GEOMETRY:

CONTENT AND STRATEGY FOR TEACHERS

SOL WEISS
West Chester State College

GEOMETRY:
CONTENT AND STRATEGY FOR TEACHERS

Bogden & Quigley, Inc., Publishers
Tarrytown-on-Hudson, New York / Belmont, California

To CHAVI

Cover and text design by Lenni Schur

Library of Congress Catalog Card No. 78–164950

Standard Book No.: 0–8005–0019–9

Printed in the United States of America

1 2 3 4 5 6 7 8 9 10—75 74 73 72

PREFACE

This book was written as a source book and guide for prospective and practicing teachers of modern mathematics. It is not a collection of prescribed lesson plans; instead, it provides a wide variety of materials from which a teacher may develop a plan reflecting his own tastes and geared to the needs and talents of his students. The heart of the book is the complete review of the content of secondary school geometry.

Commentaries on teaching strategies appear throughout the book and beside the content itself. Therefore, the book is equally useful in content and in methods courses for secondary school teachers. Supervisors of mathematics should also find this book useful for training beginning teachers and substitutes.

Each of the 13 chapters is divided into four sections. **Section One** suggests informal *methods of motivating* the study of the content. **Section Two** *reviews the content* to be taught, focusing on important ideas and relations. **Section Three** contains suggestions for *visual aids*, many of which can be made by the students, that are useful in teaching specific concepts. It also contains many suggestions for *student and teacher activities* as well as *questions and puzzles* based on the ideas in Section Two. **Section Four** provides *enrichment*: historical backgrounds, higher-level developments of the content, and other materials for teaching talented students and for enriching the course in general.

Many people and many years of experience have contributed to the development of this book. In particular, I should like to thank the following people, who read the preliminary edition of the manuscript and who were generous with their comments and suggestions: Howard Eves (University of Maine), Lorella A. McKinney (State University College, New Paltz, New York), Dora Helen Skypek (Emory University), Judith Lau (Pierce Junior High School, Milton, Massachusetts), and Ralph T. Heimer (Pennsylvania State University).

Sol Weiss
West Chester State College

CONTENTS

1 SETS

INTRODUCING THE CHAPTER TO STUDENTS

CONTENT REVIEW

Prerequisites for Learning This Chapter
Set Notation
Relations Between Sets
Operations on Sets

TEACHING SUGGESTIONS AND ACTIVITIES

Visual Aids
A Way To Illustrate Union and Intersection
Questions
Puzzles
Solutions

ENRICHMENT

Set Theory
Proving Some Properties of Sets
Cartesian Products
Partitioning a Set
Boolean Algebra

INTRODUCING THE CHAPTER TO STUDENTS

One of the fascinating things about mathematics is that we do not know today what may turn out to be useful tomorrow.

For thousands of years people have been thinking and talking about collections or sets of things—herds of cattle, sets of dishes, a baseball team, a tribe, a family, a club, a nation. But it was not until about 100 years ago that this ancient and obvious idea, the idea of a *set* of things, was recognized as very useful and important.

The concept of a set has become important because by using it we can express ideas more clearly, more exactly, and more concisely. We can use sets to express ideas about arithmetic as well as about geometry. The set ideas and language learned in connection with numbers can be applied to geometry or algebra or to any other branch of mathematics.

By using the ideas of sets we can solve problems that would be very difficult or even impossible to solve without them. For example, one of the fascinating and unanswered puzzles for a long time was the question of whether a 6-inch segment contains more points than a 1-inch segment. It was not until the mathematician Georg Cantor (1845–1918) developed ideas about sets that we finally got a clear and definite answer to this question. And what a surprising answer it was!

Today we use sets to define whole numbers; the operations of addition, subtraction, multiplication, and division with whole numbers; and the properties of whole numbers. When we count, all that we are really doing is attaching a number to a set. The study of arithmetic is based on the study of sets, and arithmetic is the beginning of mathematics.

Today we also use sets to give us clearer, more exact, and more concise definitions in geometry. We think of a *curve* as a *set of points* and of *space* as the *set of all points*. As a matter of fact, *all* the geometric figures we shall talk about are sets of points.

Sets are useful not only in mathematics but in other fields as well. The mathematics of life insurance divides the population into *sets of people* of similar age. Physicists do not think of an atom as a single object but as a member of a *set of atoms* of similar type.

Because the ideas of sets occur so frequently, a whole language of sets has been developed. In this chapter you will be introduced to the language and some uses of sets and to operations on sets. We shall later use these ideas in our geometry.

CONTENT REVIEW

PREREQUISITES FOR LEARNING THIS CHAPTER

None.

SET NOTATION

A *set* is any collection of objects. The objects in the set are called its *elements* or *members.*[1]

In dealing with sets, we adopt, for convenience, the following symbolism:

1. Capital letters such as A, B, or C are the names for sets.
2. The elements of the set are listed within braces. For example, $A = \{1, 5, 13\}$. This is read "A is the set whose elements are 1, 5, and 13."
3. The statement "5 is an element of set A" may be abbreviated $5 \in A$. To show that 3 is not an element of set A, we write $3 \notin A$.

A set must be well defined, which means that there must be a reliable way of telling whether or not any given element belongs to the given set. A set may be defined

1. By *listing* the elements of the set, for example, $A = \{2, 3, 5\}$.[2]
2. By giving a *rule* or description by which the elements of the set may be found, for example,

   $B = \{\text{all odd whole numbers less than } 7\}$

   or

   $C = \{x \mid x \text{ is a counting number greater than } 5\}$,

   which is read "C is the set whose elements are all x such that x is a counting number greater than 5."[3]

[1] 1. These "objects" need not be tangible. They can be abstract ideas, meaningless symbols, or anything at all.

2. It is our ability to think of a collection of objects *as a single entity* that makes the set concept useful. A set, as a single entity, has characteristics not possessed by any of its members. For example, the set of people who make up the U.S. Senate, *as a whole*, has the authority to pass laws. This characteristic is not possessed by any of its members.

[2] A set is distinguished from another set only by its members, not by the *order* in which its members are listed. The set denoted by $\{2, 3, 5\}$ and the set denoted by $\{3, 5, 2\}$ are exactly the same set.

[3] 1. The expression "$x \mid x$ is a counting number greater than 5" is called a *set builder* because

it selects from some larger set (the set of counting numbers) a set consisting only of certain members of the set.

2. Set C (above) can be expressed as $C = \{6, 7, 8, 9, \ldots\}$, where the three dots mean "and so on." We assume that the words "and so on" make perfectly clear which numbers are and which numbers are not members of C.

3. The *rule* method of defining a set becomes very useful when we wish to discuss a set containing a large or unlimited number of elements. For example,

a. If we are talking about the set of 30 students in Mr. Cole's class in Central High School, it would be more convenient to write

$S = \{\text{all students in Mr. Cole's class}\}$

than to list all 30 names within braces.

b. If we are talking about the set of rational numbers, there is no simple way to represent this set by the *listing* method, because any number in the set that is listed has no "next larger" number in the set.

4. Certainty as to whether a given element is or is not a member of a given set is the most important idea in defining a set. Let students see that if membership in a set depends on *opinion* (the set of "interesting" people in a room), then the set is not clearly defined. In a well-defined set there must be complete agreement on its membership.

Although it is natural to assume that a "collection" has two or more elements in it, we find it useful to refer to collections containing just *one* element, or even *no* elements, as sets.

1. The set that has *no* members is called the *empty set*, or the *null set*. We denote this set by the symbol $\varnothing$ or $\{\ \}$ (braces without any elements inside).[4]

[4] 1. We talk about *the* empty set rather than *an* empty set because there is only one empty set, the set that contains no elements.

2. The symbol $\varnothing$ is an adaptation of a letter of the Danish alphabet.

3. Students sometimes represent the empty set by $\{\varnothing\}$, which is incorrect. What is being represented by $\{\varnothing\}$ is a set containing one element, $\varnothing$. For a similar reason, $\{0\}$ is not the empty set.

2. When a set has a limited number of elements it is called a *finite set*. When a set has an unlimited number of elements it is called an *infinite set*.[5]

[5] 1. These statements are not definitions of *finite* and *infinite* sets but only intuitive descriptions.

2. The empty set is a finite set.

3. It is incorrect to say that a set is finite if it has a largest element. The set {all fractional numbers from 1 to 2} has a last element, 2, but is not a finite set.

RELATIONS BETWEEN SETS

Subsets

Definition
If every element of set A is also an element of set B, then A is a *subset* of B. We write this $A \subseteq B$.

For example,

1. If $A = \{1, 2\}$ and $B = \{1, 2, 3\}$, then $A \subseteq B$.
2. If $A = \{1, 2, 3\}$ and $B = \{1, 2, 3\}$, then $A \subseteq B$.[6]

[6] 1. Since the definition of subset does not specify that A contain *fewer* elements than B, A may have exactly the same elements as B (example 2).

2. The definition of a subset may be expressed this way:

$A \subseteq B$ if and only if $x \in A$ implies $x \in B$.

3. The definition of a subset may also be stated this way: Set A is a subset of set B if A contains no element that is not in B.

We can think of the definition of a subset in still a different way: A is *not* a subset of B if and only if A contains an element not in B.

If A is a subset of B and B has at least one element that is not an element of A, for example, $A = \{1, 2\}$, $B = \{1, 2, 3\}$, then A is said to be a *proper subset* of B, and we write $A \subset B$.

Definition
If A is a subset of B but B is not a subset of A, then we say that A is a *proper subset* of B.[7]

[7] 1. When the subset A does not contain all the elements of set B, we call A a proper subset of B. When the subset A contains exactly the same elements as set B, A is called an *improper subset* of B.

2. Just as $2 < 3$ means the same thing as $3 > 2$, so does the expression $A \subset B$ mean the same thing as $B \supset A$, read "B is a *superset* of A."

The following are important properties of subsets:

1. Every subset of B, other than B itself, is a *proper* subset of B.
2. Every set is an *improper* subset of itself.
3. The empty set is considered to be a proper subset of every set except itself. Since the empty set contains no elements, all its elements belong to *every* set.[8]

[8] The concept that the empty set is a subset of every set is a difficult one for some students to grasp. Sometimes, posing the following question to the student helps convince him: Can you name an element in the empty set that is not contained in any given set? Illustrate with specific examples. (See the last two paragraphs of comment **6**.)

4. For every set of n elements there are exactly 2^n subsets.[9]

[9] 1. For students to discover this formula, ask them to form all possible subsets for sets containing one, two, three, and four elements. Let them then tabulate the results:

Number of Elements in Set	*Number of All Possible Subsets*
1	2
2	4
3	8
4	16

Then ask them whether they see a pattern.

2. To form *all* possible subsets for any set, we consider all possible subsets containing *no* elements, *one* element, *two* elements, *three* elements, . . . , *all* the elements, and then find the sum of all these possibilities.

If two sets have no common elements, they are said to be *disjoint sets*. For example, if $A = \{1, 2, 9\}$ and $B = \{3, 5\}$, then A and B are disjoint sets.

When all the sets in a discussion are subsets of a given set, we call the given set the *universal set*, or the *universe*.[10]

[10] 1. In any discussion it is important to know what the universe is.

2. In this chapter we shall denote the universal set by U.

Definition

If $A \subseteq U$, then the elements of U that are *not* in A form another subset of U called the *complement of A*, denoted by A'.

For example, if $U = \{1, 3, 5, 7\}$ and $A = \{5\}$, then $A' = \{1, 3, 7\}$.

Venn Diagrams

A way of picturing relations between sets is to use Venn diagrams. In these diagrams we let a region, such as a rectangular region, represent the universal set (FIG. 1-1) and let subregions, such as circular regions completely contained in U, represent subsets of U (FIG. 1-2).[11]

[11] 1. Although we are using rectangles and circles to construct Venn diagrams, any simple closed curves would do. What is important is the *relationship* between the interiors of the curves used rather than their shapes.

2. Venn diagrams are so named in honor of the English mathematician John Venn (1834–1923), who popularized their use in the study of logic.

1. FIGURE 1-3 shows $A \subseteq U$, $B \subseteq U$, A and B disjoint.
2. FIGURE 1-4 shows C' represented by the shaded region.

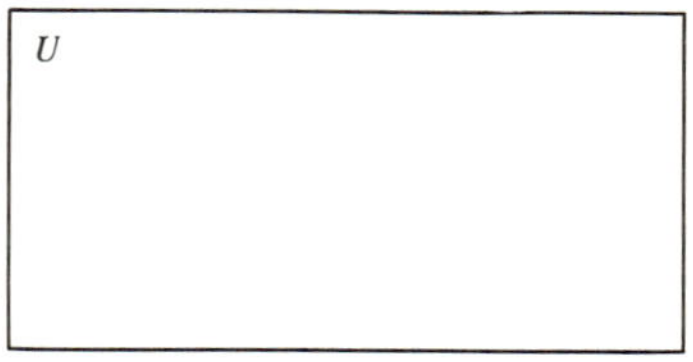

FIGURE 1-1

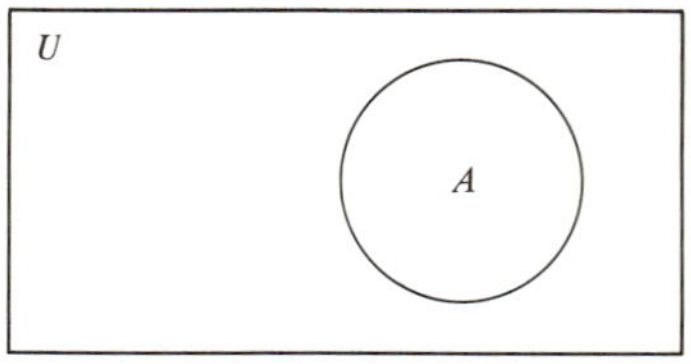

FIGURE 1-2

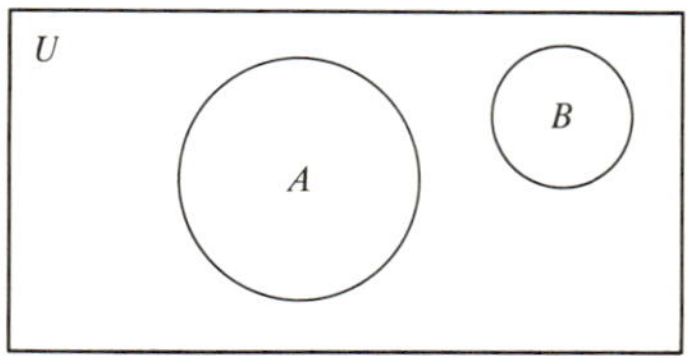

FIGURE 1-3

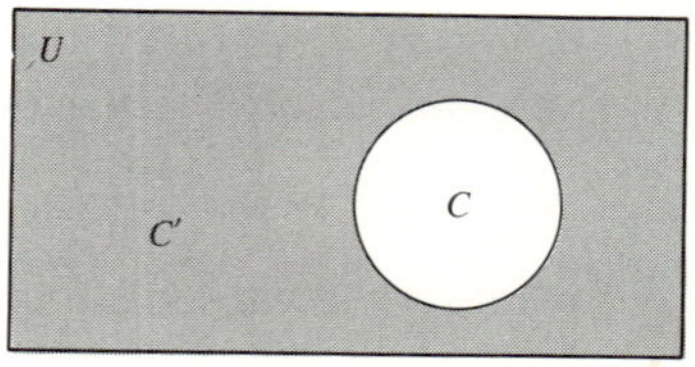

FIGURE 1-4

Equality of Sets

Definition

Two sets are *equal* if each is a subset of the other; that is, $A = B$ if $A \subseteq B$ *and* $B \subseteq A$.[12]

Equivalence of Sets

Definition

Two sets are *equivalent* if there exists a one-to-one correspondence between their elements. If set A is equivalent to set B, we write $A \sim B$.[13]

[13] 1. The *way* in which the elements of two sets are matched has no bearing on the equivalence of the sets as long as to every element in the first set there corresponds some element in the second set, and to every element in the second set there corresponds some element in the first set. For example, if $A = \{a, b, c\}$ and $B = \{1, 2, 3\}$, there are six different ways in which the elements can be matched:

(1)	(2)	(3)	(4)	(5)	(6)
$a \leftrightarrow 1$	$a \leftrightarrow 1$	$a \leftrightarrow 2$	$a \leftrightarrow 2$	$a \leftrightarrow 3$	$a \leftrightarrow 3$
$b \leftrightarrow 2$	$b \leftrightarrow 3$	$b \leftrightarrow 1$	$b \leftrightarrow 3$	$b \leftrightarrow 1$	$b \leftrightarrow 2$
$c \leftrightarrow 3$	$c \leftrightarrow 2$	$c \leftrightarrow 3$	$c \leftrightarrow 1$	$c \leftrightarrow 2$	$c \leftrightarrow 1$

A pairing that assigns to each member of each set one and only one member of the other set is called a *one-to-one correspondence.*

2. Finite sets whose elements can be paired in a one-to-one correspondence contain the same *number* of elements. This number is the same regardless of the method by which the one-to-one correspondence is established. (For a discussion of infinite sets, see pages 17–18.)

3. The *number* of elements in a set is called the *cardinal number* of the set.

4. All equivalent sets have the same cardinal number, and all sets with the same cardinal number are equivalent.

5. We now have two ways of determining the equivalence of finite sets: First, by

[12] 1. It follows from this definition that two sets are equal if they contain the same elements. The sentence "A and B contain the same elements" can be broken down into "every element of A is an element of B, and every element of B is an element of A." This is another way of saying that $A \subseteq B$ and $B \subseteq A$.

2. It also follows that two sets are equal if they contain the same elements *regardless of the order* in which they are listed.

3. Equal sets can be shown in a Venn diagram as in FIG. 1A.

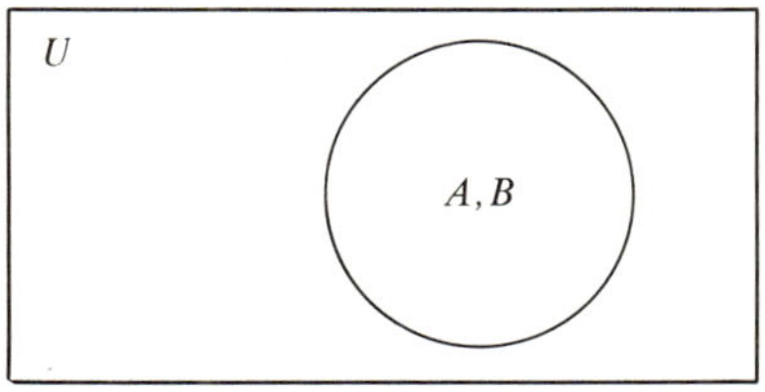

FIGURE 1A

determining whether there exists a *one-to-one correspondence* between the elements of the two sets; second, by *counting* the elements in each set. When we count the elements in two sets we not only determine whether the sets are equivalent, but we also determine *how many* elements each set contains. When we set up a one-to-one correspondence we determine only whether or not the sets are equivalent.

6. Stress the difference between *equal* and *equivalent* sets. Equal sets contain the *same* elements; equivalent sets contain only the *same number* of elements. All equal sets are equivalent, but all equivalent sets are not necessarily equal.

7. The usefulness of the equivalence property depends on the fact that it does not matter what kind of elements the sets contain. Equivalence is concerned only with the fact that we can pair the elements in such a way as to produce a one-to-one correspondence.

OPERATIONS ON SETS[14]

[14] Just as numbers may be combined by addition and multiplication to form other numbers, so may sets be combined by certain operations to form other sets. The operations we shall be concerned with here are *union*, *intersection*, and *Cartesian product*.

Union

Definition
The *union* of two sets is the set of all elements belonging to either of the two sets or to both of them. The union of sets A and B is symbolized by $A \cup B$, read "the union of A and B."

If $A = \{1, 3, 5\}$ and $B = \{2, 3, 4\}$, then $A \cup B = \{1, 2, 3, 4, 5\}$.[15]

[15] 1. The elements contained in *both* A and B are written only once.

2. Stress that the union of two sets is another *set*. Answers should be expressed in *set notation*.

3. Since the union of two sets contains all the elements of both sets, each of the original sets is a subset of the union set. That is, for all sets A and B, $A \subseteq (A \cup B)$ and $B \subseteq (A \cup B)$.

4. Another way to define the union of A and B is to say that it is the set whose members

are the elements belonging to at least one of the sets A and B.

5. "$x \in (A \cup B)$" is equivalent to "$x \in A$ *or* $x \in B$."

6. The union of sets is closely related to the meaning of the word "or." The test of membership in $A \cup B$ is whether the element in question is or is not a member of set A *or* set B (or of both A and B).

1. We can show the union of two sets with a Venn diagram (FIG. 1-5).

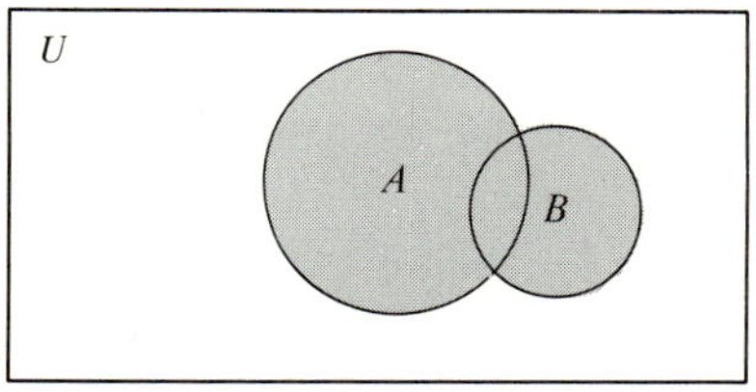

$A \cup B$ is represented by the shaded region

FIGURE 1-5

2. The following are two important properties of union (see pages 18–19):

   a. The *order* in which we form the union of two sets has no effect on the result. For example, if $A = \{1, 2\}$ and $B = \{0, 2, 4\}$, then

   $$A \cup B = \{1, 2, 0, 4\} = \{0, 1, 2, 4\}$$

   and

   $$B \cup A = \{0, 2, 4, 1\} = \{0, 1, 2, 4\}.$$

   That is, $A \cup B = B \cup A$. We call this property the *commutative property* of union and say that *union is commutative.*

   b. The way in which we *group* the sets when we take their union has no effect on the result. For example, if $A = \{1, 2\}$, $B = \{0, 2, 4\}$, and $C = \{1, 3, 5\}$, then

   $$\begin{aligned}(A \cup B) \cup C &= \{1, 2, 0, 4\} \cup \{1, 3, 5\} \\ &= \{1, 2, 0, 4, 3, 5\} \\ &= \{0, 1, 2, 3, 4, 5\}\end{aligned}$$

and

$$A \cup (B \cup C) = \{1, 2\} \cup \{0, 2, 4, 1, 3, 5\}$$
$$= \{1, 2, 0, 4, 3, 5\}$$
$$= \{0, 1, 2, 3, 4, 5\}.$$

That is, $(A \cup B) \cup C = A \cup (B \cup C)$. We call this property the *associative property* of union and say that *union is associative.*

Intersection

Definition
The *intersection* of two sets is the set of elements belonging to *both* sets.

The intersection of sets A and B is symbolized by $A \cap B$, read "the intersection of A and B." If $A = \{1, 3, 5\}$ and $B = \{2, 3, 4, 5\}$, then $A \cap B = \{3, 5\}$.[16]

1. We can show the intersection of two sets with a Venn diagram (FIG. 1-6).

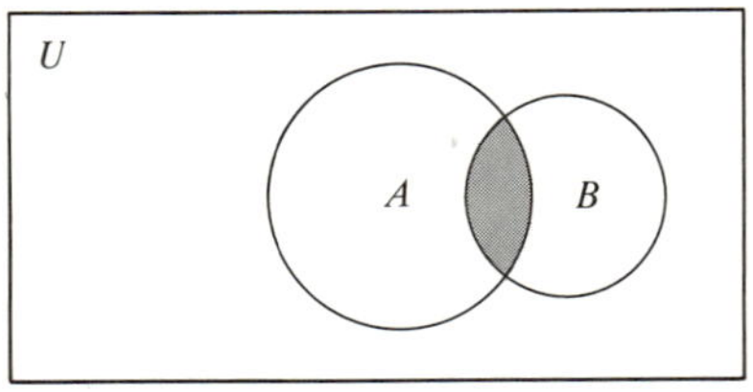

$A \cap B$ is represented by the shaded region

FIGURE 1-6

2. The following are three important properties of intersection (see pages 19–20):
   a. *Intersection is commutative;* that is, $A \cap B = B \cap A$. For example, if $A = \{1, 2\}$ and $B = \{0, 2, 4\}$, then

   $A \cap B = \{2\}$

   and

   $B \cap A = \{2\}$.

[16] 1. Stress that $A \cap B$ is another *set.*
2. Since the elements of the intersection set belong to both sets, $A \cap B$ is a subset of set A and set B. That is, for all sets A and B, $A \cap B \subseteq A$ and $A \cap B \subseteq B$.
3. The statement "$x \in (A \cap B)$" is equivalent to "$x \in A$ *and* $x \in B$."
4. The words "intersect" and "intersection" do not necessarily imply the same thing. The statement "A and B intersect" implies that A and B have at least one element in common. But the phrase "the intersection of A and B" can also refer to the empty set.
5. Disjoint sets may now be defined as sets whose intersection is the empty set.
6. The intersection of sets is closely related to the meaning of the word "and." The test of membership in $A \cap B$ is whether the element in question is or is not a member of both A and B.

b. *Intersection is associative;* that is, $(A \cap B) \cap C = A \cap (B \cap C)$. For example, if $A = \{1, 2\}$, $B = \{0, 2, 4\}$, and $C = \{1, 2, 5\}$, then

$$(A \cap B) \cap C = \{2\} \cap \{1, 2, 5\} = \{2\}$$

and

$$A \cap (B \cap C) = \{1, 2\} \cap \{2\} = \{2\}.$$

c. When the intersection and union of sets are combined, the combined operation has the properties that

(1) $A \cap (B \cup C) = (A \cap B) \cup (A \cap C)$

and

(2) $A \cup (B \cap C) = (A \cup B) \cap (A \cup C)$.

This is known as the *distributive property*. Note that each operation is *distributive* over the other.[17]

[17] 1. Illustrate both parts of this property by letting A, B, and C represent any three sets.

2. The distributive property can be verified very neatly with a Venn diagram (FIG. 1B). Let A, B, and C represent the following regions:

$A = \{1, 2, 3, 4\}$,

$B = \{2, 3, 5, 6\}$,

$C = \{3, 4, 6, 7\}$.

a. To establish part (1) of the property, show that $A \cap (B \cup C)$ and $(A \cap B) \cup (A \cap C)$ represent the same region in the diagram.

b. To establish part (2) of the property, show that $A \cup (B \cap C)$ and $(A \cup B) \cap (A \cup C)$ represent the same region in the diagram.

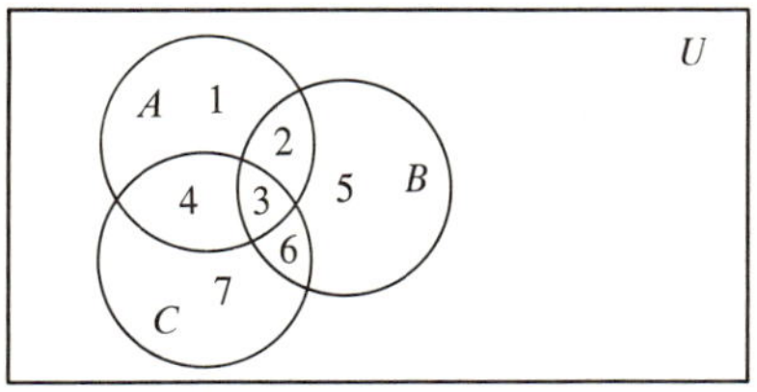

FIGURE 1B

Cartesian Product

Before talking about Cartesian products we shall define an *ordered pair*.

1. An *ordered pair* is a pair of numbers (or objects) where the *order* in which the numbers are considered is important. If 2 denotes the first number and 3 the second number, the ordered pair is expressed as (2, 3).[18]

[18] 1. In the ordered pair (2, 3), 2 is called the *first component*, and 3 is called the *second component*. Do not refer to the 2 and the 3 as the first and second "elements," because we have reserved the word "element" to mean *member of a set*, and (2, 3) is not a set in the sense in which we have used the term.

2. Ordered pairs appear throughout mathematics, although sometimes in disguised form. Consider the ordered pair (6, 2). In the operation of *addition* it is expressed as $6 + 2$, in subtraction as $6 - 2$, in multiplication as 6×2, and in division as $6 \div 2$. All these operations involve ordered pairs of numbers, because the order in which we consider the 6 and the 2 is important: $6 \div 2$ is not the same thing as $2 \div 6$.

3. Although an ordered pair is not a set, we can have a *set of ordered pairs,* that is, a set whose elements are ordered pairs. For example, consider the following test results for five job applicants:

Applicant	1	2	3	4	5
Test score	65	82	79	90	62

These data can be expressed as a set of five ordered pairs:

$$J = \{(1, 65), (2, 82), (3, 79), (4, 90), (5, 62)\}.$$

Each element of J is an ordered pair whose first component is the applicant number and whose second component is that applicant's test score.

4. Two ordered pairs are equal if and only if they have identical first components and identical second components; for example, $(a, b) = (c, d)$ if and only if $a = c$ and $b = d$.

2. Let $A = \{1, 2\}$ and $B = \{a, b, c\}$.

a. Let us now form *all possible* ordered pairs such that the *first component* is an element of A and the *second component* is an element of B. We obtain this set of ordered pairs:

$$\{(1, a), (1, b), (1, c), (2, a), (2, b), (2, c)\}.$$

b. This set of ordered pairs is called the Cartesian product of A and B and is denoted $A \times B$ (read "A cross B").

c. *Definition:* If A and B are sets, then the set $A \times B$, called the *Cartesian product of A and B,* is the set of all possible ordered pairs such that the first component of an ordered pair is an element of A and the second component is an element of B.[19]

[19] 1. "Cartesian product" takes its name from René Descartes (1596–1650), who related ordered pairs of numbers to points in a plane and thereby conceived the ideas of analytic geometry.

2. If A and B are different sets, $A \times B$ is a set different from $B \times A$. We therefore say that the formation of Cartesian products of sets is *not* commutative.

3. Unlike union and intersection, the formation of a Cartesian product produces a set *not* in the universal set. For example, let

$$U = \{1, 2, 3, 4, 5\}, \qquad A = \{1, 2\}, \qquad B = \{1, 3, 5\}.$$

Then

a. $A \cup B = \{1, 2, 3, 5\}$, a set whose elements are in U.

b. $A \cap B = \{1\}$, a set whose elements are in U.

c. $A \times B = \{(1, 1), (1, 3), (1, 5), (2, 1), (2, 3), (2, 5)\}$, a set whose elements are *not* in U.

TEACHING SUGGESTIONS AND ACTIVITIES

VISUAL AIDS

Visual 1

PURPOSE

To verify relationships between sets, by use of *flexible* Venn diagrams.

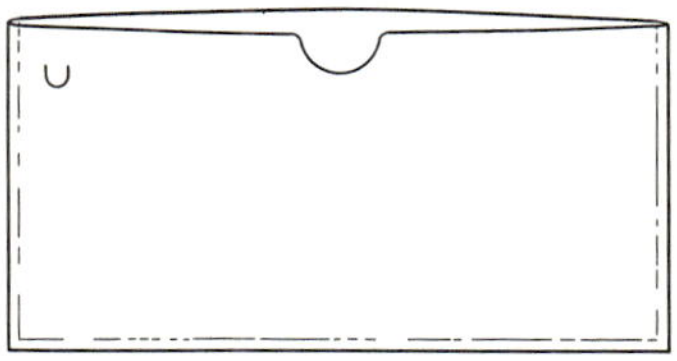

FIGURE 1-7

CONSISTS OF

1. A transparent, stiff plastic envelope,* as shown in FIG. 1-7 (to represent the universe).
2. Four plastic sheets with a circle drawn on each, as shown in FIG. 1-8 (to represent subsets).

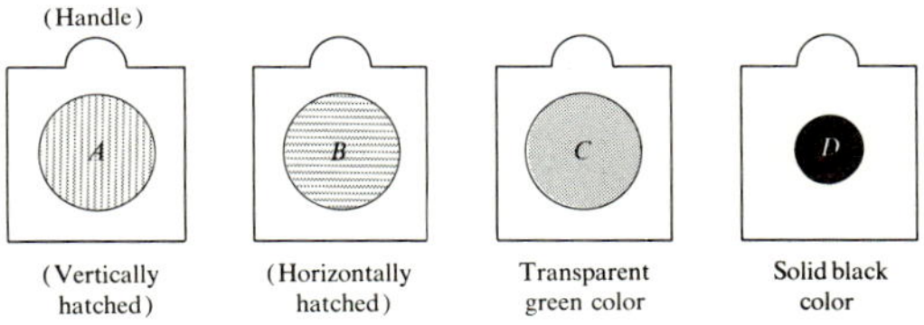

FIGURE 1-8

HOW TO USE

1. Arrange sets in the universe to show
   a. Subsets (place set D inside set C).
   b. Disjoint sets.
   c. Intersecting sets.
   d. Equivalent sets.
   e. Equal sets (superimpose set A over set B).
   f. Complement of a set.
2. Place A and B in the universe to show
   a. $A \cup B$. In FIG. 1-9 $A \cup B$ is represented by regions that are either (1) vertically hatched, (2) horizontally hatched, or (3) crosshatched. In FIG. 1-10 $A \cup B$ is represented by regions that are either (1) vertically hatched or (2) horizontally hatched.
   b. $A \cap B$. In FIG. 1-11 $A \cap B$ is represented by the crosshatched region.
   c. That *union and intersection are commutative.* Show that no matter with which set you start the union or intersection, you end up with the same region on the Venn diagram.

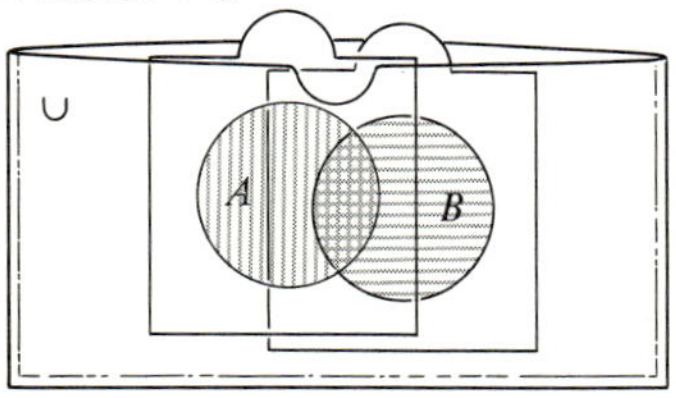

FIGURE 1-9

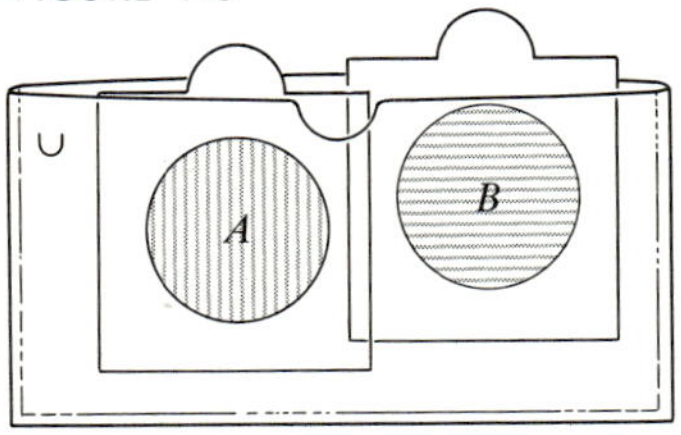

FIGURE 1-10

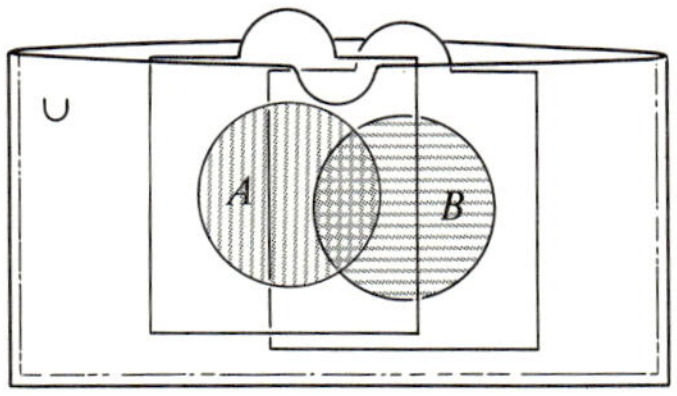

FIGURE 1-11

* The envelope serves as the universe into which to insert various combinations of subsets in various relationships.

3. Use sets A, B, and C to verify that *union and intersection are associative*. Show that no matter which way you group your sets, you end up with the same region on the Venn diagram.
4. Arrange sets A, B, and C to show regions on the Venn diagram represented by
   a. $A \cup (B \cap C)$.
   b. $A \cap (B \cup C)$.
5. Use with pages 6–11.

A WAY TO ILLUSTRATE UNION AND INTERSECTION

You can make use of your class seating arrangement to illustrate union and intersection. If your students are seated as shown in the accompanying table, label each row and column by a capital letter. Each letter will then represent a set of students who sit in a particular row or column. Then ask students to stand up if they belong to a set you write on the board. (1) C; (2) V; (3) $A \cup C$; (4) $B \cup Q$; (5) $E \cap R$; (6) $D \cap A$; (7) the complement of D; etc.

	P	*Q*	*R*	*S*	*T*	*V*
A	×	×	×	×	×	×
B	×	×	×	×	×	×
C	×	×	×	×	×	×
D	×	×	×	×	×	×
E	×	×	×	×	×	×

A seating arrangement can also be used to emphasize that an ordered pair is a single entity. For example, the student in row C, column R, is identified with (C, R).

QUESTIONS

1. Complete the following statements if neither set E nor K is the empty set:
   a. If $E \subseteq K$, then $E \cup K =$ ____________.
   b. If $E \subset (C \cup K)$, then $E \neq$ ____________.
   c. If $E \subseteq K$, then $E \cap K =$ ____________.

2. Show by means of Venn diagrams that

   a. $A \cup (B \cap C) = (A \cup B) \cap (A \cup C)$.
   b. $(A \cap B)' = A' \cup B'$.
   c. $(A \cup B)' = A' \cap B'$.

3. Let U = {all people living in the United States},
   A = {all professional actors},
   B = {all airplane pilots},
   C = {all people over 6 feet tall},
   D = {all people earning more than $50,000 a year}.

   Use the language of sets to express the following:

   a. All pilots are over 6 feet tall.
   b. No actor makes more than $50,000 a year.
   c. No pilot is an actor.
   d. All actors who are pilots are over 6 feet tall.
   e. A person who is either 6 feet tall or a pilot cannot earn more than $50,000 a year.
   f. All people who are not pilots earn more than $50,000 a year.

PUZZLES

1. At a certain hotel there are 60 guests who had traveled to London, 45 who had traveled to Paris, and 40 who had traveled to Rome. Seventeen of these guests had been to both London and Paris, 15 had been to both Paris and Rome, 13 had been to both Rome and London, and 5 visited all three cities. How many guests are there altogether at the hotel? (Use a Venn diagram to solve this problem.)

2. Three of the characteristics possessed by the inhabitants of a certain island are onx, bonx, and conx. Five of the inhabitants are onx, 6 are bonx, and 7 are conx. Two of the inhabitants are onx and bonx, 2 are bonx and conx, 2 are conx and onx, and 2 are at the same time onx, bonx, and conx. Altogether there are 15 inhabitants on the island. Why are the conditions in this problem contradictory?

SOLUTIONS

1. There are 105 guests.

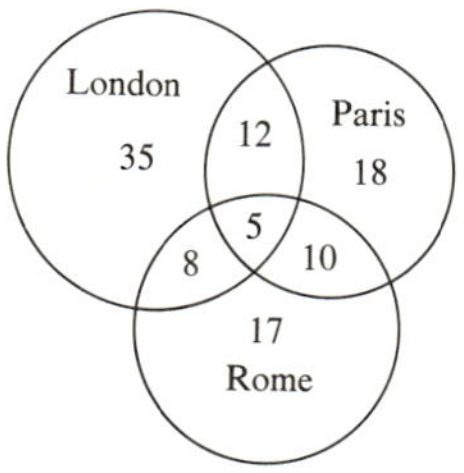

2. The total number of inhabitants under the given conditions except the last must be 14, not 15 as stated in the problem.

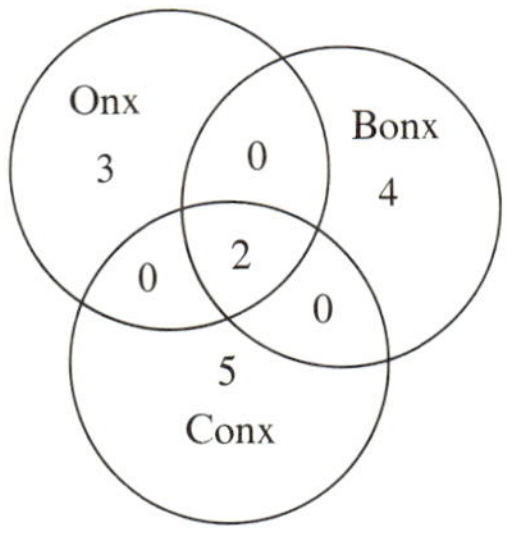

3. The number of points on each segment is the same because it is possible to set up a one-to-one correspondence between the points on the two segments (see page 18).

3. Which line segment contains more points: a segment one-millionth of an inch long or a segment stretching from the earth to the moon? How can you prove your answer?

ENRICHMENT

SET THEORY

The subject of set theory was developed by Georg Cantor (1845–1918) toward the end of the nineteenth century. As a result of his work, it has become possible to answer such questions as

1. Are there more natural numbers than even numbers?
2. Does a 5-inch segment contain more points than a 2-inch segment?

We have already seen that the equivalence of two finite sets can be established either by counting or by a one-to-one correspondence. Although we cannot count the number of elements in an infinite set, it is possible to devise techniques to ascertain whether there exists a one-to-one correspondence between the elements of two infinite sets. If such a correspondence can be shown to exist, then the two sets must be equivalent.

To answer the question whether there are more natural numbers than positive even numbers, let us denote the set of natural numbers by N and the set of positive even numbers by G and pair their elements this way:

$$\begin{array}{lcl} N & = & \{1, 2, 3, 4, 5, \ldots, n, \ldots\}, \\ & & \;\;\updownarrow\;\updownarrow\;\updownarrow\;\updownarrow\;\updownarrow \qquad\quad \updownarrow \\ G & = & \{2, 4, 6, 8, 10, \ldots, 2n, \ldots\}. \end{array}$$

This pairing establishes a one-to-one correspondence between N and G, since for every natural number there exists a positive even number and for every positive even number there exists a natural number. Therefore, G is equivalent to N, and the answer to question 1 is that there are as many positive even numbers as there are natural numbers!

But although G is equivalent to N, it is much "thinner" than N, since N also contains the positive odd numbers. Since G is a subset of N but N contains elements that are not in G, G is a proper subset of N.

We are now confronted with the curious paradox that the set of positive even numbers is a proper subset of the set of natural numbers and *at the same time* equivalent to it! In other words, the set N is no greater than a piece of itself—the set of positive even numbers. There are many such pieces of N that are as great as N, for example, the set of square numbers, the set of positive cube numbers, and the set of positive multiples of 3.

This paradox involves a property used to define finite and infinite sets.

Definition
A set is an *infinite* set if it is equivalent to a proper subset of itself. If a set is *not* an infinite set, then it is a *finite* set.

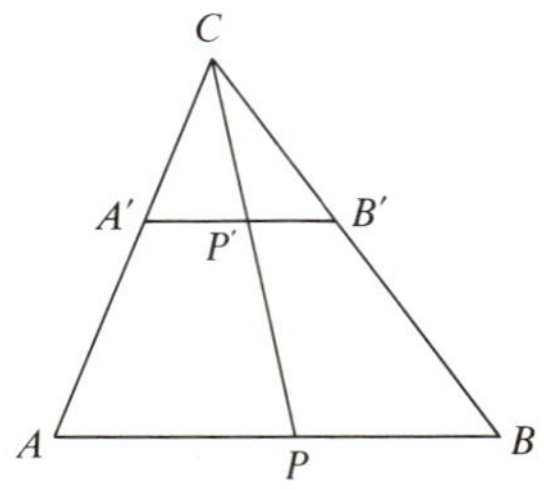

FIGURE 1-12

The famous Euclidean axiom that "the whole is greater than any of its parts" is not valid for infinite sets.

The answer to question 2, whether a 5-inch segment contains more points than a 2-inch segment, can likewise be found by using the ideas of equivalence of sets. Let $\overline{AB}$ represent a 5-inch segment and $\overline{A'B'}$ a 2-inch segment (FIG. 1-12). It is possible to establish a one-to-one correspondence between the points on $\overline{A'B'}$ and $\overline{AB}$ by this scheme: If C is the intersection of $\overrightarrow{AA'}$ and $\overrightarrow{BB'}$, where $\overline{A'B'}$ is placed parallel to and nonincident with $\overline{AB}$, then to any point P of $\overline{AB}$ there will correspond a point P' of $\overline{A'B'}$ which is on $\overline{CP}$. Similarly, to any point P' of $\overline{A'B'}$ there will correspond a point P of $\overline{AB}$ which is on $\overline{CP}$. There is, therefore, a one-to-one correspondence between the set of points on a 5-inch segment and the set of points on a 2-inch segment. This proves that the number of points on each segment is the same. By the same method we can show that the number of points on a segment measuring the thickness of a hair is exactly the same as the number of points on a segment extending from New York to Tokyo!

PROVING SOME PROPERTIES OF SETS

We have seen that union and intersection are commutative and associative. We shall now prove these properties for any sets A, B, C. In all these proofs we shall rely on the definition of equality of sets; that is, two sets are equal if each is a subset of the other.

THEOREM 1 Union is commutative.

That is, $A \cup B = B \cup A$.

Proof
1. Each element of $A \cup B$ is an element of $B \cup A$ and each element of $B \cup A$ is an element of $A \cup B$.
2. Therefore, $A \cup B \subseteq B \cup A$, and $B \cup A \subseteq A \cup B$.
3. Therefore, $A \cup B = B \cup A$, since each is a subset of the other.

THEOREM 2 Union is associative.

That is, $(A \cup B) \cup C = A \cup (B \cup C)$.

Proof

1. We need to show that
   a. $(A \cup B) \cup C \subseteq A \cup (B \cup C)$.
   b. $A \cup (B \cup C) \subseteq (A \cup B) \cup C$.
2. To prove statement a, we show that any element x of $(A \cup B) \cup C$ is an element of $A \cup (B \cup C)$.
   a. $x \in (A \cup B) \cup C$ means $x \in (A \cup B)$ or $x \in C$.
   b. $x \in (A \cup B)$ means $x \in A$ or $x \in B$.
   c. Therefore, $x \in A$ or $x \in B$ or $x \in C$.
   d. Therefore, $x \in A$ or $x \in (B \cup C)$; that is, $x \in A \cup (B \cup C)$.
   e. Therefore, $(A \cup B) \cup C \subseteq A \cup (B \cup C)$.
3. To prove statement b, we show that any element y of $A \cup (B \cup C)$ is an element of $(A \cup B) \cup C$. Follow the same procedure as for statement a.
4. Therefore, $(A \cup B) \cup C = A \cup (B \cup C)$, since each set is a subset of the other.

THEOREM 3 Intersection is commutative.

That is, $A \cap B = B \cap A$.

Proof

1. We need to show that
   a. $A \cap B \subseteq B \cap A$.
   b. $B \cap A \subseteq A \cap B$.
2. To prove statement a, we must show that any element x of $A \cap B$ is also an element of $B \cap A$.
   a. $x \in A \cap B$ means $x \in A$ and $x \in B$; $x \in A$ and $x \in B$ means $x \in B$ and $x \in A$.
   b. $x \in B \cap A$.
   c. Therefore, $A \cap B \subseteq B \cap A$.
3. To prove statement b, we show that any element y of $B \cap A$ is also an element of $A \cap B$. Follow the same procedure as for statement a.
4. Therefore, $A \cap B = B \cap A$, since each is a subset of the other.

THEOREM 4 *Intersection is associative.*

That is, $(A \cap B) \cap C = A \cap (B \cap C)$.

Proof

1. We need to show that
   a. $(A \cap B) \cap C \subseteq A \cap (B \cap C)$.
   b. $A \cap (B \cap C) \subseteq (A \cap B) \cap C$.
2. To prove statement a, we must show that any element x of $(A \cap B) \cap C$ is also an element of $A \cap (B \cap C)$.
   a. $x \in (A \cap B) \cap C$ means $x \in (A \cap B)$ and $x \in C$.
   b. $x \in (A \cap B)$ means $x \in A$ and $x \in B$.
   c. Therefore, $x \in A$ and $x \in B$ and $x \in C$, or
   d. $x \in A \cap (B \cap C)$.
   e. Therefore, $(A \cap B) \cap C \subseteq A \cap (B \cap C)$.
3. To prove statement b, we show that any element of $A \cap (B \cap C)$ is also an element of $(A \cap B) \cap C$. Follow the same procedure as for statement a.
4. Therefore, $(A \cap B) \cap C = A \cap (B \cap C)$, since each is a subset of the other.

CARTESIAN PRODUCTS

1. We use Cartesian products to define multiplication of whole numbers.

Set	*Cardinal Number*
A	2
B	3
$A \times B$	6

We define the product of 2 and 3 as follows (see the table): If 2 is the cardinal number of a set A and 3 is the cardinal number of a set B, then the *product* of 2 and 3, written 2×3, is the cardinal number of the Cartesian product $A \times B$, which is 6. More generally, if a is the cardinal number of set A and b is the cardinal number of set B, then the product of a and b, written $a \times b$, is defined to be the cardinal number of the Cartesian product $A \times B$.

Sometimes this definition is stated more briefly by using the notation $n(A)$ (read "n of A") to represent the cardinal number of A. The definition of multiplication of whole numbers then becomes $n(A) \times n(B) = n(A \times B)$.

2. We also use Cartesian products to define a *relation*. A *relation* is a subset of a Cartesian product. Any subset A of $R \times R$ is

said to be a *relation defined on the set R*. The set of all the *first* components of the ordered pairs of A is called the *domain* of A. The set of all the *second* components of the ordered pairs of A is called the *range* of A. For example, the set of ordered pairs

$A = \{(1, 2), (2, 2), (3, 1)\}$

is a relation whose domain is $\{1, 2, 3\}$ and whose range is $\{1, 2\}$.

Of special interest is the Cartesian product of a set with itself. If $R = \{1, 2, 3\}$, the $R \times R = \{(1, 1), (1, 2), (1, 3), (2, 1), (2, 2), (2, 3), (3, 1), (3, 2), (3, 3)\}$. A graphical representation of $R \times R$ is given in FIG. 1-13. If R is the set of real numbers, then $R \times R$ is the set of all ordered pairs of real numbers. Graphically, this would represent all the points of the plane called the Cartesian plane.

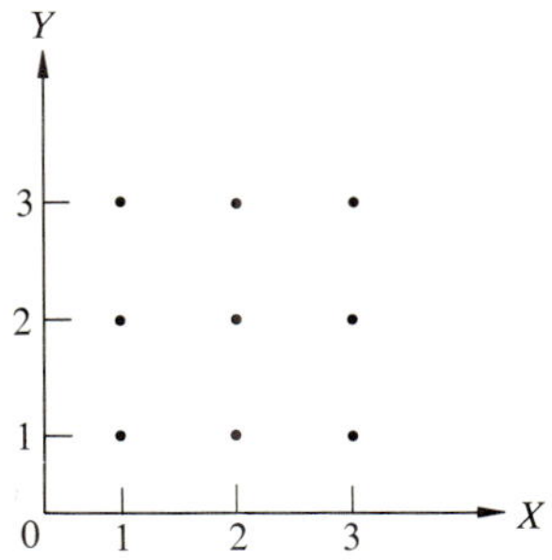

FIGURE 1-13

PARTITIONING A SET

Many problems related to sets can best be described in terms of *partitions* of a set. A *partition* is a subdivision of a set into mutually disjoint subsets. We shall soon see that a set and its complement form a partition of the universal set.

If A, B, and F are sets such that

1. $A \cup B = F$,
2. $A \cap B = \varnothing$,

then we say that A and B *partition* F.

For example, the set of positive even numbers E and the set of positive odd numbers O partition the set of natural numbers N because

1. Every natural number is either even or odd; that is, $E \cup O = N$.
2. No natural number can be both even and odd; that is, $E \cap O = \varnothing$.

A set may be partitioned into more than two sets. For example, A, B, and C partition F if and only if

1. $A \cup B \cup C = F$.
2. $A \cap B = \varnothing$, $A \cap C = \varnothing$, and $B \cap C = \varnothing$.

If A and B partition F, then we call A the *complement of B with respect to F*, and B is the complement of A with respect to F.

BOOLEAN ALGEBRA

Set operations obey certain abstract laws, some of which were stated in the Content Review and some of which were proved in this section. These laws resemble the laws of the algebra with which we are already familiar. This is why sets and their laws of operation are thought of as an algebraic system known as Boolean algebra, so named after George Boole (1815–1864), who first developed this algebra.

The following are the basic laws of Boolean algebra.

1. Given any sets A and B, there is a unique set $A \cup B$.
2. Given any sets A and B, there is a unique set $A \cap B$.
   } Closure laws

3. $A \cup B = B \cup A$.
4. $A \cap B = B \cap A$.
   } Commutative laws

5. $(A \cup B) \cup C = A \cup (B \cup C)$.
6. $(A \cap B) \cap C = A \cap (B \cap C)$.
   } Associative laws

7. $A \cap (B \cup C) = (A \cap B) \cup (A \cap C)$.
8. $A \cup (B \cap C) = (A \cup B) \cap (A \cup C)$.
   } Distributive laws

9. $U \cap A = A$ for all A.
10. $\varnothing \cup A = A$ for all A.

11. $U \cup A = U$ for all A.
12. $\varnothing \cap A = \varnothing$ for all A.

13. $A \cup A = A$.
14. $A \cap A = A$.
   } Idempotent laws

15. $A \cup (B \cup C) = (A \cup B) \cup (A \cup C)$.
16. $A \cap (B \cap C) = (A \cap B) \cap (A \cap C)$.
   } Self-distributive laws

17. $A \cup B \supset A$.
18. $A \cap B \subset A$.

19. $A \subset U$ for all A.
20. $A \supset \varnothing$ for all A.

21. If $A \subset B$, then $A \cup B = B$.
22. If $A \supset B$, then $A \cap B = B$.

23. $A \cup A' = U$.
24. $A \cap A' = \varnothing$.
   } Complementation laws

25. $(A \cup B)' = A' \cap B'$.
26. $(A \cap B)' = A' \cup B'$.
   } De Morgan's laws

The similarities and differences between the familiar laws of algebra and Boolean algebra become apparent if we replace $\cup$ by $+$ and $\cap$ by $\times$.

Notice that laws 1–7 also hold for our familiar algebra, whereas laws 8, 11, and 13–26 are different.

Also notice that the properties of the number zero that hold in ordinary algebra do not necessarily hold in Boolean algebra. For instance, one of the properties in algebra states that if a and b are numbers and $a \neq 0$ and $b \neq 0$, then $ab \neq 0$. But in Boolean algebra we can have $A \neq \varnothing$ and $B \neq \varnothing$, yet $A \cap B = \varnothing$ if A and B happen not to have any common elements.

Also notice that if we interchange $\cup$ and $\cap$, $\varnothing$ and U, or $\subset$ and $\supset$ in any of laws 1–26, we obtain another law in Boolean algebra. Thus the odd- and even-numbered laws enclosed with braces are obtainable from each other. This is known as the principle of *duality* of Boolean algebra.

Boolean algebra has application to electrical switching circuits, which are used in the construction of electronic computers.

2 POINTS, LINES, AND PLANES

INTRODUCING THE CHAPTER TO STUDENTS

CONTENT REVIEW

Prerequisites for Learning This Chapter
Undefined Terms
Postulates of Incidence
Betweenness
Congruence of Segments
Convex Sets
Separation
Simple Closed Curves

TEACHING SUGGESTIONS AND ACTIVITIES

Visual Aids
Drawing and Describing Diagrams
Random Suggestions
Questions
Puzzles
Solutions

ENRICHMENT

The Nature of Geometry
History of Geometry
Defects in Euclid's Geometry
Hilbert's Axioms
Euclid's Axioms
Finite Geometries

INTRODUCING THE CHAPTER TO STUDENTS

In the work we are about to begin, we will not do much computing with numbers. We will mostly use our "spatial" imagination.

By now we have all had some experience with geometric figures. We learned something about their characteristics, how to measure them, and, perhaps, how to find their perimeters and areas. This may have left some of us with the impression that geometry deals mostly with measurement. But this is not so.

Sometimes geometry is concerned with the study of properties that are not related to measurement. It may be concerned with reasoning, where we start with a few statements we consider to be true and, from these, discover and prove many other statements to be true.

The study of geometry goes back four thousand years, and there is no doubt that it began with measurement. The name "geometry," which means "earth measure," indicates this. Part of the early development of geometry was related to the yearly overflow of the Nile River that washed out the property boundaries of the ancient Egyptians. Every year they were faced with the practical need to remeasure and redraw their boundaries. This is how primitive surveying contributed to the beginnings of the science of measurement.

Two thousand years later the Greeks took the physical concept of geometry and developed it into an abstract science that reached its peak with the famous Greek mathematician Euclid. What Euclid did was to transform the nature of geometry from the science of measurement to a system of reasoning. In this system we arrive at conclusions, called theorems, not by measuring but by reasoning.

Before we can understand what geometry is all about, we must first understand the building blocks on which it rests—ideas such as point, space, curve, line, segment, ray, and plane. Although we use these ideas over and over again in geometry, we do not define all of them. Some of these ideas are *undefined* terms. We attach meaning to them only through a description of some of their characteristics. Other ideas are defined in terms of these undefined terms.

Why do we use undefined terms? Why don't we define *every* term? We can see the reason for this by a comparison with the way we learn a foreign language. Suppose you want to learn French, but you do not know *any* French words. Would it help you to use an all-French dictionary? If you looked up a French word in such a dictionary, you could find only a definition that uses other French words, and this would not help you very much. But if you first learn the meanings of a few French words *before* using such a dictionary, you can then use these words as a basis for learning other words.

This is also true in mathematics. To find the definition of a "triangle" we could define a triangle in terms of "line segments." But what are "line segments?" We could then define a "line segment" in terms of "points." But what are "points?"

We can see that sooner or later this process must come to an end because with each new definition we can pose the same question: What are the definitions of the terms used in the definition? We must, therefore, use *some* terms without definitions and then use these *undefined* terms to define other terms.

How do we know the meaning of a term if we do not define it? We can obtain a good idea of its meaning from a description of some of its characteristics. For example, the term "point" in geometry is often taken as undefined. But we can obtain a good idea of what a "point" is by a description of some of its properties. We say that a "point" is suggested by the tip of a pencil or by a dot on a piece of paper and represents an exact location. With these assumed properties of a "point" in mind, we proceed to *define* such terms as "segment" and "triangle."

In this chapter we shall talk about the properties of such undefined terms as "point," "line," and "plane" and then use the undefined terms to define other terms. We shall discover interesting relations between points, lines, and planes, and in the process of doing this we shall be stretching our "spatial" imagination.

CONTENT REVIEW

[1] Geometry can be developed through two fundamentally different approaches: the *metric* and the *synthetic*.

Under the *metric* approach, the concept of measurement is introduced first and then other concepts are defined in terms of measurement. Geometry is thus related to the real number system. The ideas of "betweenness" and "congruence" are *defined* in terms of distance and angular measure, and then their basic properties are proved as theorems (except the SAS postulate). This approach to geometry was proposed by G. D. Birkhoff in 1932.

Under the *synthetic* approach, "betweenness" and "congruence" can be taken as *undefined* terms, and their basic properties then stated as postulates. Distance or measurement need not be mentioned. The *postulates* of the synthetic approach turn out to be *theorems* of the metric approach. The synthetic route was used by Euclid and perfected by David Hilbert in 1898. (See pages 64–67.)

We shall use the metric approach because we believe that students will understand it better.

[2] 1. Other names for *postulate* are *axiom* and *assumption*.

2. Distinguish for your students between the two levels on which ideas will be presented: the *informal* and the *formal*. The informal level will appeal, essentially, to our intuition. Formally, the only things we shall claim to know about points, lines, and planes will be based on definitions, postulates, and theorems. The informal level will be used to make the formal ideas more "reasonable" and more understandable. This distinction must be kept in mind at all times.

PREREQUISITES FOR LEARNING THIS CHAPTER

1. An understanding of the following: set, subset, union of sets, intersection of sets, one-to-one correspondence.
2. Knowing that the real number system contains the natural numbers, the integers, the rationals, and the irrationals.
3. Knowing that the *absolute value* of a real number x is written $|x|$ and may be defined as follows:
   a. If x is positive or zero, then $|x| = x$.
   b. If x is negative, then $|x| = -x$.[1]

UNDEFINED TERMS

We begin our geometry with three undefined terms: *point*, *line*, and *plane*. We obtain an *intuitive* understanding of the meaning of these terms from an informal description of their properties. A *formal* description of their properties will come from *postulates* that will be stated later.[2]

Following is an informal description of *point*, *line*, and *plane*.

1. Point
   a. A point can be described as an exact location in space.
   b. A point is suggested by the head of a pin, a speck of dust, the tip of a pencil, a dot on the board, or the location in a room where two walls and the ceiling meet.
   c. A point does not move. If a dot on the board is erased, the point—or location—still remains.
   d. A point has no size. It has position only.
   e. A point is usually represented by a dot and is named by a capital letter.
2. Line
   a. A line is a set of points. The number of points on a line is unlimited.
   b. A line may be represented by a string that is tightly

stretched infinitely far in both directions. In all our discussions, a "line" will always mean a "straight line."

c. When we draw a picture of a line, we can draw a picture of only *part* of the line. We show that a picture represents a line by attaching arrows to both ends of the "line" (FIG. 2-1).

FIGURE 2-1

d. A line has neither width nor thickness, only length.
e. We usually name a line either by a single lowercase letter, such as *m*, or by any two points on the line, such as *A* and *B*, denoted by $\overleftrightarrow{AB}$ (FIG. 2-2).

FIGURE 2-2

f. A "piece" of a line, with definite endpoints *A* and *B*, as shown in FIG. 2-2, is called a *segment* and is denoted by $\overline{AB}$.
g. A "line" that stretches out infinitely far in only *one* direction is called a *ray*. A good physical representation of a ray is a beam of light emanating from a pin-point source. If the source is at *A* and the beam shines in the direction of *B*, we denote ray *AB* by $\overrightarrow{AB}$ (FIG. 2-3).

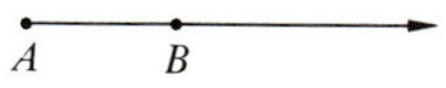

FIGURE 2-3

3. Plane
   a. A plane is a set of points.
   b. A plane is suggested by any flat surface such as a wall, floor, door, or desk top if you can imagine that it stretches out infinitely far in every direction.
   c. A plane has no thickness.

d. A plane is usually named by a capital letter or by three of its points all of which do not lie on the same line (FIG. 2-4).

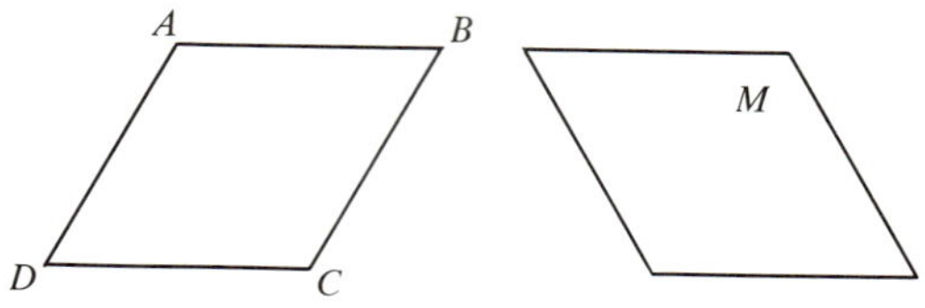

FIGURE 2-4

POSTULATES OF INCIDENCE

We now begin the formal development of our geometry by stating certain postulates about the undefined terms *points*, *lines*, and *planes*. We shall also state certain definitions related to these undefined terms. From these postulates and definitions we shall derive certain theorems.

In a formal, rigorous development of geometry, the only properties of points, lines, and planes that we have a mathematical "right" to accept are those coming from these postulates, definitions, and theorems. No others, no matter how "reasonable" or "evident," are acceptable. This is the essence of a *deductive*, or *axiomatic*, system.

POSTULATE 2-1 All lines and planes are sets of points.

Definition
The set of all points is called *space*.

POSTULATE 2-2 Given any two different points, there is exactly one line containing them.

If the points are A and B, then the line containing them is denoted by $\overleftrightarrow{AB}$.

Definitions
1. Points lying on one line are *collinear*.
2. Points lying in one plane are *coplanar*.

Let us agree that

1. If a line is a subset of a plane, we will say that the line "lies in" the plane, or that the plane "contains" the line.
2. If a point belongs to a line, we will say that the point "lies on" the line, that the line "passes through" the point, or that the line "contains" the point.
3. If a point belongs to a plane, we will say that the point "lies in" the plane, that the plane "passes through" the point, or that the plane "contains" the point.

POSTULATE 2-3 Three different noncollinear points are contained in exactly one plane (FIG. 2-5).

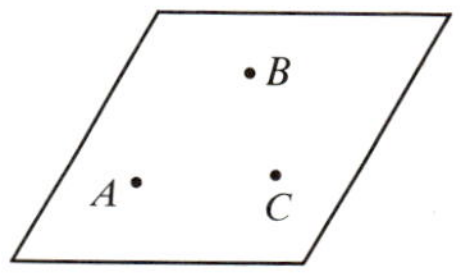

FIGURE 2-5

POSTULATE 2-4 If a plane contains two distinct points, then the line containing them lies in the plane (FIG. 2-6).

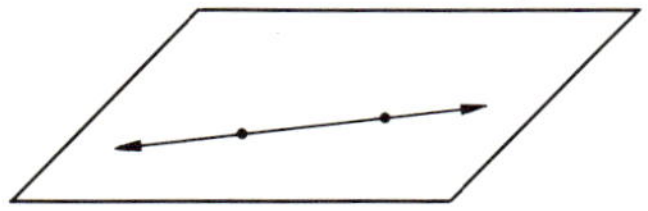

FIGURE 2-6

POSTULATE 2-5 If two distinct planes intersect, then their intersection is a line (FIG. 2-7).

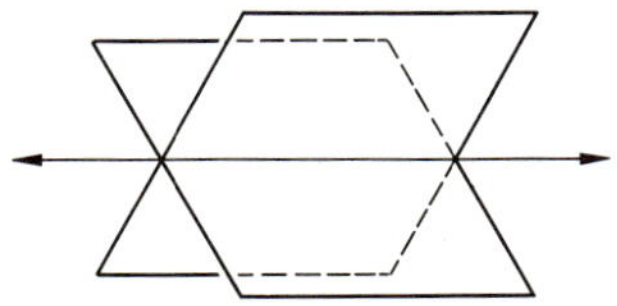

FIGURE 2-7

POSTULATE 2-6 Every line contains at least two distinct points. Every plane contains at least three noncollinear points. Space contains at least four noncoplanar points.[3]

[3] 1. These six postulates are called *postulates of incidence*. They express the basic positional relationships of points, lines, and planes. Euclid never spelled out these postulates in his list of axioms, although he assumed them in some of his proofs. In a tight deductive system, unstated assumptions are not permitted.

2. Another way in which Postulate 2-2 is often expressed is "Two distinct points determine a line."

3. Postulate 2-3 can be expressed this way: Three noncollinear points determine a plane.

4. Postulate 2-4 suggests itself intuitively by the straightness of the line and the flatness of the plane.

5. Postulate 2-5 can be "seen" in the classroom through the intersection of the floor and the walls and the intersection of the ceiling and the walls.

From the postulates and definitions we can derive several theorems that will tell us more about the behavior of points, lines, and planes.

THEOREM 2-1 *If two distinct lines intersect, then their intersection contains exactly one point* (FIG. 2-8).

FIGURE 2-8

Proof

1. Assume that the two lines intersect in two distinct points.
2. This is impossible because it contradicts Postulate 2-2, which says that there is exactly *one* line containing two distinct points.
3. Therefore, our assumption in statement 1 is false, and the theorem is proved.[4]

[4] This proof, as well as the next one for Theorem 2-2, is called an "indirect proof." In such a proof we prove that something cannot be false by showing that if it were false we would arrive at a contradiction. In this case we showed that if the theorem were false, then such a conclusion would contradict one of our postulates that we assumed to be true.

THEOREM 2-2 *If a line intersects a plane not containing it, then the intersection is a single point* (FIG. 2-9).

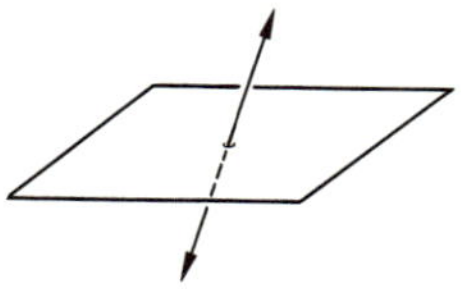

FIGURE 2-9

Proof

1. Assume that the line intersects the plane in two distinct points.
2. By Postulate 2-4, the line would have to lie in the plane.
3. Since this contradicts the hypothesis, assumption 1 is false, and the intersection is a single point.

THEOREM 2-3 *If l is a line and P is a point not on l, then there is one and only one plane containing l and P* (FIG. 2-10).

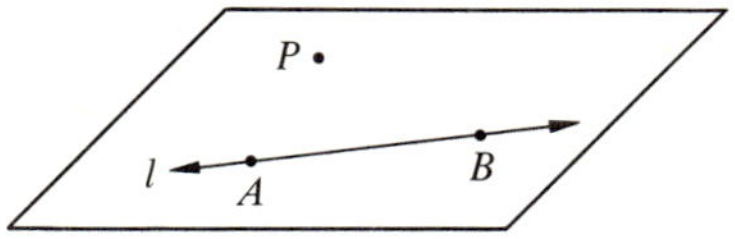

FIGURE 2-10

Proof

1. By Postulate 2-6, l contains at least two distinct points A and B.
2. A, B, and P are noncollinear since l is the only line that contains A and B (Postulate 2-2), and l does not contain P (hypothesis).
3. There is exactly one plane containing A, B, and P (Postulate 2-3), and by Postulate 2-4 the plane also contains l.
4. No other plane can contain l and P, since no other plane can contain A, B, and P.

THEOREM 2-4 If two distinct lines intersect, then there is exactly one plane containing them (FIG. 2-11).

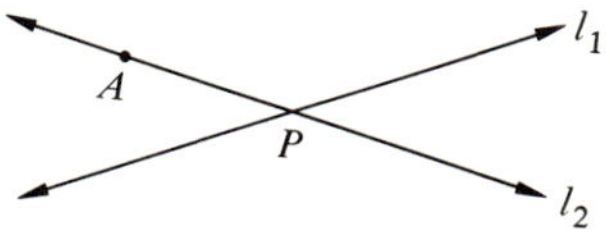

FIGURE 2-11

Proof

1. Let P be the intersection of l_1 and l_2 (Theorem 2-1), and let A be a point on l_2 but not on l_1.
2. By Theorem 2-3, there is one and only one plane containing A and l_1.
3. Since A and P are in the plane and on l_2, the plane contains l_2 by Postulate 2-4.
4. Since no other plane can contain A and l_1, no other plane can contain l_1 and l_2.[5]

Although the geometry of parallel lines will be discussed in Chapter 5, the definition of parallelism will be given here because the concept is an important one in the theory of incidence.

Definitions

See FIG. 2-12.

[5] 1. Just as Postulates 2-1—2-6 are called *incidence postulates,* so are Theorems 2-1—2-4 called *incidence theorems.* They deal with such questions as whether and how points, lines, and planes intersect and whether one contains the other.

2. By the postulates and theorems given so far, we have established the following ways of determining a plane:

a. By any three noncollinear points (Postulate 2-3).

b. By any line and any point not on the line (Theorem 2-3).

c. By any two distinct intersecting lines (Theorem 2-4).

Note: Although we proved all these theorems, we shall not always prove the theorems we state because of space limitations. Proofs can be found in the references given at the end of the book.

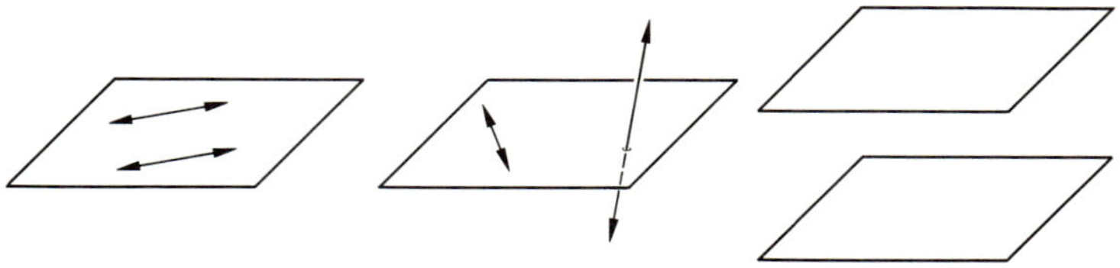

FIGURE 2-12

1. Two *lines are parallel* if they are contained in the same plane and do not intersect. If line m is parallel to line k, we write $m \parallel k$.
2. Two lines are said to be *skew lines* if and only if they are not in the same plane.
3. Two *planes are parallel* if they do not intersect.
4. *A line and a plane are parallel* if they do not intersect.[6]

An important relation between parallel planes and parallel lines is established by Theorem 2-5.

THEOREM 2-5 *If a plane intersects two parallel planes, then it intersects them in two parallel lines* (FIG. 2-13).

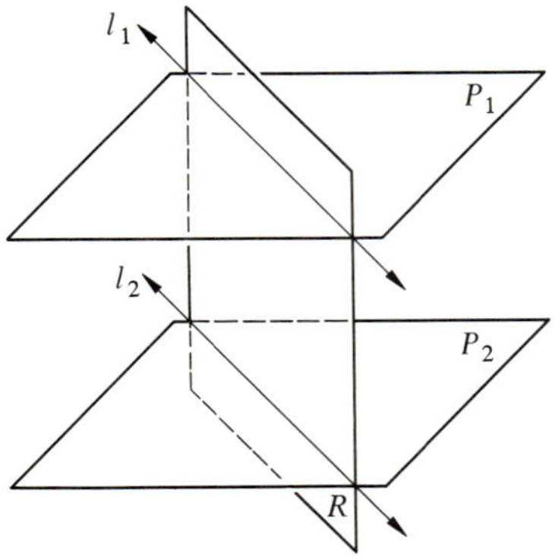

FIGURE 2-13

Proof

1. Let plane R intersect the two parallel planes P_1 and P_2.
2. By Postulate 2-5, the intersections will be lines, which we shall call l_1 and l_2.

[6] 1. In defining parallel lines, we are not assuming that they exist.

2. Bearing in mind that "do not intersect" implies an "empty intersection," we can restate Definitions 1, 3, and 4 by saying that two coplanar lines, two planes, and a line and a plane "are parallel if their intersection is empty."

3. Note that two planes either intersect or are parallel, but that two lines either intersect, or are parallel, or are skew.

3. l_1 and l_2 do not intersect, since they lie in P_1 and P_2, respectively, and P_1 and P_2 do not intersect.

4. Therefore, $l_1 \parallel l_2$ by definition.

BETWEENNESS

We said earlier that in this book we shall use the concept of *distance* to define *betweenness*. The meaning of "the distance between two points" is given by the following postulates.

POSTULATE 2-7 (Distance Postulate) *To each pair of distinct points there corresponds a unique positive number called the* distance *between the two points.*

The distance between points P and Q is denoted by PQ or QP, and it is agreed that if P and Q are the same point, then $PQ = 0$.

POSTULATE 2-8 (Ruler Postulate) *The points of a line can be placed in correspondence with the real numbers in such a way that*

1. *To every point on the line there corresponds exactly one real number.*
2. *To every real number there corresponds exactly one point of the line.*
3. *The distance between two points equals the absolute value of the difference between the corresponding numbers.*

We shall call such a correspondence a *coordinate system* for the line.[7]

We are now equipped to give a definition of "betweenness."

Definition
B is *between* A and C (FIG. 2-14) if (1) A, B, and C are distinct and collinear, and (2) $AB + BC = AC$.

We write "A-B-C" to denote that B is between A and C.

FIGURE 2-14

[7] Postulate 2-7 tells us what "distance" means, and Postulate 2-8 gives us a method for finding it. Postulate 2-8, in effect, provides us with a ruler for finding distances between any two points.

From this definition we can derive the following theorems.

THEOREM 2-6 *If A-B-C then C-B-A.*

That is, if B is between A and C, then it is also between C and A. This is true, since if $AB + BC = AC$, then $CB + BA = CA$.

THEOREM 2-7 *Of any three distinct points on a line, exactly one is between the other two.*

That is, if A-B-C, then neither B-A-C nor A-C-B is true. See FIG. 2-15.

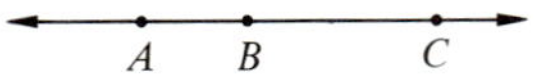

FIGURE 2-15

Proof

1. If B-A-C were true, then $BA + AC = BC$ by definition.
2. But $AB + BC = AC$ by hypothesis.
3. By adding statements 1 and 2 we get $2AB = 0$, or $AB = 0$.
4. This is false, since $A \neq B$.
5. Therefore B-A-C cannot be true.
6. We show in a similar way that A-C-B cannot be true.

THEOREM 2-8 *Given four distinct collinear points, it is possible to name them A, B, C, and D in such a way that A-B-C, A-B-D, A-C-D, and B-C-D* (FIG. 2-16).

FIGURE 2-16

In other words, given four distinct collinear points, two are "endpoints" of a particular segment and two are "inside points" of the segment.

THEOREM 2-9 *If A and B are any two distinct points, then* (1) *there is a point C such that A-B-C, and* (2) *there is a point D such that A-D-B* (FIG. 2-17).

FIGURE 2-17

THEOREM 2-10 *If A-B-C, then A, B, and C are three different points of the same line.*[8]

With the meaning of "betweenness" established, we can now define several important geometric figures that we previously described only intuitively.

Definitions

1. If A and B are two distinct points, then the set consisting of A, B, and all points between A and B is called *segment AB* and is denoted by $\overline{AB}$ (FIG. 2-18).

FIGURE 2-18

The points A and B are the *endpoints* of the segment which is said to *join A and B*.

2. Point B is called the *midpoint of* $\overline{AC}$ if (a) A-B-C, and (b) $AB = BC$ (FIG. 2-19).

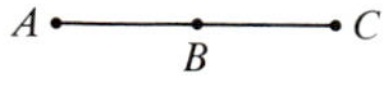

FIGURE 2-19

Point B is said to *bisect* $\overline{AC}$.

3. *Ray AB*, denoted by $\overrightarrow{AB}$, where A and B are distinct points, is the union of two sets of points: (a) the line segment $\overline{AB}$ and (b) all the points C such that point B is between points A and C, that is, A-B-C (FIG. 2-20).

[8] 1. These five theorems may be taken as the assumed "betweenness postulates" under the synthetic approach.
2. These properties are also called "order" properties because they establish the relative position of points on a line. These properties are needed to define concepts such as *opposite, interior, separation,* and *direction*.

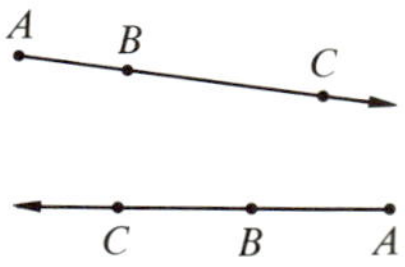

FIGURE 2-20

In other words, a ray is the set of points $\overline{AB} \cup \{C \mid A\text{-}B\text{-}C\}$.

4. If A-B-C, the rays $\overrightarrow{BA}$ and $\overrightarrow{BC}$ are called *opposite* rays (FIG. 2-21). The point B is called the *common endpoint* of the two rays.[9]

FIGURE 2-21

[9] 1. Note that a ray has only one endpoint.

2. In naming a ray, we identify its endpoint and one other point on the ray. The endpoint is always listed first. Since the order of points is important, $\overrightarrow{AB} \neq \overrightarrow{BA}$.

3. We can use the definition of a ray to define the term *direction*. We can say that "B and C (distinct from A) are in the same direction from A" means that "B and C belong to the same ray with endpoint A" (FIG. 2A).

FIGURE 2A

CONGRUENCE OF SEGMENTS[10]

Definitions

1. The *length* of a segment is the distance between its endpoints.
2. Two segments are *congruent* if they have the same length. If $\overline{AB}$ is congruent to $\overline{CD}$, we write $\overline{AB} \cong \overline{CD}$, or $AB = CD$.

The relation of congruence between segments can easily be shown to possess three important properties:

1. For any $\overline{AB}$,

   $\overline{AB} \cong \overline{AB}$ (reflexive property).

   That is, every segment is congruent to itself.

2. For any segments $\overline{AB}$ and $\overline{CD}$,

   $\overline{AB} \cong \overline{CD}$ implies $\overline{CD} \cong \overline{AB}$ (symmetric property).

   That is, if one segment is congruent to another segment, then the second segment is congruent to the first.

[10] We are now about to introduce the concept of *congruence* for geometric figures. We shall start with the simplest geometric figures, segments, and later discuss congruence for angles, for triangles, and then for any two sets of points.

The intuitive idea of congruence is that two figures are congruent if one is an exact copy of the other, that is, if both figures have the same size and shape.

How can we determine whether two figures are congruent? A good test to satisfy our intuition might be to superimpose one figure over the other and see whether they fit perfectly, that is, coincide.

But such a test has certain shortcomings. First, it appears to be more physical than mathematical. Second, how do we apply such a test to, say, two footballs? How does one place one football on top of another football

3. For any segments $\overline{AB}$, $\overline{CD}$, $\overline{EF}$,

if $\overline{AB} \cong \overline{CD}$ and $\overline{CD} \cong \overline{EF}$, then $\overline{AB} \cong \overline{EF}$

(transitive property).

That is, if one segment is congruent to a second segment and the second segment is congruent to a third segment, then the first segment is congruent to the third segment.

Relations that are reflexive, symmetric, and transitive are called *equivalence relations*. Since congruence of segments can be shown to possess these three properties, we have

THEOREM 2-11 Congruence of segments is an equivalence relation.

to see whether they coincide? Third, an implied assumption of such a test is that we can "move" a geometric figure from one place to another and yet preserve its size and shape.

Our object here is to formulate a *mathematical* theory of congruence with which we can establish properties of geometric figures. To formulate such a theory, we can use either of the two approaches mentioned on page 28, the metric or the synthetic. Our approach will be to assume the existence of the real number system and then base our theory of congruence on the notion of measurement.

We begin with congruence of segments.

CONVEX SETS

Definition
A set of points S is *convex* if for every two distinct points A and B of S, the entire segment $\overline{AB}$ is in S.

For example, the sets in FIG. 2-22 are convex, and the sets in FIG. 2-23 are not convex.[11]

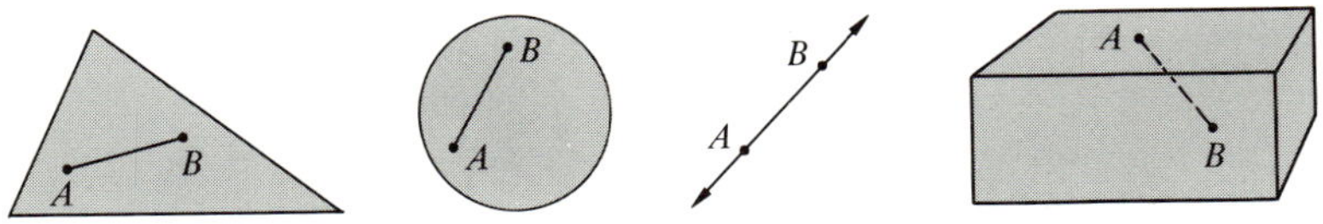

FIGURE 2-22

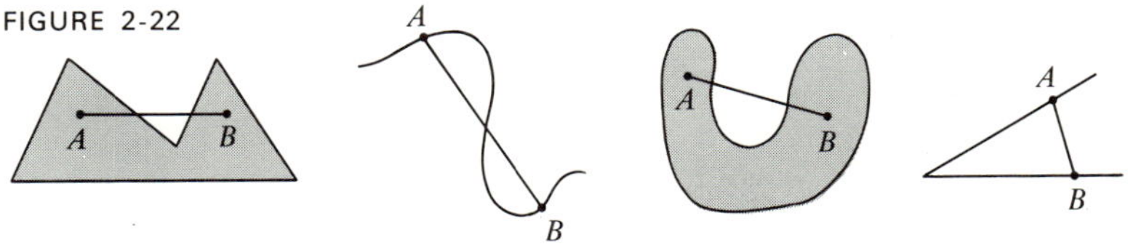

FIGURE 2-23

The following are consequences of the definition of convexity:

1. Any line is convex, since $\overline{AB} \subset \overleftrightarrow{AB}$.

[11] 1. We can restate the last definition as follows: A set of points S is said to be *convex* if $A, B \subset S$ and $A \neq B$ always imply $\overline{AB} \subset S$.

2. Examples of convex objects are a ball bearing or a cube of sugar. Examples of objects that are not convex are the surfaces of a wastepaper basket or a spoon.

2. Any plane is convex.
3. Space (the set of all points) is convex.
4. The empty set is convex since there are *no* two distinct points A and B of this set such that $\overline{AB}$ is *not* in the set.
5. A set consisting of one point is convex for the same reason as in 4.
6. Any segment is convex.
7. Any ray is convex.

THEOREM 2-12 *The intersection of any two convex sets is also a convex set.*

Proof

To prove Theorem 2-12, let S and T be two convex sets, and let distinct points A and B belong to $S \cap T$. Then

1. By the definition of intersection, A and B belong to set S and to set T.
2. Since S and T are convex sets, $\overline{AB} \subset S$ and $\overline{AB} \subset T$.
3. By the definition of intersection, $\overline{AB} \subset (S \cap T)$.
4. Therefore, by the definition of convexity, $S \cap T$ is a convex set.

A *corollary* to Theorem 2-12 is that the intersection of a finite number of convex sets is a convex set.

SEPARATION[12]

Line Separation

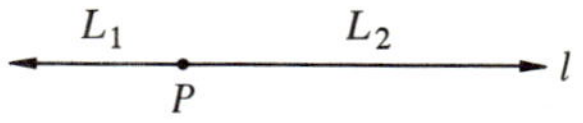

FIGURE 2-24

Consider line l and a point P on l (FIG. 2-24). We say that P *separates* l into the following two sets of points:

1. The set L_1 on one side of P.

[12] 1. We shall now see how basic geometric figures can be divided, or *separated*, into other figures. But before examining this idea, we note that the ideas of *separation* find frequent application in everyday situations. For example,

a. A room divider (screen) "separates" the room into the set of points in front of the screen and the set of points behind the screen.

b. An aisle in an auditorium "separates" the seats on the left side from the seats on the right side.

c. Any mark on the ruler, say the 5-inch mark, "separates" the set of points before the 5-inch mark from the set of points after the 5-inch mark.

2. Although the idea of *separation* is suggested by these physical illustrations, it is important to remember that *geometric separation* involving abstractions such as points, lines, and planes is not a physical process but an interrelationship among sets, as we shall see soon.

2. The set L_2 on the other side of P.

The point P itself is not in either set.

L_1 and L_2 are called *half-lines.* A half-line is like a ray without the endpoint. Like a ray, a half-line is convex.

Note that a line segment has two endpoints, a ray has one endpoint, and a half-line has no endpoints.

If two points on line l belong to the same half-line, we say they are *on the same side of P.* If they are on different half-lines, then they are *on opposite sides of P.*

We now give a more formal definition of *separation.*

Definition

Point P *separates* line l into the nonempty sets L_1 and L_2 if the following conditions are satisfied (FIG. 2-25):

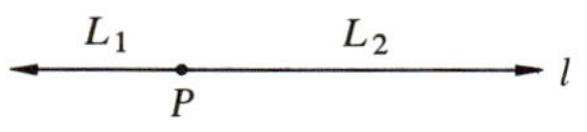

FIGURE 2-25

1. $l = L_1 \cup L_2 \cup P$, where L_1, L_2, and P are disjoint.
2. P is between each point of L_1 and each point of L_2.
3. P is not between two points of L_1 or of L_2.

Plane Separation

Intuitively, we might expect the separation of a plane by a line to be analogous to the separation of a line by a point.

Just as a point separates a line into two half-lines, so might we expect a line to separate a plane into two *half-planes.* The line would serve as a *boundary* between the two half-planes and would not belong to either. Just as the half-lines are convex, so might we expect the half-planes to be convex.

Although this analogy appears to be reasonable, we cannot accept it formally until its ideas are incorporated formally into our system. We do this by means of a *Plane-Separation Postulate.*

POSTULATE 2-9 (Plane-Separation Postulate) Given a line l and a plane X containing it, the set of all points of the plane that do not lie on l is the union of two disjoint sets such that

1. *Each of the sets is convex.*

2. *If point A belongs to one set and point B to the other, then* $\overline{AB}$ *intersects l* (FIG. 2-26).

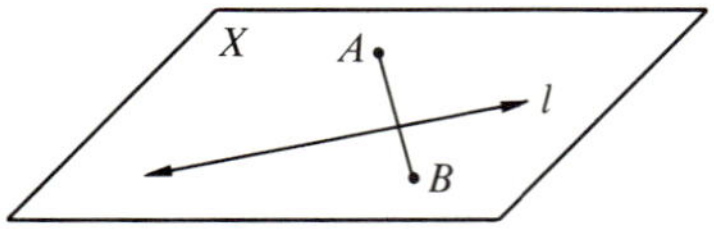

FIGURE 2-26

Definitions

1. Given a line l and a plane X containing l and the two sets P_1 and P_2, as given in Postulate 2-9, then each of the sets P_1 and P_2 is called a *half-plane* and l is called the *edge* of each of them (FIG. 2-27).

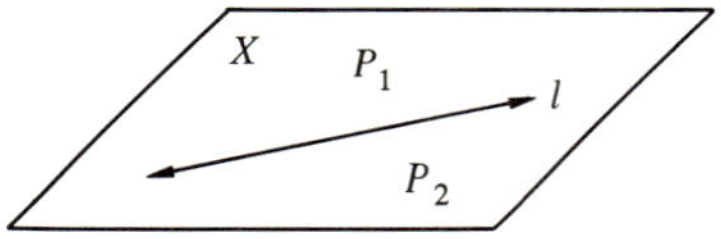

FIGURE 2-27

2. We say that l separates X into two half-planes called *opposite half-planes.*
3. If two points A and B of the plane lie in the same half-plane, we say that they lie *on the same side of l.* If A lies in one of the half-planes and B in the other, then they lie *on opposite sides of l.*

Since every half-plane is determined by a line l and a point A not on l (FIG. 2-28), we shall denote such a half-plane by l/A, read "the half-plane determined by line l and containing point A."[13]

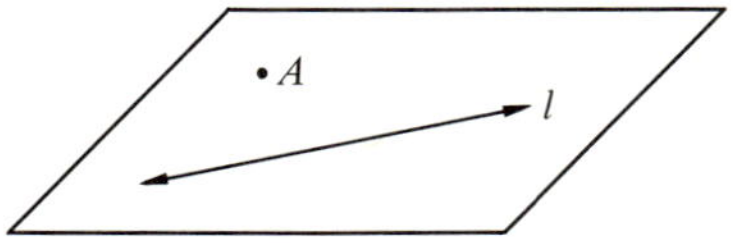

FIGURE 2-28

[13] 1. If we do not restrict ourselves to one plane, we can have infinitely many half-planes having ℓ as an edge.

Suggest to students the various positions of a revolving door or the pages of a book as representing different half-planes having the same edge.

2. Compare Postulate 2-9 with the definition of *separation* given on page 41.

3. Postulate 2-9 says, in effect, that any line of a plane separates the plane into three disjoint sets of points:

a. The set of points on one side of the line.

b. The set of points on the other side of the line.

c. The set of points of the line itself.

4. A half-plane can be seen, intuitively, as a generalization of a ray. Think of a *ray* as the set of points of a line that are on a given side of a given point of the line.

Think of a *half-plane* as the set of points of a plane that are on a given side of a given line of the plane.

Space Separation

As might have been expected, we shall now generalize the ideas of line separation and plane separation to space separation by means of a *Space-Separation Postulate.*

> *POSTULATE 2-10 (Space-Separation Postulate)* *Given a plane in space, the set of all points that do not lie in the plane is the union of two disjoint sets S_1 and S_2 such that*
>
> 1. *Each of the sets is convex.*
> 2. *If A belongs to one of the sets and B belongs to the other, then $\overline{AB}$ intersects the plane.*

Definition
S_1 and S_2 in Postulate 2-10 are called *half-spaces*, and the plane is called the *face* of each half-space.[14]

SIMPLE CLOSED CURVES[15]

A *curve* is a set of points that can be represented by a pencil drawing without lifting the pencil off the paper. Curves are, therefore, *connected*, or *continuous* sets of points.

Examples of curves are given in FIG. 2-29. Examples of sets of points that are *not* curves are given in FIG. 2-30.

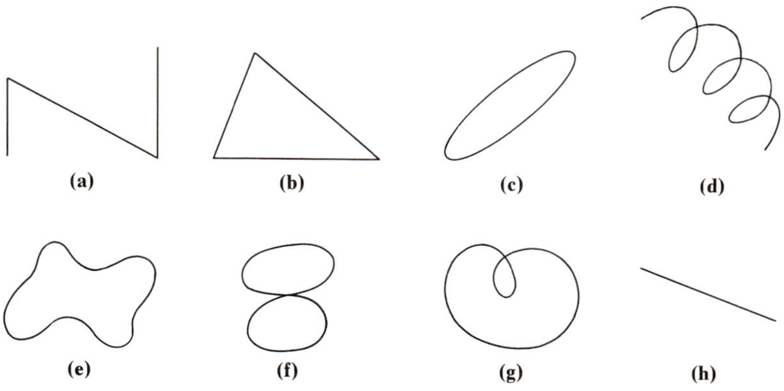

FIGURE 2-29

[14] 1. Although a line is an edge of infinitely many half-planes, a plane is a face of only two half-spaces.

2. Postulate 2-10 says, in effect, that any plane separates space into three disjoint sets of points:

a. The set of points on one side of the plane.
b. The set of points on the other side of the plane.
c. The set of points of the plane itself.

3. The Plane-Separation Postulate can be used to *prove* the Space-Separation Postulate, which would then become a theorem. For a proof, see C. R. Wylie, *Foundations of Geometry*, New York: McGraw-Hill Book Company, Inc., 1964, p. 65.

4. The concepts related to "separation," as they move from half-lines to half-planes to half-spaces, help to establish in the student's mind a sense of continuity in the study of geometry.

[15] We shall now consider, informally, special sets of points called *curves.* The ideas will be presented on an intuitive level only.

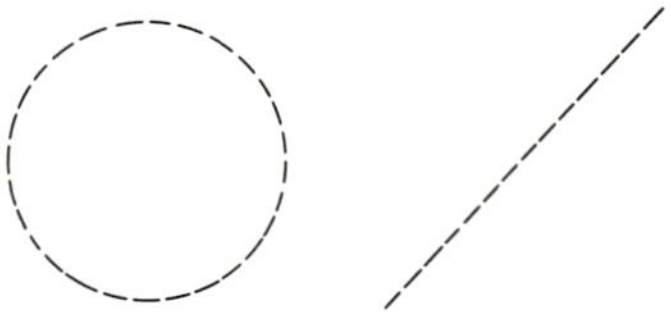

FIGURE 2-30

A *closed* curve is a curve that returns to its starting point. In the curves shown in FIG. 2-29, (b), (c), (e), (f), and (g) are closed curves, and (a), (d), and (h) are not closed curves.

A planar closed curve that does not cross itself is called a *simple closed curve*. In FIG. 2-29, (b), (c), and (e) are simple closed curves, and (a), (d), (f), (g), and (h) are not simple closed curves. Since (f) and (g) are closed but do not cross themselves, they are *non-simple* closed curves.[16]

The ideas of *separation* apply to simple closed curves. This was proved as a theorem by Camille Jordan (1838–1922) and is known as the *Jordan Curve Theorem*. The theorem states that

> *A simple closed curve C in the plane separates the set of all points in the plane not on the curve into two sets of points, an* interior *and an* exterior (FIG. 2-31), *such that*

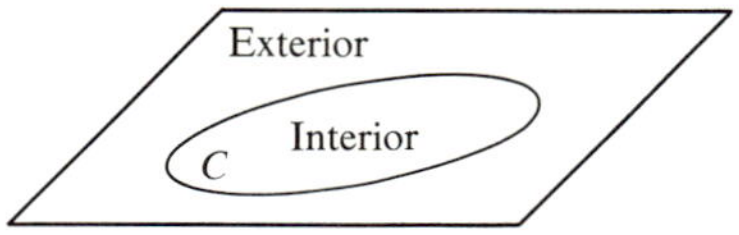

FIGURE 2-31

1. *Any pair of points of the same set can be joined by a curve (in the plane) that does not intersect C* (FIG. 2-32).

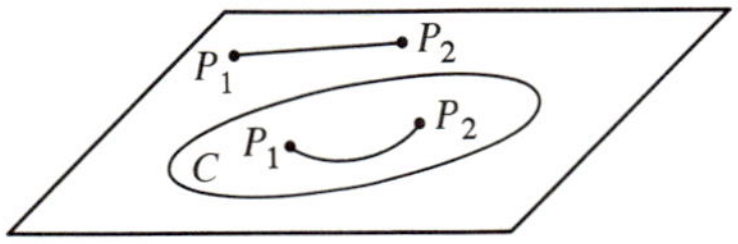

FIGURE 2-32

[16] 1. Note that any triangle, circle, parallelogram, or regular polygon is a simple closed curve. Also note that not every quadrilateral is a simple closed curve as, for example, the quadrilateral shown in FIG. 2B.

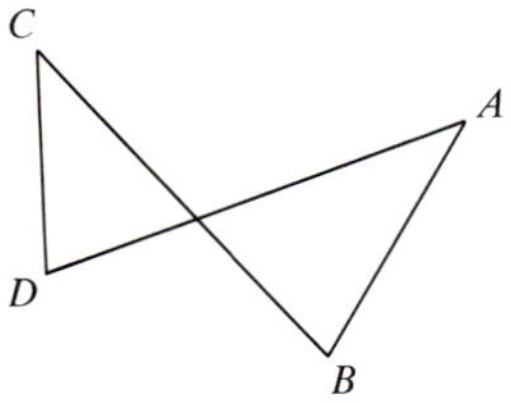

FIGURE 2B

2. Identify simple closed curves in the room, such as the boundary of the door, windows, clock, etc.

3. Exhibit more elaborate curves, for example, FIG. 2C. Then ask the class to determine whether they are simple closed curves.

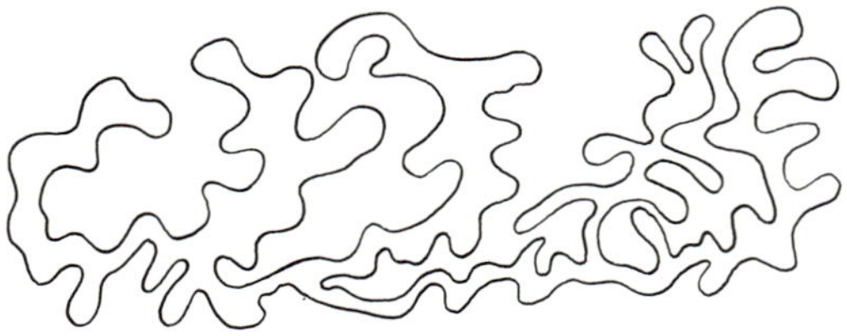

FIGURE 2C

4. The description of a *curve* given first in this section is the "naive" concept of curve. There are curves constructed by Giuseppi Peano and others that are infinitely crinkled and whose points completely fill space. Such curves cannot be represented by a pencil drawing.

5. What, *in three dimensions*, corresponds to a simple closed curve or to a nonsimple closed curve? (For an interesting discussion of this question, see E. Kasner and J. Newman, *Mathematics and the Imagination*, New York: Simon and Schuster, Inc., 1940, pp. 278–287.)

2. *Any curve (in the plane) joining a pair of points belonging to different sets must intersect C* (FIG. 2-33).

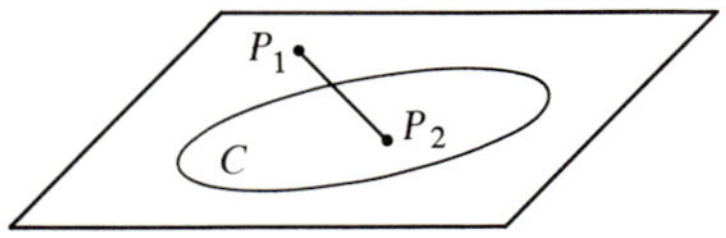

FIGURE 2-33

A planar simple closed curve *and* its interior is called a *region.* The curve itself is called the *boundary* of the region. (See FIG. 2-34.)

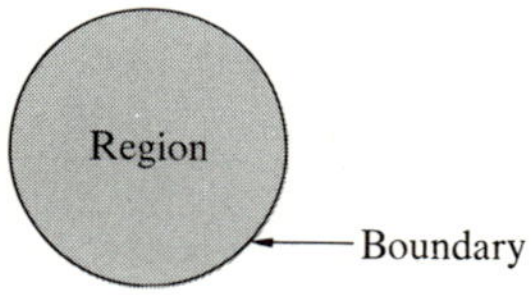

FIGURE 2-34

TEACHING SUGGESTIONS AND ACTIVITIES

VISUAL AIDS

Visual 1

PURPOSE

To illustrate point–line, point–plane, and line–plane relations.

CONSISTS OF

1. Different colored dowels, string, or wire to represent "lines."
2. Pegboard.
3. Pegs (three colors, three different lengths) to represent "points."
4. Pieces of stiff cardboard (or plastic or plywood) to represent "planes."

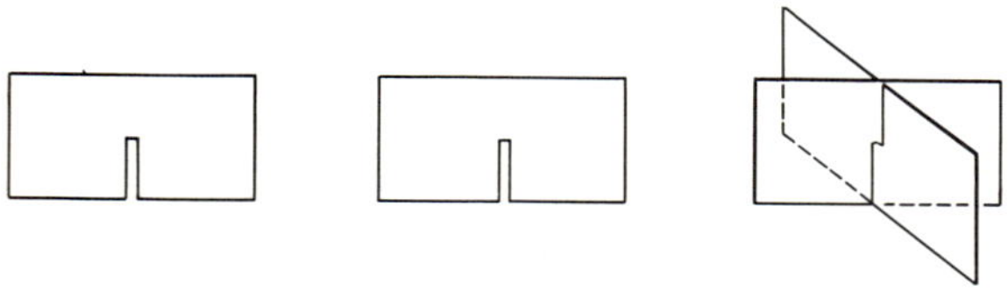

FIGURE 2-35

HOW TO USE

1. Cut two "planes" halfway through, as shown in FIG. 2-35, to illustrate that if two distinct planes intersect, then their intersection is a line (Postulate 2-5).
2. Cut three "planes" halfway through, as shown in FIG. 2-36, to illustrate that if a plane intersects two parallel planes, then it intersects them in two parallel lines (Theorem 2-5).
3. To illustrate that (1) infinitely many planes can pass through two points in space and (2) three different noncollinear points are contained in exactly one plane (Postulate 2-3),
   a. Insert two "points" into the pegboard. Take a "plane" and show there are *many* positions in which a plane can be held against two points.
   b. Now insert a third "point" into the pegboard so that the three "points" are noncollinear. Show that there is only *one* way to hold a flat surface against the three points. Although we may slide that surface, it represents just one plane. In other words, through any three noncollinear points there is only one plane.

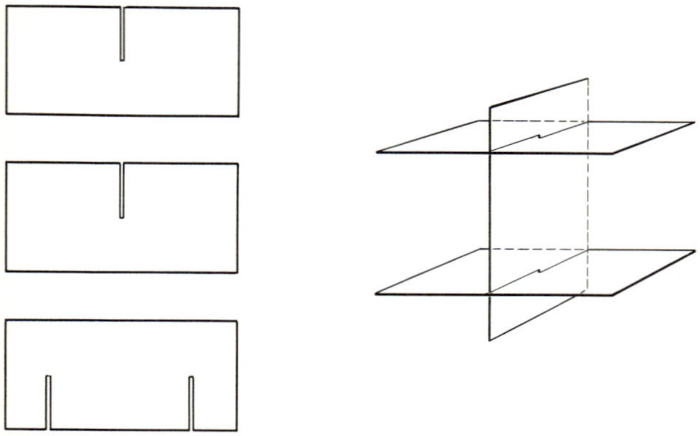

FIGURE 2-36

c. Insert three noncollinear pegs of different heights. Place a flat surface against the three pegs and see what happens. Add a fourth peg and see what happens.
d. Insert three noncollinear pegs of different heights and turn the pegboard over and let the heads of the pegs rest on top of your desk. Does the pegboard rest solidly against your desk top?

Now add a fourth peg and repeat the above procedure. What happens? What does this illustrate?

Note: Other ways to illustrate the ideas in item 3 are

1. Suggest the various positions of a revolving door or the pages of a book as representing different planes passing through the same two points.
2. Take a rectangular-shaped card and hold opposite corners by your thumb and a finger. Note that the card can spin freely and that each position of the card represents a different plane.

But, if you extend another finger to the card, the card can no longer spin. The three points represented by the tips of your thumb and two fingers determine one and only one position of the card.

4. Use these materials to illustrate Theorems 2-1–2-4.

Visual 2

PURPOSE

To provide model drawings of various combinations and positions of points, lines, and planes. Students can use these as guides for their own drawings.

CONSISTS OF

Sets of drawings like those shown on page 48.

HOW TO USE

1. As each drawing is displayed for the first time, describe the relationships among the parts in mathematical language. (See pages 48–49.)
2. Provide sufficient opportunities for students to make these and similar drawings to represent space relationships.

DRAWING AND DESCRIBING DIAGRAMS[17]

Planes are usually represented as shown in FIG. 2-37.

The fact that C is a point of line m in FIG. 2-38 can be stated in several ways:

1. C lies on m.
2. C is a point of m.
3. m passes through C.
4. $C \in m$.
5. C is contained in m.

The intersection of lines m and n in FIG. 2-39 may be stated in the following ways:

1. m and n intersect at P.
2. $m \cap n = \{P\}$.

Figure 2-40 shows the following:

1. A lies in plane R.
2. n lies in plane R.
3. Plane R contains A and n.
4. $n \subset R$.
5. R passes through n.

In FIG. 2-41,

1. T passes through r.
2. T contains r.
3. T intersects s.
4. T does not pass through s.

In FIG. 2-42,

1. S and T intersect.
2. $\overleftrightarrow{AB}$ is the line of intersection of S and T.

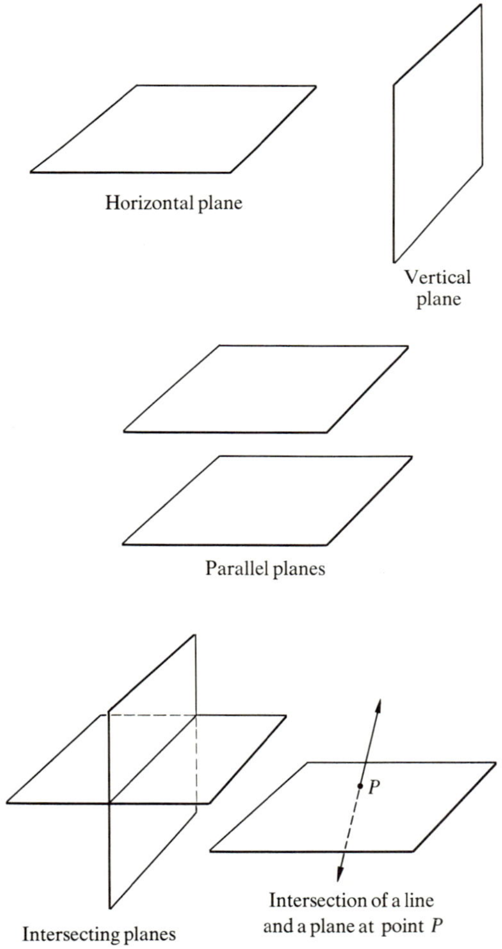

FIGURE 2-37

[17] Because many students have difficulty drawing and describing accurately diagrams that represent space relationships, it would be well to take class time to provide for adequate student experience with such diagrams. Illustrative diagrams and their descriptions follow.

3. $\overleftrightarrow{AB}$ lies in plane S and in plane T.
4. S and T pass through $\overleftrightarrow{AB}$.
5. $\overleftrightarrow{AB}$ is contained in S and T.

Note: Ask students to think of and then describe relationships contained in FIGS. 2-38 through 2-42 that have not been mentioned.

FIGURE 2-38

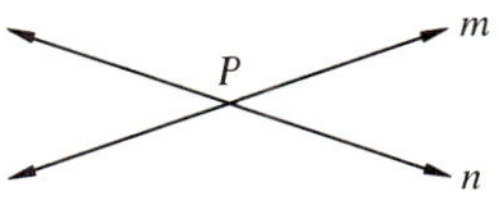

FIGURE 2-39

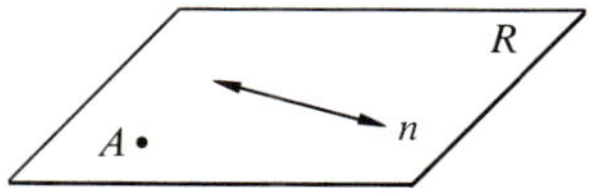

FIGURE 2-40

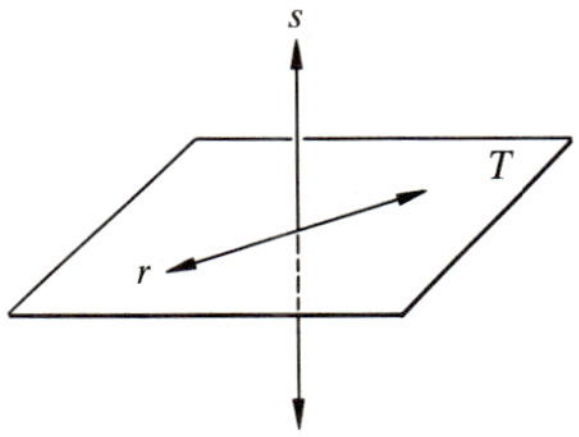

FIGURE 2-41

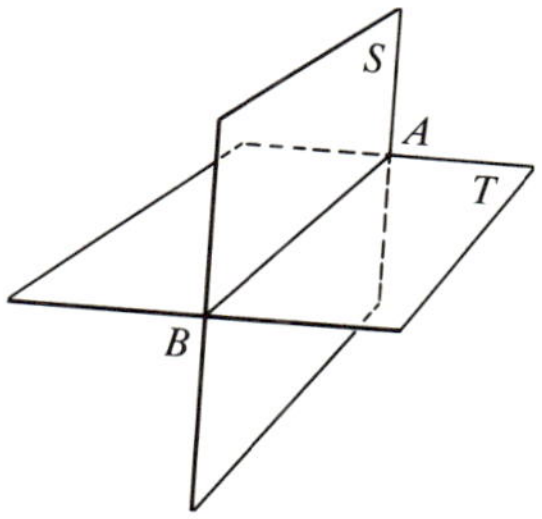

FIGURE 2-42

RANDOM SUGGESTIONS

In Teaching This Chapter

1. Use many examples from the classroom and the students' daily life to illustrate points, lines, and planes. Identify situations in the room that illustrate different cases of intersection of lines and planes.
2. Encourage students to express their ideas in clear and precise statements.
3. Encourage students to make simple models that will help them visualize spatial relationships.
4. Point out the similarity between using numerals to represent numbers and using drawings to represent points, lines, and planes. Let students see that points, lines, and planes—like numbers—are abstractions that exist only in our minds and that they have no physical existence.
5. Use a variety of visual aids, but the ultimate goal is to understand the ideas with little dependence on their physical representations.

Stating the Postulates

After students *understand* a postulate, they should be expected to be able to state it. For example, students should be able to state that "three different noncollinear points are contained in exactly one plane," or that "if two planes intersect, then their intersection is a line." To encourage understanding rather than mere memorization, ask students to reformulate these statements in their own language and then see whether the two statements of the same postulate say the same thing. Also, where diagrams are involved, change frequently the lettering on the figures.

Bulletin Board Displays

Bulletin board displays for this chapter might include

1. Pictures or models of geometric figures that can be found in architecture, advertising, nature, sports, industry, fashions, articles in the home, science, and art.
2. Materials related to the history of geometry such as photographs of famous geometers, Egyptian rope stretchers, and unsolved problems in geometry.
3. Geometry in the news: newspaper and magazine clippings that report on geometric concepts involved in such fields as space flights, medicine, and new inventions.
4. Puzzle of the week: intriguing problems and puzzles involving geometry.

The Form of a Theorem

A theorem is usually a statement that *if* something is true, *then* something else is also true. The *if* part of the theorem is called the *hypothesis*, or the given information. The *then* part is called the *conclusion*, or the thing to be proved.

Consider, for example, this theorem: "If two distinct lines intersect, then there is exactly one plane containing them." Here, the *hypothesis* is that two given distinct lines intersect. The *conclusion* is that they determine exactly one plane.

Not all theorems are stated in the *if . . . then* form. An example of a theorem that is not stated in this form is "The intersection of any two convex sets is also a convex set." In such a case, it is important for students to be able to restate the theorem in the *if . . . then* form. This will help them recognize more easily what data are given and what has to be proved. Students should, therefore, be able to restate the above theorem in this form: "If two convex sets intersect, then their intersection is also a convex set."

Abstractions and Their Representations

When we first begin talking about "points," "lines," and "planes" with students in the lower grades, we must continually make the distinction between points, lines, and planes as mathematical abstractions and their physical representations.

A way of sharpening this distinction with a "point," for example, is to mark a large dot on the board and then mark smaller and smaller succeeding dots. Then ask which dot is the best representation of a geometric "point."

The following are illustrative questions to help students understand the concept of a "point." Mark a large dot on the board, and then ask

1. Is this a "point?"
2. If I erase the dot, what happens to the point?
3. How many points are covered by this dot?
4. Does a big dot represent a larger point than a small dot?
5. Did you ever see a "point?"

QUESTIONS

1. a. If three points are coplanar, must they also be collinear?

   b. If three points are collinear, must they also be coplanar?

2. Draw a simple closed curve. Locate point A inside the curve and point B outside the curve such that the intersection of $\overline{AB}$ and the curve is three points.

3. Given four points with no three of them on a line and not all in one plane. If planes are passed through every set of three points, how many planes are there altogether?

4. What is the most number of planes determined by the ends of four legs of a table if no two legs have the same length?

5. How many different lines may contain
   a. one point?
   b. a pair of points?

6. How many different planes may contain
   a. one point?
   b. a pair of points?
   c. a set of three noncollinear points?

7. If A, B, and C are three distinct points and all are in each of two different planes, what will be true about these points?

8. Give a mathematical reason for the fact that a four-legged table sometimes rocks, but a three-legged table is always stable.

9. If two planes have a common point, does it follow that they have more than one common point? Why?

10. If a point is removed from a line, is the remaining set of points convex? Why?

11. Restate the following postulates in the *if . . . then* form:

    a. Three different noncollinear points are contained in exactly one plane.
    b. Every line contains at least two points.

12. Is the statement "If a figure does not contain at least three noncollinear points, then it is not a plane" equivalent to the statement "Every plane contains at least three noncollinear points"? Why?

13. Given the following mathematical system:

Undefined terms:	azie, beezie, woozie.
Postulate 1:	Exactly one azie contains any two distinct woozies.
Postulate 2:	If two distinct woozies lie in a beezie, then the azie containing these woozies lies in the same beezie.
Postulate 3:	Azies and beezies are sets of woozies.

    a. Prove that if two distinct azies intersect, then their intersection contains exactly one woozie.
    b. Derive as many other theorems as you can from this system.

PUZZLES

1. In the accompanying figure is a set of six squares made with 17 line segments. How can you change the six squares into three squares by removing 5 line segments?

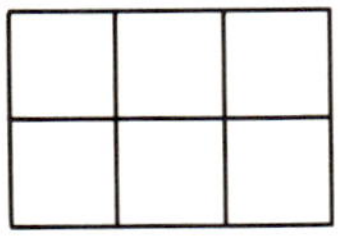

SOLUTIONS

1.

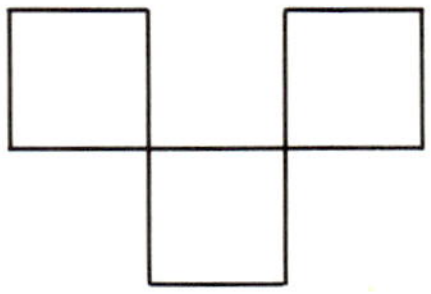

2. Without taking your pencil off the paper, draw 4 line segments that pass through every point.

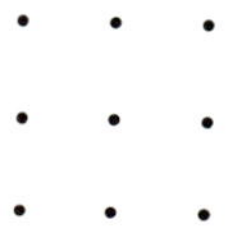

3. Without taking your pencil off the paper, draw 6 line segments that pass through every point.

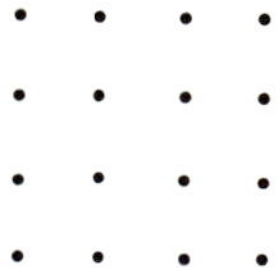

4. Starting at point *P*, draw the figure shown without lifting your pencil from the paper, without tracing any part of the figure more than once, and without crossing any part of the figure.

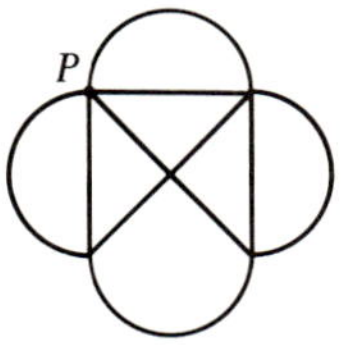

5. Connect like numerals by lines that do not cross each other or any other lines in the figures shown.

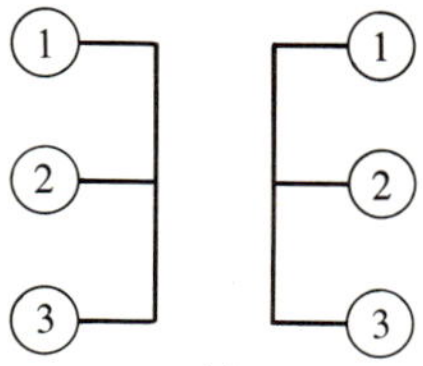

(a)

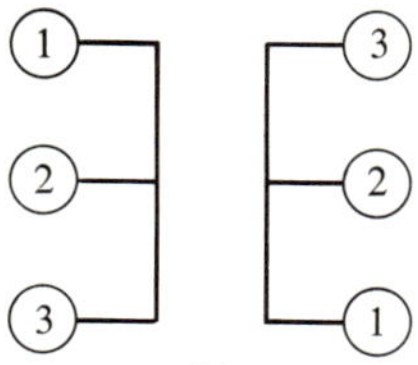

(b)

2.

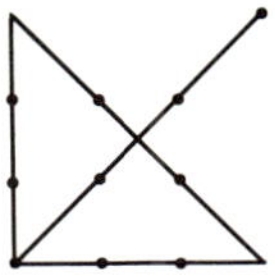

3.

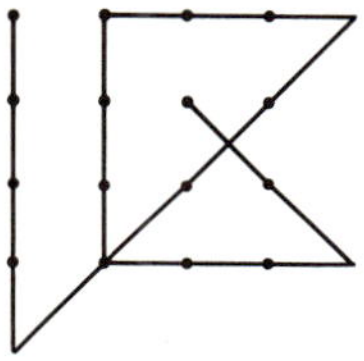

4.

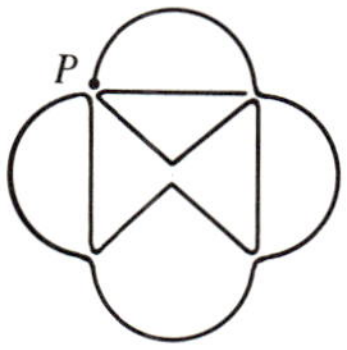

5. a.

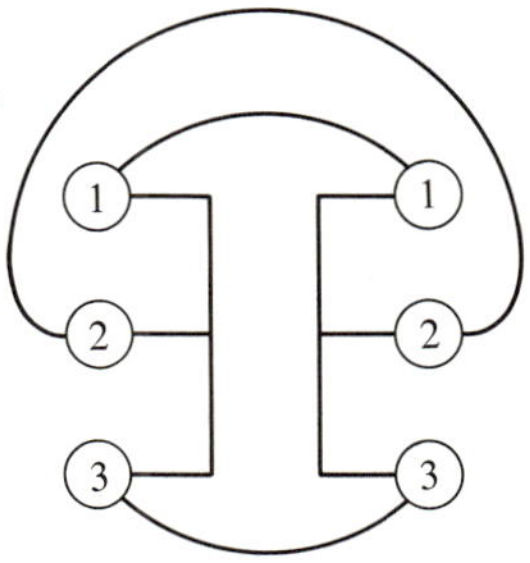

b. It is impossible to connect like numerals in this figure without crossing a line. We can see intuitively why this is so. When we connect 1 to 1 and 2 to 2, we obtain a simple closed curve. This leaves one 3 inside and the other 3 outside the simple closed curve, and by the Jordan Curve Theorem you cannot get from the inside to the outside of a simple closed curve without intersecting the curve.

(*Note:* It is not always easy to distinguish between the inside and the outside of a simple closed curve. There is the story of the intoxicated man who was found one morning going round and round a great stone pillar in front of a bank building and moaning to himself. When asked by a passerby why he was so sad he replied, "Look, I'm walled in.")

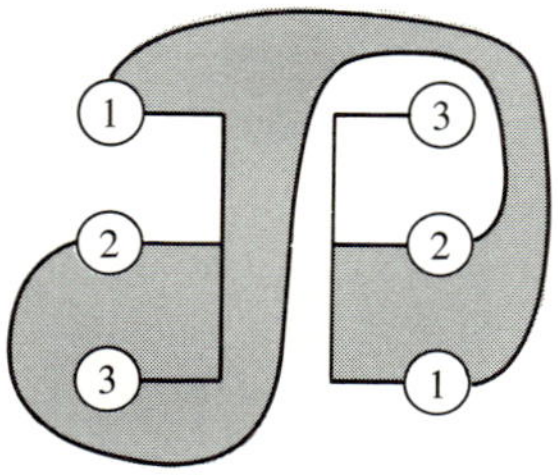

ENRICHMENT

THE NATURE OF GEOMETRY

Although geometry may have had its beginnings in land measurement, today we think of geometry as a perfect example of an abstract science. Such a system starts with a few undefined terms, and these are then used to define new terms. We then introduce some statements, called postulates, that we assume to be true. By applying deductive reasoning to these undefined terms, definitions, and postulates, we prove that other statements, called theorems, are also true. Later, these theorems help us to prove other theorems.

Although our choice of postulates may have been suggested by the physical world, this need not be their origin. The only *necessary* condition for our choices is that the system be internally *consistent*, that is, that no postulate within the system contradicts any other postulate in the system. This condition ensures us against proving two contradictory theorems from the same set of postulates.

Besides the necessary criterion of consistency, there are certain *desirable* criteria for the choice of a set of postulates:

1. *Independence*, that is, that no postulate can be derived from the other postulates. This condition ensures that we do not have any redundant postulates. Independence in our postulates is desirable aesthetically, but sometimes it is not desirable pedagogically, because the less we assume, the harder may be our proofs.

2. *Completeness*, that is, that the postulate system be sufficient to establish or disestablish any statement about the primitive terms of the discourse.

3. *Meaningfulness*, that is, that the postulates have some useful application to other branches of mathematics.

Establishing the Criteria for a Set of Postulates

How to prove that a set of postulates is *consistent* has not yet been answered to everybody's satisfaction. An approach is suggested on page 151 in connection with the consistency of non-Euclidean geometries. There we shall see that consistency of a set of postulates is established by relating it to a model that we know, or believe to be, consistent. Once a single model of the system has been found, the consistency of the system is established, and no additional models are needed.

Unlike consistency, which is required, *independence* of the postulates is only desirable, desirable because it prevents us from cluttering our system with superfluous postulates. This makes our system more economical and more elegant. For pedagogical reasons, however, independence is sometimes sacrificed to make learning easier or more interesting.

Like consistency, *independence* is also established through the use of models. But here we need as many models as there are postulates. To show that a postulate is independent of the other postulates in the system, it is sufficient to show that the other postulates, together with some contradiction of the postulate in question, form a consistent set of postulates.

For example, let us assume we have n postulates P_1, P_2, $P_3, \ldots, P_n$ in our system and we wish to test the independence of P_1. If we can exhibit a model in which $P_2, P_3, \ldots, P_n$ are true but in which P_1 is false, then P_1 must be independent of the other postulates. This is so because if P_1 were dependent, that is, if it were a consequence of the other postulates, then it, too, would have to be true in our model.

To prove *independence* of the entire system, we must construct a separate model for each postulate.

To prove *completeness* of a mathematical system, we must prove that it is impossible to enlarge it by adding any other postulate that is consistent with and independent of the other postulates. This is not easy to do. To attempt to write such a set of postulates for a mathematical system, in the first place, implies that we have a complete understanding of the essential properties of this system. It is often very difficult to prove that we have succeeded. We cope with this difficulty by proving completeness *indirectly*. It can be shown that if an axiomatic system is *categorical*, then it must be *complete*. Therefore, by proving that a system is *categorical* we prove that it is *complete*. Before defining a *categorical* system, we shall define a related concept.

We shall soon see that an abstract axiomatic system may have a variety of interpretations or models, some of which may appear to be very different from each other. It sometimes turns out, however, that the structures within these different models are really the same, differing only in the names given the undefined terms and relations.

Whenever a one-to-one correspondence exists between the elements of two models such that all true statements in the first model can be translated into true statements in the other model by the substitution of corresponding elements, then the two systems are said to be *isomorphic*. This means that for two systems to be *isomorphic*, there must exist a one-to-one correspondence between the elements of the two systems which preserves all relations existing in either system.

Now, a set of postulates is *categorical* if and only if *each* of its models is isomorphic to every other model. This means that a system is *categorical* if and only if *all* its models are structurally identical; that is, all the models are really one model with different names for the elements and relations.

Geometry and the Physical World

Because our assumptions in a mathematical system are not necessarily true in the physical world, the theorems we prove are not necessarily true in the physical world. When we prove a theorem in geometry we are saying only that the theorem is true in the mathematical system described by the postulates.

Although geometric ideas are often suggested by objects, situations, or problems in the physical world, our main interest does not lie in any one specific object, situation, or problem but in the properties *common* to all *similar* objects, situations, or problems.

For example, when we study "points" in geometry, we are not concerned with each separate representation of a point, such as the head of a pin. We are more interested in the properties *common* to all such representations, such as that they all represent exact locations in space. This approach enables us to study *all* situations involving points rather than to study each situation separately.

HISTORY OF GEOMETRY

Geometry of the Egyptians

Geometry had its beginnings in measurement. The ancient Egyptians were concerned with land measures because each year they had to restore washed-out property boundaries caused by the annual floods of the Nile River. As a result, they became very proficient in dealing with these practical aspects of geometry. They undoubtedly knew some of the fundamental formulas for measurement. In the construction of their pyramids, the ancient Egyptians showed such remarkable accuracy that to this day we do not fully know how they achieved it.

THALES - 600 B.C.
PYTHAGORAS - 540 B.C.
ARISTOTLE - (384 - 322 B.C.)
EUCLID - 300 B.C.
ARCHIMEDES (287-212 BC)

Geometry of the Greeks

The Greeks moved away from this practical concept of geometry and, instead, developed geometry as a deductive system. In about 300 B.C., Euclid was the first to organize and systematize geometry into the deductive science which we still study today. Other important Greek mathematicians were Thales (600 B.C.), who was the first to prove that any circle is bisected by its diameter; Pythagoras (540 B.C.), who is believed to have proved the Pythagorean Theorem; Aristotle (384–322 B.C.), who developed the logic that we still use today; and Archimedes (287–212 B.C.), one of the greatest mathematicians of all time, who developed some of the methods of the integral calculus and who also found a method of computing π.

DESCARTES, RENÉ (1596-1650)
FERMAT, PIERRE (1601-1655)

Analytic Geometry

In the seventeenth century, René Descartes (1596–1650) "algebratized" geometry by his invention of *analytic geometry.* What he invented was a method of combining algebra with geometry to produce what is considered one of the most useful and revolutionary inventions of the human mind.

His method makes use of two intersecting straight lines on a plane and a way of locating points on the plane with reference to these lines. Analytic geometry was the first major advancement from classical Greek geometry and became a major factor in the development of the calculus and modern mathematics.

Pierre de Fermat (1601–1665) discovered analytic geometry independently of Descartes.

DESARGUES, GÉRARD (1593-1662)
PASCAL, BLAISE (1623-1662)

Projective Geometry

Out of the experiments with perspective by seventeenth-century artists grew an entirely new kind of geometry, known as *projective geometry.* In this geometry the basic concepts are *projection* and *section.*

A *projection* may be thought of as a set of lines of light from the eye to the points of an object (FIG. 2-43). A *section* is the pattern formed by the intersection of these lines with a glass screen placed between the eye and the object.

A main goal of projective geometry is to find properties of geometric figures that remain invariant for any section of any projection of those figures. These properties have been shown to involve collinearity of points, concurrence of lines, cross-ratio, and the principle of duality according to which we can interchange "point" and "line" in a theorem and obtain a valid statement.

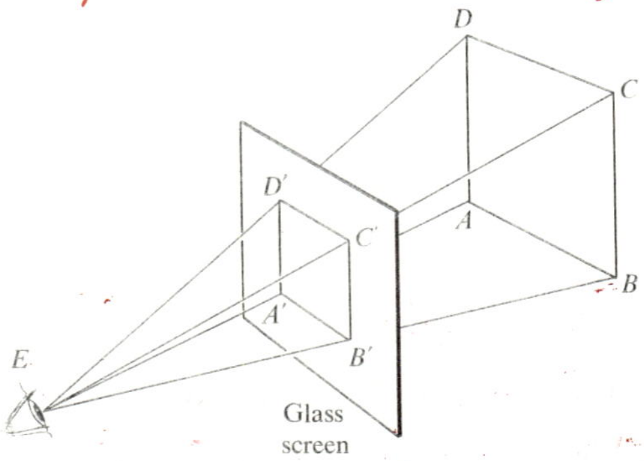

FIGURE 2-43

In contrast to projective geometry, Euclidean geometry seeks to study properties of geometric figures that remain invariant under rigid motion. These properties involve equality of lengths, angles, and areas.

Gérard Desargues (1593–1662) and Blaise Pascal (1623–1662) made important contributions to projective geometry.

Non-Euclidean Geometry

For two thousand years most mathematicians accepted as a "self-evident" truth Euclid's postulate which implied that there exists only one line parallel to a given line through a given point not on the line. They also assumed that the geometry of Euclid was the only geometry possible.

But in the last 200 years mathematicians have successfully challenged these beliefs. They have shown that it is possible, for example, to build consistent and useful geometries with a postulate that contradicts Euclid's parallel postulate. These geometries are called *non-Euclidean geometries.*

In the early part of the nineteenth century, three mathematicians, working independently, developed a geometry with a postulate that assumed that *more* than one parallel can be drawn to a given line through a given point not on the line. This geometry, later named *hyperbolic geometry*, was developed by N. I. Lobachevsky (1793–1856), Janos Bolyai (1802–1860), and K. F. Gauss (1777–1855).

Later, G. F. B. Riemann (1826–1866) assumed that *no* line can be drawn parallel to a given line through a point not on the line. This resulted in the creation of another new and consistent geometry which differed from that of Euclid and from that of Lobachevsky, Bolyai, and Gauss. Riemann, in fact, created a whole class of new geometries, one of which was the basis for Einstein's general theory of relativity.

The discovery of these equally valid and consistent non-Euclidean geometries destroyed the notion that Euclid's axioms were self-evident truths that were immutable. We now regard axioms, or postulates, as *assumptions* that need only be consistent. (For a more detailed discussion of non-Euclidean geometries, see pages 142–152.)

Topology

Since the middle of the nineteenth century, geometry has taken a new and exciting turn through the development and spectacular

MORE THAN ONE LINE ∥ THROUGH P+ OUTSIDE LINE
HYPERBOLIC – LOBACHEVSKY, N.I. (1793-1856)
BOLYAI, JANOS (1802-1860)
GAUSS, K.F. (1777-1855)

NO PARALLEL –
RIEMANN, G.F.B. (1826-1866)
⇒ EINSTEIN'S GENERAL THEORY OF RELATIVITY.

See p142-152

EULER, LEONARD (1707-1783)
MÖBIUS, A.F. (1790-1868)
RIEMANN, BERNHARD (1826-1866)
POINCARÉ, HENRI (1854-1912)
HAUSDORFF, FELIX (1868-1942)

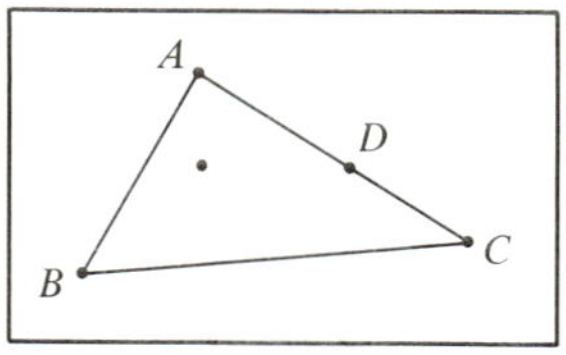

FIGURE 2-44

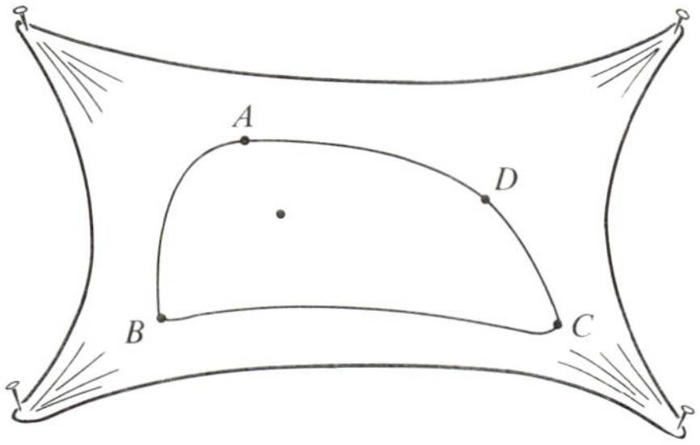

FIGURE 2-45

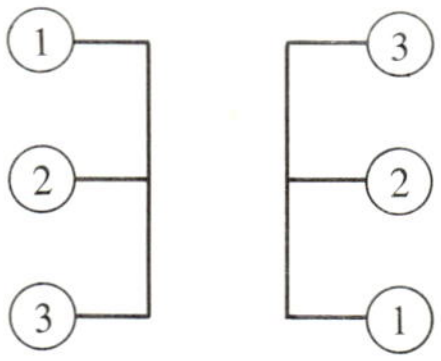

FIGURE 2-46

KOENIGSBERG BRIDGE PROBLEM.

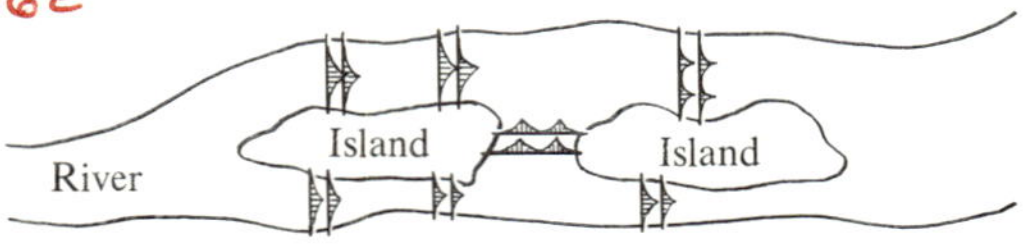

FIGURE 2-47

growth of a branch of mathematics called *topology*. Some of the men who made important contributions to this subject are Leonard Euler (1707–1783), A. F. Möbius (1790–1868), Bernhard Riemann (1826–1866), Henri Poincaré (1854–1912), and Felix Hausdorff (1868–1942).

The object of topology is to study those properties of geometric figures that remain invariant even after the size and shape of these figures have been altered. Such deformations would occur if we imagine our figures to be drawn on a rubber sheet and then allow them to be distorted by stretching the rubber sheet in any way we wish without tearing it or folding it back onto itself.

For example, suppose we draw a triangle on a rubber sheet and mark a dot inside the triangle (FIG. 2-44). No matter how we stretch or bend the rubber sheet, the dot is still *inside* the triangle (FIG. 2-45). This "inside" position of the dot remains invariant even though the triangle was subjected to so drastic a deformation that it lost its original shape and size.

Certain other properties will remain invariant under this deformation. For instance, the *order* of the points *A*, *B*, *C*, and *D* remains the same. No matter how we stretch the rubber sheet, the path from *A* to *B* to *C* to *D* to *A* remains a path that does not cross itself. No matter where we start on this path, we will return to the starting point without crossing the path.

Topology has been used to solve some fascinating puzzles and problems, such as the problem of connecting like numerals by lines that do not cross each other or any other lines in the figure (FIG. 2-46).

A famous problem that substantially led to the beginnings of topology was the Koenigsberg bridge problem. This relates to the Preger River, which ran through the town of Koenigsberg and was crossed in part by seven bridges (FIG. 2-47). The problem was how to start out from any point in the town and cross each bridge once, and only once, in returning to the starting point. Euler became interested in this problem and proved that it was impossible to cross each bridge only once and return to the starting point. Euler's solution led to the founding of topology.

A famous unsolved topological problem, known as the four-color conjecture, asks us to prove or disprove that four different colors are sufficient to color any map on a plane or sphere so that countries with a common border are colored differently. So far, only four different colors have been needed for any maps suggested, but no mathematical proof has ever been given that four colors are sufficient for *all* maps.

Although topology can be a "fun" subject and can be used to solve some fascinating problems, it is being applied more and more to many practical problems and serves to illuminate other branches of mathematics.

DEFECTS IN EUCLID'S GEOMETRY

Although Euclid's geometry, as he developed it in his *Elements*, is a monumental prototype of a deductive system, it has been found to be logically deficient in a number of important ways.

Let us recall that a deductive system consists of a collection of undefined words, other words defined in terms of the undefined words, axioms we assume to be true, and theorems we deduce from our definitions and axioms by applying our assumed laws of logic.

We shall now examine the ways in which Euclid treated each of these components in his system of geometry. Although we shall not question the truth or falsity of his theorems, we will be concerned with the way in which he proves some of them and the assumptions under which they are true.

Euclid's Definitions

Let us begin with Euclid's definitions. Euclid did not recognize the need for undefined terms. Instead of starting with a few undefined words whose meaning would come from the axioms, Euclid tried to define every term he used. This led him to some curious "definitions." For instance, he attempted to define a *point* by saying, "A point is that which has no part." But what does "part" mean? He attempted to define a *line* this way: "A line is breadthless length." But what does "breadthless length" mean? He defined a *straight line* this way: "A straight line is a line that lies evenly with the points on itself." But what does "lies evenly" or "on" mean? Such "definitions" are defective because they lack clarity and precision.

Euclid's Axioms*

More serious are the gaps in Euclid's axioms. Euclid assumed and used a number of properties which are not included in his list of axioms and which cannot be derived from them. In a rigorous, logical system, tacit but unstated assumptions are not allowed.

An example of such a gap is the complete absence of *axioms of*

---

* For a listing of these axioms, see page 67.

order in Euclid's list. Euclid's axioms say nothing about the relation "between." We therefore do not know from his axioms what the relation "point C is between points A and B" means and cannot therefore adequately define *line segment*.

It is sometimes impossible to establish with certainty the location of a particular point with respect to some other points. For example, in the well-known fallacious "theorem" which asserts that "every triangle is isosceles," the fallacy lies in the assumption that a certain point lies inside a triangle when, in fact, it lies outside the triangle. Since Euclid's axioms say nothing about the *inside* and *outside* of a triangle, we do not even know the meaning of *inside* and *outside* of a triangle. (For a "proof" of this "theorem," see C. F. Adler, *Modern Geometry*, New York: McGraw-Hill Book Company, Inc., 1958, p. 23.)

Euclid's axioms do not contain the *axioms of existence* he needs in his system. For instance, no axiom states that points and lines even *exist* in the system. While it may seem reasonable to assume the existence of points and lines in our geometry, the absence of such a statement in a rigorous mathematical system may mean that we are dealing with an empty system.

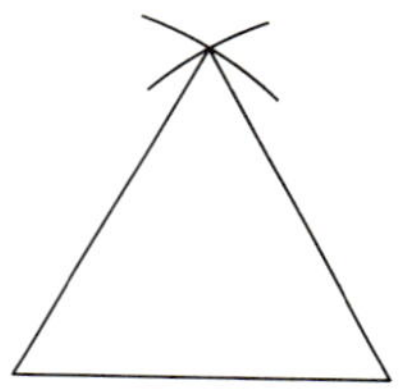

FIGURE 2-48

Another "existence" problem in Euclid relates to constructions involving the intersection of two circles or of a line and a circle (FIG. 2-48). For example, Euclid's first proposition involves the construction of an equilateral triangle on a given segment as base. The construction rests upon the intersection of two circles, and Euclid assumed the existence of their intersection point. But how do we know that the circles do, in fact, intersect? Although we are certain, intuitively, that the circles do intersect, Euclid's *axioms* do not guarantee the existence of such a point of intersection.

Proposition 16 (Book 1) states that "In any triangle, if one of the sides be produced, the exterior angle is greater than either of the interior and opposite angles." In the proof of this and other theorems, Euclid assumed that a line is infinite in extent, but he did not have an axiom that says this. The assertion that a straight line may be produced "continuously" (Axiom 2) does not necessarily imply that the line is infinite in extent. It implies only that it is *boundless*. We shall see (page 149) how Riemann built a non-Euclidean geometry by discarding this assumption by Euclid. In Riemann's geometry, a line may be produced "continuously" and still be finite in length.

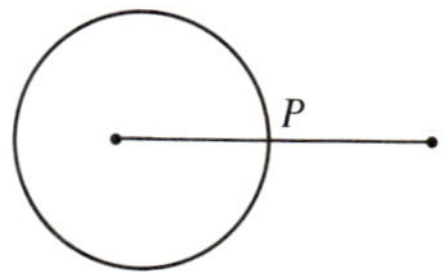

FIGURE 2-49

Euclid's axioms say nothing about "continuity." Yet, in a figure such as FIG. 2-49, Euclid assumed that the circle intersects the segment in some point P. From the axioms alone, it is conceivable that the points on the segment may be spaced like beads on a string

(FIG. 2-50). In that case, the circle could slip through one of the "holes" without ever intersecting the segment.

To guarantee the intersection of the circle and segment, we need some *axiom of continuity*.

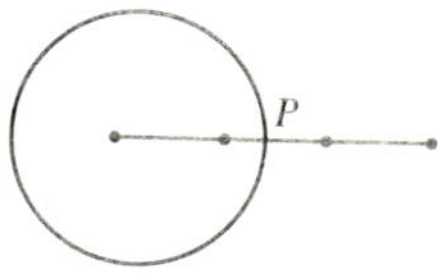

FIGURE 2-50

Proposition 21 (Book 1) states that "if on one of the sides of a triangle, from its extremities, there be constructed two straight lines meeting within the triangle, the straight lines so constructed will be less than the remaining two sides of the triangle, but will contain a greater angle" (FIG. 2-51). In the proof of this proposition, Euclid assumed that if a straight line enters a triangle at a vertex, it must, if sufficiently produced, intersect the opposite side. But, again, there is no axiom that guarantees this intersection.

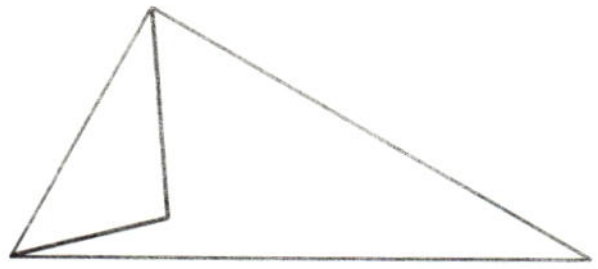

FIGURE 2-51

Euclid's first axiom states, "To draw a straight line from any point to any point." This axiom guarantees the existence of at least one straight line between any two points, but it does not guarantee that there cannot be more than one such line. Yet, Euclid frequently assumed that there is one such line.

Congruence

Euclid made congruence depend on superposition, which in turn involves concepts of motion and invariance (see page 123). He tried to prove that two triangles are congruent by the side angle side postulate (see Postulate 4-1). In his proof, he placed one triangle on top of the other and assumed that the triangle he moved remained the same triangle after it was moved. But none of his axioms assures us of this. The axioms do not deny the possibility, for example, that "moving" a triangle makes it shrink.

In summary, the major weaknesses we find in Euclid are the absence of

1. Undefined terms.
2. Axioms of incidence.
3. Axioms of order.
4. Axioms of congruence.
5. An axiom of continuity.

It was not until the end of the nineteenth and the early twentieth centuries that these weaknesses were eliminated. During this period a number of mathematicians subjected the foundations of geometry to intense scrutiny and formulated various systems that correct the deficiencies found in Euclid's axioms.

Perhaps the most widely accepted set of such axioms is that of David Hilbert (1862–1943). He first suggested his system in 1898, and in later years many improvements of his original axioms were introduced, all of which are still referred to as Hilbert's axioms.

HILBERT'S AXIOMS

Hilbert built his system of geometry on

1. Three *undefined terms*: point, line, and plane.
2. Five *undefined relations*: on, between, congruence, parallel, and continuous.
3. Twenty-one *axioms*.

Hilbert's 21 axioms are divided into six groups:

1. Axioms of incidence.
2. Axioms of order.
3. Axioms of congruence.
4. Axiom of parallels.
5. Axiom of continuity.
6. Axiom of completeness.

The 21 axioms in slightly modified form follow.

Axioms of Incidence

1. There is a straight line that contains two given distinct points.
2. There is at most one straight line that contains two given distinct points.
3. There is a plane that contains three given points which do not lie on the same straight line.
4. There is at most one plane that contains three given points which do not lie on the same straight line.
5. If two points of a straight line lie in a plane, then every point of the straight line lies in the plane.

6. If two planes have one point in common, then they have at least a second point in common.

7. On every straight line there are at least two points, in every plane there are at least three points that are not on the same straight line, and in space there are at least four points that are not on the same plane.[18]

[18] The axioms of incidence relate the undefined terms "point," "line," and "plane" to each other and postulate the existence of points, lines, and planes in the system. This assures us that the system is not empty.

Axioms of Order

8. If A, B, and C are points of a straight line and B is between A and C, then B is also between C and A (FIG. 2-52).

FIGURE 2-52

9. If A and C are two points of a straight line, then there is at least one point B that is between A and C, and there is at least one point D such that C is between A and D (FIG. 2-53).

FIGURE 2-53

10. Of any three points on a straight line there is always one and only one which is between the other two.

11. Any four points on a straight line can be labeled A, B, C, and D in such a way that B is between A and C and also between A and D and that C is between A and D and also between B and D.

12. Let A, B, and C be three points that are not on the same straight line, and let a be a straight line in the plane determined by these points and not passing through any of them (FIG. 2-54). Then if a passes through a point of the segment AB, it will also pass through either a point of the segment BC or a point of the segment AC.[19]

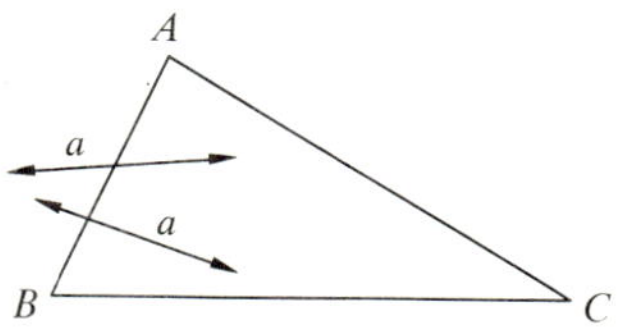

FIGURE 2-54

[19] 1. With the concept of "betweenness" now introduced, we are able to define a *segment AB* as the set of points consisting of the two points A and B and all points between them. We can now define a *polygon* as a closed chain of segments joined end to end.

2. Axiom 10 satisfies our intuitive notion of straightness and guarantees that our lines will not be closed. Consider three points A, B, and C on a line such that A-B-C. Without Axiom 10, we could have more than one betweenness relation, that is, not only A-B-C but also B-A-C and A-C-B. Our "line" could then look like FIG. 2D.

3. Although Axiom 12 is one of the Hilbert axioms, it actually predates Hilbert's system,

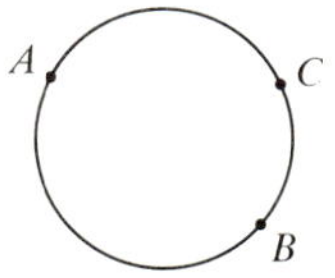

FIGURE 2D

having been formulated by M. Pasch (1843–1930) in 1882. Known as the Pasch Axiom, it implies that if a straight line enters a triangle at a vertex, it must, if sufficiently produced, intersect the opposite side.

4. Hilbert's order axioms enable us to show that a point on a line separates the line into two half-lines and that a line on a plane separates the plane into two half-planes. We can now define a *ray* and an *angle* and the *interior* and *exterior* of an angle and of a triangle.

Axioms of Congruence

13. If AB is a segment and A' is a point on a line, then there is on that line on a given side of A' one and only one point B' such that $AB \cong A'B'$. $AB \cong AB$.

14. If $AB \cong A'B'$ and $AB \cong A''B''$, then $A'B' \cong A''B''$.

15. Let AB and BC be two segments of a straight line that lie on opposite sides of B, and let $A'B'$ and $B'C'$ be two segments of a straight line that lie on opposite sides of B' (FIG. 2-55). Then if $AB \cong A'B'$ and $BC \cong B'C'$, we also have $AC \cong A'C'$.

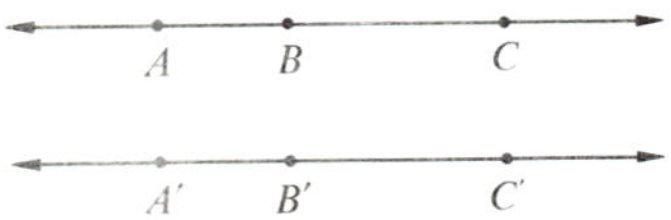

FIGURE 2-55

16. If (h, k) is an angle, a is a line on a plane, and h' is a half-line on a with vertex O, then there is on the plane on a given side of a one and only one half-line k' with vertex O such that angle $(h, k) \cong (h', k')$. Angle $(h, k) \cong$ angle (h, k).

17. If angle $(h, k) \cong$ angle (h', k') and angle $(h, k) \cong$ angle (h'', k''), then angle $(h', k') \cong$ angle (h'', k'').

18. If in two triangles ABC and $A'B'C'$, $AB \cong A'B'$, $AC \cong A'C'$, and angle $BAC \cong$ angle $B'A'C'$, then angle $ABC \cong$ angle $A'B'C'$, and angle $ACB \cong$ angle $A'C'B'$.[20]

[20] 1. These axioms apply the congruence relation to segments, angles, and triangles. Notice that Axiom 18 postulates, in essence, what Euclid tried to prove by superposition.

2. The congruence axioms lead to *measure* concepts for segment and angle by agreeing that segments or angles will have the same measure if and only if they are congruent.

Axiom of Parallels

19. In a plane containing a given straight line and a given point that is not on the line, there is one and only one straight line that does not intersect the given line.[21]

[21] This form of the postulate was popularized by J. Playfair (1748–1819) and is known as the Playfair Axiom. It can be shown that this axiom is equivalent to Euclid's fifth postulate.

Axiom of Continuity

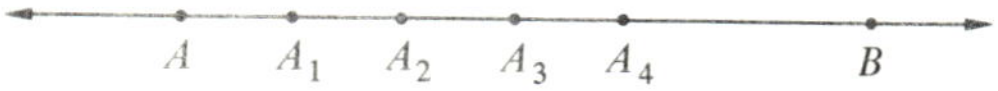

FIGURE 2-56

20. Let A_1 be any point on a straight line between the points A and B (FIG. 2-56). Take the points A_2, A_3, A_4,... so that A_1 is

between A and A_2, A_2 is between A_1 and A_3, A_3 is between A_2 and A_4, etc., and so that

$$AA_1 \cong A_1A_2 \cong A_2A_3 \cong A_3A_4 \cong \cdots.$$

Then there is a positive integer n such that B is between A and A_n.[22]

Axiom of Completeness

21. No additional points, lines, or planes can be added to the system without violating one of Axioms 1–20.[23]

Hilbert's 21 axioms remove all the defects in Euclid's *Elements* that have been pointed out and are sufficient for a rigorous development of Euclidean geometry.

EUCLID'S AXIOMS[24]

Postulates

1. To draw a straight line from any point to any point.
2. To produce a finite straight line continuously in a straight line.
3. To describe a circle with any center and distance.
4. That all right angles are equal to one another.
5. That, if a straight line falling on two straight lines makes the interior angles on the same side less than two right angles, the two straight lines, if produced indefinitely, meet on that side on which are the angles less than the two right angles.

Common Notions

1. Things that are equal to the same thing are also equal to one another.
2. If equals be added to equals, the wholes are equal.
3. If equals be subtracted from equals, the remainders are equal.
4. Things that coincide with one another are equal to one another.
5. The whole is greater than the part.

[22] 1. This axiom, known as the Archimedian Axiom, assures us that there are no "holes" in the Euclidean line by implying a one-to-one correspondence between the real numbers and the points on a line.

2. The previous 19 axioms can be satisfied by a "rational number geometry." In such a geometry, all points can be represented by rational coordinates, and all lines are expressible as linear equations with rational coefficients. The Archimedian Axiom makes Euclidean geometry a "real number geometry."

[23] This axiom says that the space under consideration is the largest one that will satisfy all the preceding axioms.

[24] The following are Euclid's ten assumptions as translated by Sir Thomas L. Heath. Euclid called five of his assumptions "postulates" and the other five "common notions." Today, no distinction is made between the two categories.

FINITE GEOMETRIES

Euclidean geometry is the most famous example of an abstract mathematical system, but it cannot be "seen" easily as a *whole*—especially by high school students—because of its many postulates and theorems. For a clearer grasp of the structure and workings of such a system we need a smaller and less complex example—based on only a few postulates that yield only a few theorems—that exposes more clearly the interrelation of its various parts. We have just such an example in the *finite geometries.*

A finite geometry is a miniature geometry which is based on a small set of postulates, undefined terms, and undefined relations and which limits the set of all points and lines to a finite number.

Criteria for Postulates

The postulates in a finite geometry must meet the same criteria as the postulates of any other mathematical system. They must be *consistent*; that is, no two postulates can contradict each other. It is desirable that they be *independent*, that is, that no postulate be a logical consequence of, or deducible from, the remaining postulates. It may also be desirable that the postulates be *complete*, that is, that it is always possible to prove every statement that can be made about the undefined terms of the system from the postulates.

Seven-Point Geometry

Seven-point geometry is an example of a finite geometry. It consists of a set of undefined elements called "points," certain undefined subsets of points called "lines," the undefined relations "contain" and "lies on," and a set of six postulates.

Although we have given specific names to our undefined terms, it is essential that we not read Euclidean or any other meanings into them. We must not confuse these "points" and "lines" with Euclidean points and lines. Our present "points" and "lines" will be completely defined by our six postulates. From them it will soon be seen, for example, that our "line," unlike the Euclidean line, is not continuous and consists of a finite number of points.

Postulates for Seven-Point Geometry

The six postulates for our geometry are

> *POSTULATE 1* *If P_1 and P_2 are any two distinct points, there is at least one line containing both P_1 and P_2.*

POSTULATE 2 *If P_1 and P_2 are any two distinct points, there is at most one line containing both P_1 and P_2.*

POSTULATE 3 *If l_1 and l_2 are any two distinct lines, there is at least one point which lies on both l_1 and l_2.*

POSTULATE 4 *Every line contains at least three points.*

POSTULATE 5 *If l is any line, there is at least one point which does not lie on l.*

POSTULATE 6 *There exists at least one line.*

Note that Postulates 1, 2, 4, 5, and 6 are also properties of Euclidean plane geometry. However, Postulate 3 contradicts Euclid's parallel postulate, since it asserts that any two lines intersect.

Models for Seven-Point Geometry

Can we find some interpretation or model for the postulates of our seven-point geometry? If in the arrangement

$$\begin{array}{ccccccc} P_1 & P_2 & P_3 & P_4 & P_5 & P_6 & P_7 \\ P_2 & P_3 & P_4 & P_5 & P_6 & P_7 & P_1 \\ P_4 & P_5 & P_6 & P_7 & P_1 & P_2 & P_3 \end{array}$$

we let the P's represent the "points" in our geometry and we let the vertical columns represent the "lines," then FIG. 2-57 can be a geometric model for our seven-point geometry.

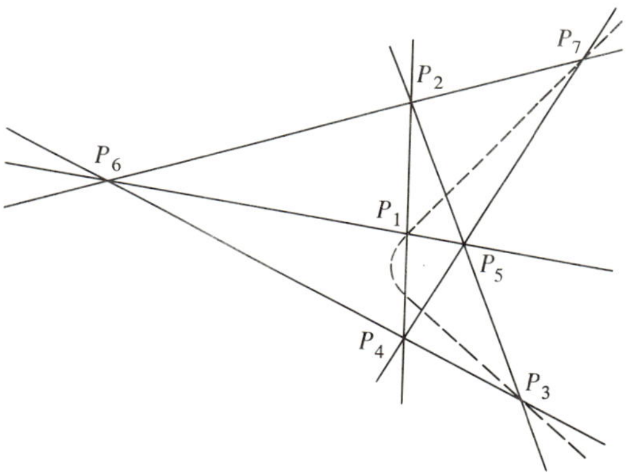

FIGURE 2-57

If we go back to the six postulates and check them against this model, we will see that the model satisfies each of them. Note that since the postulates do not say that a line must be "straight," it may be curved like line $P_7P_1P_3$.

Because we have exhibited at least one model that satisfies all our postulates, we can conclude that the six postulates are consistent.

We can construct a different geometric diagram (FIG. 2-58) that satisfies all our postulates, again letting the P's represent "points" and the vertical columns represent "lines":

$$\begin{array}{ccccccc} P_1 & P_1 & P_2 & P_1 & P_3 & P_2 & P_3 \\ P_2 & P_4 & P_4 & P_6 & P_5 & P_5 & P_4 \\ P_3 & P_5 & P_6 & P_7 & P_6 & P_7 & P_7. \end{array}$$

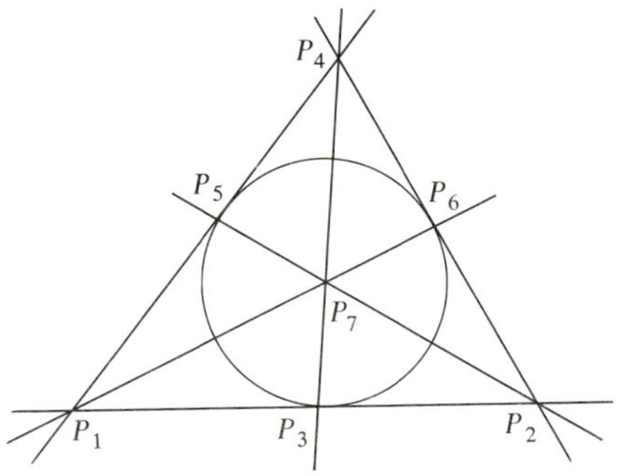

FIGURE 2-58

Here, again, inspection of the model will show that it satisfies each of the six postulates.

Geometric diagrams are not the only models we can use to represent our postulates. We can translate the postulates of seven-point geometry into statements about numbers. For instance, let us adopt these definitions:

1. The seven "points" correspond to the seven numbers 1, 2, 3, 4, 5, 6, and 7.
2. The seven "lines" correspond to the seven numbers 123, 145, 246, 167, 356, 257, and 347.
3. A point is "on" a line if and only if the digit corresponding to the point is a digit of the numeral corresponding to the line. If a point and a line have this relation, then the line is said to "contain" the point.

With these definitions, every statement of the seven-point geometry becomes a statement about certain numbers. Inspection of this model will show that it, too, satisfies all six postulates.

We can interpret the postulates of our seven-point geometry in still another way. We can relate them to people and committees. To create this new model, we adopt these definitions:

1. The seven "points" correspond to the seven people *A*, *B*, *C*, *D*, *E*, *F*, and *G*. "People" and "members" will be used interchangeably.
2. The seven "lines" correspond to seven committees, as indicated by the following seven columns:

A	*A*	*B*	*A*	*C*	*B*	*C*
B	*D*	*D*	*F*	*E*	*E*	*D*
C	*E*	*F*	*G*	*F*	*G*	*G*.

3. A line "contains" a point if and only if a person is a member of the committee. If a point and a line have this relation, then the point is also said to "lie on" the line.

With these definitions, our postulates become

POSTULATE 1 *If A and B are different people, there is at least one committee of which both A and B are members.*

POSTULATE 2 *If A and B are different people, there is at most one committee on which both A and B are members.*

POSTULATE 3 *Any two committees have at least one member in common.*

POSTULATE 4 *Every committee contains at least three members.*

POSTULATE 5 *For any committee, there is at least one person who is not a member of it.*

POSTULATE 6 *There exists at least one committee.*

An inspection of the membership of each committee will again show that all six postulates of our geometry are satisfied. The isomorphism between the point–line model and the people–committee model becomes apparent when we rename the points with people (FIG. 2-59).

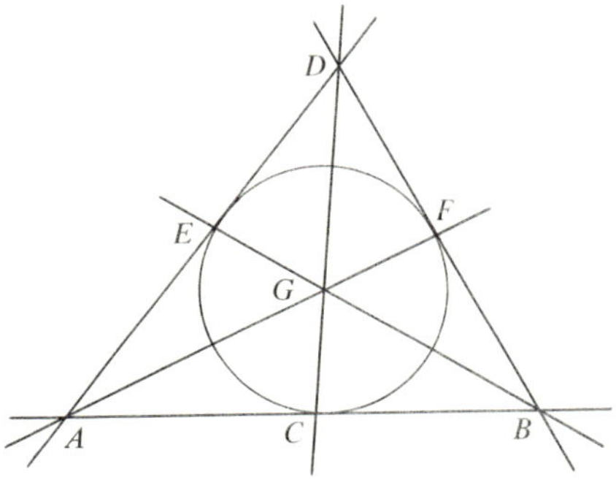

FIGURE 2-59

Theorems of Seven-Point Geometry

Now that we have described the postulates of seven-point geometry, we shall see what *theorems* can be deduced from them. To prove theorems in this system we do precisely what we do in other mathematical systems—but in this system we have only six postulates and four undefined terms to consider.

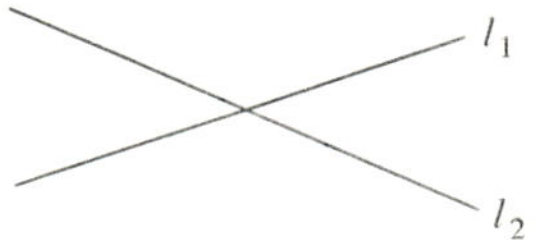

FIGURE 2-60

THEOREM 1 *If l_1 and l_2 are any two distinct lines, there is at most one point which lies on both l_1 and l_2* (FIG. 2-60).

Proof
Assume the contrary, that is, that two distinct lines have two points in common. This contradicts Postulate 2. Therefore, two distinct lines have at most one point in common.

THEOREM 2 *There exists at least one point.*

Proof
By Postulate 6 there exists at least one line, and by Postulate 4 every line contains at least three points. Therefore, there exists at least one point.

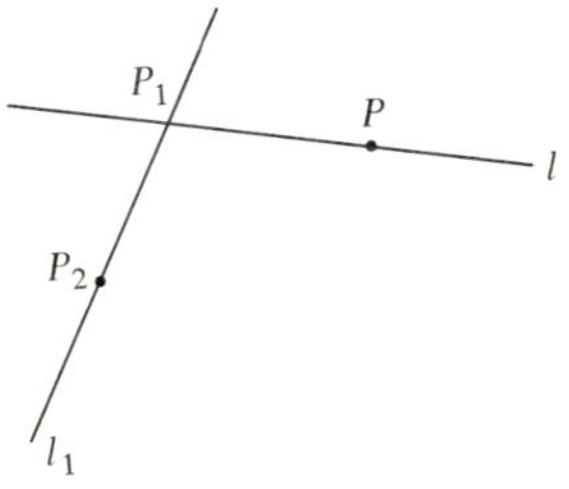

FIGURE 2-61

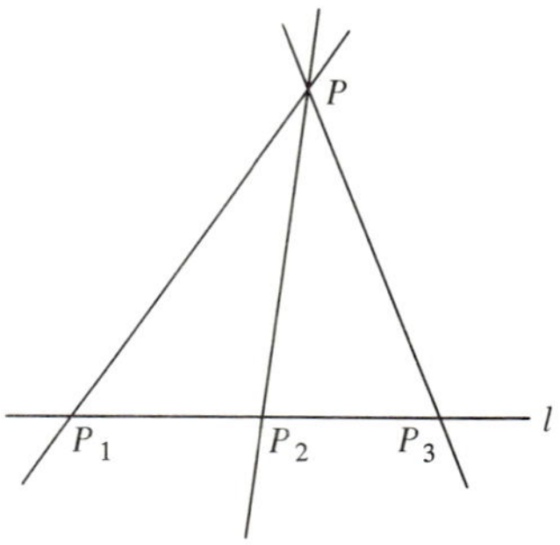

FIGURE 2-62

THEOREM 3 *If P is any point, there is at least one line which does not pass through P.*

Proof

1. By Postulate 6 there exists at least one line l.
2. If this line does not pass through P, our proof is complete.
3. Suppose l passes through P (FIG. 2-61). Then, by Postulate 4, l contains at least one point P_1 besides P.
4. By Postulate 5, there is at least one point P_2 which does not lie on l.
5. By Postulates 1 and 2, there is a unique line l_1 which contains P_1 and P_2.
6. l and l_1 are distinct, since l_1 contains P_2 and l does not.
7. By Postulate 3 and Theorem 1, l and l_1 have exactly one point in common, P_1.
8. Therefore, P, which lies on l, cannot also lie on l_1. In other words, l_1 is a line that does not pass through P.

THEOREM 4 *Every point lies on at least three lines.*

Proof

1. By Theorem 3, there is at least one line l that does not pass through a given point (FIG. 2-62).
2. By Postulate 4, l contains at least three points, P_1, P_2, and P_3.
3. By Postulates 1 and 2, each of these points determines with P a unique line.
4. These lines are distinct, since if two of them coincided, that line would have two points in common with l. This would contradict Postulate 3 and Theorem 1.
5. Therefore, there are at least three lines passing through every point P.

Duality

By proving these four theorems we have established a very interesting and important property of seven-point geometry—that it has *duality*.

According to the principle of *duality*, any true statement about lines and points is also true when "line" and "point" are interchanged, and when a corresponding change is also made in the connective used. For instance, "all lines do not pass through the same point" implies, by duality, "all points do not lie on the same line."

If we examine the theorems and postulates of our seven-point geometry, we find that

1. Postulate 1 is the dual of Postulate 3.
2. Theorem 1 is the dual of Postulate 2.
3. Theorem 2 is the dual of Postulate 6.
4. Theorem 3 is the dual of Postulate 5.
5. Theorem 4 is the dual of Postulate 4.

This proves that our seven-point geometry has duality.

In general, the importance of this property lies in the fact that in any system that has duality, each proof actually establishes two theorems—the theorem we set out to prove and its dual.

The following are two other theorems in this geometry:

THEOREM 5 *Seven-point geometry contains exactly seven points.*

THEOREM 6 *Seven-point geometry contains exactly seven lines.*

For a proof of these theorems, see B. E. Meserve, *Fundamental Concepts of Geometry*, Reading, Mass.: Addison-Wesley Publishing Company, Inc., 1955, p. 16.

Note that Theorems 5 and 6 are duals of each other; therefore, one theorem implies the other.

Theorems 5 and 6 have been generalized to yield information about the number of points and lines contained in any finite projective geometry:

1. If a line contains exactly n points, then the system contains exactly $n^2 - n + 1$ points and $n^2 - n + 1$ lines.
2. If a line contains exactly n points, then exactly n lines pass through every point.

For example, in a finite projective geometry where a line contains 4 points,

1. Exactly 4 lines will pass through every point.
2. The system will contain $n^2 - n + 1$ points and lines, or 13 points and 13 lines.

The study of finite geometries goes back to 1892, when G. Fano first considered a three-dimensional geometry containing 15 points and 35 lines, each plane containing 7 points and 7 lines. About 15 years later, O. Veblen and J. W. Young developed seven-point geometry in their book *Projective Geometry*, Vol. 1, Boston: Ginn & Company, 1910.

3 ANGLES AND TRIANGLES

INTRODUCING THE CHAPTER TO STUDENTS

CONTENT REVIEW

- Prerequisites for Learning This Chapter
- What Is an Angle?
- Interior and Exterior of an Angle
- Theorems About Angles
- Measurement of Angles
- Congruence of Angles
- What Is a Triangle?
- Interior and Exterior of a Triangle
- Theorems About Triangles

TEACHING SUGGESTIONS AND ACTIVITIES

- Visual Aids
- Random Suggestions
- Questions
- Puzzles
- Solutions

ENRICHMENT

- Famous Theorems Related to the Triangle

INTRODUCING THE CHAPTER TO STUDENTS

If we hinge four strips of wood to form a quadrilateral, we will find that the shape of the quadrilateral can be changed without changing the lengths of its sides. But if we hinge three strips of wood to form a triangle, we will find that the shape of the triangle cannot be changed without changing the lengths of its sides. This is why we say that a triangle is a *rigid* figure. No other polygon has this property. The quadrilateral we just formed does not have this property.

To make the quadrilateral rigid, we can place a brace across two opposite corners, forming two triangles out of the quadrilateral. The figure will now be rigid (FIG. 3-1).

Because a triangular structure, if correctly constructed, cannot change its shape under pressure or strain without being pulled apart or breaking one of its sides, it is used in construction work of all kinds—in the construction of buildings, airplanes, bridges, and "hurricane-resistant" cottages.

Triangles often help us in indirect measurement of heights and distances when it is impossible or difficult to make a direct measurement, such as finding the height of a mountain or the distance to the moon.

Even more basic than the triangle in its wide-range practical applications is the *angle*. Angles are involved in countless situations and activities. They are used in drafting, designing, and construction. We talk about the angle of incline of a road, an angle of reflection, the angle at which you hold your razor while shaving, and the angle of flight of a rocket.

Angles and triangles are not only important in the physical world, but they play a most important role in the study of geometry. In this chapter we shall deal with the *geometry* of angles and triangles rather than with their physical applications. We shall define in precise mathematical language some basic ideas about these geometric figures—such as what is meant by the "inside" of an angle—and sharpen our understanding of some of their basic properties—such as the "separation" properties.

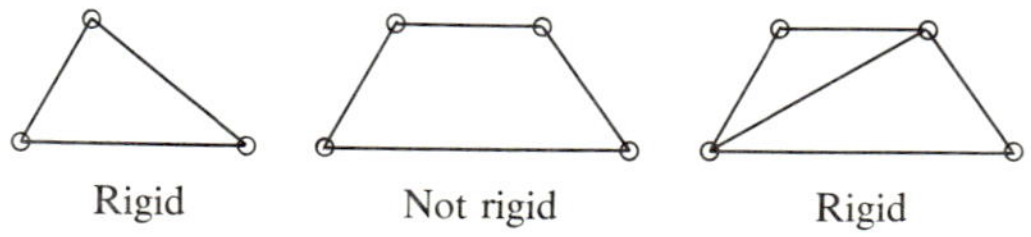

FIGURE 3-1

CONTENT REVIEW

PREREQUISITES FOR LEARNING THIS CHAPTER

An understanding of the following terms and concepts: ray, endpoint, half-plane and its designation, convex set, one-to-one correspondence, collinearity of points, ideas of separation, union and intersection of sets.

WHAT IS AN ANGLE?[1]

Definition
An *angle* is the union of two distinct rays that have the same endpoint but do not lie on the same line (FIG. 3-2).

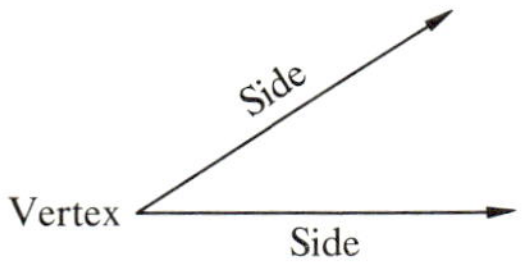

FIGURE 3-2

The endpoint is called the *vertex* of the angle, and the two rays are called the *sides* of the angle. The angle formed by $\overrightarrow{BA}$ and $\overrightarrow{BC}$ (FIG. 3-3) is called angle ABC, denoted by $\angle ABC$, or angle CBA, denoted by $\angle CBA$. Note that the middle letter always names the vertex. Other ways of naming the angle shown are $\angle B$ and $\angle 1$.[2]

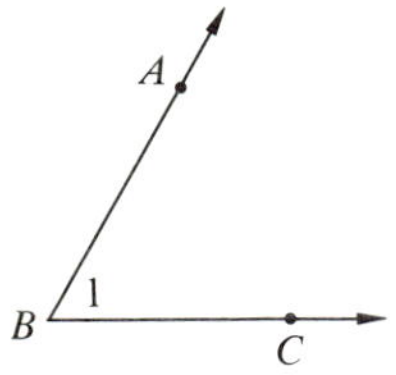

FIGURE 3-3

We can easily extend our definition of an angle on a plane to define a *dihedral angle* in space. Two rays with the same endpoint,

[1] Past experience with angles may lead us to associate an *angle* with the amount of rotation or difference of rotation between two rays, or with a region of the plane bounded by two intersecting lines. But the idea of *rotation* is really related to *measurement* of an angle, while the idea of a *region* between two intersecting lines really relates to the *interior* of an angle, as we shall soon see. But what is an *angle*?

We shall make a careful distinction among an *angle*, its *measure*, and its *interior*. We shall think of an angle as a geometric figure, that is, as a set of points, rather than as an amount of rotation. Although such a view of an angle is inadequate in trigonometry and in analytic geometry, it is adequate for Euclidean geometry.

[2] 1. The sides of an angle extend indefinitely in two directions because the sides are rays rather than segments.

2. We always have $\angle ABC = \angle CBA$ because each is the union of the rays $\overrightarrow{BA}$ and $\overrightarrow{BC}$.

3. Points *between* the rays are not part of the angle.

4. Note that "zero" angles and "straight" angles are not angles according to our definition. Although they are needed in trigonometry, they are not needed in elementary geometry.

5. If our definition of an angle included straight angles, then these would be the only angles not contained in a unique plane. (See Theorem 3-1.)

not on the same line, define a planar angle. Two noncoplanar half-planes with a common edge form a *dihedral angle.* More precisely, a *dihedral angle* is the union of a line and two noncoplanar half-planes having the line as *edge* (FIG. 3-4).

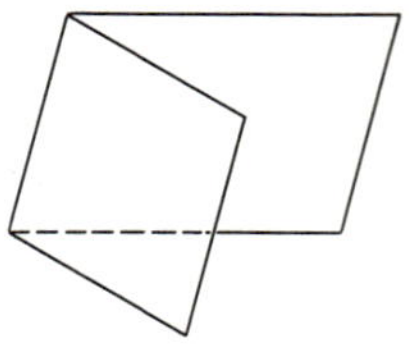

FIGURE 3-4

INTERIOR AND EXTERIOR OF AN ANGLE[3]

Definitions

1. *The interior of* $\angle ABC$ is the intersection of the two half-planes $\overleftrightarrow{AB}/C$ and $\overleftrightarrow{BC}/A$ (FIG. 3-5).

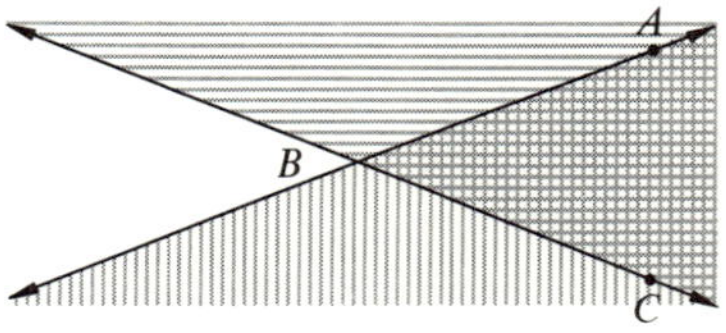

FIGURE 3-5

2. *The exterior of* $\angle ABC$ is the set of all points that are in the plane of the angle but are neither on the angle nor in the interior of the angle (FIG. 3-5).

Therefore, for a point P to lie in the interior of $\angle ABC$ (FIG. 3-6),

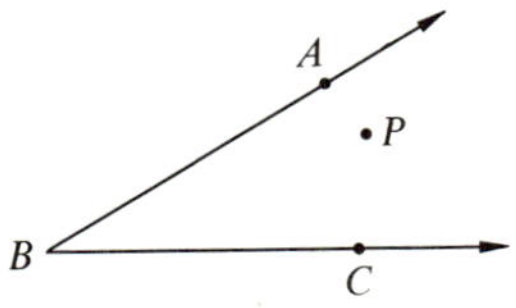

FIGURE 3-6

[3] Just as a planar simple closed curve divides the plane into the interior and the exterior of the curve, so does an angle divide the plane into the interior and the exterior of the angle (FIG. 3A).

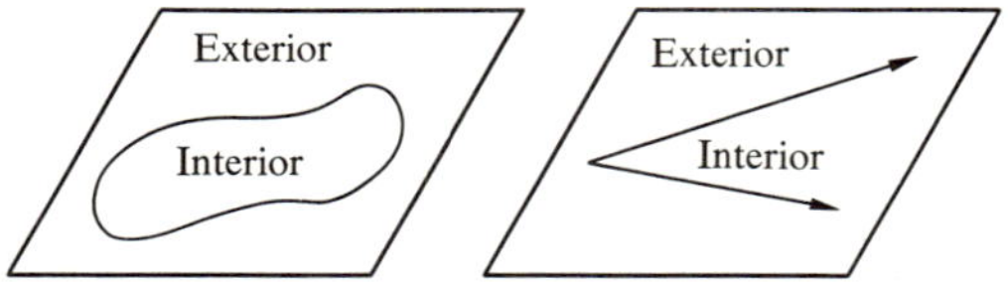

FIGURE 3A

We shall now define this idea precisely.

1. P must be in the plane of $\angle ABC$.
2. P and A must be in the same half-plane with edge $\overleftrightarrow{BC}$.
3. P and C must be in the same half-plane with edge $\overleftrightarrow{AB}$.[4]

THEOREMS ABOUT ANGLES

Following are four basic theorems about angles.

THEOREM 3-1 Every angle is contained in a unique plane.

Proof

1. Since $\overrightarrow{BA}$ and $\overrightarrow{BC}$ are noncollinear by the definition of an angle, the points A, B, and C are noncollinear (FIG. 3-7).

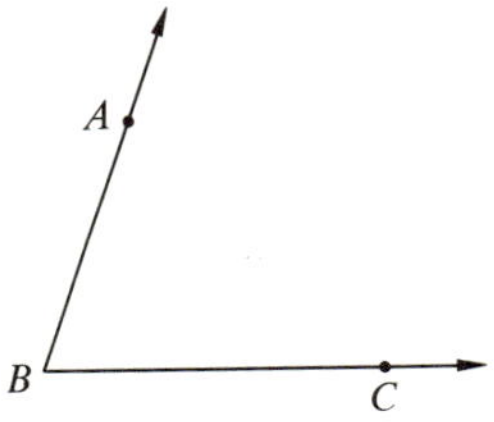

FIGURE 3-7

2. By Postulate 2-3, three different noncollinear points are contained in exactly one plane.

THEOREM 3-2 For any two points of an angle not on the same ray of the angle, every point between these two points is in the interior of the angle.

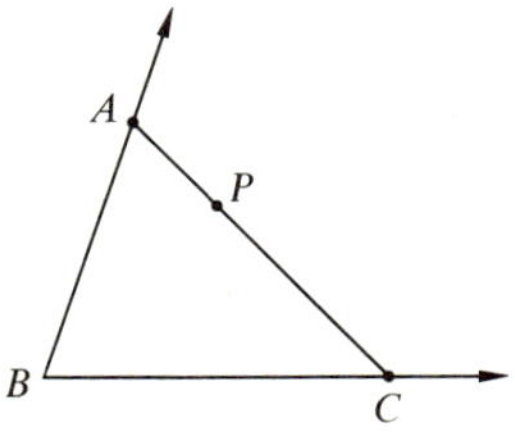

FIGURE 3-8

That is, if A-P-C, then P is in the interior of $\angle ABC$ (FIG. 3-8). To prove that P is in the interior of $\angle ABC$, we need to show that

[4] 1. Note that the interior of an angle is not a subset of the angle.

2. If our definition of an angle included straight angles, we would face the difficulty that a straight angle does not have an interior or an exterior (FIG. 3B). We think of a *straight angle* merely as a pair of opposite rays. Two rays $\overrightarrow{BA}$ and $\overrightarrow{BC}$ are called *opposite* if A-B-C.

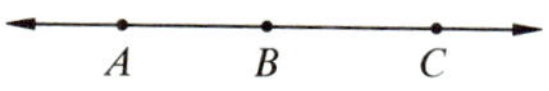

FIGURE 3B

1. P is in the plane of $\angle ABC$.
2. P and A are in the same half-plane with edge $\overleftrightarrow{BC}$.
3. P and C are in the same half-plane with edge $\overleftrightarrow{AB}$.

Proof

1. P is in the plane of $\angle ABC$ because, by Postulate 2-4, $\overleftrightarrow{AC}$ is in the plane of $\angle ABC$.
2. P and A are in the same half-plane with edge $\overleftrightarrow{BC}$ because
   a. No point of $\overleftrightarrow{CB}$ is between A and P, since $\overline{AC}$ and $\overleftrightarrow{CB}$ intersect in only one point C.
   b. P and A must be in the same half-plane with edge $\overleftrightarrow{CB}$ by a theorem which states that

   Given any plane, any line $\overleftrightarrow{CB}$ in the plane, and any two points P and A that are in the plane but not on $\overleftrightarrow{CB}$, if no point of $\overleftrightarrow{CB}$ is between P and A, then P and A are in the same half-plane with edge $\overleftrightarrow{CB}$.
3. Similarly, P and C are in the same half-plane with edge $\overleftrightarrow{AB}$.

THEOREM 3-3 *Two distinct intersecting lines form exactly four angles* (FIG. 3-9).

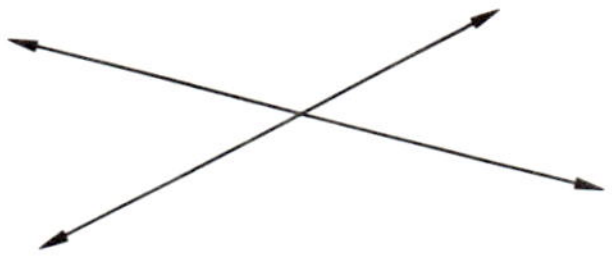

FIGURE 3-9

For a proof of this theorem, see W. Prenowitz and M. Jordan, *Basic Concepts of Geometry*, Waltham, Mass.: Blaisdell Publishing Company, 1965, pp. 256–257.

THEOREM 3-4 *The interior of an angle is a convex set.*

The proof of this theorem depends on another theorem which says that any half-plane is convex.

Proof

1. Let P_1 and P_2 be any two points in the interior of $\angle ABC$ (FIG. 3-10).

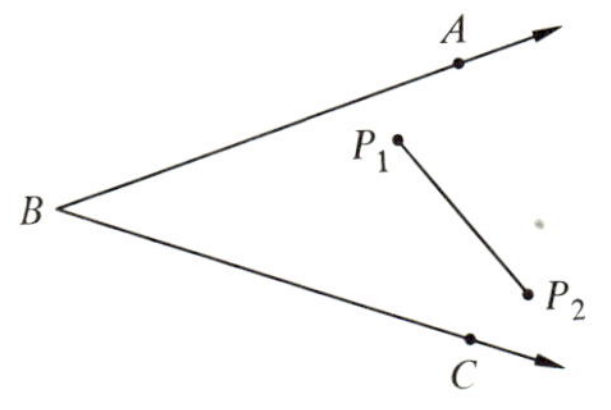

FIGURE 3-10

2. Then, by the definition of "interior," P_1 and P_2 are in the half-planes $\overleftrightarrow{AB}/C$ and $\overleftrightarrow{BC}/A$.
3. Since $\overleftrightarrow{AB}/C$ and $\overleftrightarrow{BC}/A$ are convex by the theorem mentioned above, it follows that the entire segment $\overline{P_1P_2}$ is in the two half-planes, that is, in the interior of $\angle ABC$.
4. Therefore, by the definition of convexity, the interior of $\angle ABC$ is convex.[5]

[5] 1. Theorem 3-2 suggests another definition of the *interior* of an angle. We can now think of the *interior* of $\angle ABC$ as the set of *all* points that lie between the endpoints of *all* line segments determined by two points not on the same ray of $\angle ABC$ (FIG. 3C).

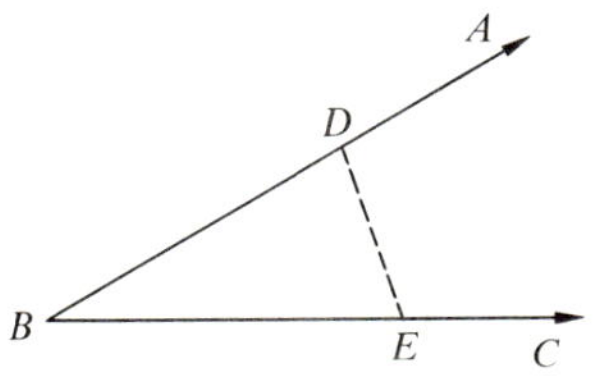

FIGURE 3C

2. According to Theorem 3-4, the interior of an angle is a convex set. It can be shown that the *exterior* of an angle is *not* a convex set.

MEASUREMENT OF ANGLES[6]

Measure of an Angle: Informal Ideas

1. To measure an angle, we first select an arbitrary angle, say $\angle a$ (FIG. 3-11), to serve as a *unit* and agree that its measure is the number 1. We write this as $m\angle a = 1$, which we read as "the measure of $\angle a$ is 1."

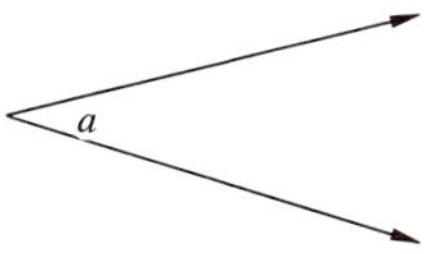

FIGURE 3-11

2. We can now form an $\angle ABC$ with measure 2 by laying off the unit angle $\angle a$ twice about a common vertex B (FIG. 3-12). In terms of the unit $\angle a$, the measure of $\angle ABC$ is 2. We write this as $m\angle ABC = 2$.
3. In a similar way we can form angles whose measures are 3, 4, etc.

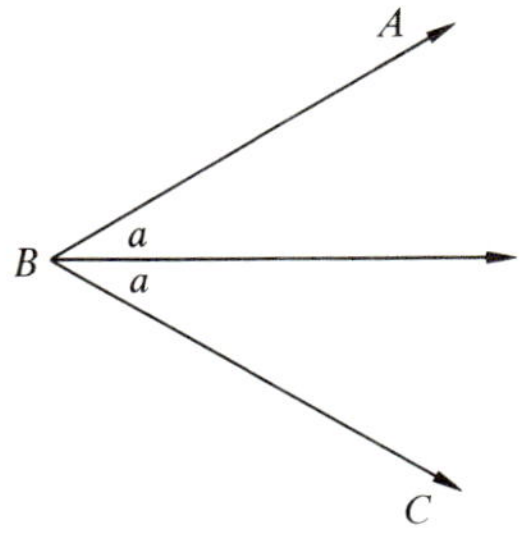

FIGURE 3-12

[6] A basic problem in mathematics is to devise a method for using numbers to measure geometric figures, which can then be compared by means of their measurements. We now wish to devise such a method for *angles*.

4. A *standard* unit of measure is as useful in measuring angles as in measuring segments. The most common *standard* unit of angle measure is called a *degree*. One degree is written 1°.

5. If we lay off 360 unit angles about a common vertex *B*, then they, together with their interiors, cover the entire plane (FIG. 3-13).

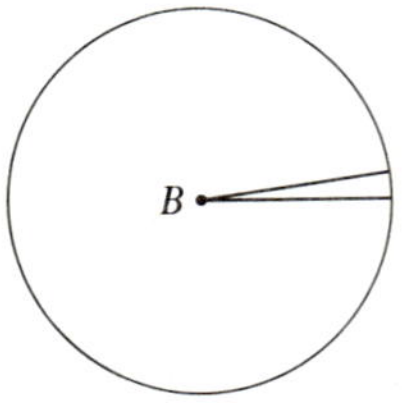

FIGURE 3-13

6. The number of degrees in an angle is called its *measure*.

7. An instrument used to determine the measure of an angle is called a *protractor*. The protractor is a scale of degree units marked, usually in a semicircle, from 0 to 180. In FIG. 3-14,

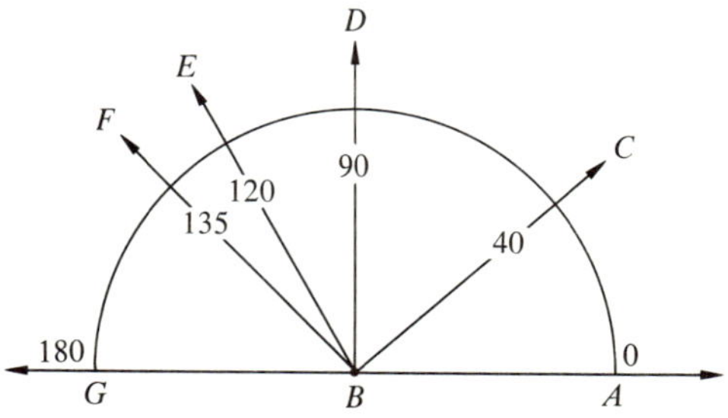

FIGURE 3-14

$m\angle ABC = 40,$

$m\angle ABD = 90,$

$m\angle ABE = 120,$

$m\angle ABF = 135,$

$m\angle CBD = 90 - 40 = 50.$[7]

[7] 1. Note that an *angle* is a *set of points*, while its *measure* is a *number*.

2. A protractor is used for finding the measures of given angles, but it is also used for drawing angles with a given measure.

3. We write $m\angle ABC = 40$, not $m\angle ABC = 40°$, because the measure is a *number*. But in labeling figures, we use the degree sign (°) to indicate that the letters or numbers are meant to be the degree measures of the angle rather than the name of the angle (FIG. 3D).

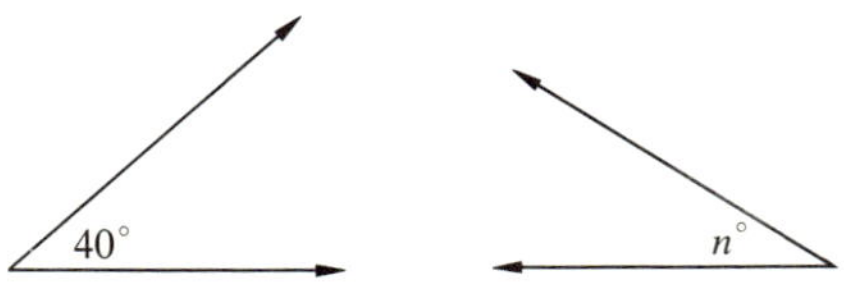

FIGURE 3D

Definitions

1. A *right angle* is an angle whose measure is 90 (FIG. 3-15). We sometimes indicate a right angle by a little square at the vertex (FIG. 3-15).

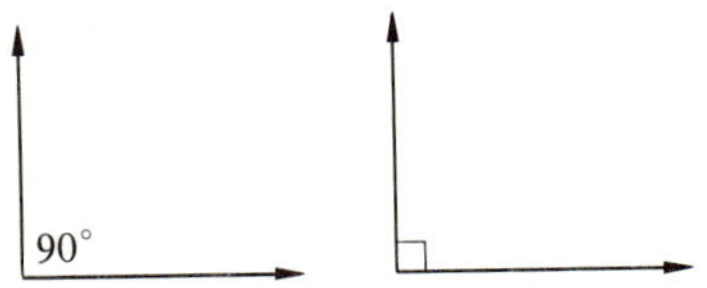

FIGURE 3-15

2. An *acute angle* is any angle whose measure is less than 90 (FIG. 3-16).

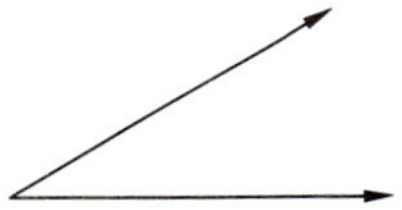

FIGURE 3-16

3. An *obtuse angle* is any angle whose measure is greater than 90 and less than 180 (FIG. 3-17).

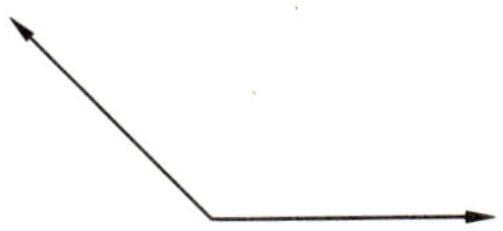

FIGURE 3-17

4. Two lines are *perpendicular* to each other if their intersection forms a right angle (FIG. 3-18). If $\overleftrightarrow{AB}$ and $\overleftrightarrow{CD}$ are perpendicular, then we write $\overleftrightarrow{AB} \perp \overleftrightarrow{CD}$.

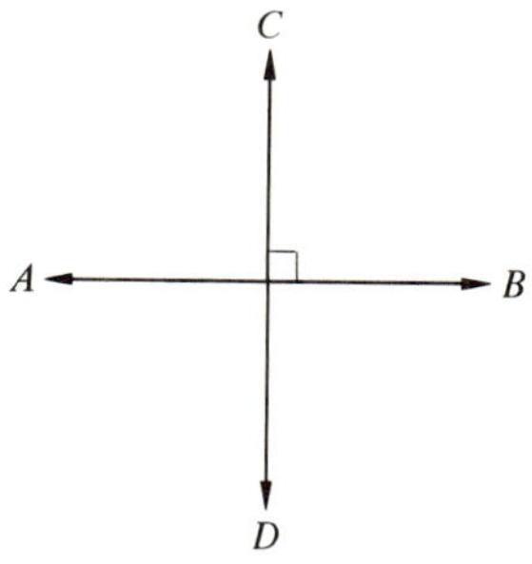

FIGURE 3-18

Note: The definition of perpendicularity applies to any two sets, either or both of which is a line, a ray, or a segment.

Angular Measure Postulates[8]

POSTULATE 3-1 (Angle Measurement Postulate) *For every angle there is one and only one real number n, called its measure. When the angle is measured in degrees, n is between* 0 *and* 180*; that is,* $0 < n < 180$.[9]

[8] The measurement of angles is based on three postulates (Postulates 3-1, 3-2, and 3-3) that are analogous to the postulates for the measurement of distance.

Just as the distance postulates are abstract descriptions of the ruler, so are the following postulates abstract descriptions of the behavior of the protractor.

POSTULATE 3-2 (Angle Construction Postulate) *Let $\overrightarrow{AB}$ be a ray on the edge of the half-plane H. For every number n such that* $0 < n < 180$, *there is one and only one ray $\overrightarrow{AP}$ with P in H such that* $m\angle PAB = n$ (FIG. 3-19).[10]

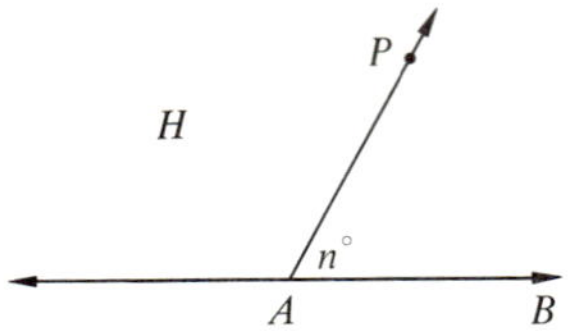

FIGURE 3-19

POSTULATE 3-3 (Angle Addition Postulate) *If D is in the interior of* $\angle ABC$, *then* $m\angle CBA = m\angle CBD + m\angle DBA$ (FIG. 3-20).[11]

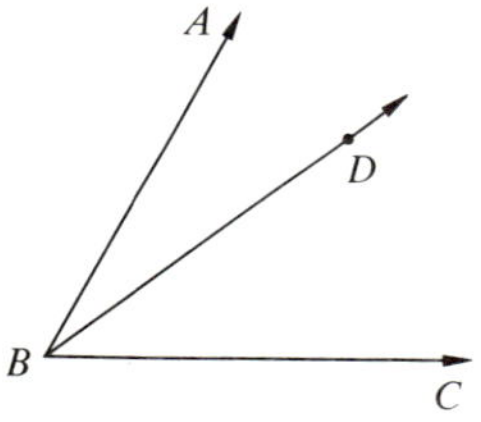

FIGURE 3-20

Angle Relationships

Definitions

1. If the sum of the measures of any two angles is 90, then the angles are *complementary*, and each angle is said to be the complement of the other.
2. If $\overrightarrow{AB}$ and $\overrightarrow{AC}$ are opposite rays and $\overrightarrow{AD}$ is any third ray, then $\angle DAB$ and $\angle DAC$ form a *linear pair* (FIG. 3-21).

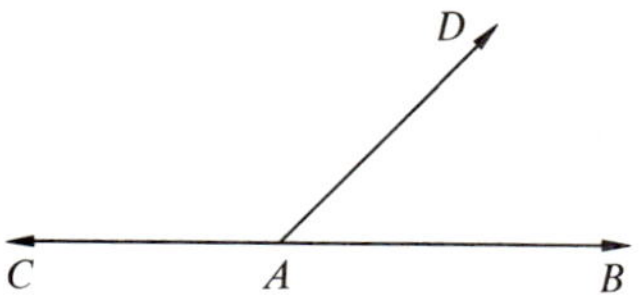

FIGURE 3-21

[9] 1. We denote the measure of an angle by the notation we adopted previously. For example, $m\angle ABC = 25$.

2. The decision that the measure of an angle be between 0 and 180 is arbitrary. We could use the range 0 to r, where r is any positive number.

[10] 1. Postulate 3-2 assures us that there is one and only one angle that can be drawn in the half-plane H, having n as its measure and $\overrightarrow{AB}$ as one of its sides.

2. Because the real number n is associated with a ray, the size of the protractor is not important in measuring angles.

3. Just as the Ruler Postulate for distance establishes a one-to-one correspondence between the real numbers and the points on a line, so does this postulate establish a one-to-one correspondence between the real numbers between 0 and 180 and the rays in the half-plane H having the endpoint A.

4. From Postulates 3-1 and 3-2 we deduce that there are exactly two angles in the plane that have $\overrightarrow{AB}$ as one side and have a measure of n (FIG. 3E).

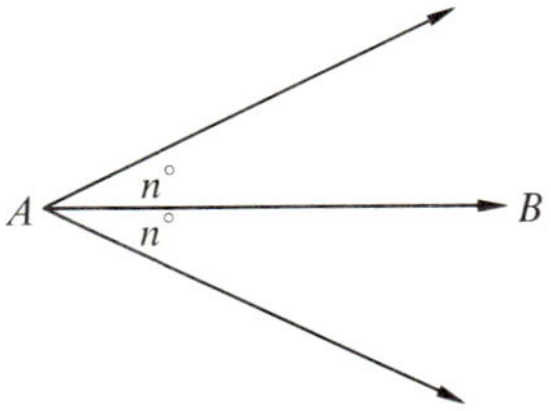

FIGURE 3E

[11] We use this postulate when we compute the measure of an angle by subtraction; for example, $m\angle DBA = m\angle CBA - m\angle CBD$.

3. If the sum of the measures of two angles is 180, then the angles are called *supplementary*, and each angle is said to be the supplement of the other.

4. Two angles in a plane are said to be *adjacent* if they have a side in common and their interiors do not intersect (FIG. 3-22).[12]

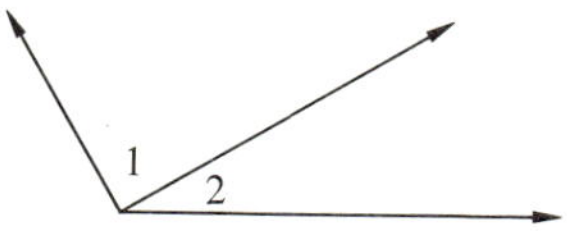

$\angle 1$ and $\angle 2$ are adjacent angles

FIGURE 3-22

5. Two angles are *vertical* angles if their union is the union of two intersecting lines (FIG. 3-23).[13]

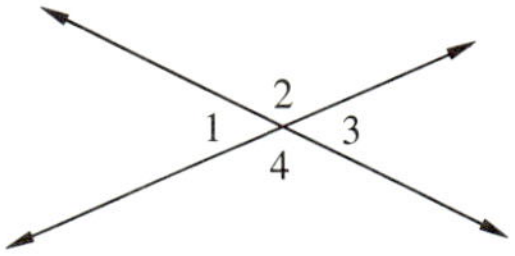

$\angle 2$ and $\angle 4$, $\angle 1$ and $\angle 3$ are vertical angles

FIGURE 3-23

POSTULATE 3-4 *If two angles form a linear pair, then they are supplementary* (FIG. 3-24).[14]

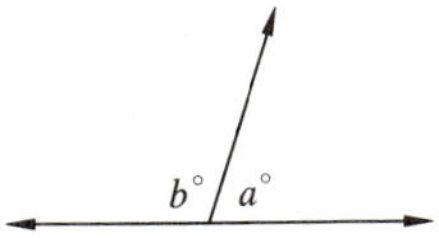

$a + b = 180$

FIGURE 3-24

CONGRUENCE OF ANGLES[15]

Definition

Two angles are *congruent* if they have the same measure. We use the symbol $\cong$ to mean "is congruent to."[16]

[12] If two angles form a linear pair, then they are adjacent, but if two angles are adjacent, they do not necessarily form a linear pair.

[13] By Theorem 3-3, two distinct intersecting lines form exactly four angles. Any two of these that are not adjacent angles are vertical angles.

[14] Postulate 3-4 eliminates the need for "straight" angles.

[15] Just as we defined congruence for segments in terms of distance, so do we define congruence for angles in terms of measure. That is, if $m\angle ABC = m\angle DEF$, then the angles are *congruent*, and we write $\angle ABC \cong \angle DEF$. This leads us to the following definition.

[16] 1. The two statements $m\angle ABC = m\angle DEF$ and $\angle ABC \cong \angle DEF$ are equivalent.

2. It is important to make the distinction between *equal* and *congruent* angles.

a. $\angle APB = \angle CPD$ refers to the same angle with different names (FIG. 3F).

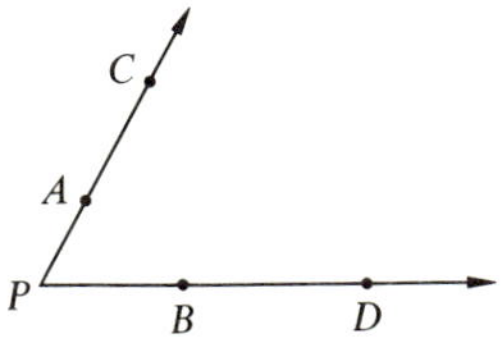

FIGURE 3F

b. $\angle ABC \cong \angle DEF$ refers to two angles having the same measure, say 30 (FIG. 3G).

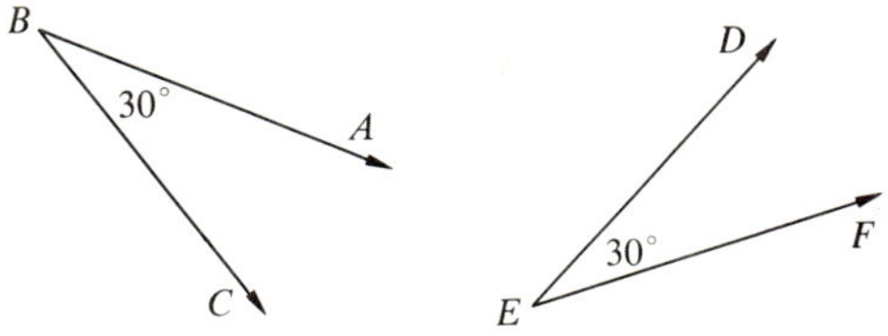

c. Note also that $\angle APB = \angle BPA$ as well as $\angle APB \cong \angle BPA$.

The definition of congruent angles leads to the following basic theorems.

THEOREM 3-5 *Congruence of angles is an equivalence relation.*

This is true because

1. For any $\angle A$,

   $\angle A \cong \angle A$ (reflexive property).

2. For any $\angle A$ and $\angle B$,

   if $\angle A \cong \angle B$, then $\angle B \cong \angle A$ (symmetric property).

3. For any $\angle A$, $\angle B$, and $\angle C$,

   if $\angle A \cong \angle B$ and $\angle B \cong \angle C$,

   then $\angle A \cong \angle C$ (transitive property).

THEOREM 3-6 *Any two right angles are congruent.*

THEOREM 3-7 *If two angles are both congruent and supplementary, then each of them is a right angle.*

THEOREM 3-8 *Supplements of congruent angles are congruent.*

THEOREM 3-9 *Complements of congruent angles are congruent.*

THEOREM 3-10 *Vertical angles are congruent.*

WHAT IS A TRIANGLE?

Definition

If A, B, and C are three noncollinear points, then the set $\overline{AB} \cup \overline{BC} \cup \overline{AC}$ is called a *triangle* (FIG. 3-25).

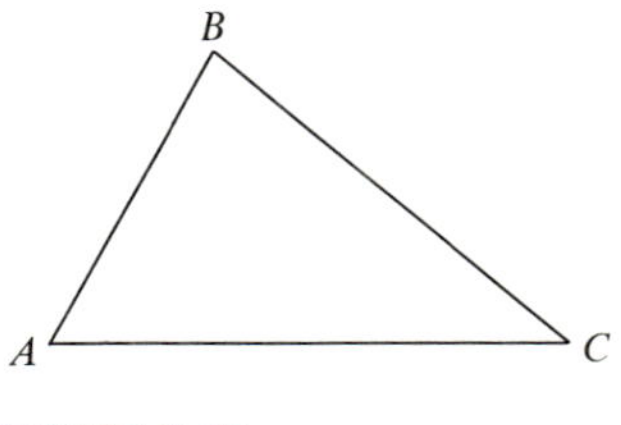

FIGURE 3-25

1. The triangle is usually denoted by the symbol $\triangle ABC$.
2. The points A, B, and C are called the *vertices* of the triangle. (Each point is called a *vertex*.)
3. The segments $\overline{AB}$, $\overline{BC}$, and $\overline{CA}$ are called the *sides* of the triangle.
4. $\angle A$, $\angle B$, and $\angle C$ are called the *angles* of the triangle.[17]

[17] 1. Since three noncollinear points determine a unique plane, a triangle also determines a unique plane.
2. Note that while $\triangle ABC$ *determines* three angles, it does not actually *contain* them because the sides of a triangle are segments while the sides of an angle are rays. Although a triangle does not contain its angles, it does contain its sides.
3. Note that the definition of a triangle is based on the idea of betweenness.

We sometimes talk about "included" and "opposite" angles and sides of a triangle. We shall now give precise definitions of these terms.

Definitions

1. The *included angle* of two sides of a triangle is the angle whose rays contain those two sides.

For example, $\angle A$ *is included* between $\overline{AB}$ and $\overline{AC}$ (FIG. 3-26).

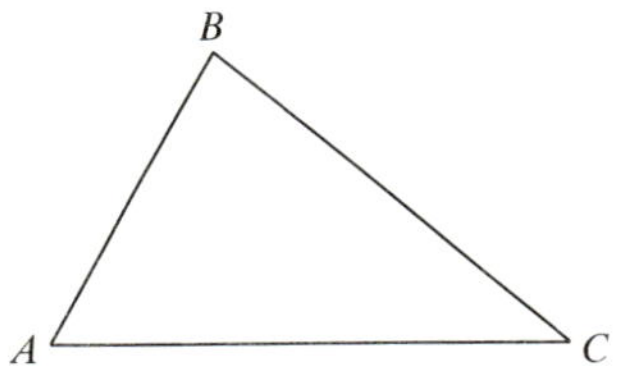

FIGURE 3-26

2. The *included side* of two angles of a triangle is the side whose endpoints are the vertices of the two angles.

For example, $\overline{AC}$ *is included* between $\angle A$ and $\angle C$.

3. The *angle opposite* a side of a triangle is the angle whose vertex is not on that side.

For example, $\angle B$ is opposite $\overline{AC}$.

4. The *side opposite* an angle of a triangle is the side that does not contain the vertex of that angle.

For example, $\overline{BC}$ *is opposite* $\angle A$.

Triangles can be classified by comparing their *sides* or their *angles*.

1. Comparing the *sides* of a triangle (FIG. 3-27),

[18] The definition of the *interior of an angle* suggests the definition of the *interior of a triangle*.

The intersection of the two half-planes $\overleftrightarrow{AB}/C$ and $\overleftrightarrow{AC}/B$ yields all the "inside" points of $\angle BAC$ (FIG. 3H).

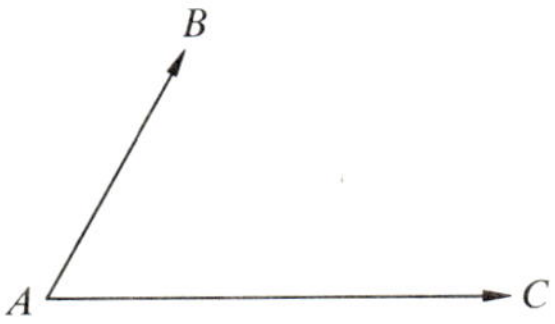

FIGURE 3H

We might, therefore, expect that the intersection of the three half-planes $\overleftrightarrow{AB}/C$, $\overleftrightarrow{BC}/A$, and $\overleftrightarrow{AC}/B$ would yield all the "inside" points of $\triangle ABC$ (FIG. 3I). This, in fact, is the way we define the *interior* of a triangle.

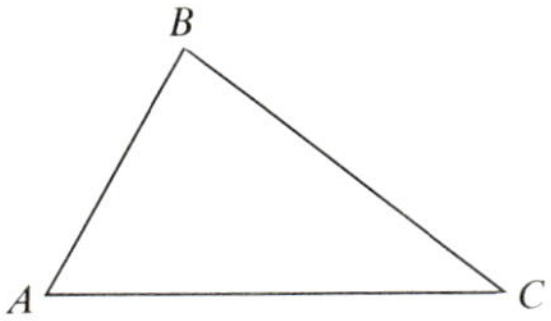

FIGURE 3I

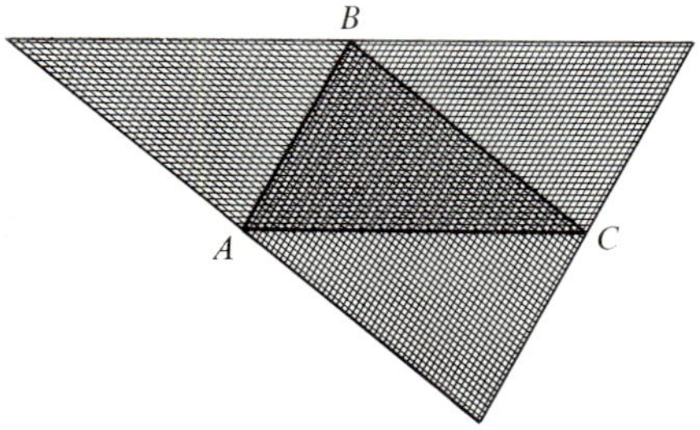

FIGURE 3-29

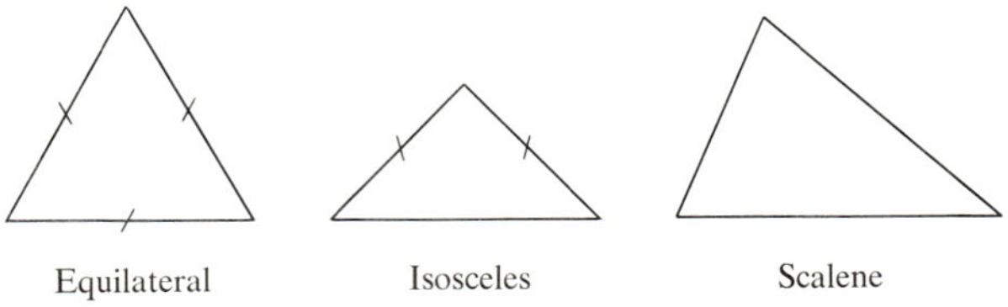

FIGURE 3-27

a. If all three sides are congruent, the triangle is *equilateral.*
b. If two sides are congruent, the triangle is *isosceles.*
c. If no two sides are congruent, the triangle is *scalene.*

2. Comparing the *angles* of a triangle (FIG. 3-28),

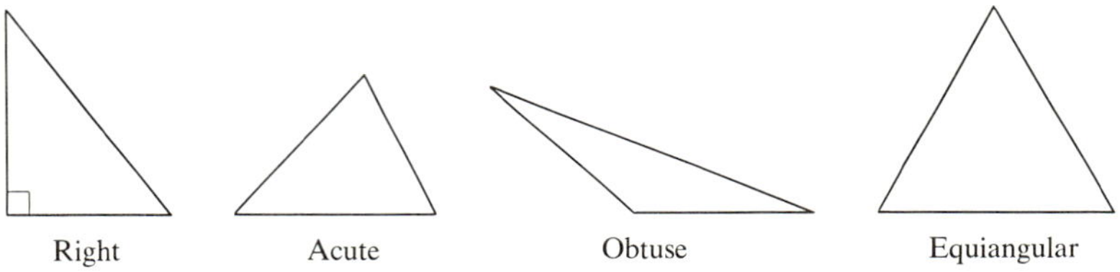

FIGURE 3-28

a. One angle in the triangle may be a right angle. We call such a triangle a *right* triangle.
b. All the angles in the triangle may be acute. We call such a triangle an *acute* triangle.
c. One angle in the triangle may be obtuse. We call such a triangle an *obtuse* triangle.
d. All the angles in the triangle may be congruent. We call such a triangle an *equiangular* triangle.

INTERIOR AND EXTERIOR OF A TRIANGLE[18]

Definitions

1. The *interior of a triangle ABC* is the intersection of the half-planes $\overleftrightarrow{AB}/C$, $\overleftrightarrow{BC}/A$, and $\overleftrightarrow{AC}/B$ (FIG. 3-29). That is, $\overleftrightarrow{AB}/C \cap \overleftrightarrow{BC}/A \cap \overleftrightarrow{AC}/B$.
2. The *exterior of triangle ABC* is the set of all points in the plane of the triangle that are not on the triangle and not in the interior of the triangle.[19]

THEOREMS ABOUT TRIANGLES

THEOREM 3-11 *The interior of a triangle is always a convex set.*

This follows from the definition of the "interior" of a triangle and from the fact that the intersection of several convex sets is convex.

THEOREM 3-12 *If a line in the plane of a triangle intersects one side of the triangle, then the line intersects at least one other side of the triangle.*

That is, if l intersects $\overline{AC}$, then it intersects $\overline{AB}$ or $\overline{BC}$ or both (FIG. 3-30).

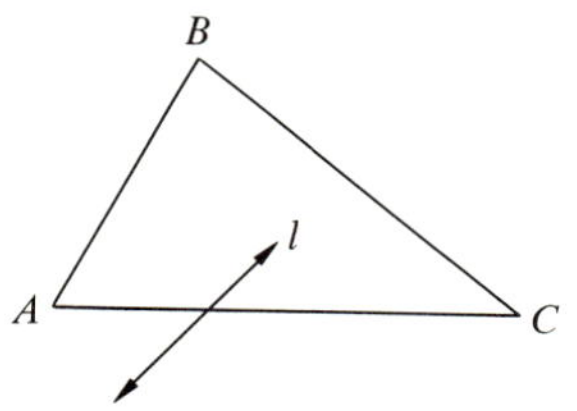

FIGURE 3-30

Before stating the next important theorem we need the following

Definitions

1. An *exterior angle of a triangle* is an angle that forms a linear pair with one of the angles of the triangle.

In FIG. 3-31, $\angle 1$ is an exterior angle of $\triangle ABC$.

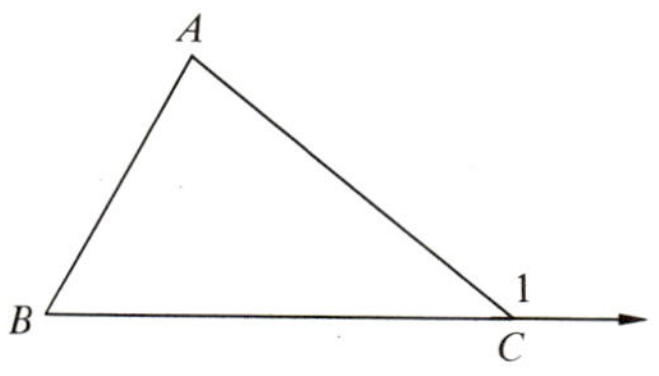

FIGURE 3-31

2. The two angles of the triangle not adjacent to the exterior angle are called the *remote interior angles.*

19 1. We can think of the interior of a triangle as the set of all points that are interior to each of the angles of the triangle.

2. The ideas of *separation* apply to a triangle the same way they do to an angle. A triangle separates the plane into three disjoint sets of points:

a. The set of points in the interior of the triangle.
b. The set of points in the exterior of the triangle.
c. The set of points on the triangle itself.

THEOREM 3-13 *The measure of an exterior angle of a triangle is greater than the measure of either remote interior angle.*

That is (FIG. 3-32), $m\angle 1 > m\angle A$ and $m\angle 1 > m\angle B$.

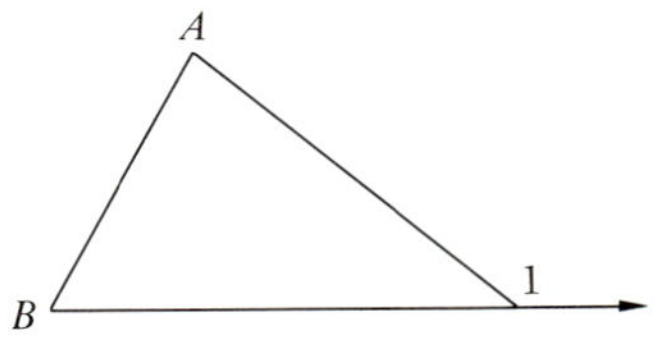

FIGURE 3-32

THEOREM 3-14 *If the lengths of two sides of a triangle are unequal, the measures of the angles opposite these sides are unequal, and the angle with the larger measure is opposite the longer side* (FIG. 3-33).

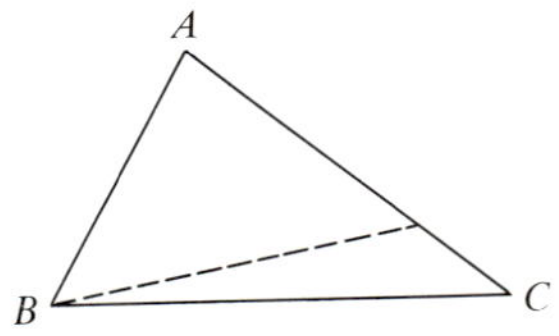

FIGURE 3-33

That is, if $AC > AB$, then $\angle B > \angle C$.

THEOREM 3-15 *If the measures of two angles of a triangle are unequal, the lengths of the sides opposite these angles are unequal, and the longer side is opposite the angle with the greater measure* (FIG. 3-34).

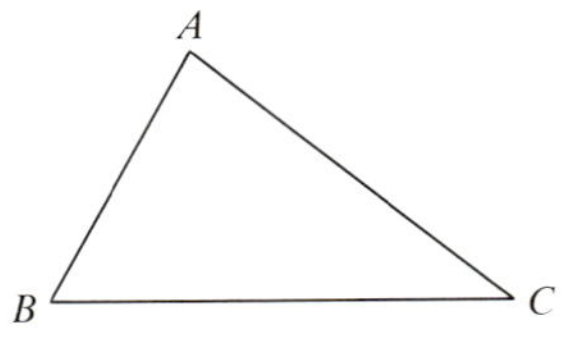

FIGURE 3-34

That is, if $\angle B > \angle C$, then $AC > AB$.

THEOREM 3-16 *The shortest segment joining a point to a line is the perpendicular segment* (FIG. 3-35).

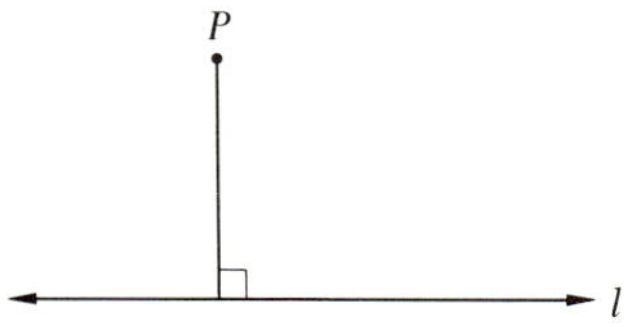

FIGURE 3-35

Definition

The *distance* between a line and a point not on the line is the length of the perpendicular segment from the point to the line. The distance between a line and a point on the line is defined to be zero.

THEOREM 3-17 *The sum of the lengths of two sides of a triangle is greater than the length of the third side.*

TEACHING SUGGESTIONS AND ACTIVITIES

VISUAL AIDS

Visual 1

PURPOSE

1. To show the distinction between an angle, its measure, and its interior.
2. To illustrate right, acute, and obtuse angles and perpendicularity.
3. To illustrate complementary and supplementary angles, linear pairs, adjacent angles, vertical angles.
4. To illustrate classification of triangles.

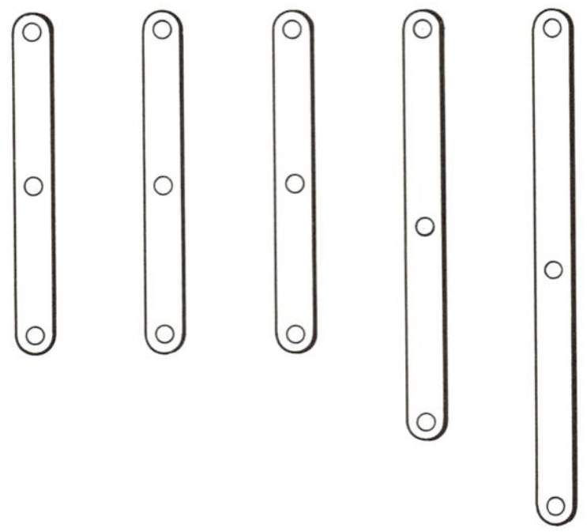

FIGURE 3-36

CONSISTS OF

1. Five strips (made of any stiff material), three of them the same size, punched with holes at the ends and midpoints (FIG. 3-36).
2. Brass fasteners.

HOW TO USE

1. Fasten strips either at midpoints or at ends, as required, to illustrate ideas listed under Purpose.
2. Use with the following Content Review: definition of "angle," page 77; comment **2**; Theorem 3-3; Measure of an Angle, page 81; the definitions on page 83; Postulate 3-2; Postulate 3-3; Angle Relationships, page 84; Postulate 3-4; and pages 87–88.

Visual 2

PURPOSE

1. To illustrate that the interior of an angle is the intersection of two half-planes.

2. To illustrate that the interior of a triangle is the intersection of three half-planes.

CONSISTS OF

1. A card with $\angle ABC$ drawn on one side and $\triangle ABC$ drawn on the other side (FIG. 3-37).
2. Three transparent overlays marked as in FIG. 3-38. Note that marked on the top of each overlay is the name of the half-plane represented by the shading.

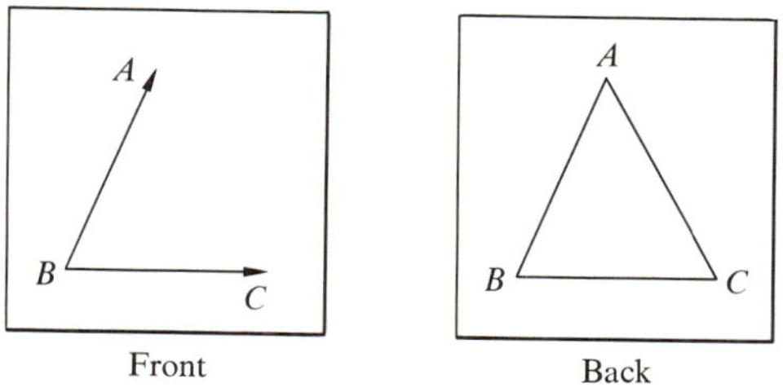

FIGURE 3-37

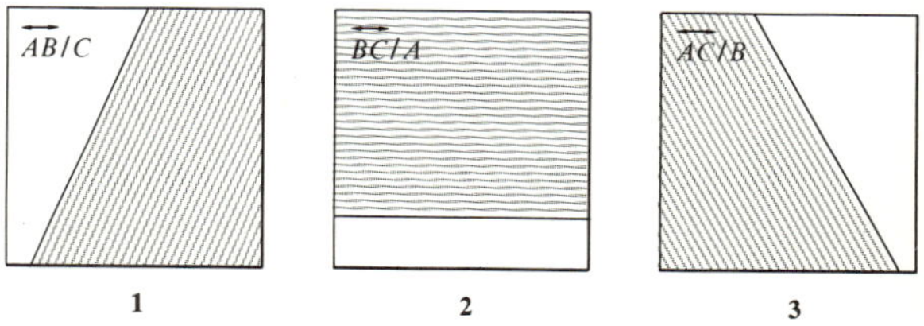

FIGURE 3-38

HOW TO USE

1. To illustrate that the interior of an angle is the intersection of two half-planes,
   a. Place overlays 1 and 2 over the front of the card.
   b. The region of intersection of the two sets of stripes is the interior of $\angle ABC$.
2. To illustrate that the interior of a triangle is the intersection of three half-planes,
   a. Place overlays 1, 2, and 3 over the back of the card.
   b. The region of intersection of the three sets of stripes is the interior of $\triangle ABC$.
3. Use with Definitions 1 and 2, page 78, and Definitions 1 and 2, page 88.

Visual 3

PURPOSE

To illustrate Theorems 3-13, 3-14, 3-15, and 3-16.

CONSISTS OF

An adjustable triangle with three 180° protractors attached as shown in FIG. 3-39.

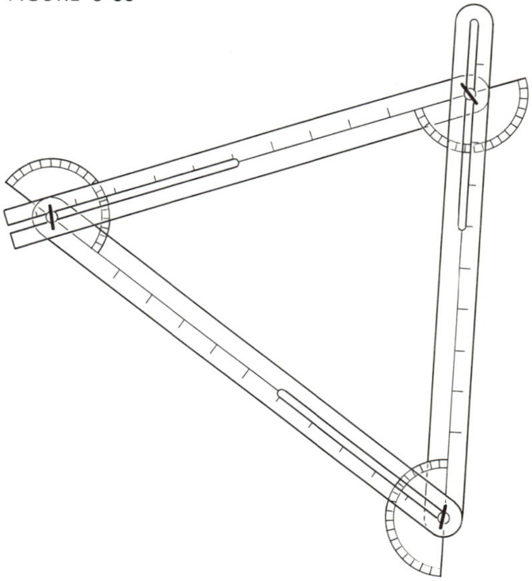

FIGURE 3-39

Note:

1. The lengths of the sides can be changed by sliding an end of a "side" along a groove in an intersecting side. Ruler markings along sides indicate their lengths.
2. This visual aid is available commercially from a number of manufacturers.

HOW TO USE

Set up the triangle to meet the conditions of each theorem. Then check the results with protractor readings and/or lengths-of-sides readings.

RANDOM SUGGESTIONS

Another View of Angles

We defined an angle as the union of two distinct rays that have the same endpoint but do not lie on the same line. This definition of an angle is not accepted by everyone, especially in Europe. Some Europeans define an angle to be what we call an "angular region," that is, the union of an angle with its interior. Some of the pedagogical reasons given for this preference are as follows:

1. In physical models of angles, if one side of an angle is rotated about the vertex to another position, the pupil sees the same angle—i.e., he sees the same two rays—yet one is "larger" and one is "smaller." This is a source of confusion to many students.
2. The union of two of our adjacent angles is a set of three rays with a common endpoint, but the union of the two related angular regions is still an angular region.
3. The intersection of two adjacent angles is a single ray, but the intersection of the two related angular regions is still an angular region.

Reasons 2 and 3 amount to the argument that it is less confusing to students to have an operation on an angle result in an angle.

Another Unit of Measure

It is essential to point out to students that the "degree" is not the only unit of angular measure used in mathematics. The use of the degree as a standard unit goes back 4000 years to the ancient Babylonians. Their selection of a unit angle which fits exactly 360 times into the plane when placed side by side and with the same vertex may have been related to their calculation of the number of days in a year as 360.

For theoretical purposes there is a more useful unit of measure for angles. This unit is called a *radian.*

A *radian* is an angle which when placed with its vertex at the center of a circle intercepts an arc equal in length to the radius of the circle (FIG. 3-40). The relationship between radian and degree measure is as follows:

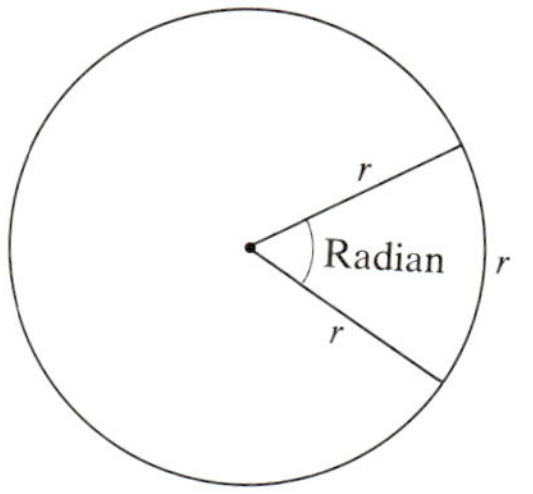

FIGURE 3-40

$$2\pi \text{ radians} = 360°,$$

$$\pi \text{ radians} = 180°,$$

$$1 \text{ radian} = \frac{180°}{\pi} = 57.3°,$$

$$1 \text{ degree} = 0.017 \text{ radian}.$$

For theoretical purposes, the radian is a more useful unit of measure for angles because many important formulas involving the trigonometric functions are simpler when expressed in radians than in degrees. Radians also involve smaller numbers; for example, instead of dealing with 100°, we can use 1.7 radians. The superiority of the radian, as a unit of measure, appears in calculus when differentiating the trigonometric functions.

Teaching Techniques

The importance of an intuitive approach to the teaching of mathematics, especially in the lower grades, cannot be overemphasized. Such an approach seeks to *persuade* the student that a theorem is "reasonable" rather than *prove* to him that it is true. The predominant procedures in such an approach are informal, nonrigorous, experimental, and investigative. After the student has satisfied his intuition, he can proceed to more formal considerations.

There are many places in geometry where such an approach can be used. In this chapter, students in the lower grades can first verify experimentally, by use of the protractor, that vertical angles are congruent. Later they can prove it.

At the beginning of this chapter we talked of triangles as rigid figures and showed that a quadrilateral can be made rigid by inserting a diagonal. A good laboratory-type lesson can ask the students to determine experimentally how many diagonals are required to convert plane polygons with any number of sides into rigid figures. How many diagonals are required in a six-sided or n-sided polygon?

Ask the students to complete the following data through actual experimentation, and then look for a pattern:

Number of Sides	3	4	5	6	7	8	...	n
Number of Diagonals Needed to Make Rigid								

At least some students will discover the pattern to be $n - 3$.

In guiding students in this type of discovery lesson, make sure they understand what they are to investigate, but do not "over-guide" them to the point where they are merely following your directions.

QUESTIONS

1. Name all the angles determined by the accompanying figure.

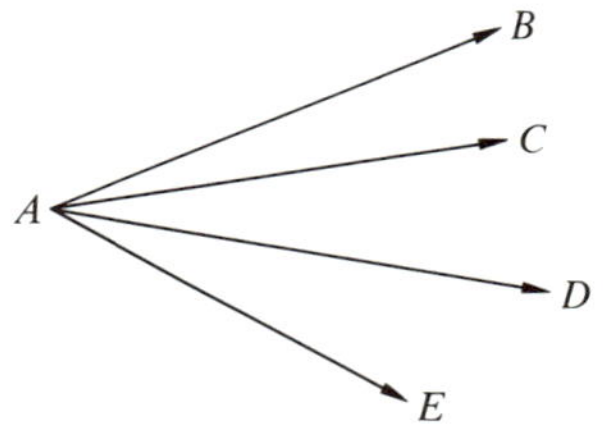

2. Make *three* true statements that begin with "If two lines intersect . . . ."

3. If $\angle A$ and $\angle B$ are complementary and $\angle A$ has a measure of 30 more than $\angle B$, what is the measure of $\angle A$ and $\angle B$?

4. If $\angle X$ and $\angle Y$ are supplementary and $\angle X$ has twice the measure of $\angle Y$, what is the measure of $\angle X$ and $\angle Y$?

5. True–false. Give a reason for your answer.
   a. If two angles are complementary, then each of them is acute.
   b. Not every angle is congruent to itself.
   c. If two angles are supplementary, then one of them must be obtuse.
   d. An angle is the union of any two rays.
   e. Vertical angles need not be congruent.
   f. Vertical angles cannot be supplementary.

6. What is the measure of an angle whose measure is twice the measure of its complement?

7. The measure of an angle is five times the measure of its supplement. Find the measure of each angle.

8. If the measure of an angle is a, what is the measure of its supplement?

9. If the measure of one of a pair of vertical angles is m, what are the measures of the other three angles formed?

10. Describe the intersection of

    a. $\triangle XYZ$ and $\triangle TXZ$.
    b. $\triangle XYZ$ and $\triangle RSZ$.
    c. $\triangle TXZ$ and $\triangle RSZ$.

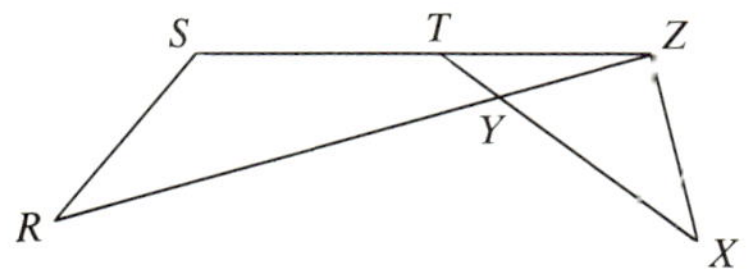

11. Why is the union of an angle and its interior a convex set?

12. Given $\triangle XYZ$, describe the following sets of points:

    a. $\triangle XYZ \cap \overrightarrow{ZY}$.
    b. $\overline{XZ} \cap \angle Y$.
    c. (Interior of $\triangle XYZ$) $\cap$ (exterior of $\triangle XYZ$).

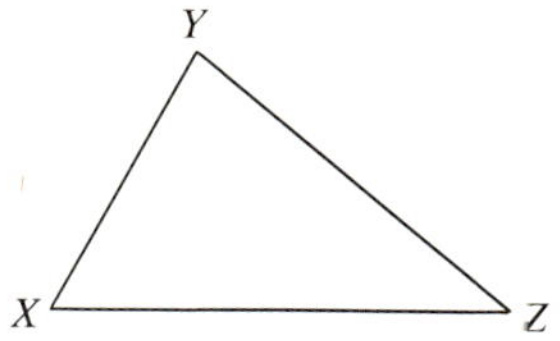

13. How are the interiors of an angle and a triangle alike? How are they different?

14. a. Can a point be in the interior of an angle and in the exterior of a triangle at the same time?

    b. Can a point be in the exterior of a triangle but not in the interior of any angle of the triangle?

15. Given $\angle DEF$, describe the following sets of points:

a. (Interior of $\angle DEF$) $\cap$ $\overrightarrow{EF}$.
b. (Exterior of $\angle DEF$) $\cap$ $\overrightarrow{ED}$.
c. (Interior of $\angle DEF$) $\cap$ (exterior of $\angle DEF$).
d. (Interior of $\angle DEF$) $\cap$ (exterior of $\angle DEF$) $\cap$ $\angle DEF$.
e. (Interior of $\angle DEF$) $\cup$ (exterior of $\angle DEF$) $\cup$ $\angle DEF$.

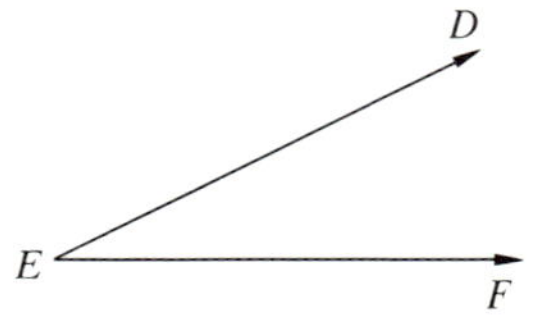

SOLUTIONS

1. Connect A, B, and C. Then (a) draw a line through A parallel to $\overline{BC}$, (b) draw a line through B parallel to $\overline{AC}$, (c) draw a line through C parallel to $\overline{AB}$.

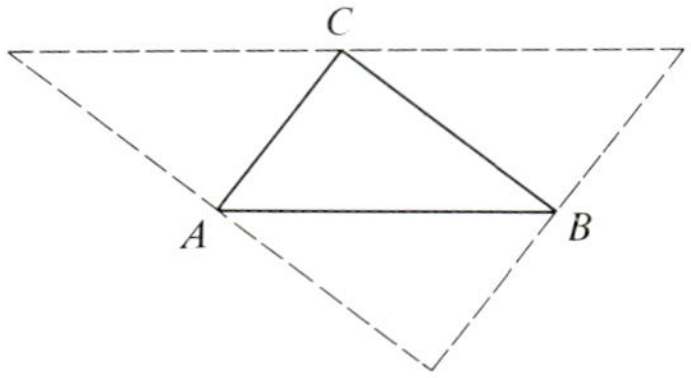

2. 35.

3. Arrange the toothpicks in the shape of a tetrahedron.

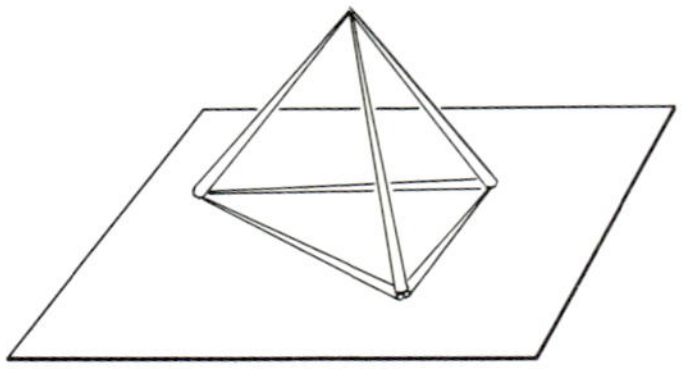

PUZZLES

1. Draw a triangle such that the points A, B, and C are the midpoints of the sides of the triangle.

2. What is the total number of triangles in the figure?

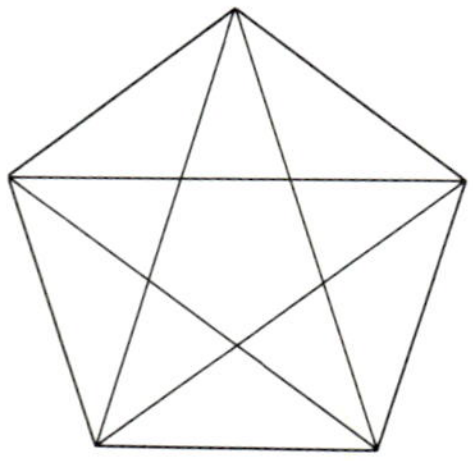

3. You are given six toothpicks of equal length. Arrange them into four equilateral triangles, the sides of each triangle being the length of a toothpick.

FAMOUS THEOREMS RELATED TO THE TRIANGLE

Some of the most interesting theorems in geometry relate to the triangle, and prominent among these are the nine-point circle theorem and the theorems of A. Miquel, Menelaus of Alexandria, and G. Ceva. [*Note:* To follow the proofs of the Miquel and Menelaus theorems, the reader is assumed to be familiar with the following ideas that are developed later in the text: (1) the measure of an inscribed angle (p. 171), and (2) similar triangles and the ratio of their corresponding sides (pp. 219–220).]

The Nine-Point Circle Theorem

Given $\triangle A_1A_2A_3$, with circumcenter O and orthocenter H (FIG. 3-41). Let O_1, O_2, and O_3 be the midpoints of the sides; H_1, H_2, and H_3 be the feet of the altitudes; and C_1, C_2, and C_3 be the midpoints of the segments $\overline{HA_1}$, $\overline{HA_2}$, $\overline{HA_3}$. Then

1. The nine points O_1, O_2, O_3, H_1, H_2, H_3, C_1, C_2, and C_3 lie on a circle known as the *nine-point circle* of the given triangle.
2. The center F of the nine-point circle is at the midpoint of $\overline{OH}$.
3. The radius of the nine-point circle is equal to half the radius of the circumcircle.

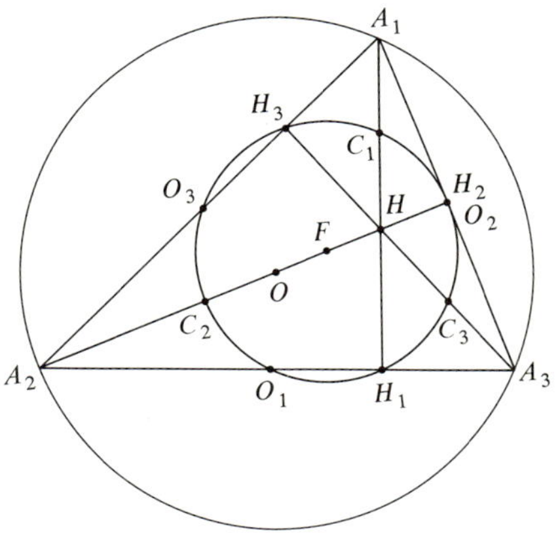

FIGURE 3-41

For a proof of this theorem, see Nathan Altshiller-Court, *College Geometry*, New York: Barnes & Noble, Inc., 1952, pp. 93 and 94.

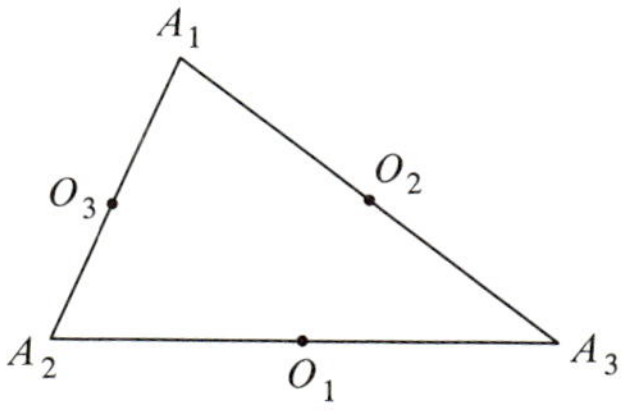

FIGURE 3-42

There are many triangles related to $\triangle A_1A_2A_3$ that have the same nine-point circle as $\triangle A_1A_2A_3$ (FIG. 3-42). For example, if a triangle is so drawn that the feet of its altitudes are at the points O_1, O_2, and O_3 of triangle $A_1A_2A_3$, then it has the same nine-point circle as $\triangle A_1A_2A_3$.

To construct $\triangle B_1B_2B_3$ such that the feet of its altitudes are at O_1, O_2, and O_3, we use the fact that the sides of $\triangle B_1B_2B_3$ can be constructed by constructing the external bisectors of $\triangle O_1O_2O_3$ (that is, the bisectors of the exterior angles of $\triangle O_1O_2O_3$) (FIG. 3-43). Therefore, join O_1, O_2, and O_3 and construct the external bisectors of $\triangle O_1O_2O_3$. These bisectors will determine $\triangle B_1B_2B_3$. Since

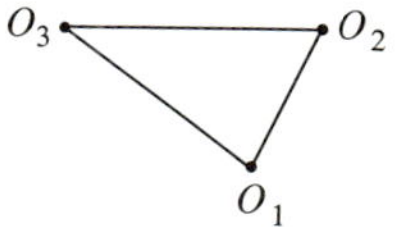

FIGURE 3-43

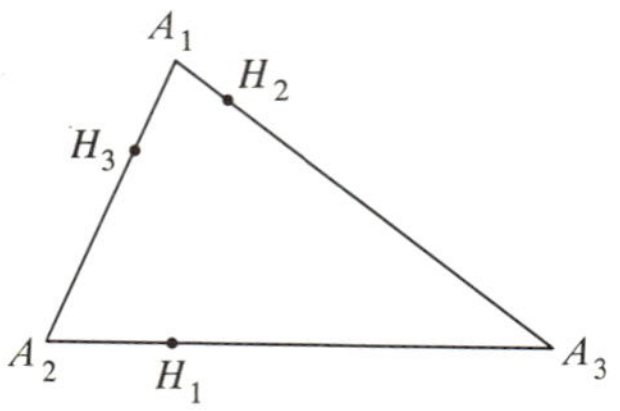

FIGURE 3-44

$\triangle B_1B_2B_3$ may in turn be replaced by a third $\triangle C_1C_2C_3$ (the feet of whose altitudes are the midpoints of the sides of $\triangle B_1B_2B_3$), $\triangle C_1C_2C_3$ by a fourth $\triangle D_1D_2D_3$, etc., we get an unlimited sequence of triangles having a common nine-point circle.

A different sequence of triangles may be obtained if a triangle is so drawn that the midpoints of its sides are at the points H_1, H_2, and H_3 of $\triangle A_1A_2A_3$ (FIG. 3-44). Both triangles have the same nine-point circle. Here again we get an unlimited sequence of triangles having a common nine-point circle.

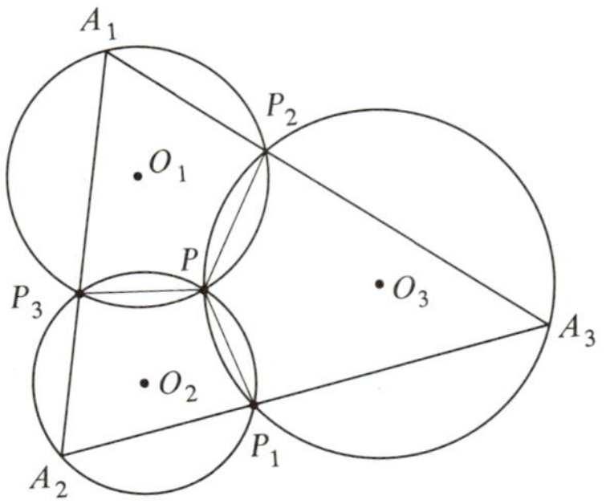

FIGURE 3-45

The Theorem of Miquel

The Miquel Theorem (1838) is an elegantly simple theorem with wide applicability:

> *If an arbitrary point is taken on each side of a triangle, the three circles determined by each vertex and the two points on the adjacent sides have a common point of intersection* (FIG. 3-45).

Let P_1, P_2, and P_3 be the arbitrary points in the given $\triangle A_1A_2A_3$, and let P be the point of intersection of circles O_2 and O_3. Then

$$m\angle P_3PP_1 = 180 - m\angle A_2,$$

$$m\angle P_1PP_2 = 180 - m\angle A_3.$$

Therefore,

$$\begin{aligned} m\angle P_2PP_3 &= m\angle A_2 + m\angle A_3 \\ &= 180 - m\angle A_1. \end{aligned}$$

This means that quadrilateral $PP_3A_1P_2$ is cyclic (that is, can be inscribed in a circle) and that point P lies on circle O_1 as well as on circles O_2 and O_3.

If P lies outside $\triangle A_1A_2A_3$, we can prove similarly that quadrilateral $PP_3A_1P_2$ is cyclic.

The point P is called the *Miquel point*, $\triangle P_1P_2P_3$ is a *Miquel triangle*, and O_1, O_2, and O_3 are *Miquel circles*.

The usefulness of this theorem is enhanced by the fact that P_1, P_2, and P_3 are completely arbitrary and so yield infinitely many Miquel points for any triangle. Incidentally, the centers of any set of Miquel circles are vertices of a triangle similar to the given triangle.

A Miquel triangle of special interest is a *pedal triangle.* A *pedal triangle* of a point with regard to a triangle is that triangle whose vertices are the feet of the perpendiculars from the point to the sides of the given triangle. In FIG. 3-46, the pedal triangle of P with respect to $\triangle A_1A_2A_3$ is $\triangle P_1P_2P_3$.

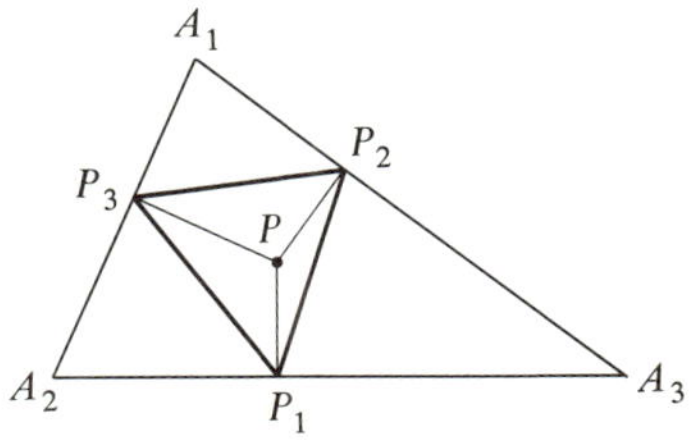

FIGURE 3-46

The Theorem of Menelaus

The theorem of Menelaus and the theorem of Ceva are related historically and conceptually. Menelaus of Alexandria discovered his theorem about 100 B.C., which was forgotten until Ceva rediscovered it. He then discovered his own theorem and published both in 1678.

One theorem deals with the collinearity of three points, one on each side of a triangle; the other theorem deals with the concurrency of three lines, one through each vertex of a triangle. Each theorem provides a general and powerful method of proving a group of seemingly unrelated theorems.

Before stating the Menelaus Theorem, let us recall that when dealing with directed line segments, $AB = -BA$ (FIG. 3-47). Also the ratio AP/PB will be positive or negative depending on whether P is an internal or external point of division. For example, in FIG. 3-48, AP/PB is positive, since both AP and PB are positive; whereas in FIG. 3-49, AP/PB is negative, since AP is positive, but PB is negative.

FIGURE 3-47

FIGURE 3-48

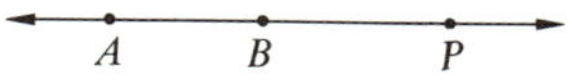

FIGURE 3-49

The Menelaus Theorem *Three points, P_1, P_2, and P_3, one on each side line of a $\triangle A_1A_2A_3$* (FIG. 3-50) *but not at any vertex of the triangle, are collinear if and only if*

$$\frac{A_1P_3}{P_3A_2}\cdot\frac{A_2P_1}{P_1A_3}\cdot\frac{A_3P_2}{P_2A_1} = -1. \tag{3-1}$$

The outline of a proof follows:

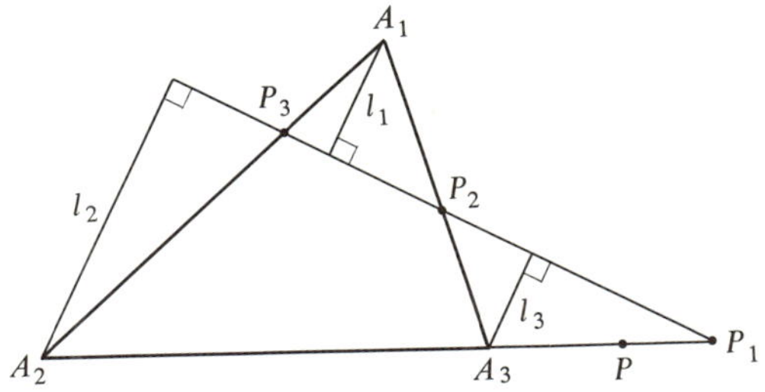

FIGURE 3-50

1. Assume P_1, P_2, and P_3 to be collinear and denote the distances from A_1, A_2, and A_3 to line $\overleftrightarrow{P_1P_2P_3}$ by l_1, l_2, and l_3. We shall show that (3-1) holds. By similar triangles, we get

$$\frac{A_1P_3}{P_3A_2} = \frac{l_1}{l_2}, \qquad \frac{A_2P_1}{P_1A_3} = \frac{l_2}{-l_3}, \qquad \frac{A_3P_2}{P_2A_1} = \frac{l_3}{l_1}.$$

Therefore,

$$\frac{A_1P_3}{P_3A_2}\cdot\frac{A_2P_1}{P_1A_3}\cdot\frac{A_3P_2}{P_2A_1} = \frac{l_1}{l_2}\cdot\frac{l_2}{-l_3}\cdot\frac{l_3}{l_1} = -1.$$

2. Assume that

$$\frac{A_1P_3}{P_3A_2}\cdot\frac{A_2P_1}{P_1A_3}\cdot\frac{A_3P_2}{P_2A_1} = -1. \tag{3-2}$$

We shall show that P_1, P_2, and P_3 are collinear. Let line $\overleftrightarrow{P_3P_2}$ intersect side line $\overleftrightarrow{A_2A_3}$ in P. Then, by step 1,

$$\frac{A_1P_3}{P_3A_2}\cdot\frac{A_2P}{PA_3}\cdot\frac{A_3P_2}{P_2A_1} = -1. \tag{3-3}$$

Dividing (3-3) by (3-2), we get

$$\frac{A_2P}{PA_3} = \frac{A_2P_1}{P_1A_3}$$

and

$$\frac{A_2P + PA_3}{PA_3} = \frac{A_2P_1 + P_1A_3}{P_1A_3}. \tag{3-4}$$

Since $A_2P + PA_3 = A_2A_3$ and $A_2P_1 + P_1A_3 = A_2A_3$, both numerators in (3-4) are equal and different from zero, and $PA_3 = P_1A_3$. Since points P and P_1 lie on the same line, P and P_1 coincide, and the three points P_1, P_2, and P_3 are collinear.

The Menelaus Theorem can be used to prove such theorems as

The bisectors of two interior angles of a triangle and the external bisector of the third angle meet the opposite sides in three collinear points (FIG. 3-51).

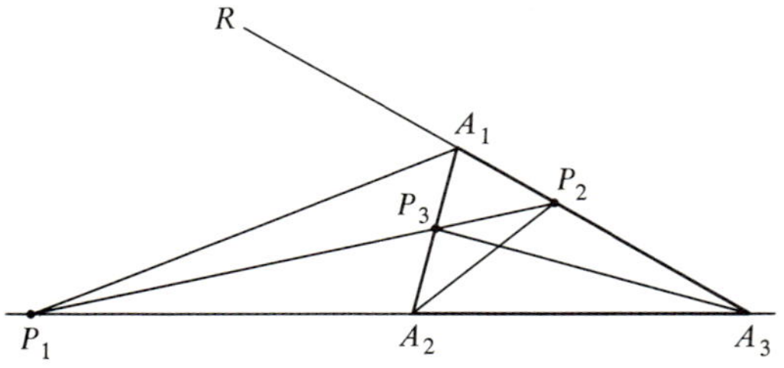

FIGURE 3-51

The Pappus Theorem *If P_1, P_2, and P_3 are three distinct points of a line l_1 and P_4, P_5, and P_6 any three distinct points of another line l_2, then the three points of intersection of the pairs P_1P_4, P_3P_5; P_2P_4, P_3P_6;* and *P_2P_5, P_1P_6 are collinear* (FIG. 3-52).

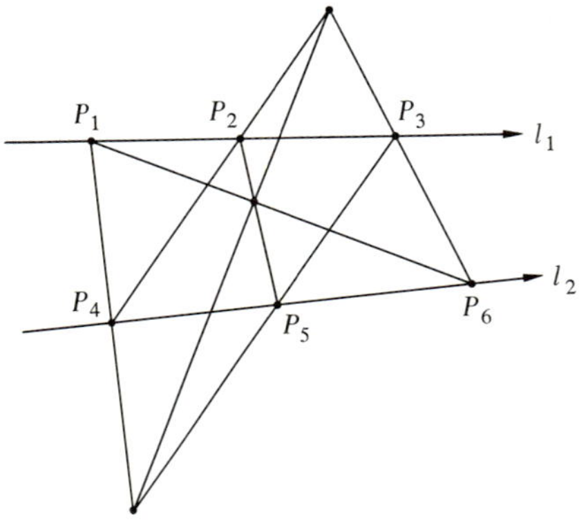

FIGURE 3-52

The Ceva Theorem *Given $\triangle A_1A_2A_3$ and points P_1, P_2, and P_3, one on each side line of the triangle, but not at any vertex of the triangle* (FIG. 3-53), *the three segments that join the points P_1, P_2, and P_3 to the opposite vertices are concurrent if and only if*

$$\frac{A_2P_3}{P_3A_1}\cdot\frac{A_1P_2}{P_2A_3}\cdot\frac{A_3P_1}{P_1A_2} = 1. \tag{3-5}$$

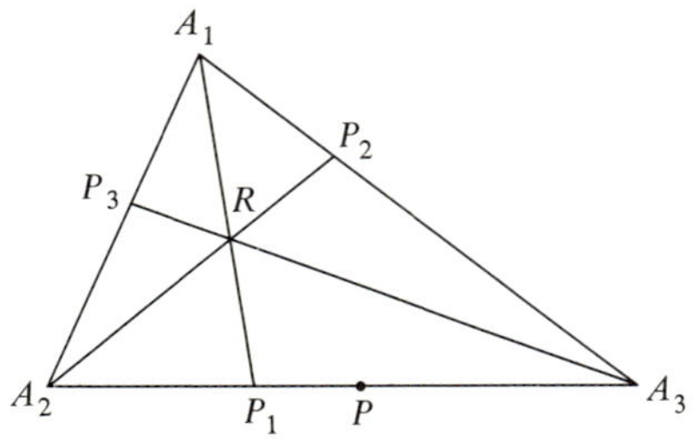

FIGURE 3-53

The following is an outline of the proof:

1. Assume that $\overline{A_1P_1}$, $\overline{A_2P_2}$, and $\overline{A_3P_3}$ intersect in point R. We shall show that (3-5) holds. In $\triangle A_3P_1A_1$, by the Menelaus Theorem,

$$\frac{A_3A_2}{A_2P_1}\cdot\frac{P_1R}{RA_1}\cdot\frac{A_1P_2}{P_2A_3} = -1, \tag{3-6}$$

and in $\triangle A_2A_1P_1$,

$$\frac{A_2P_3}{P_3A_1}\cdot\frac{A_1R}{RP_1}\cdot\frac{P_1A_3}{A_3A_2} = -1. \tag{3-7}$$

Multiplying (3-6) by (3-7), we get

$$\frac{A_2P_3}{P_3A_1}\cdot\frac{A_1P_2}{P_2A_3}\cdot\frac{A_3P_1}{P_1A_2} = 1.$$

2. Assume that (3-5) is true. We shall now show that $\overline{A_1P_1}$, $\overline{A_2P_2}$, and $\overline{A_3P_3}$ are concurrent. Let $\overline{A_2P_2}$ and $\overline{A_3P_3}$ intersect in R, and let $\overrightarrow{A_1R}$ intersect $\overline{A_2A_3}$ in P. Then

$$\frac{A_2P_3}{P_3A_1}\cdot\frac{A_1P_2}{P_2A_3}\cdot\frac{A_3P}{PA_2} = 1. \tag{3-8}$$

Dividing (3-8) by (3-5), we get

$$\frac{A_3P}{PA_2} = \frac{A_3P_1}{P_1A_2},$$

which means that

$$\frac{A_3P + PA_2}{PA_2} = \frac{A_3P_1 + P_1A_2}{P_1A_2}. \tag{3-9}$$

Since $A_3P + PA_2 = A_3A_2$ and $A_3P_1 + P_1A_2 = A_3A_2$, both numerators in (3-9) are equal and different from zero, and $PA_2 = P_1A_2$. Since P and P_1 lie on the same line, P and P_1 must coincide, and the segments $\overline{A_1P_1}$, $\overline{A_2P_2}$, and $\overline{A_3P_3}$ are concurrent.

The following well-known theorems are consequences of the Ceva Theorem:

1. *The medians of a triangle are concurrent.*
2. *The altitudes of a triangle are concurrent.*
3. *The bisectors of the interior angles of a triangle are concurrent.*

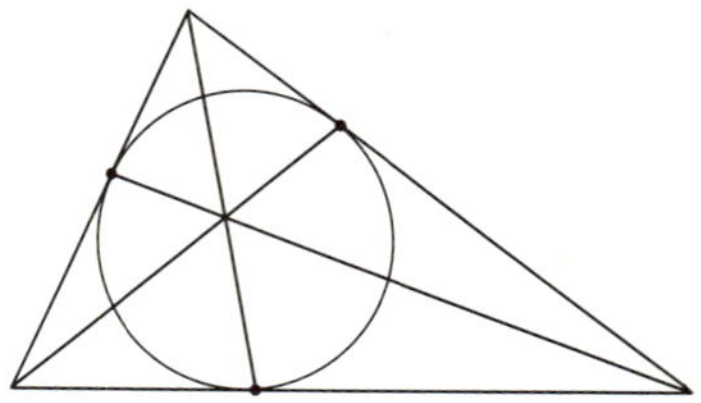

FIGURE 3-54

4. *The lines joining the vertices of a triangle to the points of contact of the opposite sides with the inscribed circle of the triangle are concurrent* (FIG. 3-54).

4 CONGRUENCE

INTRODUCING THE CHAPTER TO STUDENTS

CONTENT REVIEW

Prerequisites for Learning This Chapter
Congruence of Triangles: Informal Ideas
Congruence of Triangles: Formal Treatment
A General Concept of Congruence

TEACHING SUGGESTIONS AND ACTIVITIES

Visual Aids
Geometry by Experimentation
Introducing Congruence
Questions
Puzzles
Solutions

ENRICHMENT

"Moving" Geometric Figures
Isometries

INTRODUCING THE CHAPTER TO STUDENTS

If we take a pile of nickels and stack one on top of the other, they would match perfectly because all the nickels have the same size and shape. A deck of cards, too, can be stacked perfectly without any corners sticking out because all the cards in the deck have the same size and the same shape.

Objects that have the same size and shape are of interest and importance not only in the world of mathematics but in the physical world around us. In mathematics we use a special word to describe this condition of "same size and shape." We say that such objects are *congruent* to each other. Assuming that all our nickels have the same thickness and all our cards have the same thickness, then all the nickels in our pile are congruent to each other, and all the cards in our deck are congruent to each other.

We see many examples of congruent figures all around us. When a manufacturer engages in "mass production" he is producing congruent objects. The same model cars are congruent to each other and so are the same make and model radios and basketballs.

We just suggested that a practical test of congruence is to place one object over the other (superimpose) and see whether the two objects match. If the fit is perfect, then they are congruent. We can easily do this with a deck of cards or a pile of nickels. But how can we match, by this method, two automobiles or a couple of basketballs?

Obviously, we must have better ways of telling whether figures are congruent than placing one on top of the other and seeing whether the two figures match.

We shall see that certain properties of congruence make it unnecessary to superimpose physically one figure on the other to test their congruence. We shall see that we do not even have to *measure* all the parts of two figures to find out whether they are congruent.

Our object here is to deal with congruence *mathematically* rather than through its physical properties and applications. We shall do this by first defining congruence in terms of distance and angular measure and then stating one basic postulate, which will be followed by some theorems. We used this approach with segments and angles and will now use it with triangles.

CONTENT REVIEW

PREREQUISITES FOR LEARNING THIS CHAPTER

1. An understanding of the following: meaning of congruence of segments and angles; one-to-one correspondence; equivalence relation; Theorem 3-2; midpoint; "included" angle and sides; isosceles and equilateral triangles.
2. The ability to use a ruler and protractor to draw and measure segments and angles.

CONGRUENCE OF TRIANGLES: INFORMAL IDEAS[1]

Consider $\triangle ABC$ and $\triangle DEF$ in FIG. 4-1. Either triangle can be moved onto the other in such a way as to fit exactly.

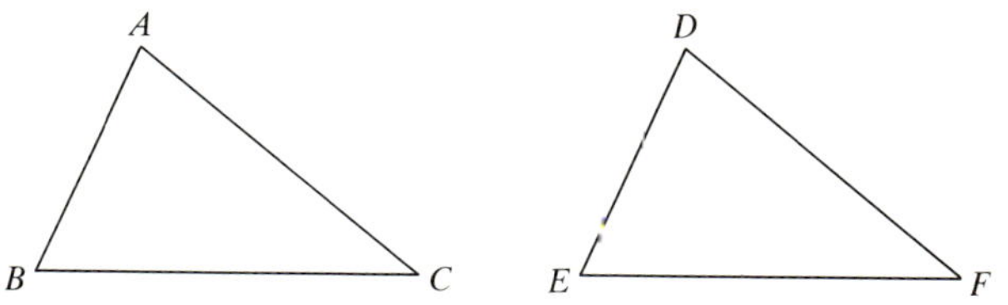

FIGURE 4-1

1. To move $\triangle ABC$ onto $\triangle DEF$ so as to fit exactly, we move $\triangle ABC$ in such a way that point A falls on point D, point B on point E, and point C on point F. That is, we must set up the following one-to-one correspondence between the vertices of the two triangles:

   $A \leftrightarrow D, \quad B \leftrightarrow E, \quad \text{and} \quad C \leftrightarrow F.$

2. A shorter way to show this correspondence is to write

   $ABC \leftrightarrow DEF,$

   where the first letter on the left (A) corresponds to the first letter on the right (D), the second (B) to the second (E), and the third (C) to the third (F).

[1] The intuitive idea of congruence is the same for triangles, or any other geometric figures, as for segments and angles. It is that two figures are congruent if the first figure can be moved so as to coincide with the second figure. Two triangles are congruent if one triangle can be moved onto the other triangle in such a way that the triangles fit exactly.

Although this is not a definition of congruence, it will soon lead us to such a formal definition.

Given a correspondence between the vertices of two triangles, $ABC \leftrightarrow DEF$ (FIG. 4-2), we speak of the *corresponding sides* as

$$\overline{AB} \leftrightarrow \overline{DE}, \qquad \overline{BC} \leftrightarrow \overline{EF}, \qquad \overline{AC} \leftrightarrow \overline{DF},$$

and we speak of the *corresponding angles* as

$$\angle A \leftrightarrow \angle D, \qquad \angle B \leftrightarrow \angle E, \qquad \angle C \leftrightarrow \angle F.$$

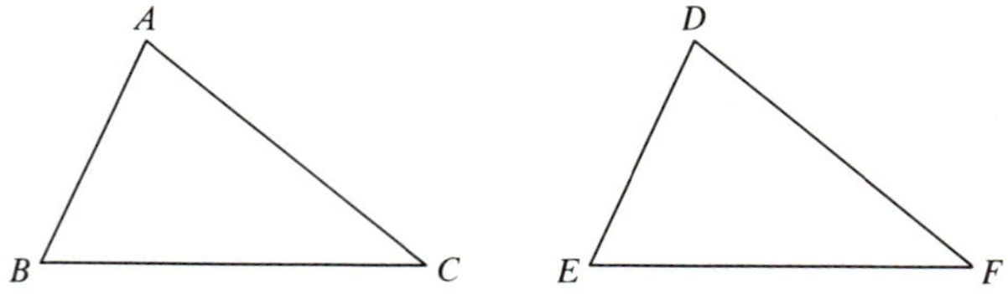

FIGURE 4-2

Note that sides of two triangles correspond if their endpoints are corresponding vertices.

Not all correspondences describe congruences. For example, in the triangles in FIG. 4-2, the correspondence $ABC \leftrightarrow EFD$ does not describe a congruence because matching the vertices in this fashion will not produce a perfect fit.

To test whether a correspondence between the vertices will produce a perfect fit, we need to check whether the corresponding sides and angles are congruent. Our formal definition of congruence, given below, is based on this idea.

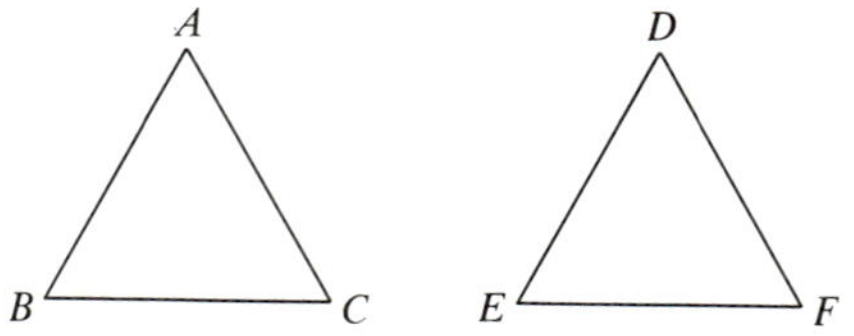

FIGURE 4-3

CONGRUENCE OF TRIANGLES: FORMAL TREATMENT

Definition
Given $\triangle ABC$, $\triangle DEF$, and a one-to-one correspondence $ABC \leftrightarrow DEF$ between their vertices (FIG. 4-3), if every pair of corresponding sides is congruent and every pair of corresponding angles is congruent, then the correspondence is a *congruence*. We then say that $\triangle ABC$ and $\triangle DEF$ are congruent, and we write $\triangle ABC \cong \triangle DEF$.

The expression $\triangle ABC \cong \triangle DEF$ (FIG. 4-4) is a short way of saying that

1. $\overline{AB} \cong \overline{DE}$ or $AB = DE$
2. $\overline{BC} \cong \overline{EF}$ or $BC = EF$

3. $\overline{CA} \cong \overline{FD}$ or $CA = FD$

4. $\angle A \cong \angle D$ or $m\angle A = m\angle D$

5. $\angle B \cong \angle E$ or $m\angle B = m\angle E$

6. $\angle C \cong \angle F$ or $m\angle C = m\angle F$[2]

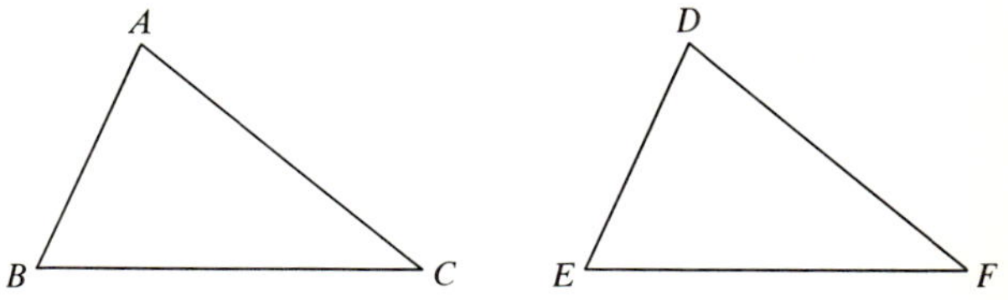

FIGURE 4-4

[2] 1. Note that although each congruence statement above on the left is interchangeable with the corresponding equality on the right, the two statements do not say exactly the same thing.

For example, $\overline{AB} \cong \overline{DE}$ expresses a *congruence* between two geometric figures. It says that *segment AB* and *segment CD* are congruent. But $AB = DE$ expresses an *equality* between two *numbers*. It says that the *distance AB* and the *distance DE* are exactly the same.

2. The expression $\triangle ABC \cong \triangle DEF$ tells us not only that the two triangles are congruent but that they are congruent in a particular way, that is, under the correspondence $ABC \leftrightarrow DEF$. This identifies the corresponding vertices and, thereby, identifies the corresponding sides and angles.

3. There is not necessarily only *one* correspondence between the vertices of two triangles that will produce a congruence. For instance, there is more than one correspondence between two equilateral triangles that can produce a congruence.

Consider the two equilateral triangles $\triangle ABC$ and $\triangle DEF$ (FIG. 4A), and suppose $AB = DE$.

Any of the following correspondences of the sets of vertices describes a congruence of the triangles:

$ABC \leftrightarrow DEF$, $ABC \leftrightarrow EFD$,

$ABC \leftrightarrow FDE$, $ABC \leftrightarrow EDF$.

So the condition for a congruence is that there is *some* correspondence between the vertices of two triangles which satisfies the six conditions given above.

FIGURE 4A

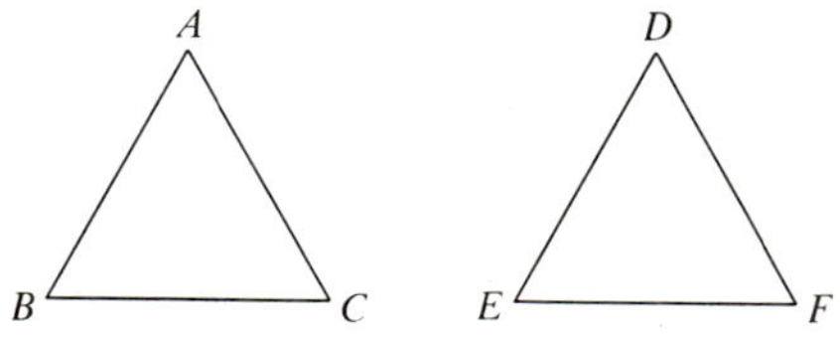

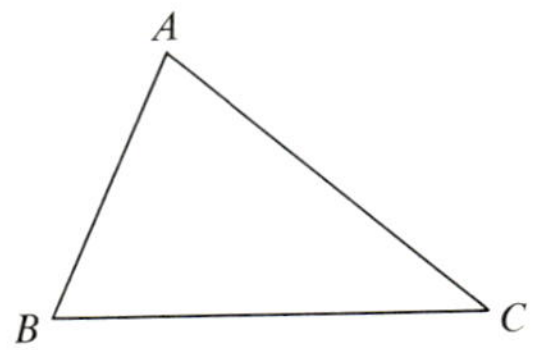

FIGURE 4B

4. In the $\triangle ABC$ (FIG. 4B), the correspondence $ABC \leftrightarrow ABC$ describes a congruence of the triangle with itself. This congruence is called the *identity congruence.*

Since the congruence of segments and the congruence of angles are equivalence relations, we can easily prove the following theorem.

THEOREM 4-1 Congruence between triangles is an equivalence relation.

That is, for any triangles $\triangle ABC$, $\triangle DEF$, and $\triangle JHK$,

$\triangle ABC \cong \triangle ABC$ (reflexive).

If $\triangle ABC \cong \triangle DEF$, then $\triangle DEF \cong \triangle ABC$ (symmetric).

If $\triangle ABC \cong \triangle DEF$ and $\triangle DEF \cong \triangle JHK$, then $\triangle ABC \cong \triangle JHK$ (transitive).

The Basic Congruence Postulate[3]

[3] Although the six pairs of corresponding parts of two triangles must be congruent for the triangles to be congruent, it is not necessary to *test* the congruence of *all* six pairs before we can say that the triangles are congruent. Sometimes, if we know only *three* pairs of corresponding parts to be congruent, we can nevertheless conclude that all the remaining corresponding parts are also congruent—if the three pairs are the right combination of parts.

One such "right" combination of three parts is any two sides and the included angle, that is, if two sides and the included angle of one triangle are congruent to the corresponding parts of another triangle (FIG. 4C).

FIGURE 4C

We can check informally whether this combination is "right" to produce a congruence by a simple experiment.

1. Draw two 40° angles on a sheet of paper (FIG. 4D).

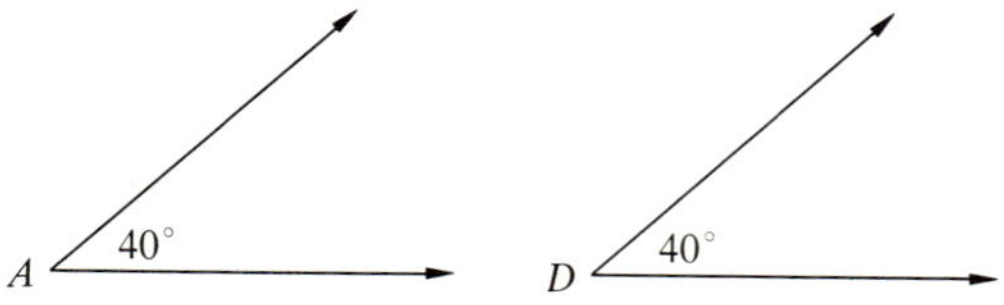

FIGURE 4D

2. Mark off on $\angle A$ points B and C such that $AB = 3$ and $AC = 4$, and mark off on $\angle D$

points E and F such that $DE = 3$ and $DF = 4$ (FIG. 4E).

3. Draw $\overline{BC}$ and $\overline{EF}$ and cut out both triangles from the sheet of paper (FIG. 4F).

4. Superimpose $\triangle ABC$ onto $\triangle DEF$ so that $ABC \leftrightarrow DEF$.

5. The two triangles will be seen to coincide.

We formalize this idea by stating it as our basic congruence postulate.

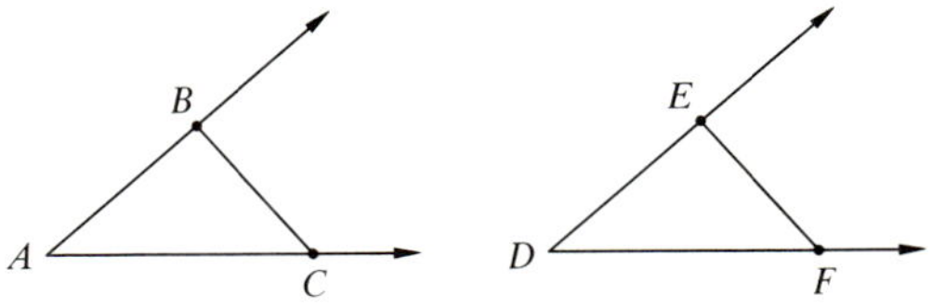

FIGURE 4E

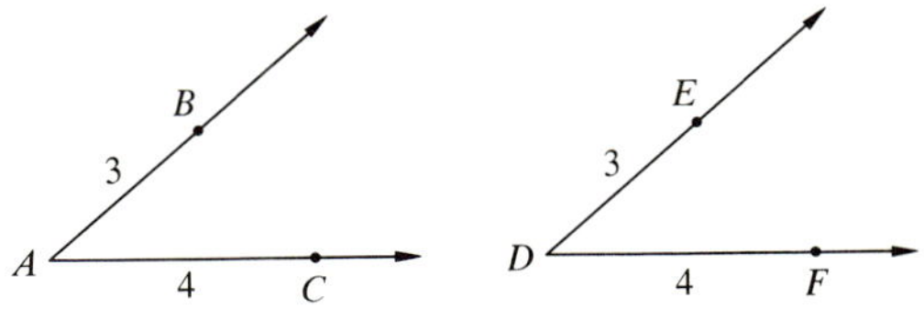

FIGURE 4F

POSTULATE 4-1 (Side Angle Side Postulate) *Given a correspondence between two triangles (or between a triangle and itself), if two sides and the included angle of the first triangle are congruent to the corresponding parts of the second triangle, then the correspondence is a congruence.*

That is, if $\overline{AB} \cong \overline{DE}$, $\overline{BC} \cong \overline{EF}$, and $\angle B \cong \angle E$, then $\triangle ABC \cong \triangle DEF$ (FIG. 4-5). We sometimes abbreviate the statement of this postulate by using the letters SAS (side angle side).

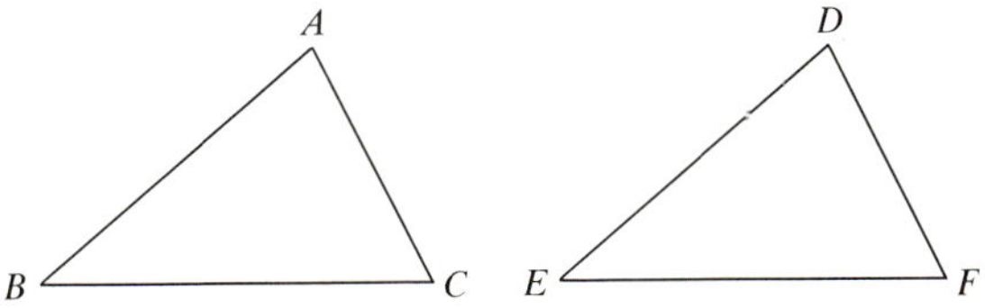

FIGURE 4-5

Basic Congruence Theorems

THEOREM 4-2 (Isosceles Triangle Theorem) *If two sides of a triangle are congruent, then the angles opposite them are congruent.*

That is, if $\overline{AB} \cong \overline{AC}$, then $\angle B \cong \angle C$ (FIG. 4-6).
The following is a simple and elegant proof of this theorem.

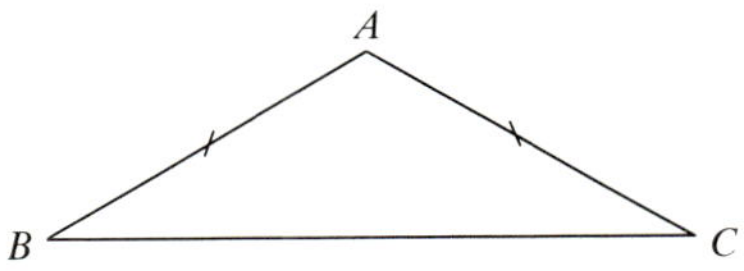

FIGURE 4-6

Proof

1. Consider the correspondence $ABC \leftrightarrow ACB$ between $\triangle ABC$ and itself. Then

$AB \leftrightarrow AC, \quad AC \leftrightarrow AB, \quad \text{and} \quad \angle A \leftrightarrow \angle A.$

2. By the SAS Postulate, $\triangle ABC \cong \triangle ACB$.
3. Therefore, $\angle B \cong \angle C$, since they are corresponding angles of congruent triangles.[4]

THEOREM 4-3 (Angle Side Angle Theorem) Given a correspondence between two triangles (or between a triangle and itself), if two angles and the included side of the first triangle are congruent to the corresponding parts of the second, then the correspondence is a congruence.

That is, if $\angle B \cong \angle E$, $\angle C \cong \angle F$, and $\overline{BC} \cong \overline{EF}$, then $\triangle ABC \cong \triangle DEF$ (FIG. 4-7). We refer to this theorem as the ASA (angle side angle) Theorem.[5]

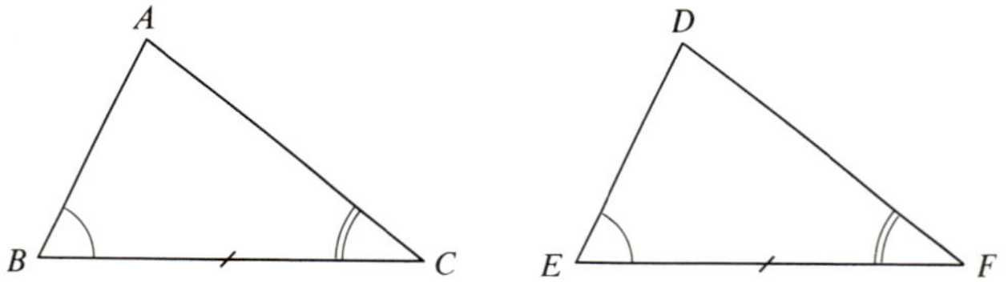

FIGURE 4-7

THEOREM 4-4 (Side Side Side Theorem) Given a correspondence between two triangles (or between a triangle and itself), if all three pairs of corresponding sides are congruent, then the correspondence is a congruence.

That is, if $\overline{AB} \cong \overline{DE}$, $\overline{BC} \cong \overline{EF}$, and $\overline{AC} \cong \overline{DF}$, then $\triangle ABC \cong \triangle DEF$ (FIG. 4-8). We refer to this theorem as the SSS (side side side) Theorem.[6]

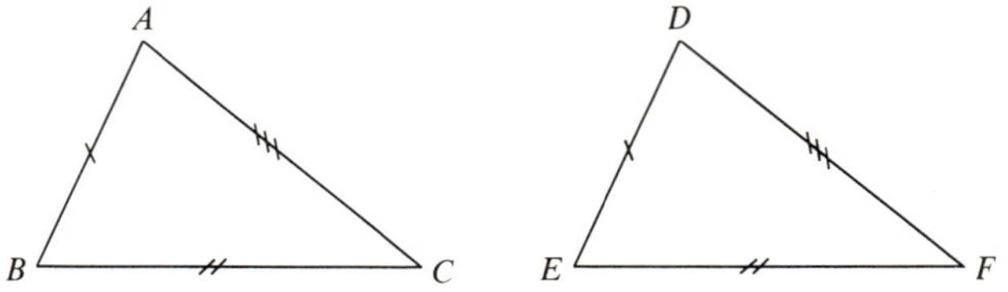

FIGURE 4-8

[4] 1. This is the famous *pons asinorum* theorem. It gets its name from the meaning of the phrase (asses' bridge) and from the figure that accompanied Euclid's proof (FIG. 4G).

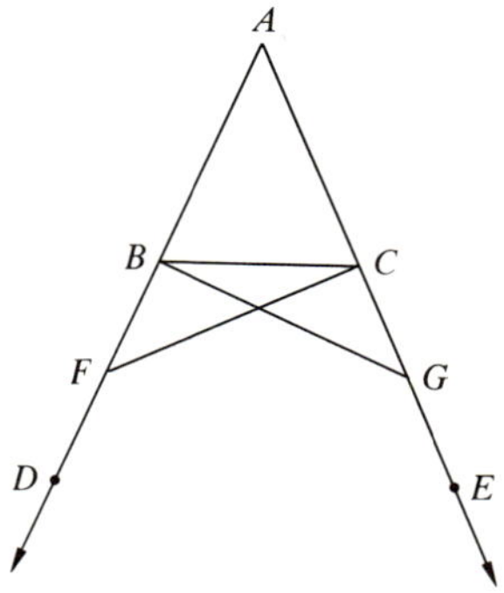

FIGURE 4G

2. A corollary of this theorem is that *in an equilateral triangle, all three angles are congruent.*

[5] A corollary of this theorem is that *if two angles of a triangle are congruent, the sides opposite these angles are congruent.* This is the converse of Theorem 4-2.

[6] 1. The ASA Theorem and the SSS Theorem can be proved on the basis of the SAS Postulate. For proofs of these theorems, see any standard geometry text.

2. From the SAS Postulate and the ASA and SSS Theorems we see that it is not necessary to test the congruence of all *six* pairs of corresponding parts before we can say that two triangles are congruent. Three "right" pairs are sufficient.

But what about congruent *quadrilaterals*? Do we need to test all *eight* pairs of corresponding parts before we can establish congruence, or can we prove congruence with fewer than the eight pairs?

It can be shown that in the case of two convex quadrilaterals, congruence of only *five* "right" pairs is sufficient to guarantee

Before proving the next theorem, we shall state another theorem and give a definition.

THEOREM 4-5 *Every segment has one and only one midpoint.*

Definition
A ray $\overrightarrow{BD}$ *bisects* $\angle ABC$ (FIG. 4-9) if (1) D is in the interior of $\angle ABC$, and (2) $\angle ABD \cong \angle DBC$.

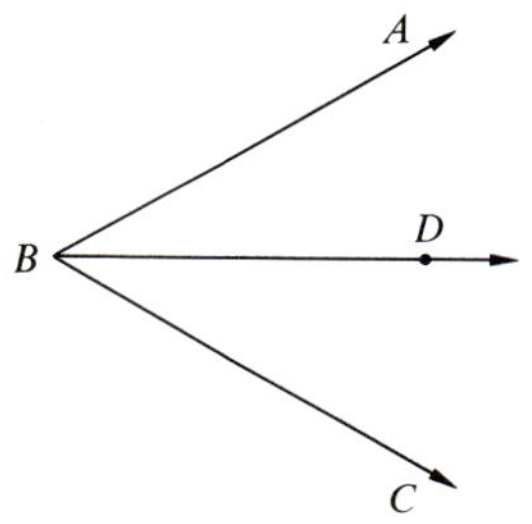

FIGURE 4-9

THEOREM 4-6 (Angle Bisector Theorem) *Every angle has exactly one bisector.*

Proof
1. We first prove that there is at least one bisector.
   a. Choose points A and C on different sides of $\angle B$ so that $\overline{AB} \cong \overline{BC}$, and let D be the midpoint of $\overline{AC}$ (FIG. 4-10).

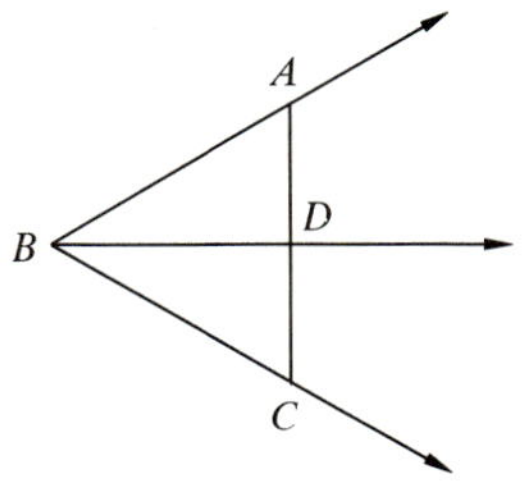

FIGURE 4-10

   b. Then, by Theorem 3-2, D is in the interior of $\angle ABC$, and by the SSS Theorem, $\triangle ABD \cong \triangle CBD$.

congruence of the quadrilaterals. Such "right" pairs are

1. Two adjacent sides and any three angles.
2. Three sides and the included angles.
3. Four sides and one angle.

c. Therefore, $\angle ABD \cong \angle CBD$, making BD a bisector of $\angle ABC$.

2. We now prove that $\angle ABC$ has at most one bisector.
   a. Suppose there is another bisector $\overrightarrow{BE}$. Then E is in the interior of $\angle ABC$ and $\overrightarrow{BE}$ intersects $\overline{AC}$ at some point D' (FIG. 4-11).
   b. By the SAS Postulate, $\triangle ABD' \cong \triangle CBD'$ and $\overline{AD'} \cong \overline{D'C}$, making D' the midpoint of $\overline{AC}$.
   c. By the theorem stated above, $D' = D$. This means that $\overrightarrow{BE}$ and $\overrightarrow{BD}$ are the same ray.

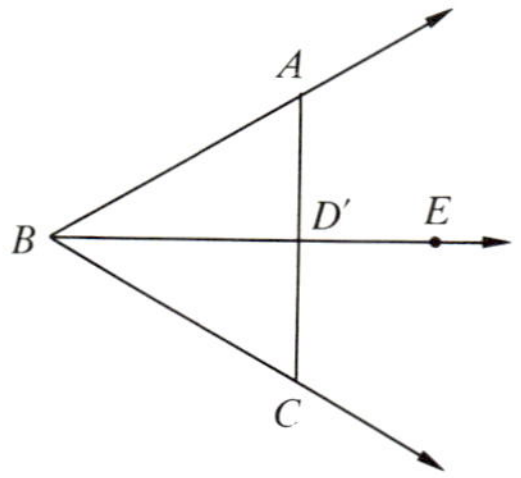

FIGURE 4-11

A GENERAL CONCEPT OF CONGRUENCE

We said that two figures are congruent if they have the same size and shape. They will have the same size and shape if their corresponding sides and angles are congruent. There is no difficulty accepting this definition of congruence when we are talking about figures like segments, angles, or triangles.

But suppose we are talking about figure A and its replica B (FIG. 4-12). What is the "size" of A? What is the "shape" of B? Here, we obviously need a more precise definition of "size" and "shape." We need to define precise conditions that determine when two figures, or any two sets of points, have the same size and shape.

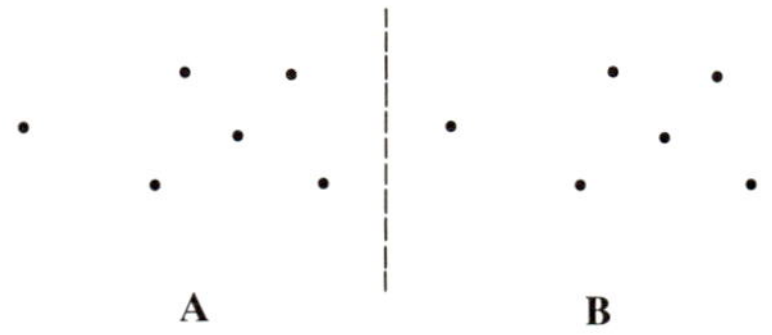

FIGURE 4-12

This need has been met by the following *general* definition of congruence that applies to any finite or to any infinite set of points:

> *Two sets of points A and B are congruent if and only if there exists a one-to-one correspondence between their points such that the segment determined by any two points of one set is congruent to the segment determined by the corresponding points of the other set.*

In other words, two figures—or sets of points—will have the same size and shape if the distance between any two points of the first figure is the same as the distance between their corresponding points of the second figure. Or, a correspondence between points that *preserves distance* assures us the same size and shape. These correspondences are called *isometries.*

TEACHING SUGGESTIONS AND ACTIVITIES

VISUAL AIDS

Visual 1

PURPOSE

1. To demonstrate the meaning of congruence.
2. To help verify the SAS, ASA, and SSS properties of congruent triangles.
3. To demonstrate that not *all* combinations of three parts will establish congruence.

CONSISTS OF

Two congruent

1. Circular discs.
2. Rectangles.
3. Triangles.
4. Irregular two-dimensional shapes.
5. Three-dimensional objects: balls.

HOW TO USE

1. To demonstrate the meaning of congruence, superimpose the congruent figures in each set and show that they fit perfectly.
2. To help verify the SAS property,
   a. Place one of the congruent triangles on a magnetic board.
   b. Trace with chalk any two sides and the included angle of the triangle.
   c. Remove the triangle from the tracing and complete construction of the triangle.
   d. Superimpose the triangle on the drawing and show that the two figures coincide *if the corresponding parts are matched.*
   e. Stress the significance of this experiment: that SAS

determines a unique triangle that can have only *one* size and shape, that is, that SAS is sufficient to determine congruence of two triangles.

f. Repeat the SAS experiment using another pair of sides and their included angle.

3. To help verify the ASA and SSS properties, follow a procedure similar to that followed in step 2.

4. Show by the same procedure that neither SSA nor AAA necessarily produces congruent triangles.

Visual 2

PURPOSE

To provide another way for students to verify the SAS, ASA, and SSS properties experimentally.

CONSISTS OF

Paper (stiff), ruler, compass, protractor, scissors.

HOW TO USE

1. To verify SAS property experimentally, ask the students to
   a. Draw any triangle on their paper.
   b. Copy on another part of their paper any two sides and the included angle of the triangle.
   c. Complete construction of the triangle in step b.
   d. Cut out of their paper both triangles, superimpose one on the other, and see whether they fit exactly.
   e. Measure all corresponding parts and see whether they are congruent.
   f. Repeat the experiment, using another pair of sides and their included angles.

2. Ask the students to perform a similar experiment to test the ASA and SSS properties.

Visual 3

PURPOSE

To introduce an appreciation of the need for proof in mathematics by exhibiting a set of optical illusions. These will illustrate the danger of drawing hasty conclusions from the evidence of our senses.

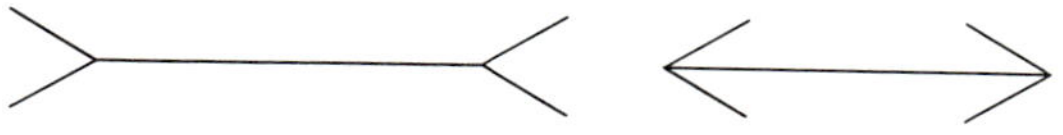

(a) (b)

FIGURE 4-13

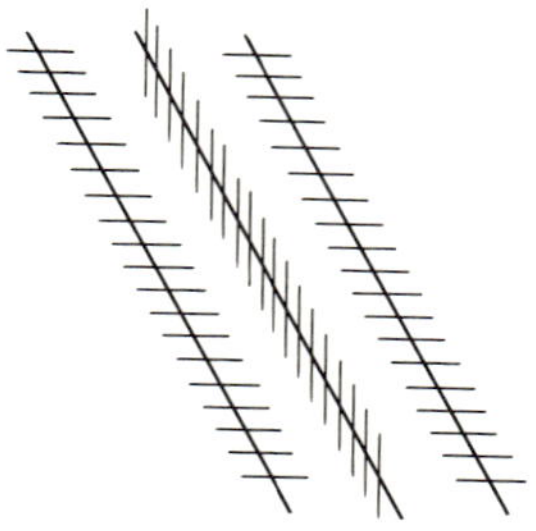

(a) (b) (c)

FIGURE 4-14

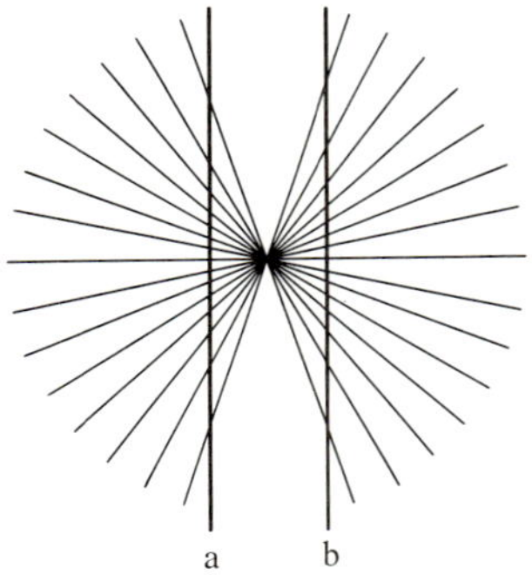

FIGURE 4-15

FIGURE 4-16

CONSISTS OF

Five charts containing the following drawings:

1. Which segment in FIG. 4-13 is longer, (a) or (b)?
2. Are lines a, b, and c in FIG. 4-14 parallel?
3. Are lines a and b in FIG. 4-15 curved or straight?
4. Look at FIG. 4-16. What do you see? Close one eye and look again. What do you see now?
5. Which segment is longer in FIG. 4-17, $\overline{AB}$ or $\overline{CD}$?

HOW TO USE

1. Display each optical illusion and ask the class to answer the question on each display.
2. Check answers:
   1. Measure both segments in FIG. 4-13.
   2. Measure the distance at several points between any two of these lines in FIG. 4-14.
   3. Apply a straightedge to both lines in FIG. 4-15.
   4. A vase; two faces "facing" each other in FIG. 4-16.
   5. Measure both segments in FIG. 4-17.

Note: These and other optical illusions are available commercially.

GEOMETRY BY EXPERIMENTATION

By means of accurate drawings, good visual aids, and careful measurements we can test or discover many geometric ideas.

It would be well, however, to reemphasize with our students the difference between conclusions based on experimentation and conclusions based on reasoning and to stress that conclusions reached in pure mathematics are derived from reasoning rather than from experimentation.

A good way to do this is to analyze an actual proof, say, of the theorem that states that vertical angles are congruent (FIG. 4-18).

1. $m\angle 1 = 180 - m\angle 3$.
2. $m\angle 2 = 180 - m\angle 3$.
3. $\angle 1 \cong \angle 2$.

In analyzing this proof with students we observe that

1. Before we could prove this theorem we needed at our disposal
   a. *Definitions:* angle, supplementary angles, vertical angles, linear pairs, and congruent angles which were defined in terms of the *undefined terms:* point and line.
   b. *A postulate: For any quantities, a, b, and c, if $a = b$ and $b = c$, then $a = c$.*

   The proof then consists of a chain of simple deductions from these definitions and the postulate.
2. The conclusion reached is true for *any* vertical angles, no matter what their measure.
3. The conclusion was obtained, not by experimentation or measuring, but by reasoning.
4. Although experimentation can open our minds to certain possibilities and lead to certain conjectures, such conclusions are only a guide to our intuition and, at best, only *probable* conclusions.
5. Optical illusions illustrate the danger of drawing hasty conclusions from the evidence of our senses.

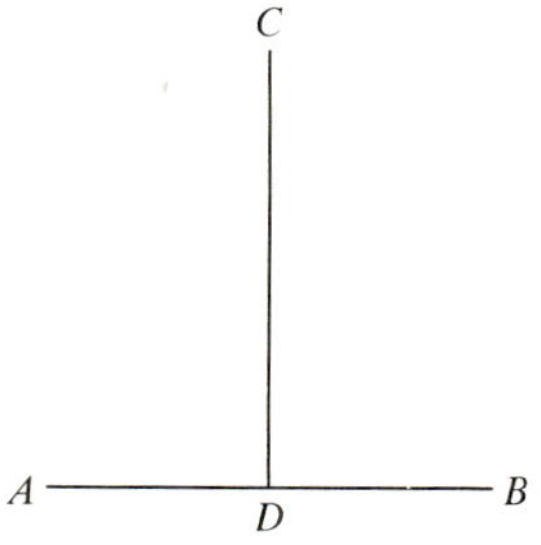

FIGURE 4-17

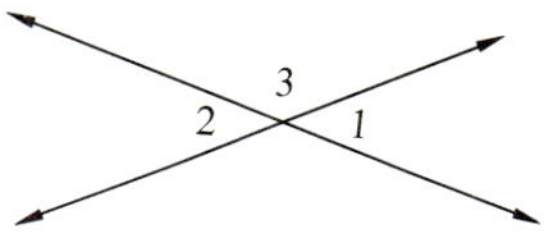

FIGURE 4-18

INTRODUCING CONGRUENCE

In introducing the concept of congruence to a class, the following guidelines may be of help:

1. Introduce the concept by intuition and experimentation (Visuals 1 and 2).
2. Define congruence informally: same size and shape, perfect fit.
3. Begin the more formal definition of congruence with the definition of congruent segments and congruent angles. From here go to a definition of congruent triangles and then to a general concept of congruence that can be used with any set of points.

In discussing the basic congruence properties of triangles, we should point out that two sides and the included angle, two angles and the included side, or three sides determine a triangle of only *one* size and shape (FIG. 4-19).

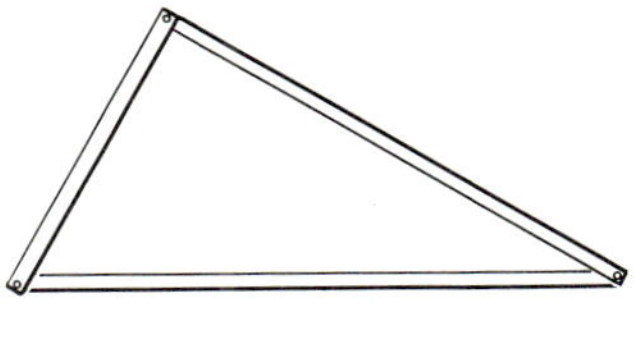
FIGURE 4-19

For instance, if we take three strips of cardboard and fasten them together at their ends, we get a unique triangle. The lengths of these three strips determine the triangle. They determine the measure of its angles, its area, its perimeter, and all its other mathematical properties.

If we were to fasten together three other strips having the same lengths as the first three, we would obtain a triangle having exactly the same size and shape as the first triangle.

QUESTIONS

1. Write down all the congruences between
   a. An equilateral triangle and itself.
   b. An isosceles triangle and itself.

2. Describe eight congruences between a square $ABCD$ and itself.

3. Prove that a triangle is equilateral if and only if it is equiangular.

4. Prove that every figure is congruent to itself.

5. Prove that if the hypotenuse and one leg of one right triangle are congruent to the corresponding parts of a second triangle, then the triangles are congruent.

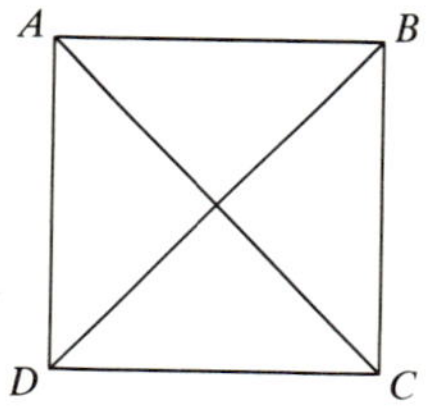

FIGURE 4-20

6. Prove that if in a triangle, two angles and a side opposite one of them are congruent to the corresponding parts of a second triangle, then the triangles are congruent.

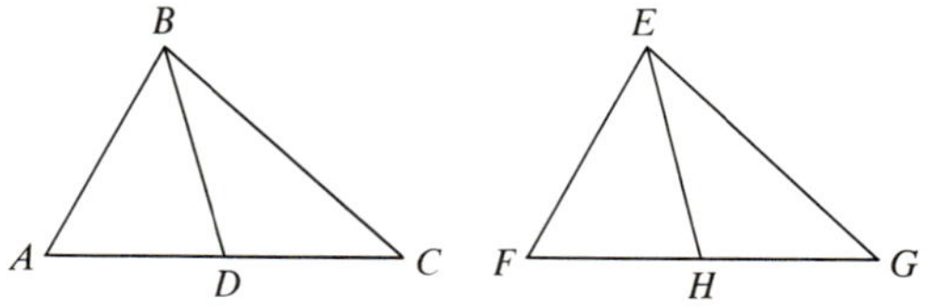

FIGURE 4-21

7. *Given:* $\angle A$, $\angle B$, $\angle C$, and $\angle D$ are right angles and $AB = BC = CD = AD$ (FIG. 4-20).

   Prove: a. $AC = BD$.
   b. $\overline{AC}$ and $\overline{BD}$ bisect each other.
   c. $\overline{AC} \perp \overline{BD}$.

8. *Given:* $AB = EF$, $BC = EG$, and median $\overline{BD} \cong$ median $\overline{EH}$ (FIG. 4-21).

   Prove: $\triangle ABC \cong \triangle FEG$.

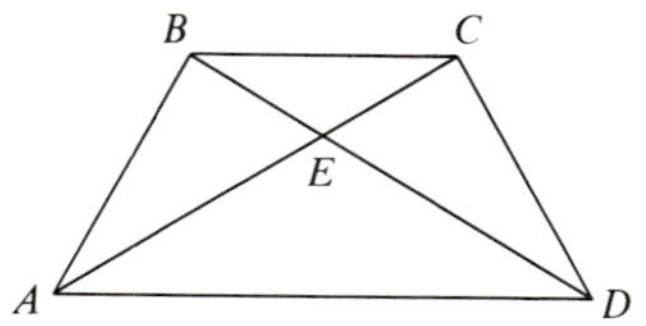

FIGURE 4-22

9. *Given:* $AB = CD$ and $AC = BD$ (FIG. 4-22).

   Prove: $\triangle AED$ is isosceles.

10. *Given:* $\angle 1 \cong \angle 2$, $\overline{BC} \cong \overline{DE}$, and $\overline{AD} \cong \overline{CF}$ (FIG. 4-23).
Prove: $\angle A \cong \angle F$.

11. *Given:* $\overline{BC} \perp \overline{AC}$, $\overline{BC} \perp \overline{CD}$, and $\overline{AC} \cong \overline{CD}$ (FIG. 4-24).
Prove: $\overline{AB} \cong \overline{BD}$.

12. Prove that if two triangles are congruent, then each pair of corresponding angle bisectors is congruent.

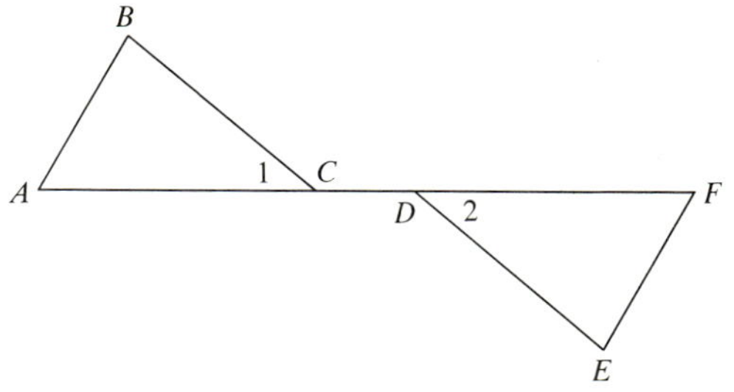

FIGURE 4-23

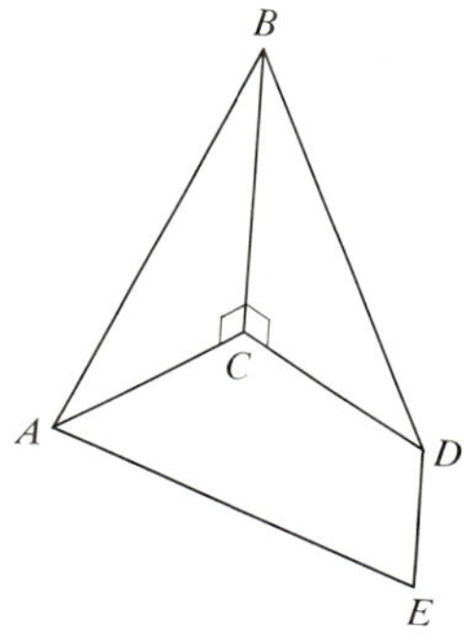

FIGURE 4-24

PUZZLES

1. You are given eight segments, four of which are half the length of the other four. Can you form three congruent squares with these eight segments?

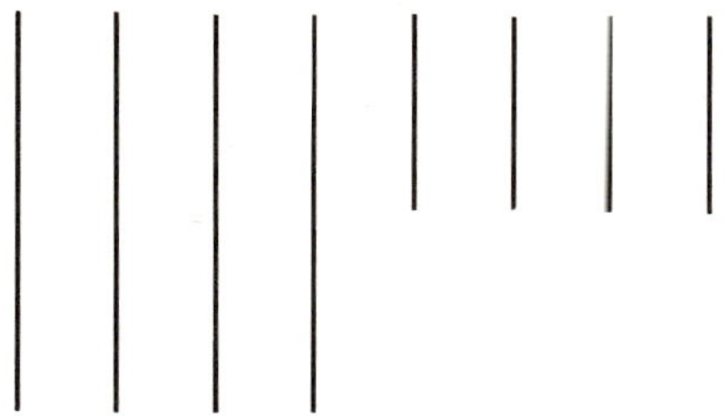

SOLUTIONS

1.

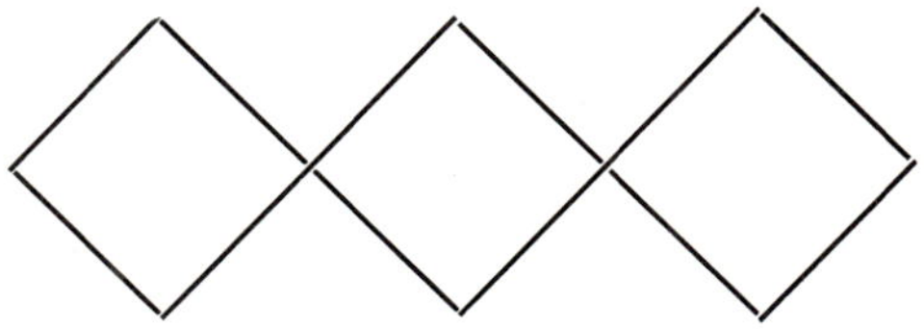

2. Two congruent triangles are formed, making the alternate interior angles congruent (see Theorem 5-3).

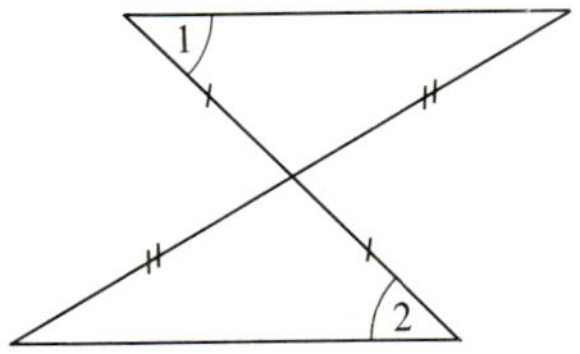

3. a. Measure $BD = BE$.
   b. Place the square so that $DF = EF$.
   c. Draw $\overrightarrow{BF}$.
   d. $m\angle 1 = m\angle 2$ because they are corresponding angles of congruent triangles.

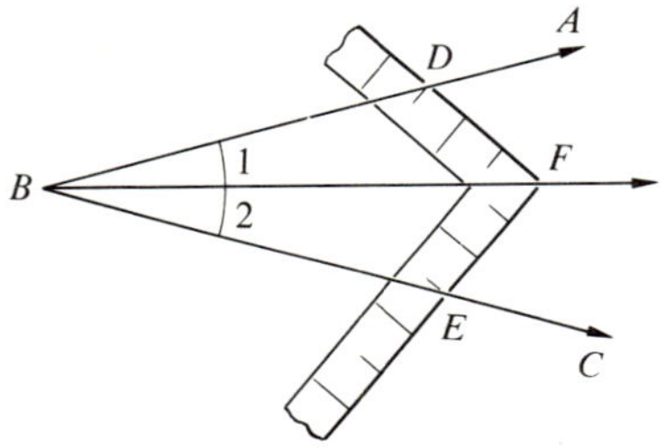

2. Why is the plane of the ironing board always parallel to the plane of the floor if the legs bisect each other?

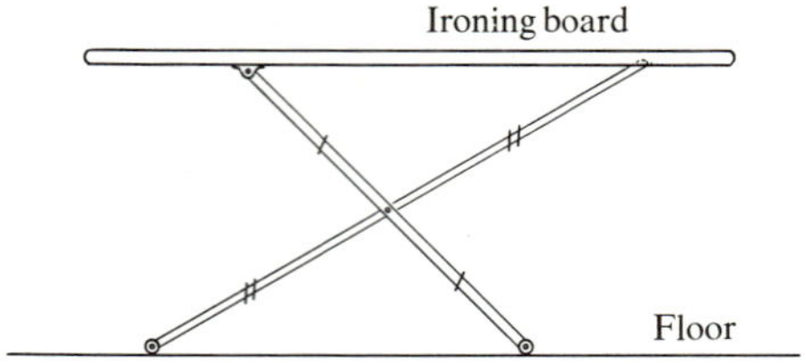

3. How can you bisect $\angle ABC$ by using the carpenter's square?

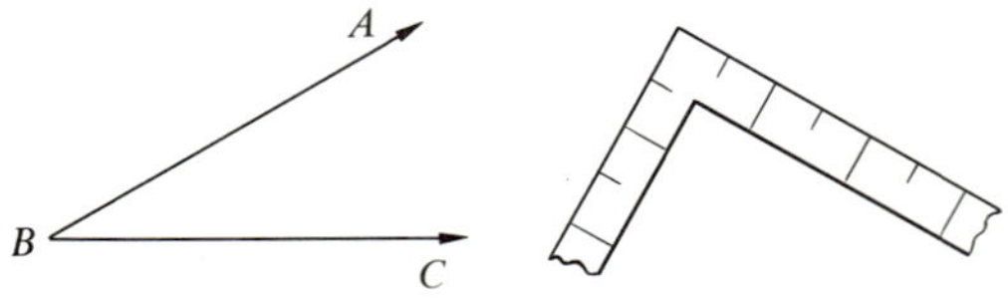

ENRICHMENT

"MOVING" GEOMETRIC FIGURES

We have seen that two figures are congruent if, by a suitable motion, one may be superimposed on the other in a perfect fit. What do we mean by "moving" a geometric figure?

The idea of "moving" abstract figures presents some difficulties. *Can* an abstract figure be moved? If it can, does it undergo change while in motion? And if it changes in motion, when it coincides with another figure at the end of its journey is it the same figure we started with?

These difficult questions have been answered by making a distinction between "moving" physical objects and "moving" abstract geometric figures.

Motion of a physical body involves different positions in space at different times between the initial and final positions of the body. But geometric "motion" of two congruent figures is not really motion in the physical sense. It is merely a *relation* or a *correspondence* between the points of two congruent figures which preserves distance. It is a mapping of one figure on the other. Geometric "motion," then, is a change that takes place only in our minds as we think of a point P in one figure and then shift to the corresponding point P' in another figure.

This concept of geometric "motion" is really the definition of congruence given on pages 114–115. A "motion" between two figures A and B is a correspondence between A and B such that if P corresponds to P' and Q to Q', then the distances PQ and $P'Q'$ are equal.

ISOMETRIES

We have seen some of the problems that arise if we try to prove congruence by "superposition." We have also suggested a way out of this difficulty: thinking of "superposition" as an abstract rather than as a physical idea. We shall now examine more closely the notion of "superposition" as an abstract idea.

To develop an abstract definition of "motion" of an object from one place to another, we rely on the concept of a *transformation.*

Imagine a block of wood resting on a table. As we slide the block along the table top, every point of the block changes position. Let P be any point on the table top which was initially occupied by a particular point of the block, and let P' be the new point on the table top to which the point of the block has been moved. Sliding the block across the table associates the initial point P with the terminal point P'. This association of each point of the table top with another point of the table top is called a *transformation* of the table top. Note that the transformation is on the set of points of the table top to the same set of points. Because no two points of the table top are carried to the same new position, we call such a transformation a *one-to-one* transformation. So by a *transformation* we mean a rule, operation, or function that assigns to every point P of the plane a unique point P' called the *image* of P.

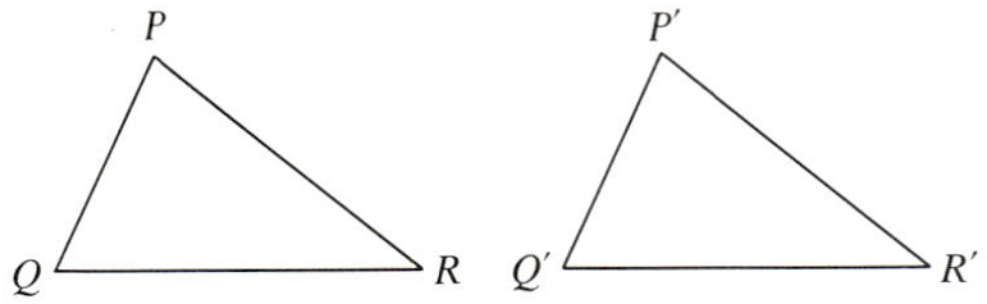

FIGURE 4-25

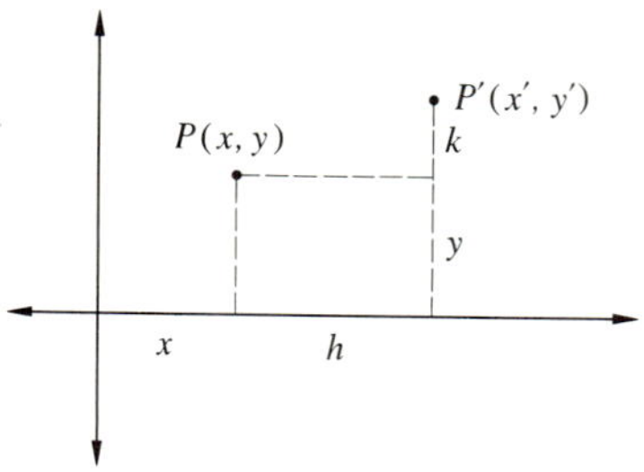

FIGURE 4-26

Now, instead of sliding a block across a table top, let us imagine that the table top is a plane and the plane itself is in motion. Such motion will carry any point P to a new position P'. If the plane moves as if it were made of stiff cardboard, then the length of any segment in the plane is preserved by the motion of the plane. The length of every segment before and after the transformation remains the same. This kind of transformation, or one-to-one mapping of the plane onto itself, "moves" every segment into a congruent segment. We call such a distance-preserving transformation a *rigid motion*, or an *isometry* ("same measure").

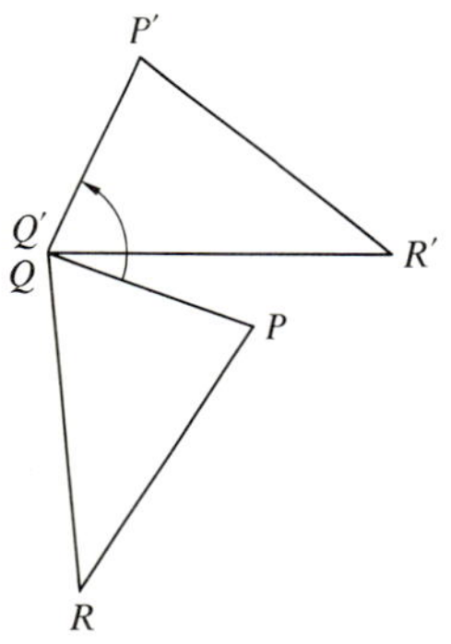

FIGURE 4-27

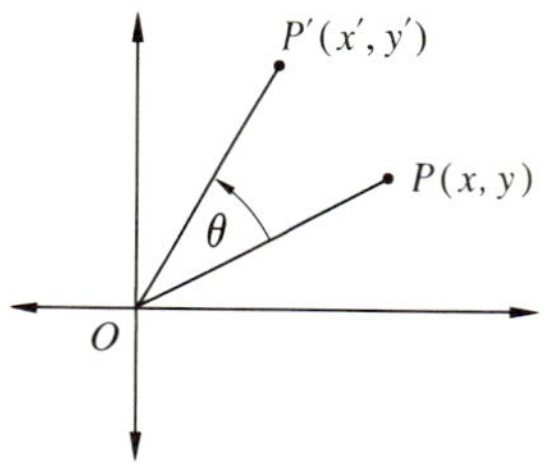

FIGURE 4-28

Examples of Isometries

1. A *translation* of the plane, which moves every point the same distance in the same direction, as in FIG. 4-25. Let $P(x, y)$ be any point in the plane, and $P'(x', y')$ be the position to which it is moved by a translation (FIG. 4-26). If the horizontal displacement is h and the vertical displacement is k, then the translation is described by

   $$x' = x + h, \qquad y' = y + k.$$

2. A *rotation* of the plane about a fixed point through a given angle, as in FIG. 4-27. If $P(x, y)$ is any point in the plane and $P'(x', y')$ is the position to which it is moved by a rotation about the origin through an angle θ (FIG. 4-28), then the rotation is described by

   $$x' = x \cos\theta - y \sin\theta,$$

   $$y' = x \sin\theta + y \cos\theta.$$

   If the angle of rotation is 180°, then O is the midpoint of $\overline{PP'}$ (FIG. 4-29). Such a rotation is called a *reflection in O*.

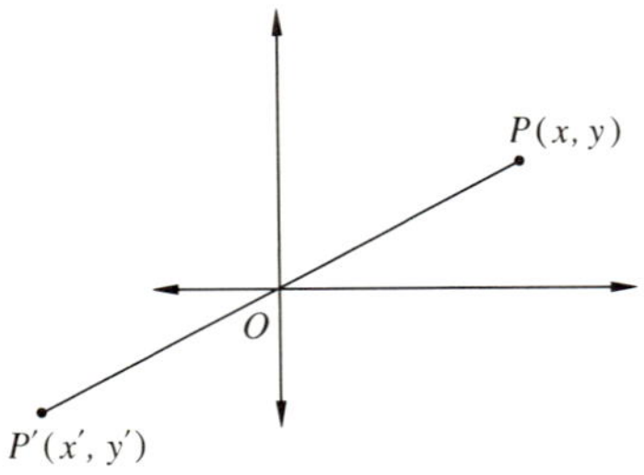

FIGURE 4-29

3. A *line reflection* in line l, as in FIG. 4-30, which moves each point $P(x, y)$ to a point $P'(x', y')$ on the other side of l so that l is the perpendicular bisector of $\overline{PP'}$. If the line reflection is across the X axis (FIG. 4-31), then the reflection is described by

$$x' = x, \qquad y' = -y.$$

Note that in a reflection one figure cannot be superimposed on the other without lifting one of the figures out of the plane.

4. A *glide reflection*, which is a combination of a translation and a reflection, as in FIG. 4-32.

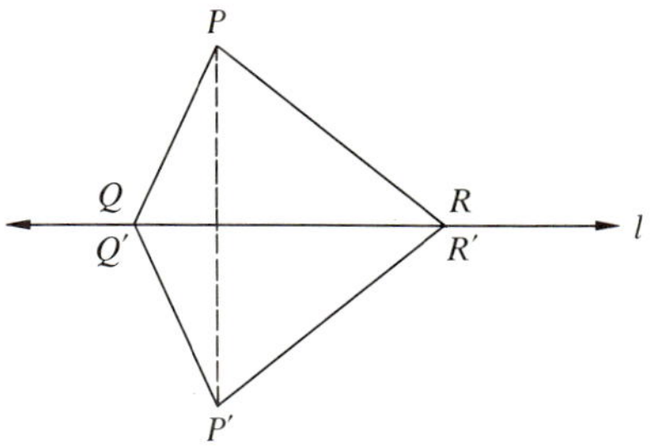

FIGURE 4-30

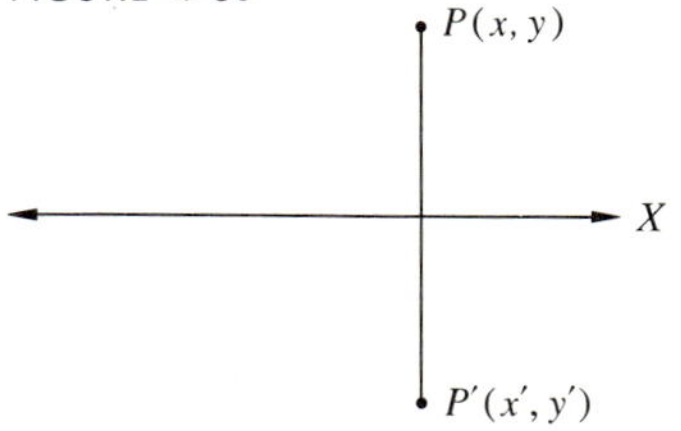

FIGURE 4-31

Product of Two Isometries

If T_1 is an isometry that moves the point P to P' and if T_2 is an isometry that moves P' to P'', the combined effect of performing T_1 and T_2 in succession is an isometry that moves P to P''. We call the resulting isometry the *product* of T_1 and T_2 and denote it by T_1T_2.

Since it can be proved that the examples given are the only four possible isometries, every isometry must be the result of the successive application of a finite number of translations, rotations, and reflections. These, then, are the only rigid motions needed to make any two congruent figures coincide.

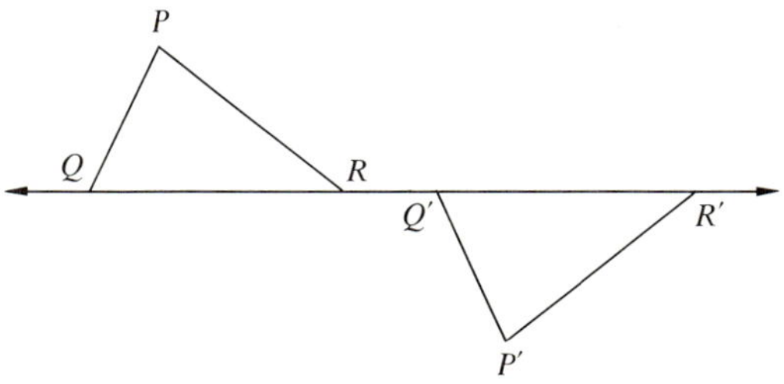

FIGURE 4-32

Isometries as Reflections

A closer look at these four isometries will show that each can be accomplished by the successive application of a finite number of reflections. Let us see why this is so.

The product of a reflection in line l_1 and a reflection in line l_2, when l_1 is parallel to l_2, is actually a translation through twice the distance from l_1 to l_2 in a direction perpendicular to the two lines (FIG. 4-33).

Conversely, if we begin with a translation, we can always find a sequence of reflections in parallel lines which is equivalent to the translation.

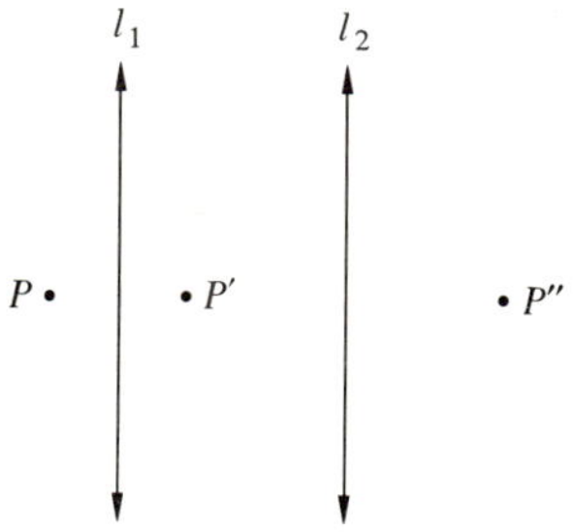

FIGURE 4-33

The product of a reflection in line l_1 and a reflection in line l_2, when l_1 and l_2 intersect, is actually a rotation about their point of intersection O through twice the angle formed by the intersection (FIG. 4-34). If l_1 is perpendicular to l_2, then the product is a rotation of 180°, or a reflection in the point O.

Conversely, if we start with a rotation, we can always find a sequence of reflections in intersecting lines which is equivalent to the rotation.

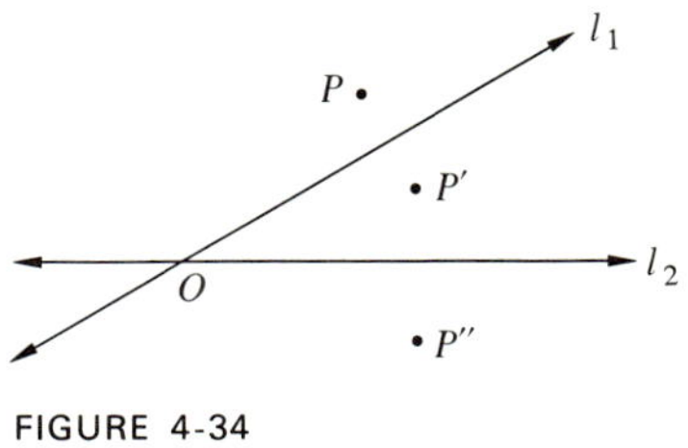

FIGURE 4-34

Properties of Isometries

1. We have already seen that the *product* of two isometries T_1 and T_2 is also an isometry denoted by $T_1 T_2$. This means that the set of isometries is closed with respect to multiplication.

2. The transformation that "moves" each point P of the plane into itself is called the *identity* transformation I. The effect of this transformation is to leave every point of the plane fixed. The identity transformation is an isometry with the property that for every isometry T, $TI = IT = T$.

3. If the isometry T carries point P to P', then the isometry that carries P' back to its original position P is called the *inverse* of T and is denoted by T^{-1}. Since the product of T and its inverse T^{-1} carries the transformed point back to its original position, we can say that $TT^{-1} = T^{-1}T = I$.

4. Isometries are associative with respect to multiplication; that is, if T_1, T_2, and T_3 are isometries, then $T_1(T_2 T_3) = (T_1 T_2)T_3$. This is true because if T_1 moves P to P', T_2 moves P' to P'', and T_3 moves P'' to P''', then $T_1(T_2 T_3)$ first moves P to $P'[T_1]$ and then P' to P''' $[T_2 T_3]$. The product $T_1(T_2 T_3)$ therefore moves P to P'''.
   $(T_1 T_2)T_3$ first moves P to P'' $[T_1 T_2]$ and then P'' to P''' $[T_3]$. The product $(T_1 T_2)T_3$ therefore moves P to P''', as does the product $T_1(T_2 T_3)$.

5. Note that the product of two isometries is not commutative; that is, $T_1 T_2 \neq T_2 T_1$. For instance, the product of a translation and a rotation does not yield the same result as the product of first the rotation and then the translation. For example, if T_1 is a translation that moves P a units to the right and T_2 is a clockwise rotation through angle θ about the point P, then the product of the two transformations would look like FIG. 4-35.

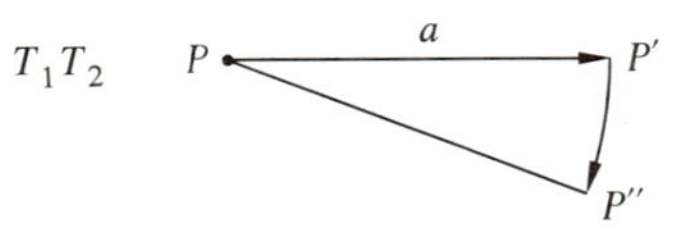

T_1 first carries P to P' and then T_2 carries P' to P''

T_2 first leaves P fixed, and then T_1 carries P to P'

FIGURE 4-35

The Set of Isometries Forms a Group

Let us now take special notice of the first four properties. These properties assert that for the operation of "multiplication," as we defined it earlier, the set of isometries

1. Is closed.
2. Is associative.
3. Contains an identity.
4. Contains an inverse for each isometry.

A set that satisfies these four properties is called a *group*. The set of isometries of a plane is, therefore, a *group*.

Groups are of special interest because they appear in so many places: in algebra, in geometry, with infinite sets. Besides the isometries, the set of all one-to-one transformations of any space onto itself is a group. The symmetries of a cube form a group. A rubber band, held in a straight line between two fixed points, may be stretched in many different ways along the line. All such deformations form a group. The set of all permutations of a finite set is a group. The nonzero real numbers under multiplication form a group, as do the integers under addition. There are many other examples of groups not only throughout mathematics but in the physical world as well.

Rigid Motions and Congruence

We said that two figures A and B are congruent if there exists a rigid motion that brings A to B. Precisely which properties of rigid motions produce the properties of congruence?

On page 110 we saw that congruence of triangles is an equivalence relation. By that we meant that

1. Congruence is *reflexive*; that is, a figure is congruent to itself.
2. Congruence is *symmetric*; that is, if A is congruent to B, then B is congruent to A.
3. Congruence is *transitive*; that is, if A is congruent to B and B is congruent to C, then A is congruent to C.

Let us now look back to the properties of rigid motions.

The existence of the *identity* isometry—whereby every figure is mapped onto itself—makes possible the *reflexive* property of congruence.

The existence of an *inverse* for every isometry—whereby for The isometry T, which maps figure A onto figure B, there exists an inverse T^{-1}, which maps figure B back onto figure A—makes possible the *symmetric* property of congruence.

If the isometry T_1 maps A onto B, and the isometry T_2 maps B onto C, then the product T_1T_2 maps A into C. Therefore, the existence of the *product* of any two isometries makes possible the *transitive* property of congruence.

Isometries and Euclidean Geometry

Felix Klein (1849–1925) once stated that any geometry can be defined in terms of the properties that are left invariant under a group of transformations. To define a geometry, we would, therefore, select some basic elements such as points, a space such as a plane, and a group of transformations such as isometries. We then look for the properties of points in a plane that are left invariant under isometries.

The system which studies properties that are unchanged under all isometries is Euclidean metric geometry. We have seen that congruence remains invariant under all isometries. They preserve all distances and measures of angles and therefore preserve both the size and shape of figures.

5 PARALLEL LINES

INTRODUCING THE CHAPTER TO STUDENTS

CONTENT REVIEW

Prerequisites for Learning This Chapter
Existence of Parallels
Corresponding and Alternate Interior Angles
Theorems About Parallel Lines
Parallels in Space

TEACHING SUGGESTIONS AND ACTIVITIES

Visual Aids
Questions
Puzzle: Find the Fallacy
Solution: The Fallacy

ENRICHMENT

Non-Euclidean Geometries
Ideal Points

INTRODUCING THE CHAPTER TO STUDENTS

In Chapter 2 we defined two parallel lines as lines that lie in the same plane and do not intersect. We commented that in defining parallel lines we were not necessarily assuming that they exist.

The question of the existence of parallel lines intrigued, baffled, and frustrated mathematicians for 2000 years and ultimately led to one of the crucial milestones in the history of mathematics—the creation of non-Euclidean geometries.

In this chapter we shall prove that parallel lines actually exist, state an essential postulate, and then show an important consequence of that postulate.

CONTENT REVIEW

PREREQUISITES FOR LEARNING THIS CHAPTER

An understanding of the following: definition of parallel lines, exterior angle of a triangle, supplementary angles, linear pair, congruent segments, Postulate 3-3. Postulate 3-4, Theorem 3-13.

EXISTENCE OF PARALLELS

THEOREM 5-1 *If l is any line and P any point not on l, then there exists at least one line that contains P and is parallel to l* (FIG. 5-1).

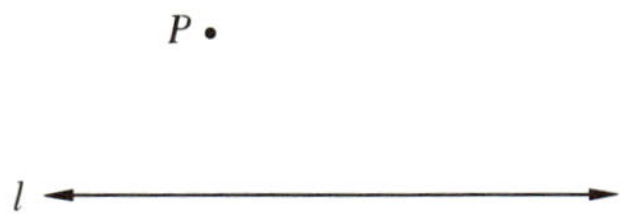

FIGURE 5-1

Proof

1. Let Q be any point on l, and let m be the line through P and Q.
2. Let A be another point on l, and let B be a point on the same side of m as A such that $\angle 1 \cong \angle 2$ (FIG. 5-2).

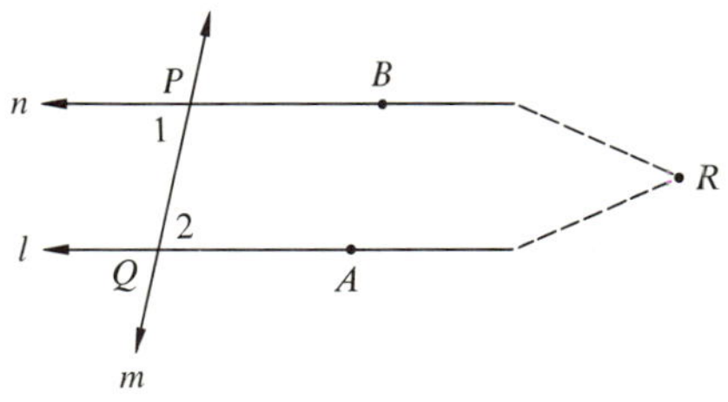

FIGURE 5-2

3. Assume that no line exists which contains P and is parallel to l. Then l and n will intersect at some point R.

4. By Theorem 3-13, $m\angle 1 > m\angle 2$, since $\angle 1$ is an exterior angle of $\triangle PQR$. But this contradicts our hypothesis in step 2.

5. Therefore, our assumption in step 3 is false, and there exists at least one line that contains P and is parallel to l.[1]

[1] Now that we proved that through a point *P* not on *l* there exists at least one line parallel to *l*, we ask whether there is *more* than one such line.

By intuition it would seem that there was one and only one such parallel line. But how can we be sure that this is so?

In the nineteenth century it was shown that from Euclid's postulates it is impossible to *prove* that there is *exactly one* such parallel. To establish the uniqueness of such a parallel we need the following postulate.

POSTULATE 5-1 (Parallel Postulate) Through any point not on a given line, there is no more than one line parallel to the given line.

Since Theorem 5-1 states that there exists at least one such parallel and Postulate 5-1 states that there is no more than one such parallel, we conclude that there is exactly one such parallel.

CORRESPONDING AND ALTERNATE INTERIOR ANGLES

A line intersecting two other lines forms certain pairs of angles that are of special interest. We shall now define these special angles.

Definitions

1. A line l that intersects lines l_1 and l_2 in exactly two points is called a *transversal* of l_1 and l_2 (FIG. 5-3). We sometimes say that l_1 and l_2 are *cut* by l.

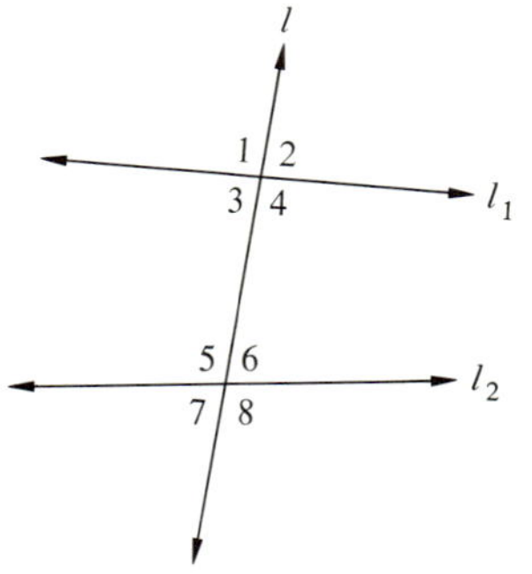

FIGURE 5-3

2. Angles 3, 4, 5, and 6 are called *interior angles.*

3. Angles 1, 2, 7, and 8 are called *exterior* angles.

4. Angles 3 and 6 are called *alternate interior* angles. So are angles 4 and 5.

5. Angles 2 and 6 are called *corresponding angles.* So are angles 1 and 5, 3 and 7, and 4 and 8.[2]

[2] Note that alternate interior angles have the points of their interiors on *opposite* sides of the transversal, whereas corresponding angles have their interiors on the *same* side of the transversal.

THEOREMS ABOUT PARALLEL LINES

Theorems Relating to Corresponding and Alternate Interior Angles

THEOREM 5-2 *If two parallel lines are cut by a transversal, then any pair of alternate interior angles is congruent.*

That is, if $l_1 \parallel l_2$, then $\angle 1 \cong \angle 2$ (FIG. 5-4).

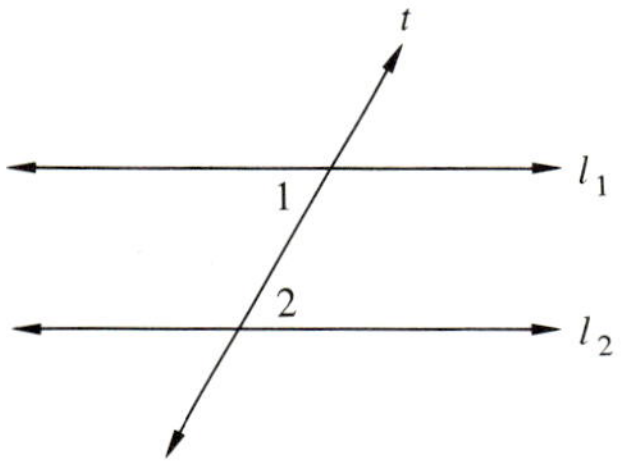

FIGURE 5-4

THEOREM 5-3 (Converse of Theorem 5-2) *If two lines are cut by a transversal and if a pair of alternate interior angles is congruent, then the lines are parallel.*

That is, if $\angle 1 \cong \angle 2$, then $l_1 \parallel l_2$.

THEOREM 5-4 *If two parallel lines are cut by a transversal, then any pair of corresponding angles is congruent.*

That is, if $l_1 \parallel l_2$, then $\angle 3 \cong \angle 4$ (FIG. 5-5).

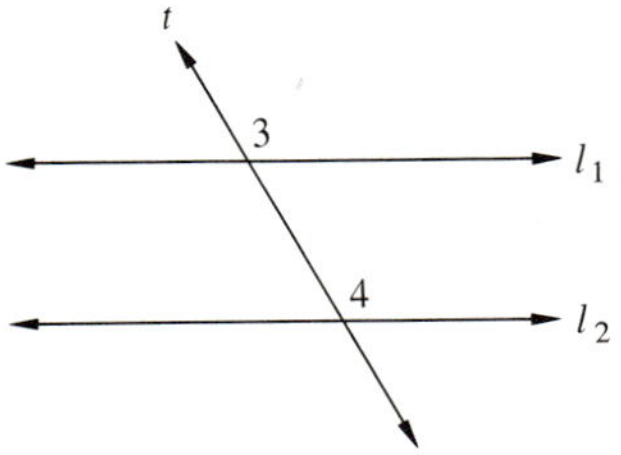

FIGURE 5-5

THEOREM 5-5 (Converse of Theorem 5-4) *If two lines are cut by a transversal and if a pair of corresponding angles is congruent, then the lines are parallel.*

That is, if $\angle 3 \cong \angle 4$, then $l_1 \parallel l_2$.

THEOREM 5-6 *If two parallel lines are cut by a transversal, interior angles on the same side of the transversal are supplementary.*

That is, if $l_1 \parallel l_2$, then $\angle 5$ is supplementary to $\angle 6$ (FIG. 5-6).

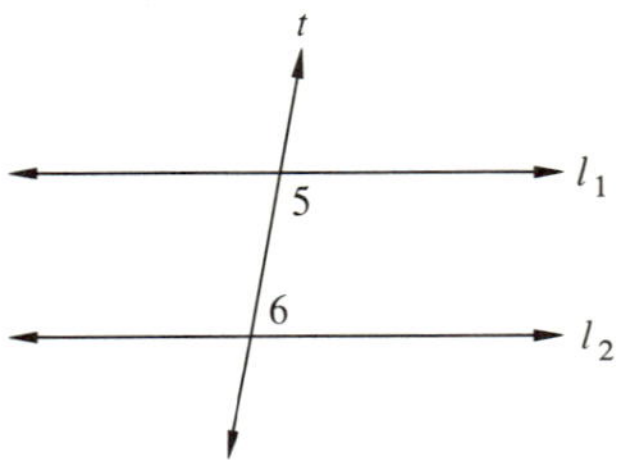

FIGURE 5-6

The following theorem is one of the most interesting and useful consequences of the Parallel Postulate.

THEOREM 5-7 *The sum of the measures of the angles of any triangle is 180.*

That is, if l is the line parallel to $\overline{BC}$ through A, and D is a point on l and is in the half-plane $\overleftrightarrow{AB}/C$, then $m\angle y + m\angle x + m\angle z = 180$ (FIG. 5-7).

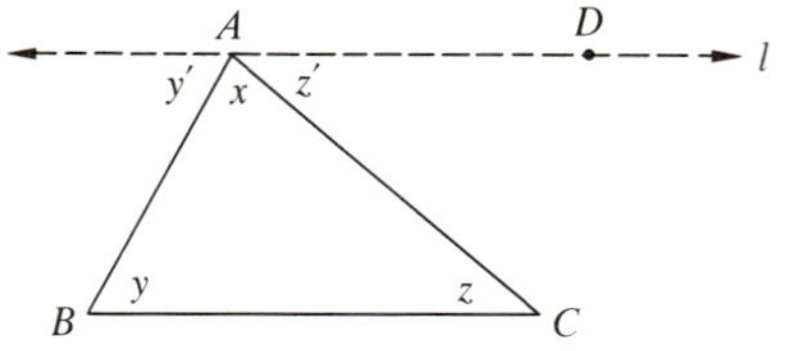

FIGURE 5-7

Proof

1. $m\angle y = m\angle y'$ and $m\angle z = m\angle z'$, since they are alternate interior angles.

2. $m\angle BAD = m\angle x + m\angle z'$ by the Angle Addition Postulate.

3. $m\angle y' + m\angle BAD = 180$, because the two angles form a linear pair and, by Postulate 3-4, two angles that form a linear pair are supplementary.

4. Therefore, $m\angle y' + m\angle x + m\angle z' = 180$.

5. $m\angle y + m\angle x + m\angle z = 180$.

Two important corollaries of Theorem 5-7 are

1. Given a correspondence between two triangles, if two pairs of corresponding angles are congruent, then the third pair of corresponding angles is also congruent.
2. The acute angles of a right triangle are complementary.

The following is an important theorem relating to transversals to three or more parallel lines. We shall start with two definitions.

Definitions

1. If two lines are cut by a transversal in points A_1 and A_2, we say that the lines *intercept* the segment A_1A_2 on the transversal (FIG. 5-8).

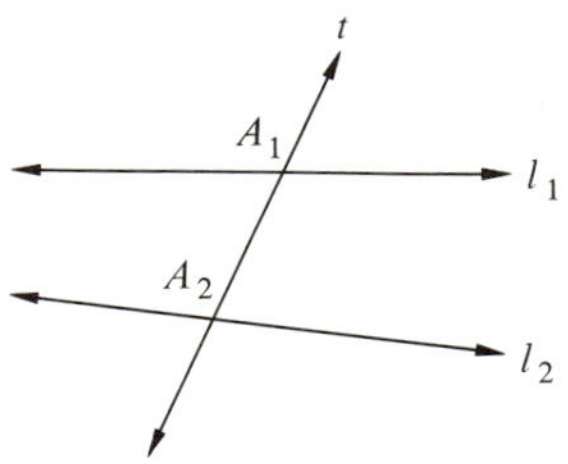

FIGURE 5-8

2. If three lines are cut by a transversal in points A_1, A_2, and A_3 such that $A_1A_2 = A_2A_3$, we say that the lines *intercept congruent segments* on the transversal (FIG. 5-9).

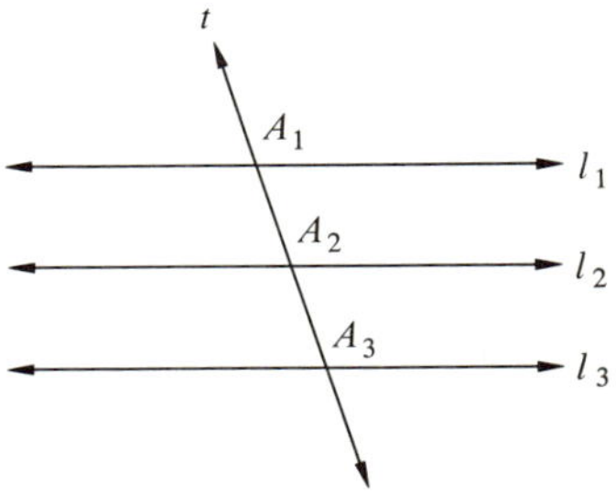

FIGURE 5-9

THEOREM 5-8 If three parallel lines intercept congruent segments on one transversal, then they intercept congruent segments on any other transversal.

[3] Theorem 5-8 can be generalized to any finite number of parallel lines.

That is, if $A_1A_2 = A_2A_3$, then $B_1B_2 = B_2B_3$ (FIG. 5-10).[3]

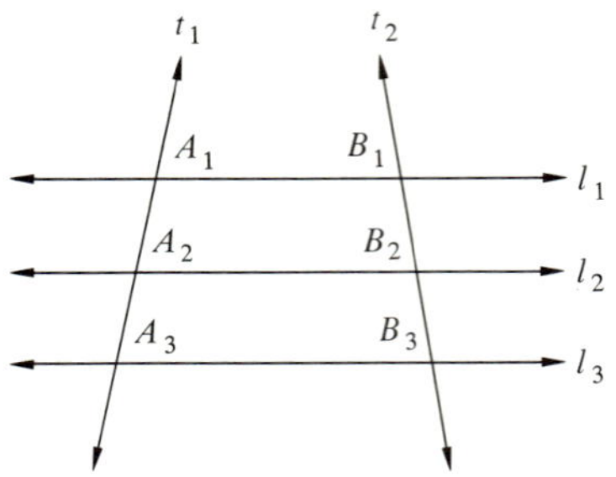

FIGURE 5-10

PARALLELS IN SPACE

We extend the discussion of parallel lines to a consideration of parallels in space. We shall state several theorems that relate parallel lines to planes and that disclose some properties of parallel planes. In addition to Theorem 2-5—which stated that if a plane intersects two parallel planes, then it intersects them in two parallel lines—we have the following theorems.

THEOREM 5-9 *Two parallel lines lie in one and only one plane.*

That is, two parallel lines determine a plane.

THEOREM 5-10 *If a line is perpendicular to one of two parallel planes, then it is perpendicular to the other* (FIG. 5-11).

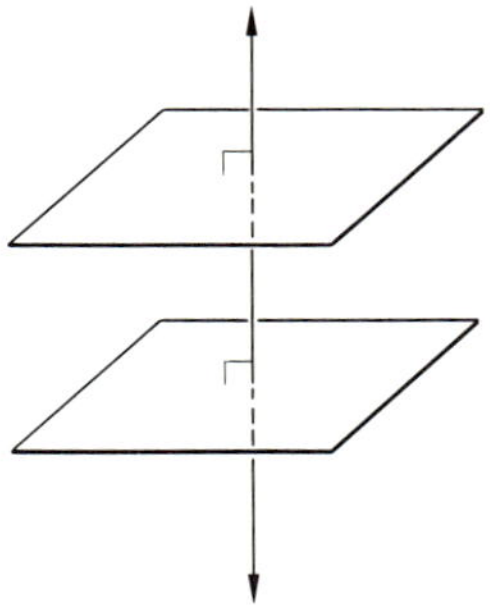

FIGURE 5-11

THEOREM 5-11 *Two lines perpendicular to the same plane are parallel* (FIG. 5-12).

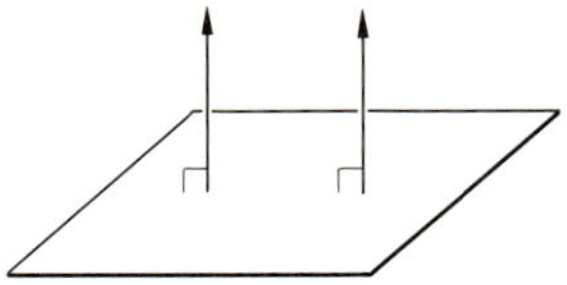

FIGURE 5-12

THEOREM 5-12 *Planes parallel to the same plane are parallel to each other* (FIG. 5-13).

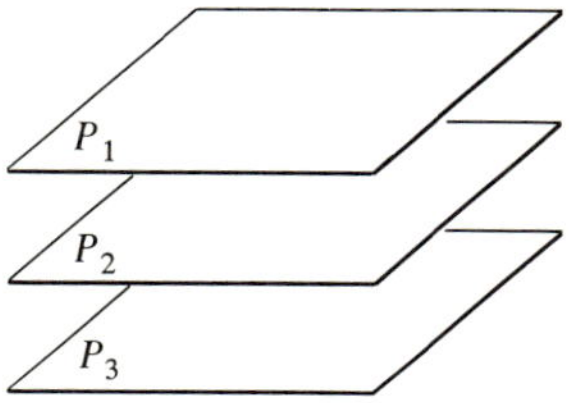

FIGURE 5-13

That is, if $P_1 \parallel P_3$ and $P_2 \parallel P_3$, then $P_1 \parallel P_3$.

THEOREM 5-13 *Planes perpendicular to the same line are parallel* (FIG. 5-14).[4]

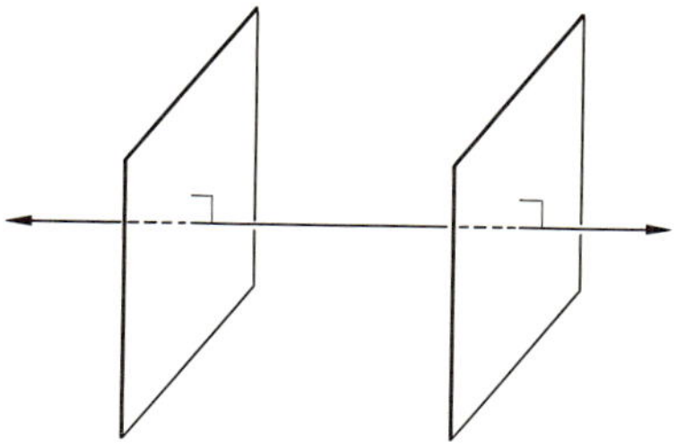

FIGURE 5-14

[4] Note that although parallel planes behave a great deal like parallel lines, there are some important differences. For example,

1. Two planes parallel to the same line are not necessarily parallel to each other.
2. Two planes perpendicular to a third plane are not necessarily parallel to each other (FIG. 5A).

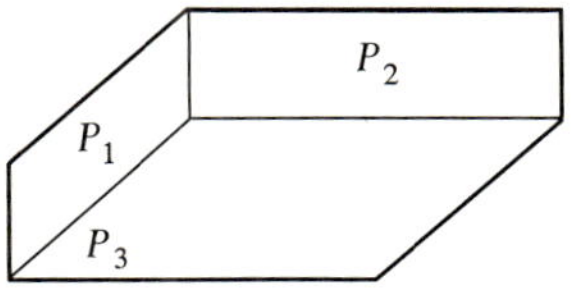

FIGURE 5A

TEACHING SUGGESTIONS AND ACTIVITIES

VISUAL AIDS

Visual 1

PURPOSE

To illustrate angle relationships when parallel lines are cut by a transversal.

CONSISTS OF

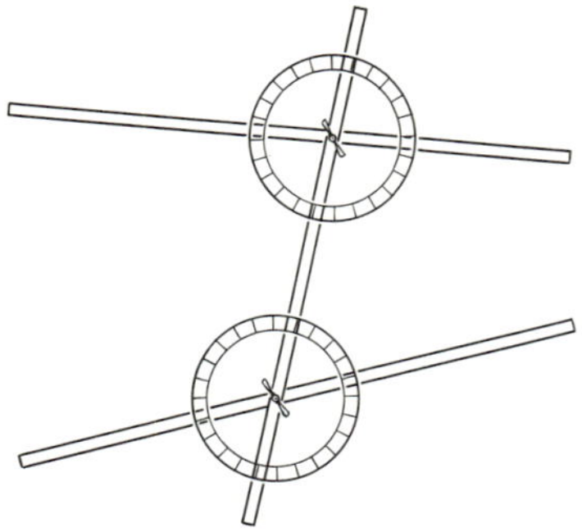

FIGURE 5-15

Three strips (cut from any stiff material) and two 360° protractors attached by bolts and wing nuts as shown in FIG. 5-15.

HOW TO USE

Set up conditions for each of Theorems 5-2–5-6. Check with protractor readings.

Visual 2

PURPOSE

To illustrate that the sum of the measures of the angles of any triangle in a plane is 180 (Theorem 5-7).

CONSISTS OF

The "adjustable triangle" with 180° protractors attached to each vertex, used in Visual 3, page 93.

HOW TO USE

Form a variety of triangles by varying the lengths of the sides. Let students see that in every case the sum of the measures of the three angles is 180.

Note: There are three other ways to illustrate Theorem 5-7. The first two are particularly good with younger children. Both ways require only a triangle cut out of paper.

1. Take the paper triangle, cut off two angles, and arrange the pieces as shown in FIG. 5-16. It appears that the three angles lie along a straight line; that is, the sum of their measures is 180. Repeat this experiment with other triangles.

2. Take the paper triangle and label it as shown in FIG. 5-17.
   a. Fold *AC* over upon itself so that the crease passes through *B*, and let the crease intersect *AC* at *D* (FIG. 5-18). (Note that *D* is the foot of the altitude from *B* to *AC*.)
   b. Fold the vertices *A*, *B*, and *C* over to *D* as shown in FIG. 5-19.
   c. It appears that the three angles lie along a straight line; that is, the sum of their measures is 180.
   d. Repeat this experiment with other triangles.

3. A third method, said to have been devised by Pascal when a mere child, is the following: Fold *A*, *B*, and *C* into the incenter of the triangle (FIG. 5-20). Then, about the incenter we have two *A*'s, two *B*'s, and two *C*'s.

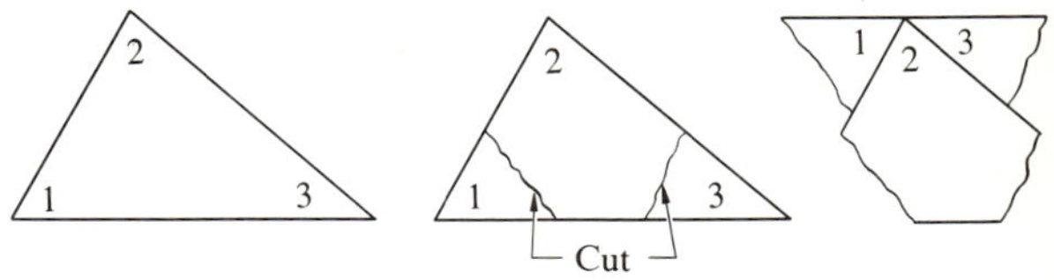

FIGURE 5-16

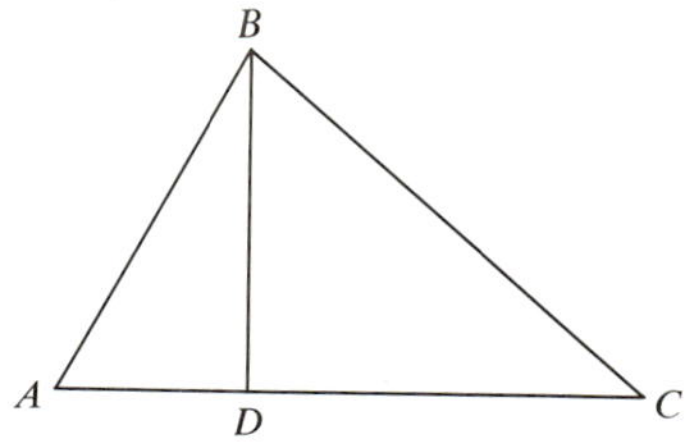

FIGURE 5-17

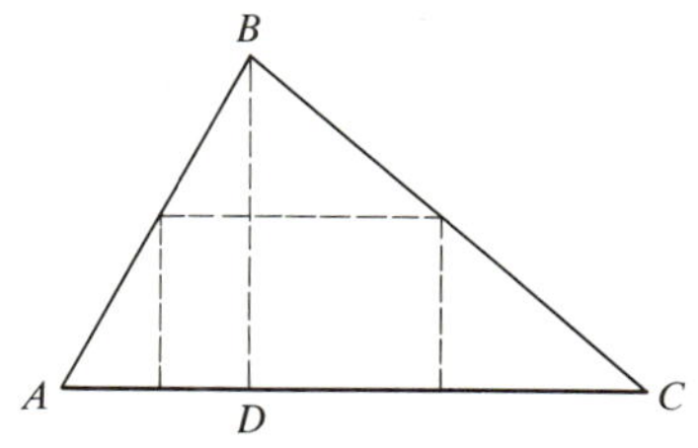

FIGURE 5-18

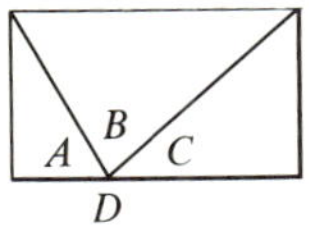

FIGURE 5-19

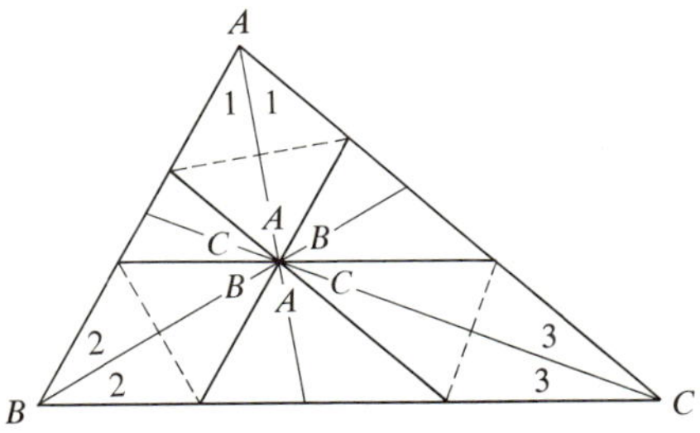

FIGURE 5-20

QUESTIONS

1. Why must the acute angles of a right triangle be complementary?

2. Prove that a transversal which is perpendicular to one of two parallel lines is also perpendicular to the other line.

3. For what value of c will l_1 be parallel to l_2?

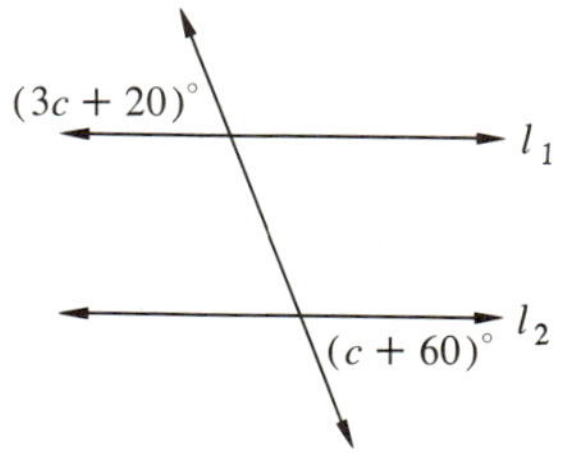

4. Prove that lines perpendicular to the same line are parallel.

5. Prove that if a pair of corresponding angles is congruent, then a pair of alternate interior angles is congruent.

6. Prove that if two lines are cut by a transversal and that if one pair of alternate interior angles is congruent, then the other pair of alternate interior angles is also congruent.

7. Write three true statements that begin with "If two lines are parallel. . . ."

8. Prove that if two angles of one triangle are congruent, respectively, to two angles of another triangle, then the third angles are congruent.

9. Use the Parallel Postulate to prove that if $l_1 \parallel l_2$ and $l_2 \parallel l_3$, then $l_1 \parallel l_3$.

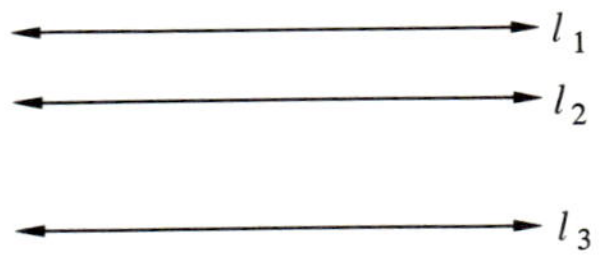

10. Prove that if two lines are cut by a transversal and that if one pair of corresponding angles is congruent, then the other three pairs of corresponding angles are congruent.

11. Prove that if the sides of one angle are parallel, respectively, to the sides of a second angle, then either the angles are congruent or they are supplementary.

12. Given that $\angle A$, $\angle C$, and $\angle D$ are right angles, prove that $\angle B$ is also a right angle.

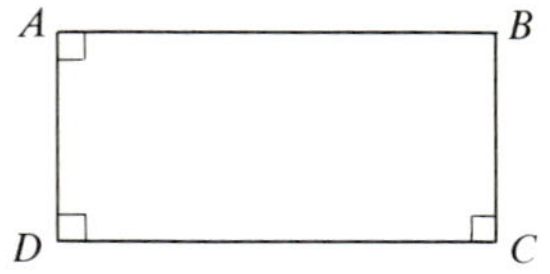

13. If $m \parallel n$, $\angle 1 \cong \angle 2$, and $\angle 3 \cong \angle 4$, why is $\overline{AB} \perp \overline{CB}$?

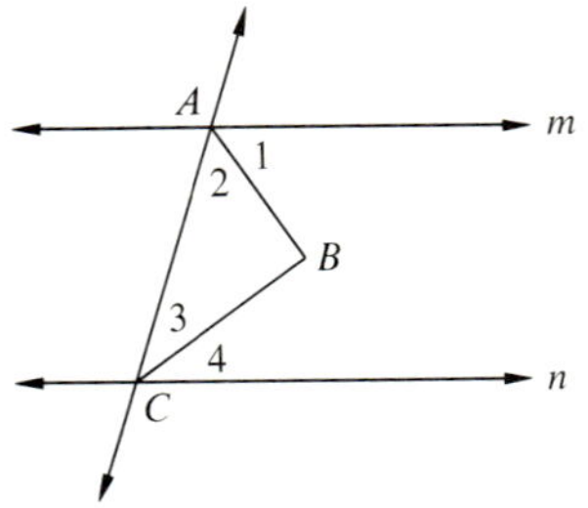

14. Prove that for any triangle, the measure of each exterior angle is equal to the sum of the measures of its two remote interior angles.

15. What is the sum of the measures of ∠1, ∠2, ∠3, ∠4, and ∠5?

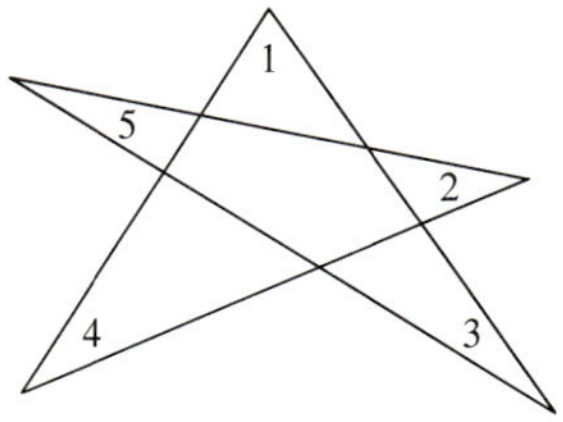

PUZZLE: FIND THE FALLACY

What is wrong with this "proof" of the Parallel Postulate?

Given line l and point P not on l, we want to prove that through P only one straight line, n, can be drawn parallel to l.

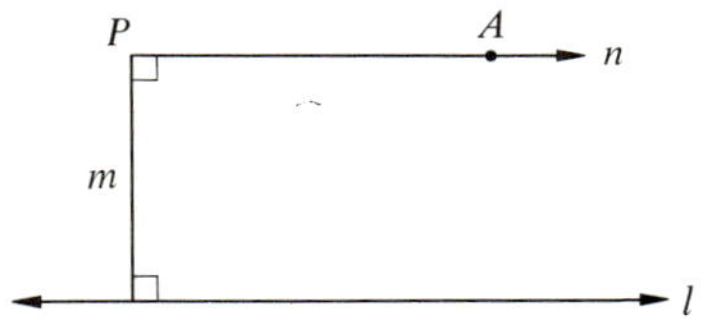

Proof

1. Drop perpendicular m from point P to line l.
2. Construct another perpendicular $\overrightarrow{PA}$ to line m at point P.
3. $n \parallel l$ because two perpendiculars to the same straight line are parallel.
4. Since we can drop only *one* perpendicular from a given point to a given line and since it is possible to construct only *one* perpendicular to a straight line from any point on the line, it follows that n is the one and only line that can be drawn through P parallel to l.

Note: The reasons given in steps 3 and 4 are theorems that can be proved without using the Parallel Postulate.

SOLUTION: THE FALLACY

In the proof given, all that is shown is that *by the method used*, that is, by the construction of two perpendiculars, we can obtain only one parallel line.

But what about the parallel lines that can be obtained by other methods? For example, we can construct n' parallel to l by making the alternate interior angles, ∠1 and ∠2, congruent.

How do we know that n' will be the same parallel as n?

To say that every possible method of constructing such a parallel will yield the same line is to assert something that has not been shown but is precisely what needs to be shown.

ENRICHMENT

NON-EUCLIDEAN GEOMETRIES

We have seen how the rejection of Euclid's fifth postulate led to the development of non-Euclidean geometries. We now wish to go more deeply into the ideas and implications of these geometries.

Euclid's Fifth Postulate

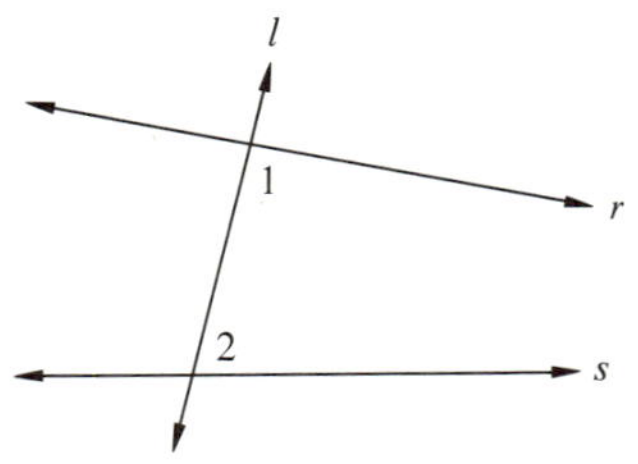

FIGURE 5-21

Let us take a closer look at the fifth postulate (page 67). It asserts that if two lines *r* and *s* are intersected by a third line *l* so as to make the sum of angles 1 and 2 less than 180°, then the lines *r* and *s meet* on the same side of line *l* as are angles 1 and 2 (FIG. 5-21). This assertion implies the existence of parallel lines.

An equivalent axiom, known as *Playfair's axiom*, expresses Euclid's fifth postulate this way:

> *Through a point not on a given straight line there is one and only one straight line that does not intersect the given straight line.*

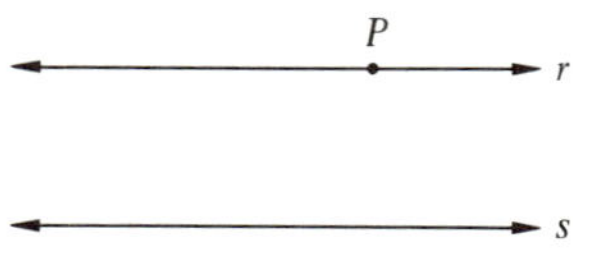

FIGURE 5-22

This says that, given line *s* and point *P* not on *s*, there is one and only one line *r* in the plane of *P* and *s* which passes through *P* and does not meet *s* (FIG. 5-22).

We say that Playfair's axiom and Euclid's fifth postulate are equivalent because from Euclid's postulates we can derive Playfair's axiom, and from Euclid's postulates with the Playfair axiom replacing Euclid's fifth postulate we can derive Euclid's fifth postulate.

The fifth postulate aroused controversy almost from the beginning. Postulates were supposed to be "self-evident truths," but the fifth is more complex than any of the others and is not quite as self-evident as, say, the fourth, which says that "all right angles are equal to one another." The fifth, unlike the other nine postulates, makes an assertion about lines that are "produced indefinitely." This transcends our experience, is not subject to experimental verification, and is, therefore, unlike the other postulates that deal only with finite segments of lines. Besides, the *converse* of the fifth postulate can be proved from the other four postulates and axioms, so why cannot the fifth be proved in a similar way? (Euclid's fifth postulate is the converse of Proposition 17 of Book I of the *Elements*.)

So for 2000 years mathematicians tried to *prove* that the fifth postulate is a necessary *consequence* of the other nine postulates. But all their attempts resulted in failure. Perhaps the most famous of these attempts was the one made by Girolamo Saccheri (1667–1733), who used the indirect method of proof to try to show that the fifth postulate is a consequence of the other nine.

The fifth postulate asserts the existence of one and only one line parallel to a given line and passing through a given point not on the given line. There are two alternatives to this assertion:

1. There is *no* line parallel to the given line and passing through a given point not on the given line.
2. There is *more* than one such line.

Saccheri's plan was to show that by assuming either of the two alternatives he would end up with a contradiction. Therefore, the only other possibility—Euclid's fifth postulate—must be true. Saccheri was able to show that the assumption that there are *no* parallels leads to a contradiction. He was not able to show a contradiction by assuming that there is *more* than one parallel.

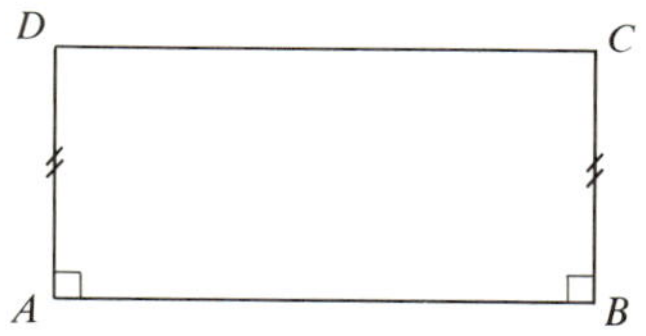

FIGURE 5-23

Saccheri's efforts centered around a special quadrilateral, now called a *Saccheri quadrilateral,* in which two equal sides are both perpendicular to a third side. Let $ABCD$ in FIG. 5-23 be a Saccheri quadrilateral with $AD = BC$ and angles A and B right angles. $\overline{AD}$ and $\overline{BC}$ are called the *legs* of the quadrilateral, $\overline{AB}$ is called the *base* of the quadrilateral, $\overline{DC}$ is called the *summit*, and angles C and D are called the *summit angles.*

After proving that $\angle C \cong \angle D$,* Saccheri considered three different hypotheses concerning these two summit angles:

1. The summit angles are right angles; that is, $m\angle C = m\angle D = 90$.
2. The summit angles are obtuse angles; that is, $m\angle C = m\angle D > 90$.
3. The summit angles are acute angles; that is, $m\angle C = m\angle D < 90$.

Saccheri was able to prove that the assumption that the summit angles are obtuse leads to a contradiction. When he assumed the

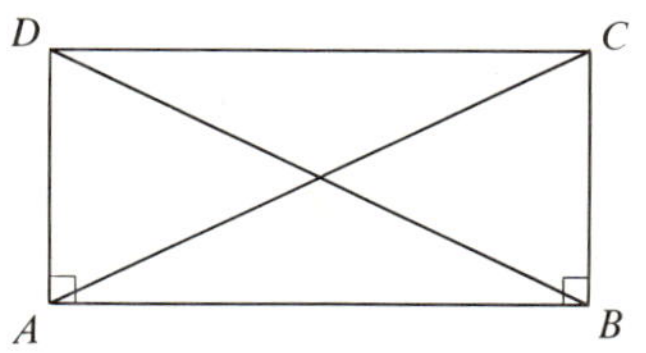

* Join D to B and C to A. $\triangle ABD \cong \triangle ABC$, making $BD = AC$. Therefore, $\triangle ACD \cong \triangle BDC$ and $\angle ADC \cong \angle BCD$.

summit angles to be acute, he ended up with several "strange" theorems, such as

1. The sum of the measures of the angles of any triangle is less than 180.
2. Two triangles are congruent if the three angles of one are congruent to the three angles of the other.

He concluded that this assumption, too, leads to a contradiction.

Believing that he had proved that hypothesis 2 and hypothesis 3 lead to contradictions, Saccheri concluded that hypothesis 1, which is equivalent to Euclid's parallel postulate, must be true.

Although Saccheri believed that he had proved the hypothesis that the summit angles are acute leads to a contradiction, he actually did not. He only proved a number of theorems in the non-Euclidean geometry that were developed a century later by Bolyai and Lobachevsky.

Later attempts by other mathematicians to prove Euclid's fifth postulate—J. H. Lambert (1728–1777), A. M. Legendre (1752–1833), and many others—were equally unsuccessful.

The inability of mathematicians to prove the fifth postulate as a consequence of Euclid's other nine postulates finally convinced them that their failure was due to the fact that the fifth is *independent* of the other nine.

This meant that it is possible to construct a consistent geometry from a set of postulates in which the parallel postulate was replaced with a contradictory postulate. Such a geometry is called a *non-Euclidean geometry.*

Why is the parallel postulate so important? What role does it play in geometry? It is no exaggeration to say that this single postulate leads to some of the most basic, useful, and important results in Euclidean geometry. It leads to

1. The conclusion that the sum of the measures of the angles of a triangle is 180.
2. The existence of rectangles and squares.
3. The existence of similar figures.
4. The Pythagorean relation.

If we change just this one postulate but keep all the other postulates, then we are creating a new geometry in which *none* of these results hold, as we shall soon see.

It was not until the nineteenth century, after centuries of failure to prove the parallel postulate, that mathematicians began to develop geometries based on a contradiction of Euclid's fifth postulate. By 1830, J. Bolyai (1802–1860), N. I. Lobachevsky (1793–1856), and K. F. Gauss (1777–1855) independently developed a geometry, called *hyperbolic geometry*, in which they assumed that there is *more* than one line parallel to a given line through an external point.

In 1854, G. F. B. Riemann (1826–1866) introduced a different non-Euclidean geometry, called *elliptic geometry*, in which he assumed that there are *no* parallel lines.

Euclidean, hyperbolic, and elliptic geometries have very different properties, and we shall now see what some of these properties are.

Hyperbolic Geometry

In *hyperbolic geometry*, also called *Lobachevskian geometry*, all the Euclidean postulates are assumed except the parallel postulate, which is replaced with another that asserts that there is *more than one* line parallel to a given line through an external point.

Two consequences immediately follow.

1. Every theorem of Euclidean geometry can also be a theorem of hyperbolic geometry as long as it does not depend on Euclid's parallel postulate.
2. There are infinitely many parallels to a given line through an external point. The reason for this conclusion is as follows:
   a. Suppose there are two distinct lines l_1 and l_2, both parallel to line m and passing through point P (FIG. 5-24).
   b. Then *any* line $\overleftrightarrow{PC}$, where C is in the interior of $\angle APB$, cannot intersect m because to do so $\overleftrightarrow{PC}$ would have to cross l_1 or l_2 and therefore meet l_1 or l_2 in a second point. But since two lines can intersect in at most one point, it follows that all lines through P which lie in the interior of $\angle APB$ will not intersect m.
   c. Therefore, the assumption that there are at least two parallels to m passing through an external point P implies that there are infinitely many parallels to m passing through P.

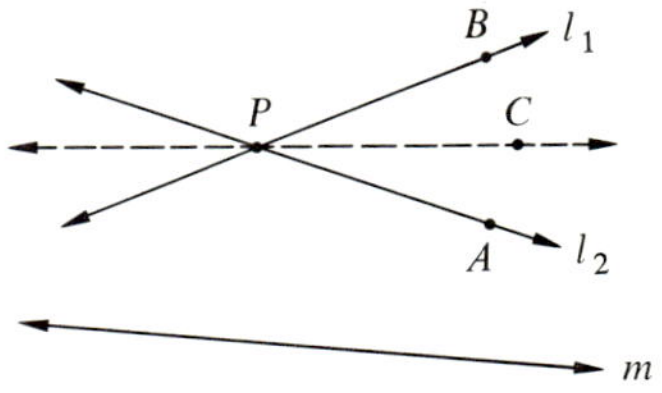

FIGURE 5-24

A Model for Hyperbolic Geometry

We use the interior of a Euclidean circle to serve as a model for hyperbolic geometry. The points of the circle itself are not included in this geometry.

The "points" in this geometry are the points *inside* the circle, and the "lines" are chords without endpoints. "Segments" are represented by Euclidean segments whose endpoints are inside the circle. The Lobachevskian "plane" is represented by the interior of the circle (FIG. 5-25).

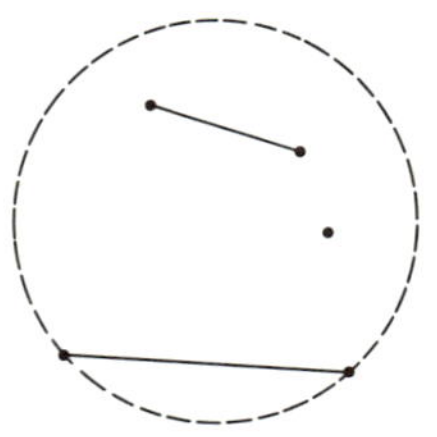

FIGURE 5-25

From the model in FIG. 5-26 we can see that it satisfies the "more-than-one-parallel" postulate of hyperbolic geometry, since neither m nor n intersects line l, making both m and n parallel to l. It is also apparent that lines parallel to the same line are not necessarily parallel to each other. But the other postulates must be verified in the model, and this requires definitions of measures of line segments and of angles.

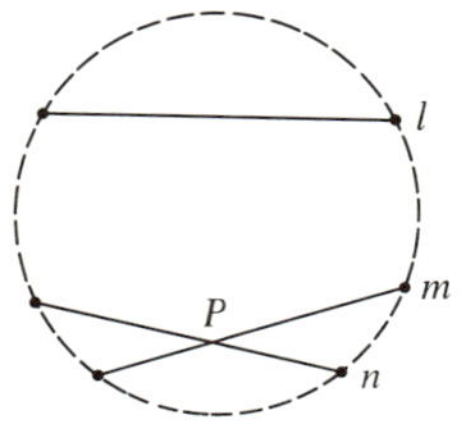

FIGURE 5-26

The Sum of the Angle Measures of a Hyperbolic Triangle

Before discussing a theorem about the sum of the angle measures of a hyperbolic triangle, we shall state but not prove another theorem in hyperbolic geometry:

Let n be a line, A a point not on n, and B a point on n, and let a side of $\overleftrightarrow{AB}$ *be given* (FIG. 5-27). *Then there exists a point C of n, on the given side of* $\overleftrightarrow{AB}$, *such that the measure of* $\angle ACB$ *is as small as we please.*

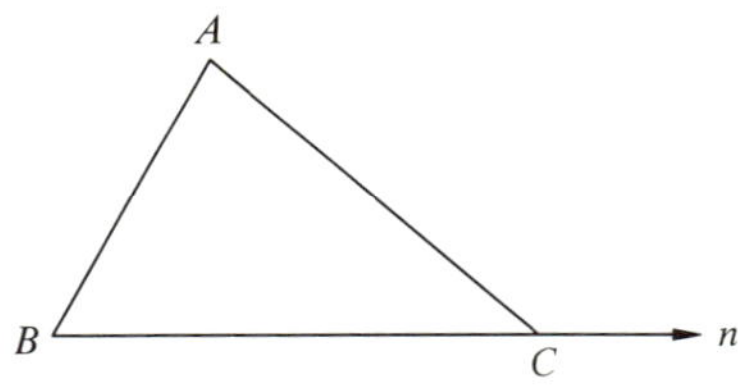

FIGURE 5-27

An important theorem in hyperbolic geometry states that the sum of the measures of the angles of every triangle is less than 180.

Proof

1. Let n be a line and A a point not on n. Also, let $\overleftrightarrow{AB}$ be perpendicular to n at B and let m be perpendicular to $\overleftrightarrow{AB}$ at A (FIG. 5-28).

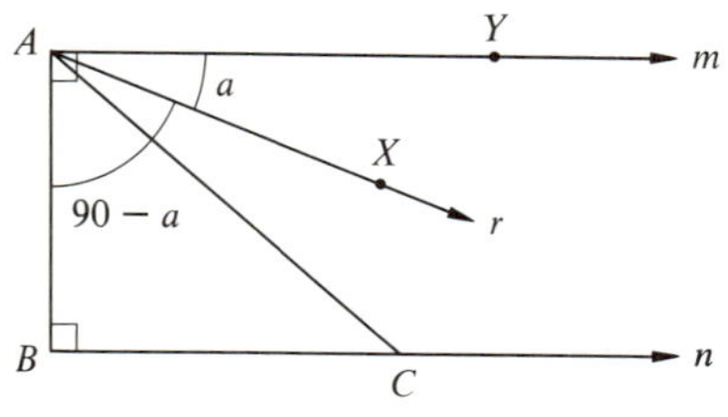

FIGURE 5-28

2. By the Lobachevskian parallel postulate, there is another line r through A parallel to n.
3. One of the angles that r makes with $\overleftrightarrow{AB}$ must be acute. Let X be a point of r such that $\angle BAX$ is acute.
4. Let Y be a point of m on the same side of $\overleftrightarrow{AB}$ as X. Let $m\angle XAY = a$. Then $m\angle BAX = 90 - a$.
5. By the theorem stated above, let C be a point of n, on the side of $\overleftrightarrow{AB}$ containing X, such that $m\angle ACB < a$.

6. In $\triangle ABC$, $m\angle ABC = 90$, $m\angle ACB < a$, and $m\angle BAC < m\angle BAX$, or $m\angle BAC < 90 - a$.

7. By adding, we obtain

   $$m\angle ABC + m\angle ACB + m\angle BAC < 90 + a + 90 - a$$

   or

   $$m\angle ABC + m\angle ACB + m\angle BAC < 180.$$

   That is, the sum of the measures of the angles of $\triangle ABC$ is less than 180.

8. What we have shown, so far, is that there exists a triangle ($\triangle ABC$) whose angle sum is less than 180°. But by another theorem we know that if one triangle has an angle sum that is less than 180°, then *every* triangle has an angle sum less than 180°.

9. Therefore, in hyperbolic geometry, the sum of the measures of the angles of every triangle is less than 180.

Properties of Hyperbolic Geometry

Unlike the Euclidean property, in hyperbolic geometry the angle sum of a triangle is *not* constant. Here the angle sum can be any value between 0° and 180°.

As a consequence of this theorem, the following properties hold in hyperbolic geometry:

1. The angle sum of every quadrilateral is less than 360°.
2. No two parallel lines are everywhere equidistant.
3. No parallelograms, rectangles, or squares exist.
4. If two triangles have their corresponding angles congruent, then the triangles are congruent. In other words, no triangles are similar unless they are congruent. This property suggests that in hyperbolic geometry exact scale models are impossible.
5. The Pythagorean relation does not hold.

Area in Hyperbolic Geometry

Since squares do not exist in hyperbolic geometry, we can no longer define area in terms of square units. We must, therefore, formulate a new definition of area for this geometry.

Any definition of area must satisfy three properties:

1. To each region there must correspond a uniquely determined positive real number, called its *area.*
2. Congruent regions must have equal areas.
3. If a region R is split into any finite number of subregions $R_1, R_2, \ldots, R_n$, then the area of R is the sum of the areas of $R_1, R_2, \ldots, R_n$.

A function that assigns to each region a specific real number in such a way that properties 1, 2, and 3 are satisfied is called an *area function* or *area measure.*

We need one more definition before we can define area in hyperbolic geometry. For the sake of simplicity, we shall consider only the area of a triangle.

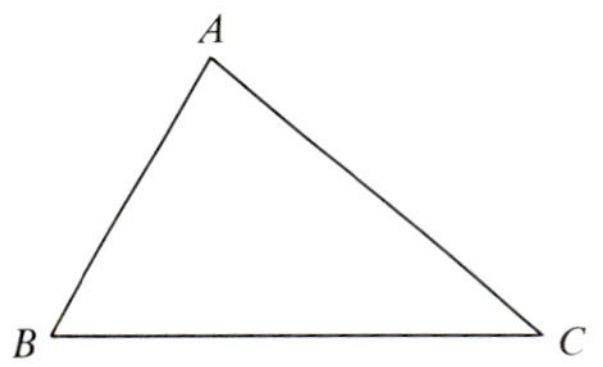

FIGURE 5-29

Definition
The amount by which the sum of the angles of a triangle differs from 180° is called the *defect* of the triangle.

That is, if d is the defect of $\triangle ABC$, then $d = 180 - (\angle A + \angle B + \angle C)$ (FIG. 5-29). Note that in this definition, $\angle A$, $\angle B$, and $\angle C$ represent the *degree measures* of the angles. Therefore, the defect of a triangle is a *real number* rather than a number of degrees.

It can be shown that the defect of a triangle satisfies the three properties of an area function. We may therefore regard the defect as an area measure for triangles in hyperbolic geometry. The closer that the sum of the angle measures is to 180, the smaller the area.

Congruence in Hyperbolic Geometry

The Congruence Theorems in Euclidean geometry apply equally well to hyperbolic geometry since their proofs do not depend on Euclid's parallel postulate.

There is, however, an additional method for proving triangles congruent in hyperbolic geometry. This method is based on property 4 on page 147; that is, two triangles in the hyperbolic plane are congruent if the three angles of one triangle are, respectively, congruent to the three angles of the other triangle.

Elliptic Geometry

We have already seen that in the selection of an axiom of parallels there are three choices open to us: Through a point not on a given line there is

1. One and only one parallel.
2. More than one parallel.
3. No parallel.

The first choice leads to Euclidean geometry in which the angle sum of a triangle is 180°.

The second choice leads to hyperbolic geometry in which the angle sum of a triangle is less than 180°.

The third choice leads to the elliptic geometry of Riemann in which the angle sum of a triangle is greater than 180°.

Unlike hyperbolic geometry where all Euclidean postulates except the parallel postulate are retained, elliptic geometry requires abandoning not only Euclid's parallel postulate but other postulates as well. Since there are no parallels in elliptic geometry, any postulate that implies the existence of parallel lines has to be abandoned.

Elliptic geometry is incompatible with Euclid's assumption that there exist straight lines of infinite extent. Therefore, in this geometry a distinction is made between "infinite" and "boundless." If we move on a circle, we can keep going around the circle endlessly, yet its length is finite. Lines may be finite though *boundless*. Riemann therefore replaced the Euclidean assumption of the infinitude of a straight line by the postulate that a straight line is *finite and boundless.*

Since the assumption of the infinitude of a straight line is related to the axioms of betweenness, these axioms, too, have to be abandoned.

If two lines cannot be parallel in elliptic geometry, how many points of intersection will they have?

Two answers have been given. The first is that any two lines intersect in exactly *one* point, as in the Euclidean plane. This leads to an elliptic plane called the Cayley–Klein plane, first suggested by Arthur Cayley (1821–1895) and Felix Klein (1849–1925). The geometry of the Cayley–Klein plane is sometimes referred to as *single elliptic geometry.*

The second answer is that two lines intersect in exactly *two* points. This leads to an elliptic plane called the Riemann plane. The geometry of this plane is sometimes referred to as *double elliptic geometry.*

A Model for Elliptic Geometry

A model of a Riemann plane is the geometry on a sphere. The points

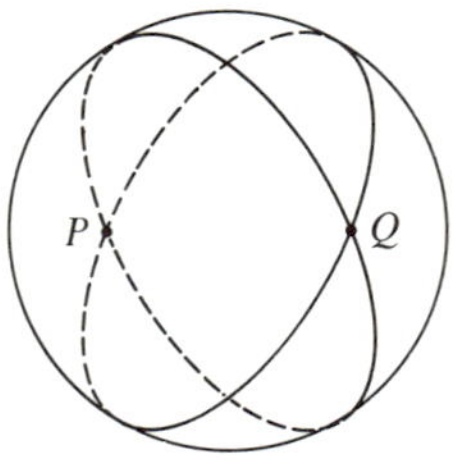

FIGURE 5-30

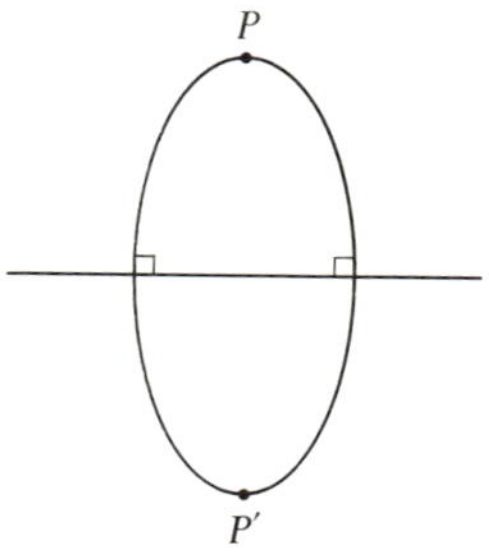

FIGURE 5-31

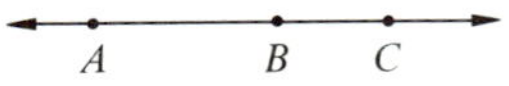

FIGURE 5-32

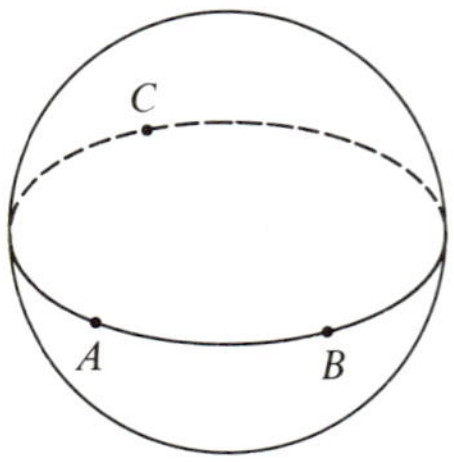

FIGURE 5-33

on a sphere correspond to the "points" in elliptic geometry. The "lines" of this geometry correspond to the great circles of the sphere. "Segments" are represented by arcs of great circles. "Distance" between two points is represented by the length of the shortest arc of a great circle joining the two points. "Angles" are represented by the spherical angles formed by two great circles, and the measure of an angle is represented by the measure of a spherical angle.

Every true statement about a sphere can be converted into a true statement about a Riemann plane by replacing the term "great circle" with the term "straight line." This is interesting when you consider that elliptic geometry and Euclidean geometry contradict each other in some important ways.

Properties of Elliptic Geometry

1. We have seen that there are no parallel lines in this geometry and that any two lines in the Riemann plane meet in exactly two points. These points are opposite ends of a diameter (FIG. 5-30).

2. Two points do not necessarily determine a unique line. For example, points P and Q (FIG. 5-30) lie on infinitely many great circles.

3. All the lines perpendicular to a given line meet at two particular points associated with the given line and known as its *poles* (FIG. 5-31). This means that the perpendicular to a line from an external point is not necessarily unique.

4. Although the "lines" in this geometry are boundless in the sense that they never come to an end at any point, they are nevertheless finite in extent. If the radius of our model sphere is 1, then the maximum possible distance between any two points is π.

5. We said earlier that some of the order properties of Euclidean geometry have to be abandoned in elliptic geometry. For example, if $\overleftrightarrow{AC}$ is a straight line in Euclidean geometry (FIG. 5-32) and if B is between A and C, then A is not between B and C, and C is not between A and B. However, on the great circle ABC (FIG. 5-33), not only is B between A and C, but A is between B and C, and C is between A and B.

6. We have already said that the sum of the measures of the angles of a triangle in this geometry is greater than 180. Therefore, it is possible for a triangle to contain more than one

right angle. In FIG. 5-34, if P is the North Pole and A and B are points on the equator that are 90° of longitude apart, then $\triangle ABP$ has three right angles.

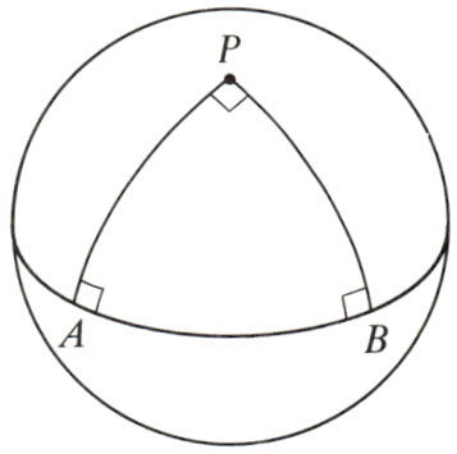

FIGURE 5-34

7. Since there exist triangles with more than one right angle, the Pythagorean relation does not hold in this geometry.

8. If the sum of the angle measures of a triangle is always greater than 180, then the sum of the angle measures of a quadrilateral will always be greater than 360. This means that we can have no rectangles or squares in this geometry.

9. The amount by which the sum of the angles of a triangle exceeds 180 is called the *excess* of the triangle. The excess of a triangle is a measure of its area, just as in hyperbolic geometry the *defect* of the triangle is a measure of its area.

10. Two similar triangles in this geometry, as in hyperbolic geometry, are necessarily congruent. This means that two figures cannot have exactly the same shape unless they also have exactly the same size.

Consistency of Non-Euclidean Geometries

We said that the non-Euclidean geometries are "consistent." How do we know that they are consistent? Can we *prove* that they are?

The way consistency is established is to relate the non-Euclidean geometries to their "models," which satisfy all the axioms of Euclidean geometry except the parallel postulate.

The argument then proceeds as follows: Hyperbolic geometry has a model in Euclidean geometry—the interior of a Euclidean circle. If Euclidean geometry is consistent, then hyperbolic geometry must also be consistent. Similarly, elliptic geometry has a model on a Euclidean sphere. If elliptic "lines" contain a contradiction, then there must be a corresponding contradiction for Euclidean great circles.

What we are really saying is that the consistency of one geometric system implies the consistency of another. Either *both* are consistent or *both* are inconsistent.

We should note that such a "proof" of consistency is not absolute but *relative*. Consistency in one system is contingent on consistency in another system. This means that we cannot really be sure that the non-Euclidean geometries are consistent. We can only be sure that they are *as consistent* as Euclidean geometry. To be able to say even this much, we must first assume the consistency of the theory of logical inference used in developing these geometries.

Some General Comments

The creation of non-Euclidean geometries had profound implications not only for geometry but for all mathematics. It demolished the notion that Euclid's axioms were immutable and that axioms are "self-evident truths." It liberated mathematics from dependence on the physical world and revealed most dramatically the abstract nature of mathematics. It established mathematics as a purely abstract science and led to the development of such abstract concepts as complex numbers, n-dimensional spaces, postulate sets and their properties (axiomatics), and symbolic logic.

We should note that even though Euclidean, hyperbolic, and elliptic geometries have different properties, they almost coincide if we confine ourselves to a small portion of the plane. The geometry of a sphere in the neighborhood of a point P nearly coincides with the geometry of the point P on the plane tangent to the sphere at P. Very small triangles in any of these non-Euclidean geometries have angle sums that almost equal 180°. So even if astronomical space may be best described as non-Euclidean, physical space in our immediate environment may be described as Euclidean.

IDEAL POINTS

In Euclidean geometry, two lines in a plane either intersect or are parallel. Therefore, any complete discussion involving two lines in a plane requires us to draw conclusions first, say, when the lines intersect and then when they are parallel. Such a procedure is not as convenient as it would be if both cases could be treated by one formulation. If we could modify our geometry so that *every* pair of lines intersects, this inconvenience would be eliminated.

We can so modify our geometry if we add to the ordinary points on each line just one more point called an *ideal point*, or a *point at infinity*. This point is considered to belong to all the lines parallel to a given line and to no other lines. By postulating the existence of *ideal points* we have modified our geometry so that every pair of lines in the plane intersects in one point. If the lines are not parallel, they intersect in an "ordinary" point. If the lines are parallel, they intersect in an *ideal point*.

Our willingness to lessen our difficulties in what must appear to be an arbitrary fashion should not surprise us. We often do this in mathematics. When we found the system of natural numbers inadequate, didn't we arbitrarily decide to enlarge it by the addition of new numbers? Isn't this the way all our number systems were created? Now we find the points in the Euclidean plane inadequate, so we decide to enlarge it by the addition of ideal points. The only restriction on our freedom to do this is that the properties we choose to assign to these ideal points must be consistent in their relations to ordinary points and to each other.

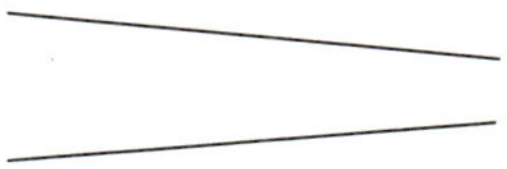

FIGURE 5-35

From the standpoint of intuition, the existence of ideal points should not be difficult to accept.

Think of a straight railroad track on a level plane (FIG. 5-35). Although we know the tracks to be parallel, they still appear to converge in a common point at the horizon. This common point at the horizon is what we defined to be an *ideal point*, or a *point at infinity*. All lines running in the same direction as these two tracks pass through the same ideal point. Every direction in the plane has a different ideal point.

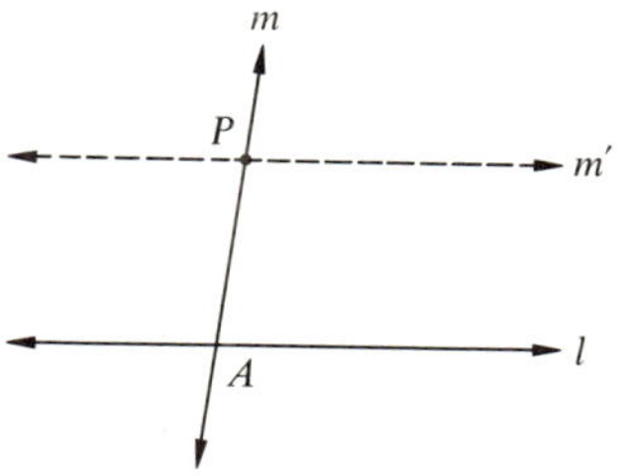

FIGURE 5-36

Drawing in another way on our intuition, think of a line *l*, a point *P* not on *l*, and a line *m* that passes through *P* and intersects *l* at *A* (FIG. 5-36). Rotate *m* clockwise about *P*. As *m* approaches the parallel position *m′* as a limit, the point of intersection *A* will recede to infinity. This suggests the idea that parallel lines meet at a *point at infinity*.

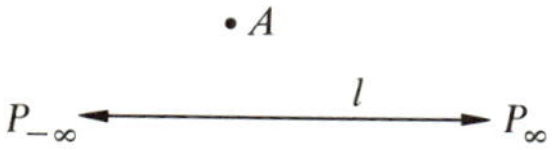

FIGURE 5-37

Since a line recedes to infinity in two directions, should we not add *two* ideal points to each line—one for each direction? Let us see what happens if we do this.

Assume that every line *l* has two ideal points $P_{-\infty}$ and P_{∞}, and let *A* be a point not on *l* (FIG. 5-37). Then a line parallel to *l* and passing though *A* would intersect *l* at two points, $P_{-\infty}$ and P_{∞}. In fact, since *l* would share the two points $P_{-\infty}$ and P_{∞} with *every* line parallel to it, infinitely many lines would pass through these two points. But this contradicts the property that through any two distinct points one and only one line may be drawn. Therefore, every line can have one and only one ideal point.

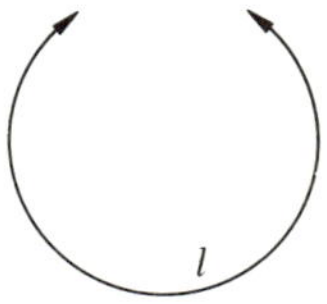

FIGURE 5-38

This means that if you look down a straight road in each direction, say east and west, the two horizons are at the same point. It also means that no matter in which direction you travel on a straight road, east *or* west, you will ultimately arrive at the same point, its ideal point. This situation can be visualized if we think of a straight line as a huge circle with its ends connected at the ideal point (FIG. 5-38).

Just as we added to the ordinary points on each line a single ideal point, we shall now add to the ordinary lines in a plane a single *ideal line*. This *ideal line* contains all the ideal points in the plane, but it contains no ordinary points.

By adding ideal points and an ideal line to the Euclidean plane, we have created the *real projective plane* and a new kind of geometry —*projective geometry*. In this geometry, we have simplified certain incidence relations by removing the exceptions caused by parallelism. Thus, in projective geometry, any two distinct lines always meet in a unique point—without exception.

The concepts of an ideal point and an ideal line are not new. They go back to Johannes Kepler (1571–1630), who postulated their existence in connection with his *principle of continuity*. He thought of a line as closed at infinity, and two parallel lines as intersecting at infinity. Although these ideas did not receive wide and immediate acceptance when Kepler first suggested them, they eventually led to the creation of *projective geometry*.

6

PLANE AND SPACE FIGURES

INTRODUCING THE CHAPTER TO STUDENTS

CONTENT REVIEW

Prerequisites for Learning This Chapter
Plane Figures
Space Figures

TEACHING SUGGESTIONS AND ACTIVITIES

Visual Aids
Using the Compass
Questions
Puzzles
Solutions

ENRICHMENT

Another View of Conic Sections
Applications of Conic Sections
More on the Sphere
Platonic Solids

INTRODUCING THE CHAPTER TO STUDENTS

All of us have had some experience with geometric figures before. When you look at these figures (Visual 1) you undoubtedly recognize some of them.

What are their mathematical names? Can you describe any one of them so exactly that if someone were to construct it according to your description he could come out only with the figure you are describing, and none other?[1]

The purpose of this chapter is to give precise definitions of certain geometric figures that are especially interesting and useful, and to examine some of their properties.

We start at the very beginning with the simplest and most basic geometric figure—a *point*. Points are the building blocks with which we build all geometric figures. Sets of points that satisfy various conditions result in different geometric figures. Certain sets of points are given special names, such as triangle, circle, prism, and trapezoid.

To make it easier to classify the endless variety of geometric figures possible, we separate all geometric figures into two groups:

1. Those whose points all lie in one plane.
2. Those whose points do not all lie in one plane.

The figures in the first group are called *plane figures*, and those in the second group are called *space figures*.

In this chapter we shall study some of the more common plane and space figures. We shall see what makes these figures look the way they do and how they are related to each other.

[1] Try to get an exact definition from the class of, say, a circle. Draw a figure on the board in accordance with the definition given and let class see whether this definition yields a circle. Let class appreciate the need, and the difficulties one encounters in trying, to define these figures precisely.

CONTENT REVIEW

PREREQUISITES FOR LEARNING THIS CHAPTER

1. Knowing the meaning of: interior of an angle; convex set; medians, altitudes of a triangle; perpendicular bisectors of the sides of a triangle; consecutive angles of a parallelogram; diagonals of a quadrilateral.
2. Knowing that: the *distance* from a point to a line is the length of the perpendicular segment from the point to the line; the perpendicular segment is the shortest segment joining a point to a line.

PLANE FIGURES

Polygons: Informal Remarks

An important classification of figures all of whose points lie in one plane are the *polygons.* These are closed figures consisting entirely of segments and look as shown in FIG. 6-1. The segments of a polygon are called its *sides*, and the endpoints of the segments are called the *vertices* of the polygon.

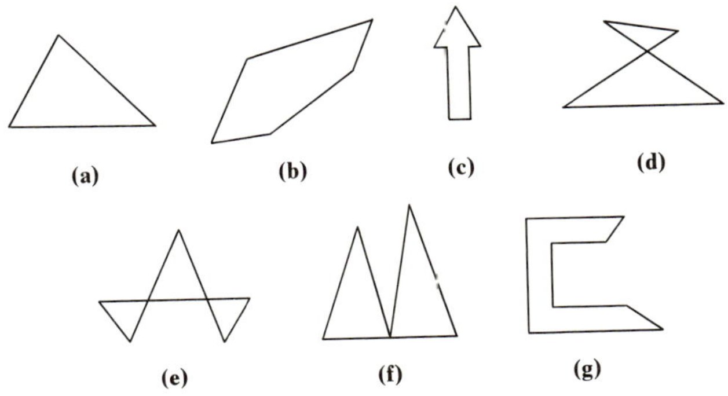

FIGURE 6-1

We wish to distinguish between two types of polygons: *simple* and *nonsimple.* A polygon is *simple* if it does not cross itself, such as in (a), (b), (c), and (g). If the polygon crosses itself, such as in (d), (e), and (f), we call it a *nonsimple* polygon. [We shall agree to think of (f) as "crossing" itself even though it does not "cross" itself the same way as (d) and (e) do.] In this book we shall confine our attention to simple polygons.

Polygons are often classified according to their number of sides:

Name of Polygon	*Number of Sides*
Triangle	3
Quadrilateral	4
Pentagon	5
Hexagon	6
Heptagon	7
Octagon	8
Nonagon	9
Decagon	10
Dodecagon	12

Polygons: Formal Treatment

We shall now give a formal definition of a polygon.

Definitions

Let $A_1, A_2, A_3, \ldots, A_n$ be a sequence of n distinct coplanar points, with $n \geq 3$ (FIG. 6-2). Let the n segments $\overline{A_1A_2}$, $\overline{A_2A_3}, \ldots, \overline{A_nA_1}$ have the following properties:

1. No two segments intersect, except for common endpoints.
2. No two segments with a common endpoint are collinear.

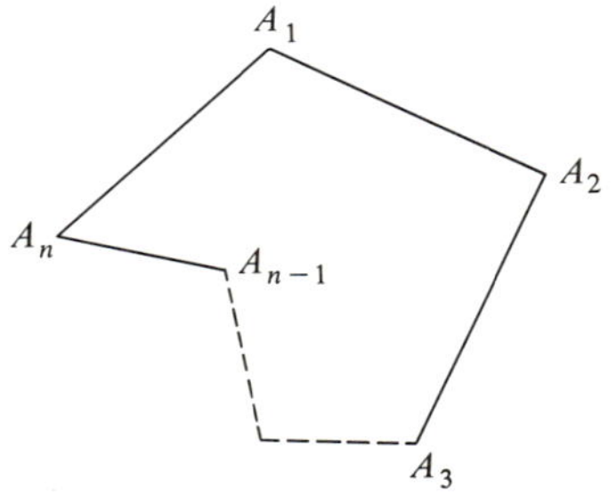

FIGURE 6-2

Then the union of the n segments is a *polygon*.
The n points $A_1, A_2, A_3, \ldots, A_n$ are called the *vertices* of the polygon.

The n segments $\overline{A_1A_2}, \overline{A_2A_3}, \ldots, \overline{A_nA_1}$ are called the *sides*.

The *angles* of the polygon are the n angles

$\angle A_nA_1A_2, \angle A_1A_2A_3, \ldots, \angle A_{n-1}A_nA_1$.

A *diagonal* of a polygon is a segment joining two vertices that are not endpoints of the same side of the polygon.

The sum of the lengths of the sides of a polygon is called the *perimeter* of the polygon.[2]

Definitions

1. A polygon is called a *regular polygon* if all its sides are congruent and all its angles are congruent.
2. A polygon is a *convex polygon* if and only if each side lies on a line determining a half-plane that contains the remainder of the polygon (FIG. 6-3).[3] Any polygon which is not convex is *concave*.

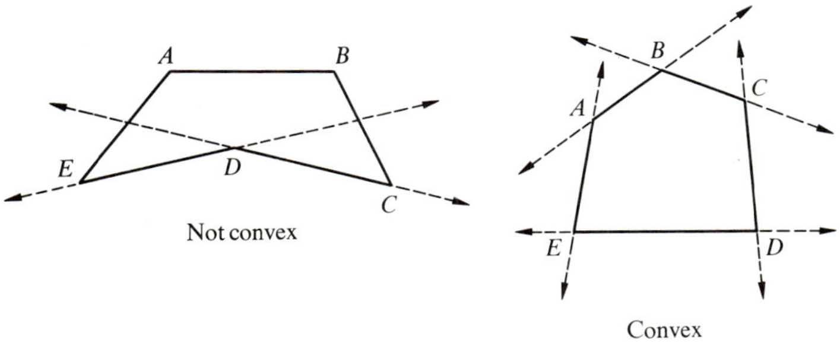

FIGURE 6-3

3. The intersection of the interiors of all the angles of a convex polygon is the *interior of the polygon*.
4. The union of a polygon and its interior is called a *closed polygonal region*.

THEOREM 6-1 *The interior of every convex polygon is a convex set.*

This theorem is a direct consequence of the definition of "the

[2] 1. A polygon must have n vertices, n sides, and n angles. A polygon of n vertices and sides is called an *n-gon*.

2. Property 1 of the definition prevents two segments from crossing and, therefore, makes our definition applicable only to simple polygons.

[3] 1. If a polygon is convex, then the polygon plus its interior forms a *convex set* as we defined it in Chapter 2. A convex polygon, by itself, is not a convex set.

2. Note that all the diagonals of a convex polygon (except their endpoints) are contained in the interior of the polygon (FIG. 6A).

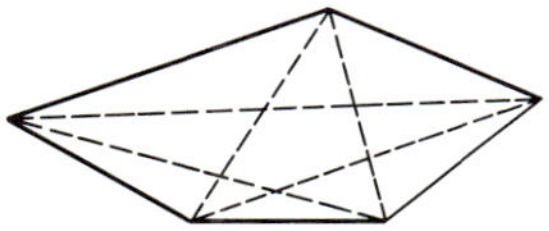

FIGURE 6A

This is not true of a concave polygon (FIG. 6B).

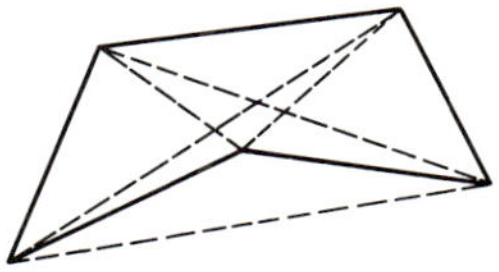

FIGURE 6B

interior of a polygon" and Theorem 2-12, which states that the intersection of any two convex sets is also a convex set.

Triangles

The following theorems describe interesting and important properties of triangles.

> THEOREM 6-2 *The segment joining the midpoints of two sides of a triangle is parallel to the third side, and the length of the segment is half the length of the third side* (FIG. 6-4).

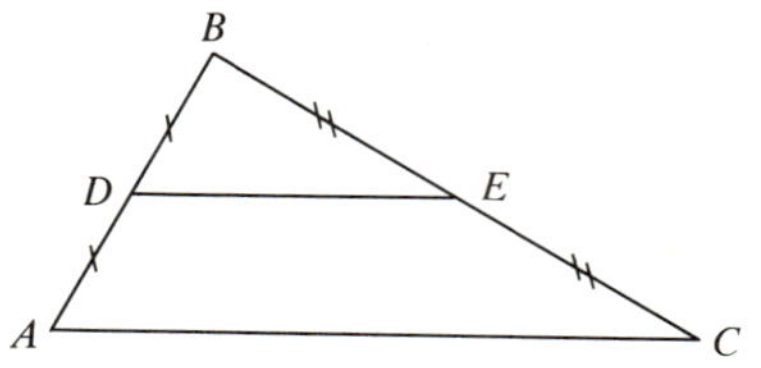

FIGURE 6-4

That is, if D and E are the midpoints of $\overline{AB}$ and $\overline{BC}$, respectively, then $\overline{DE} \parallel \overline{AC}$ and $DE = \frac{1}{2}AC$.

> THEOREM 6-3 *If a line is parallel to one side of a triangle and contains a midpoint of another side, then the line contains the midpoint of the third side* (FIG. 6-5).

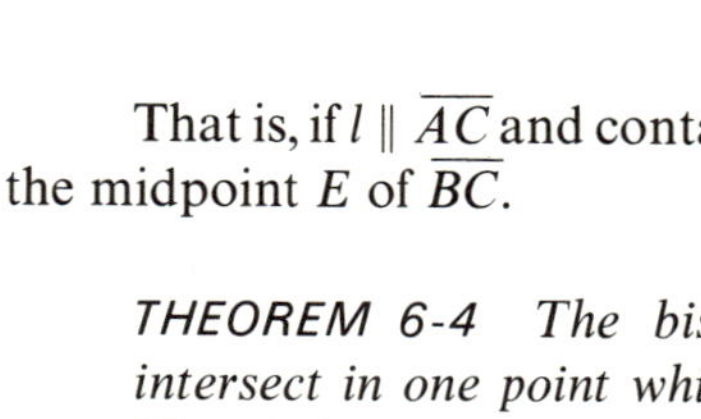

FIGURE 6-5

That is, if $l \parallel \overline{AC}$ and contains midpoint D of $\overline{AB}$, then l contains the midpoint E of $\overline{BC}$.

> THEOREM 6-4 *The bisectors of the angles of a triangle intersect in one point which is equidistant from the three sides* (FIG. 6-6).

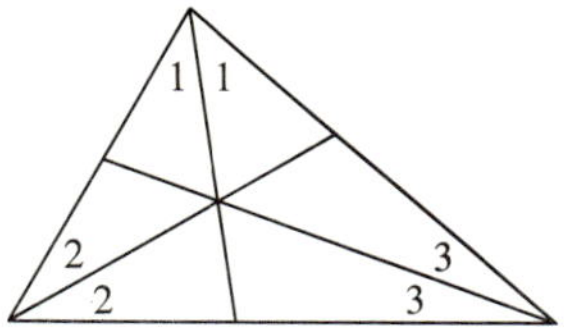

FIGURE 6-6

The point is called the *incenter* of the triangle because it is the center of the circle inscribed in the triangle.[4]

THEOREM 6-5 The perpendicular bisectors of the sides of a triangle intersect in one point which is equidistant from the three vertices (FIG. 6-7).

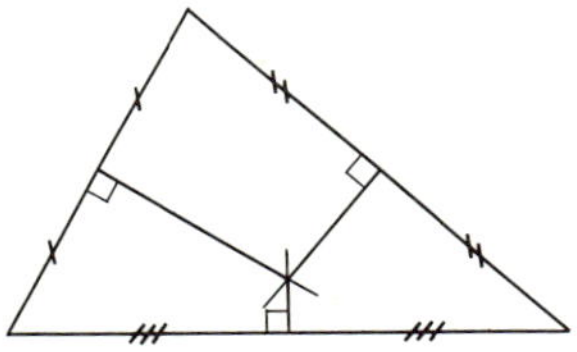

FIGURE 6-7

The point is called the *circumcenter* of the triangle because it is the center of the circle circumscribed about the triangle.

THEOREM 6-6 The medians of a triangle intersect in one point which is two thirds of the distance from each vertex to the midpoint of the opposite side (FIG. 6-8).

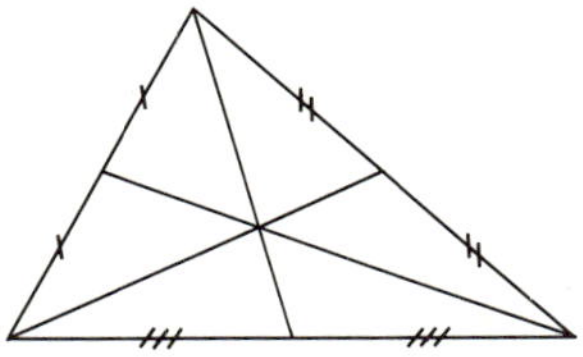

FIGURE 6-8

The point is called the *centroid* of the triangle and is the "point of balance" or the "center of gravity" of an object, of uniform thickness, in the shape of a triangle. Such an object would balance if it were supported at the centroid.

[4] Theorem 6-4 speaks about the three angle bisectors of a triangle. What about the *trisectors* of the angles of a triangle?

It is possible to prove that the *adjacent* trisectors of the angles of any triangle intersect at three points which are the vertices of an equilateral triangle (FIG. 6C). This is known as *Morley's Theorem*.

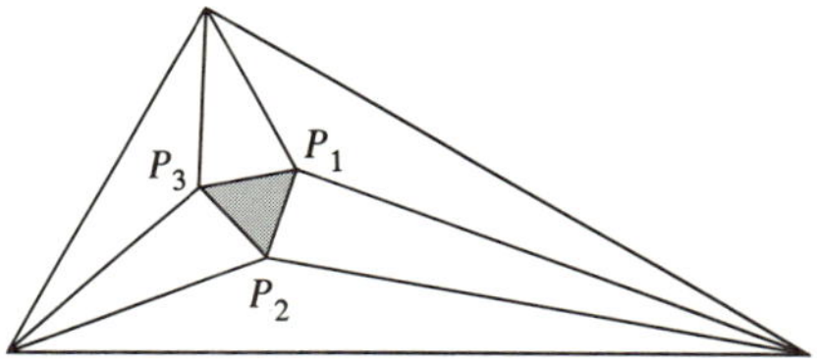

FIGURE 6C

THEOREM 6-7 *The altitudes of a triangle intersect in one point* (FIG. 6-9).

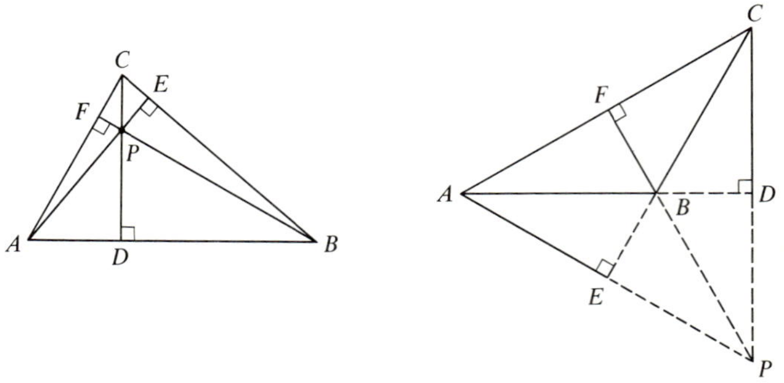

FIGURE 6-9

The point is called the *orthocenter* of the triangle.[5]

THEOREM 6-8 *If two internal angle bisectors are congruent, the triangle is isosceles* (FIG. 6-10). (This theorem is known as the Steiner–Lehmus problem.)

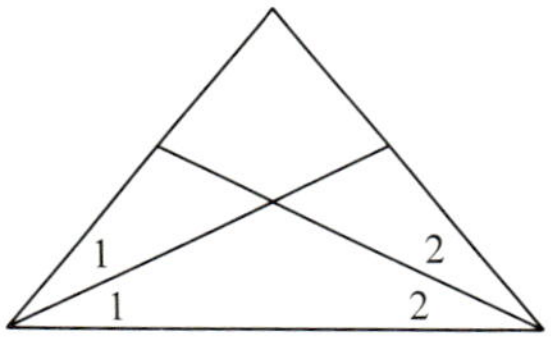

FIGURE 6-10

THEOREM 6-9 *The sum of the distances from any point within an equilateral triangle to its sides is a constant equal to the length of any altitude* (FIG. 6-11).

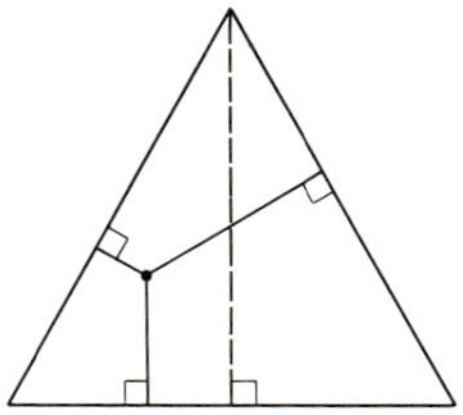

FIGURE 6-11

[5] 1. The three altitudes of a right triangle meet at the vertex of the right angle (FIG. 6D). Why?

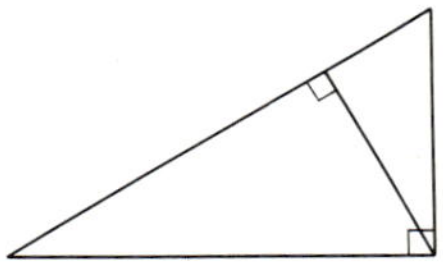

FIGURE 6D

2. The median, the altitude, and the angle bisector to the base of any isosceles triangle are the same segment (FIG. 6E). Why?

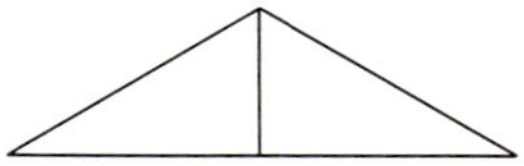

FIGURE 6E

Quadrilaterals[6]

Definitions
See Fig. 6-12.

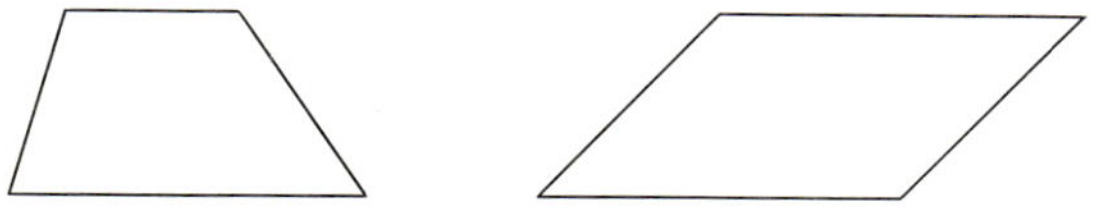

FIGURE 6-12

1. A quadrilateral that has *at least one* pair of parallel sides is called a *trapezoid.*
2. A quadrilateral that has two pairs of parallel sides is called a *parallelogram.*[7]

Theorems About Parallelograms

THEOREM 6-10 *In a parallelogram, any two opposite sides are congruent.*

THEOREM 6-11 *Any two opposite angles of a parallelogram are congruent.*

THEOREM 6-12 *Any two consecutive angles of a parallelogram are supplementary.*

THEOREM 6-13 *The diagonals of a parallelogram bisect each other.*

Conditions That Make a Quadrilateral a Parallelogram

THEOREM 6-14 *If in a given quadrilateral, both pairs of opposite sides are congruent, then the quadrilateral is a parallelogram.*

THEOREM 6-15 *If two sides of a quadrilateral are both parallel and congruent, then the quadrilateral is a parallelogram.*

THEOREM 6-16 *If the diagonals of a quadrilateral bisect each other, then the quadrilateral is a parallelogram.* (This is the converse of Theorem 6-13.)

[6] Certain quadrilaterals are of great importance because of special properties they possess. One such property relates to parallelism.

We shall now define these special quadrilaterals.

[7] These definitions make a parallelogram a special type of trapezoid.

Some people prefer to define a trapezoid as a quadrilateral in which *exactly one pair* of opposite sides are parallel. This eliminates parallelograms from the class of trapezoids.

Our definition of a trapezoid is more general and, therefore, carries the advantage that theorems about trapezoids automatically apply to parallelograms.

Definitions
See FIG. 6-13.

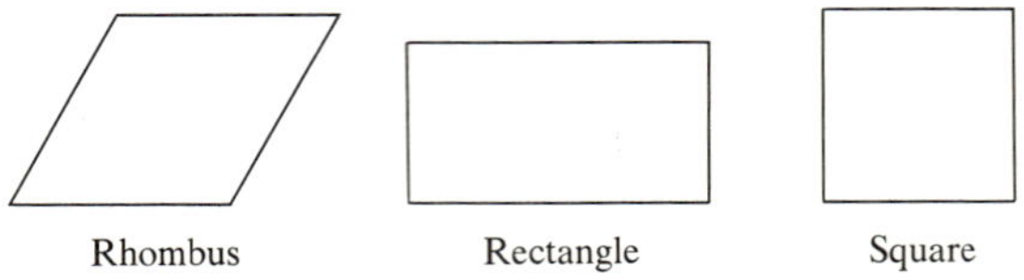

FIGURE 6-13

1. A *rhombus* is a parallelogram all of whose sides are congruent. (Or, if two adjacent sides of a parallelogram are congruent, then the quadrilateral is a *rhombus.*)
2. A parallelogram containing at least one right angle is called a *rectangle*.
3. A rhombus with at least one right angle is called a *square.* (Or, a *square* is a rectangle all of whose sides are congruent.)

Related Theorems

THEOREM 6-17 *The diagonals of a rhombus are perpendicular to each other and bisect each other.*

THEOREM 6-18 (Converse of Theorem 6-17) *If the diagonals of a quadrilateral bisect each other and are perpendicular to each other, then the quadrilateral is a rhombus.*

THEOREM 6-19 *If a quadrilateral has three right angles, it is a rectangle.*

Circles

Definitions
1. A *circle* is the set of all points in a plane at a given distance from a given point in the plane. The given point P (FIG. 6-14) is called the *center*, and the given distance r is called

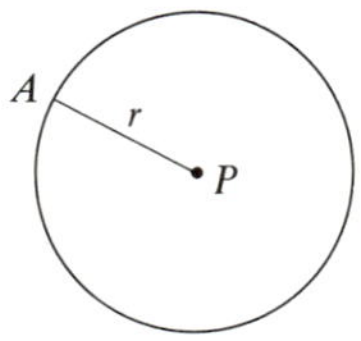

FIGURE 6-14

ircle. If A is any point of the circle, then s a radius of the circle and A is called its

nent having both endpoints on a circle, $\overline{CE}$ (FIG. 6-15).

6-15

3. If the chord contains the center of a circle, it is called a *diameter* of the circle, for example $\overline{CE}$ (FIG. 6-15). The length of a diameter is the number $2r$.

4. A *tangent* to a circle is a line in the plane of the circle which intersects the circle in only one point, for example $\overleftrightarrow{AB}$ (FIG. 6-16). This point is called the *point of tangency*, or the *point of contact*. If the line intersects the circle at more than one point, it is called a *secant* line, for example $\overleftrightarrow{CD}$ (FIG. 6-16).

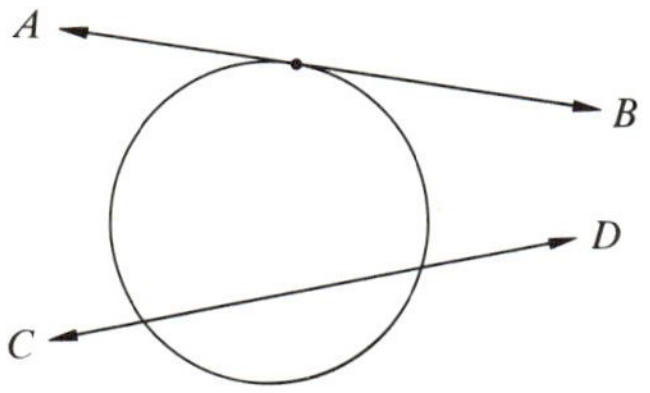

FIGURE 6-16

5. Circles that lie in the same plane and have the same center are called *concentric circles* (FIG. 6-17).

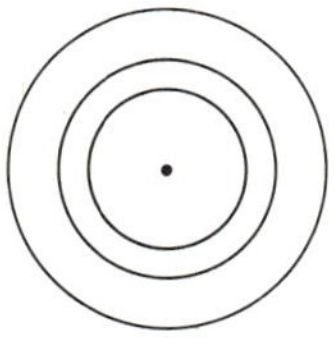

FIGURE 6-17

6. The *interior of a circle* is the set of all points of the plane whose distance from the center is less than the radius (FIG. 6-18). The *exterior of a circle* is the set of all points of the plane whose distance from the center is greater than the radius.

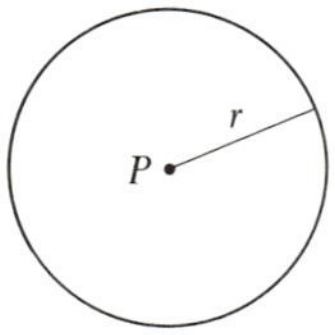

FIGURE 6-18

7. The union of a circle and its interior is a *closed circular region*.

8. Two circles are *congruent* if and only if they have congruent radii.[8]

[8] 1. Unlike previous figures we talked about, a circle is an example of a set of points *not* made up of lines, rays, and segments.

2. Note that the center is *not* a point of the circle, and a radius is *not* a part of the circle.

3. Note that we use the word "radius" in two senses. It may mean either a number r or a segment $\overline{PA}$ (FIG. 6-14). We use the word "diameter" in the same two senses: either a number $2r$ or a segment $\overline{CE}$ (FIG. 6-15).

4. Circles play a particularly important role in the history of civilization. Wheels, for example, make possible our present-day machines.

Related Theorems

THEOREM 6-20 *If a line is perpendicular to a radius of a circle at its outer end, then the line is a tangent.*

Proof

1. Let P be the center, $\overline{PA}$ be a radius, and let l be perpendicular to $\overline{PA}$ at A (FIG. 6-19).

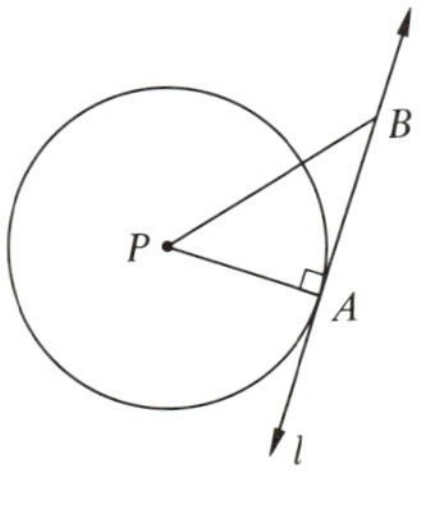

FIGURE 6-19

2. If B is any other point of l, then $PB > PA$ because the shortest segment joining a point to a line is the perpendicular segment.

3. Therefore, B is in the exterior of the circle.
4. Therefore, l intersects the circle only at A and is, by definition, a tangent line.

THEOREM 6-21 (Converse of Theorem 6-20) *Every tangent to a circle is perpendicular to the radius drawn to the point of contact.*

THEOREM 6-22 *Any perpendicular from the center of a circle to a chord bisects the chord* (FIG. 6-20).

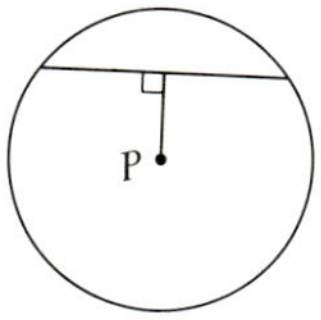

FIGURE 6-20

THEOREM 6-23 *The perpendicular bisector of a chord passes through the center of the circle* (FIG. 6-21). (This is the converse of Theorem 6-22.)

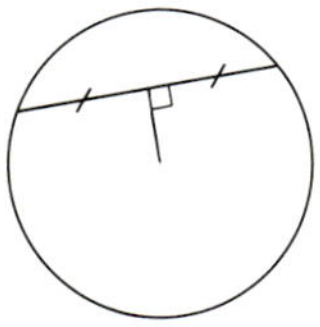

FIGURE 6-21

THEOREM 6-24 *In the same or congruent circles, chords equidistant from the center are congruent* (FIG. 6-22).

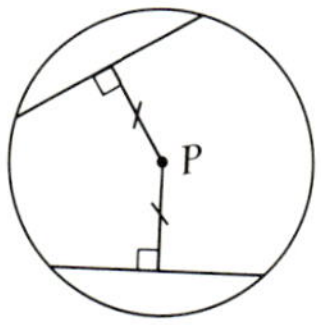

FIGURE 6-22

THEOREM 6-25 In the same circle or in congruent circles, any two congruent chords are equidistant from the center (FIG. 6-23). (This is the converse of Theorem 6-24.)

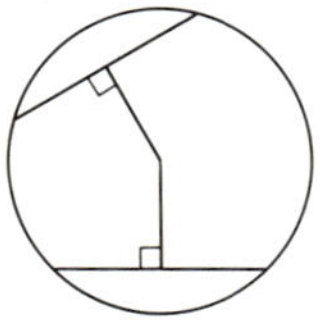

FIGURE 6-23

Arcs and Inscribed Angles

Definitions

1. A *central angle* of a given circle is an angle whose vertex is the center of the circle (FIG. 6-24).

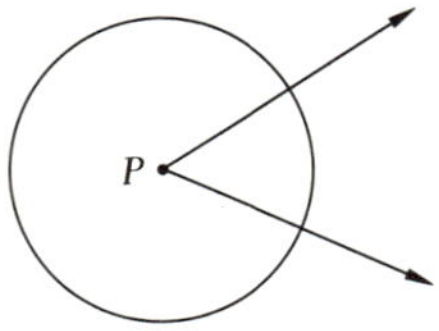

FIGURE 6-24

2. In the central angle $\angle APB$, the set of points consisting of A and B together with all points of the circle that are in the interior of $\angle APB$ is called a *minor arc* of the circle. We denote this arc by $\widehat{AB}$ (FIG. 6-25). The set consisting

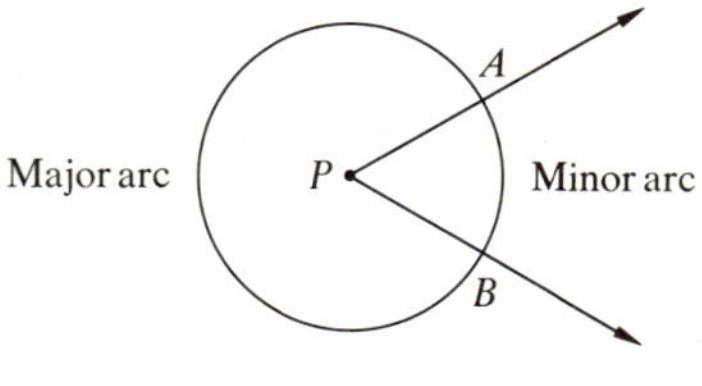

FIGURE 6-25

of A, B, together with all points of the circle that are in the exterior of $\angle APB$, is called a *major arc* of the circle. To distinguish between two arcs that have the same endpoints

A and B, we take any third point X in the arc and denote the arc by $\widehat{AXB}$ (FIG. 6-26).

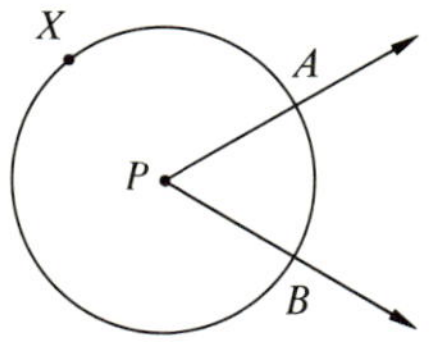

FIGURE 6-26

3. If A and B are the endpoints of a diameter, then the set of points consisting of A and B together with all points of the circle that lie on a given side of the line is called a *semicircle*. There are two such semicircles, $\widehat{AYB}$ and $\widehat{AXB}$ (FIG. 6-27).

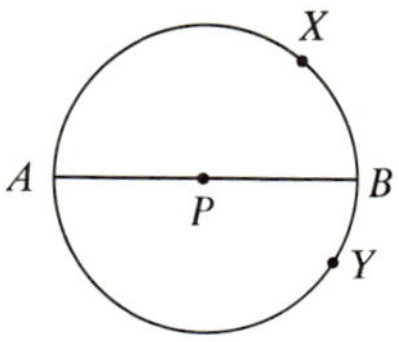

FIGURE 6-27

4. An *arc* is any minor arc, major arc, or semicircle.

The *degree measure* $m\widehat{AXB}$ of an arc $\widehat{AXB}$ is defined as follows:

1. If $\widehat{AXB}$ is a minor arc, then $m\widehat{AXB}$ is the measure of the corresponding central angle (FIG. 6-28).

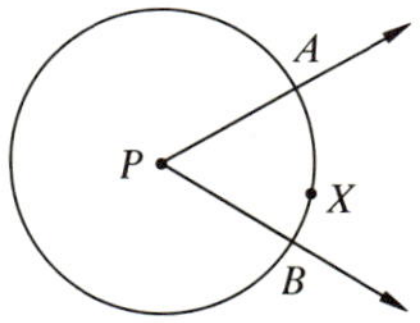

FIGURE 6-28

2. If $\widehat{AXB}$ is a semicircle, then $m\widehat{AXB} = 180$.
3. If $\widehat{AXB}$ is a major arc and $\widehat{AYB}$ is the corresponding minor arc, then $m\widehat{AXB} = 360 - m\widehat{AYB}$ (FIG. 6-29).[9]

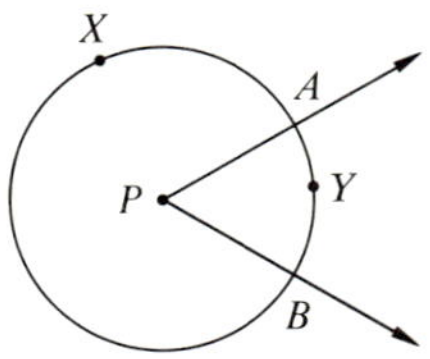

FIGURE 6-29

[9] Note that the degree measure of an arc does not depend on the radius of the circle. The degree measure of an arc is less than or greater than 180, depending on whether it is a minor or a major arc (FIG. 6F).

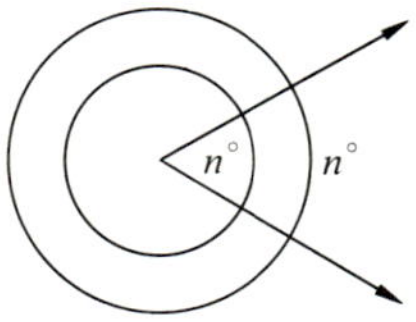

FIGURE 6F

Related Theorems

The first theorem says that degree measure for arcs is additive.

THEOREM 6-26 *If $\widehat{AB}$ and $\widehat{BC}$ are arcs of the same circle, having only the point B in common, then*

$m\widehat{AB} + m\widehat{BC} = m\widehat{AC}$

(FIG. 6-30).

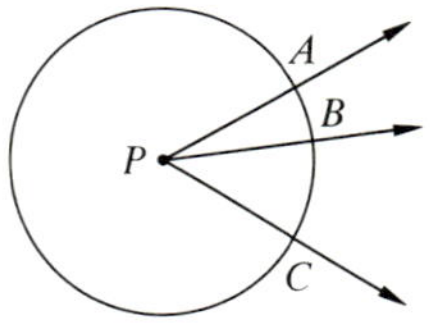

FIGURE 6-30

Before proceeding to the next theorem, we need two definitions.

Definitions
See FIG. 6-31.

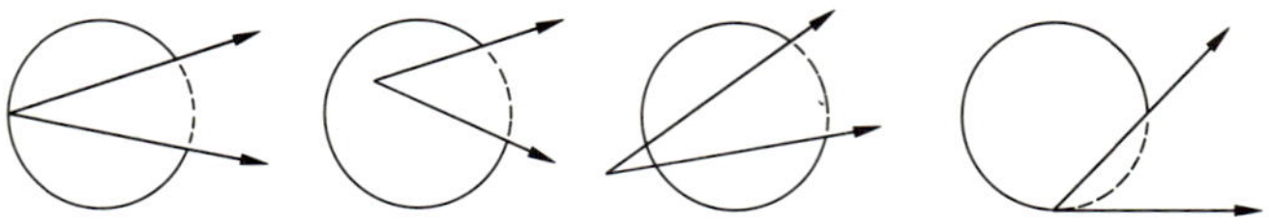

FIGURE 6-31

1. An angle *intercepts* an arc if

   a. The endpoints of the arc lie on different rays of the angle.
   b. Except for its endpoints, the arc lies in the interior of the angle.

2. An angle is *inscribed* in an arc if the intersection of the angle and the arc consists of three points: the vertex of the angle and the two endpoints of the arc (FIG. 6-32).[10]

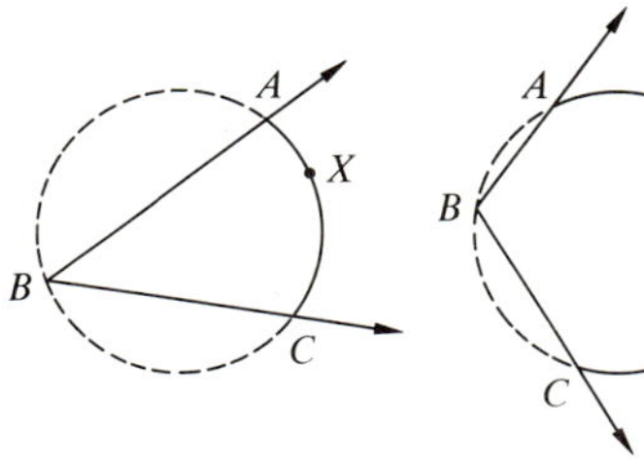

$\angle ABC$ is inscribed in $\widehat{ABC}$

FIGURE 6-32

THEOREM 6-27 The measure of an inscribed angle is half the measure of its intercepted arc (FIG. 6-33).

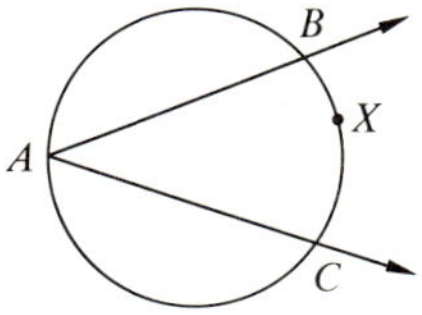

FIGURE 6-33

That is, if $\angle A$ intercepts $\widehat{BXC}$, then $m\angle A = \frac{1}{2} \cdot m\widehat{BXC}$.

Two important consequences of this theorem are the following corollaries:

1. An angle inscribed in a semicircle is a right angle.
2. All angles inscribed in the same arc are congruent.

The Two-Circle Theorem

THEOREM 6-28 (Two-Circle Theorem) Let C and C′ be two coplanar circles with radii a and b, and let d be the distance between their centers (FIG. 6-34). *Then*

[10] 1. Note that $\angle ABC$, which is inscribed in $\widehat{ABC}$, *intercepts* $\widehat{AXC}$.

2. An angle may be inscribed in a minor arc, a major arc, or in a semicircle.

3. It is possible for two angles to be inscribed in the same arc. For example, in FIG. 6G, $\angle ABD$ and $\angle ACD$ are both inscribed in $\widehat{AXD}$, and both intercept $\widehat{AYD}$.

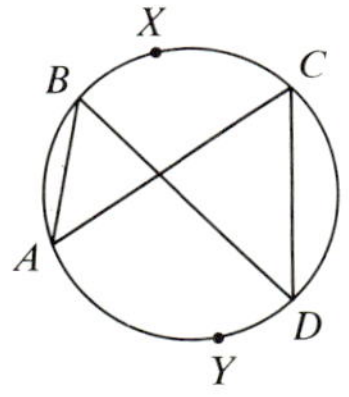

FIGURE 6G

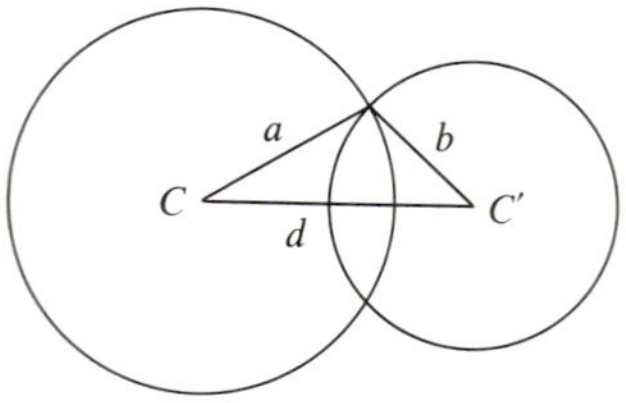

FIGURE 6-34

1. *C and C′ intersect in exactly two points if and only if each of the numbers a, b, and d is less than the sum of the other two.*
2. *If C and C′ intersect in exactly two points, then these points lie on opposite sides of the line of centers.*

For a proof of this theorem, see E. Moise, *Elementary Geometry from an Advanced Standpoint*, Reading, Mass.: Addison-Wesley Publishing Company, Inc., 1963, pp. 197–201.[11]

SPACE FIGURES[12]

Polyhedra

Definitions

1. A *polyhedron* is a solid whose surface consists entirely of polygonal regions (FIG. 6-35). The polygonal regions are called *faces* of the polyhedron. The intersection of two adjacent faces of the polyhedron is a line segment called an *edge* of the polyhedron. The points of intersection of the edges are the *vertices* of the polyhedron. (The plural of *polyhedron* is *polyhedra*.)

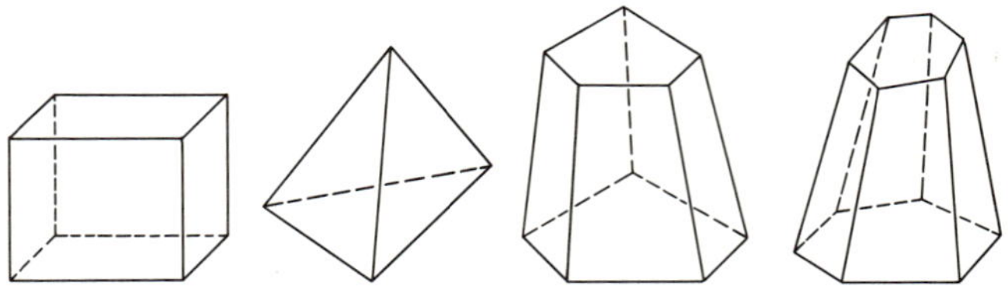

FIGURE 6-35

2. The figure formed by the intersection of a solid and a plane is called a *section* of the solid.

[11] This is an important theorem because, among other reasons, it is the basis for many construction problems. We have seen (page 61) that Euclid assumed it in his geometry but never stated it either as a postulate or as a theorem.

The theorem gives a necessary and sufficient condition for two circles to intersect in exactly two points. Obviously, the nature of the intersection depends on the relationship between the radii and the distance between the centers.

It can be shown that two circles will not intersect at all if at least one of the lengths *a*, *b*, and *d* is greater than the sum of the other two; and will intersect in only one point if at least one of the three lengths is equal to the sum of the other two. It can similarly be proved that two distinct coplanar circles cannot intersect in more than two points.

[12] The geometry of space figures is sometimes called *solid* geometry. The figures in solid geometry, like those in plane geometry, are sets of points. If a set of points separates space into three distinct parts—an interior, an exterior, and the set of points itself—then we call the union of this set of points and the points of the interior a *solid*. We may, therefore, think of a *solid* as a portion of space bounded by plane or curved closed surfaces or by a combination of such surfaces. Physical objects such as a sugar cube, a block of wood, a dowel, or an ice cream cone are examples of solids. Geometric solids are abstractions of such physical objects.

A *solid*, we have just said, consists of a "closed surface" and an interior. But what is a "closed surface"? Because the concept of *surface* is difficult to define precisely, we shall discuss it here only informally. Let us think of a *closed surface* in terms of a rubber balloon.

An inflated rubber balloon divides space into three sets of points: the points in the

Polyhedra are classified according to the number of their faces. The smallest number of faces possible is four.

Name	*Number of Faces*
Tetrahedron	4
Pentahedron	5
Hexahedron	6
Octahedron	8
Decahedron	10
Dodecahedron	12
Icosahedron	20

Prisms

Definitions

1. A *prism* is a polyhedron of which two faces, called *bases*, are congruent polygons in parallel planes (*ABC* and *DEF*), and the other faces, called *lateral faces*, are parallelograms (FIG. 6-36). Each edge determined by the intersection of two lateral faces is called a *lateral edge* ($\overline{AD}$, $\overline{BE}$, $\overline{CF}$).

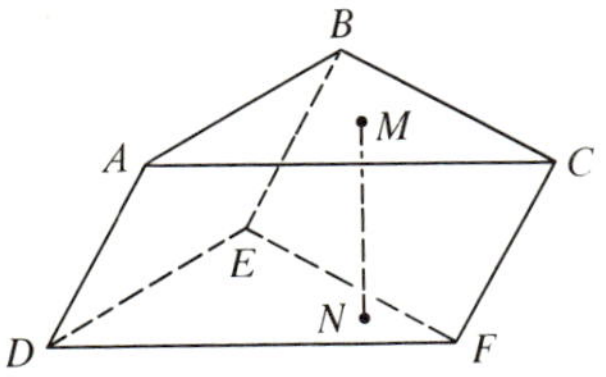

FIGURE 6-36

2. An *altitude* of the prism is any segment that has endpoints in the two bases and is perpendicular to them ($\overline{MN}$). We call the length of any altitude the *height* of the prism.

3. If the lateral edges of a prism are perpendicular to its bases, the prism is called a *right prism* (FIG. 6-37).

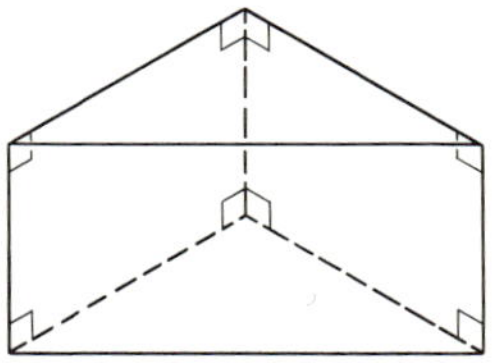

Right prism

FIGURE 6-37

interior of the balloon, the points in the exterior of the balloon, and the points of the balloon itself.

Now think of a point *A* outside the balloon and a point *B* inside the balloon (FIG. 6H). Then the segment $\overline{AB}$ contains points that lie outside, points that lie inside, and points that lie on the balloon ("rubber points").

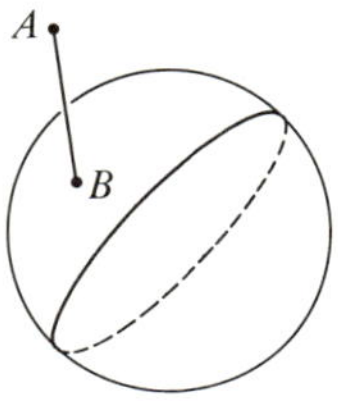

FIGURE 6H

If we now imagine the balloon to be so thin that $\overline{AB}$ contains exactly one "rubber point" and that every segment joining a point inside the balloon to a point outside the balloon contains exactly one "rubber point," then we can think of such a balloon as an example of a "closed surface."

Perhaps the most elementary geometric solids are the polyhedra, which we shall now define.

4. A *parallelepiped* is a prism whose faces are parallelograms. If each of the faces is a rectangle, then the prism is a *rectangular parallelepiped* or a *rectangular solid* or box (FIG. 6-38).

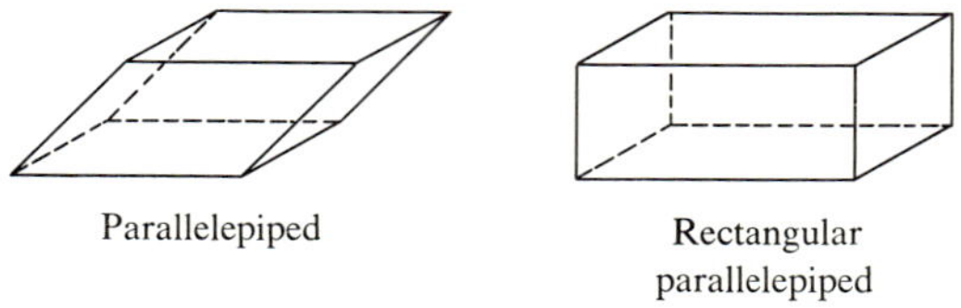

FIGURE 6-38

Prisms are classified according to the shape of their bases, that is, triangular, quadrangular, pentagonal, etc. (see FIG. 6-39).

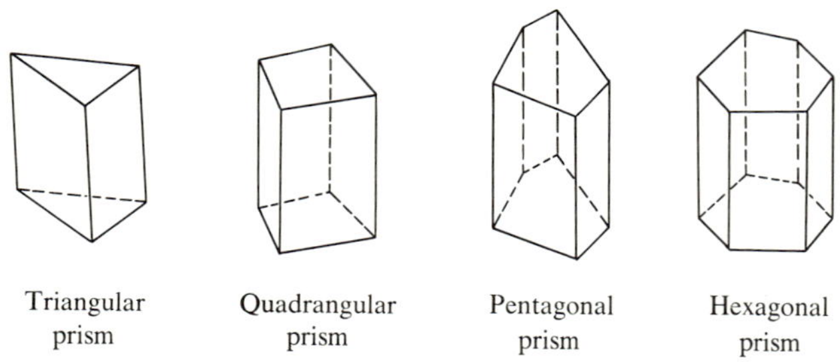

FIGURE 6-39

Theorems About Prisms

THEOREM 6-29 *The lateral edges of a prism are parallel and congruent.*

THEOREM 6-30 *The lateral faces of a right prism are rectangles.*

THEOREM 6-31 *An altitude of a right prism is congruent to each of its lateral edges.*

Pyramids

Definitions

1. A *pyramid* is a polyhedron of which one face is called

the *base* (*ABCD*), and the other faces, called *lateral faces*, are triangles (*PAD*, *PCD*, *PBC*, *PAB*) having a common vertex called the *apex* (*P*) (FIG. 6-40).

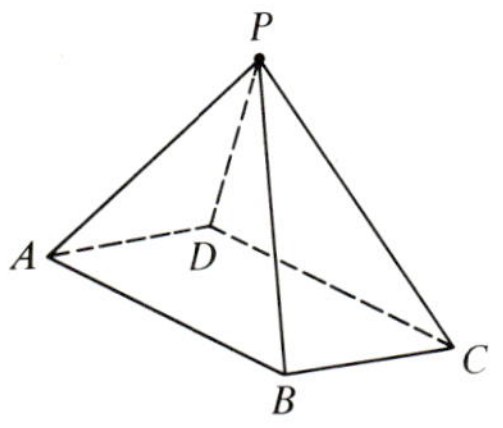

FIGURE 6-40

2. If the base of a pyramid is a regular polygon, the line through the vertex of the pyramid and the center of the base is the *axis* of the pyramid. If the base of a pyramid is a regular polygon and if its axis is perpendicular to its base, then the pyramid is called a *regular pyramid* (FIG. 6-41).

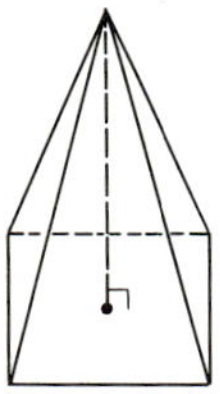

FIGURE 6-41

3. If a plane parallel to the base of a pyramid intersects all its lateral edges, then the polyhedron formed between the parallel plane and the base is called a *frustum* (FIG. 6-42). The lateral faces of a frustum are trapezoids.

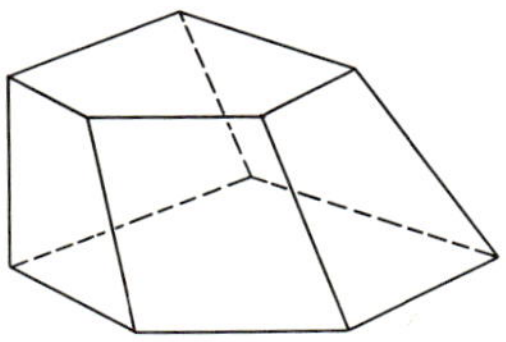

FIGURE 6-42

Pyramids are classified according to the shape of their base, that is, triangular, quadrangular, pentagonal, etc. (see FIG. 6-43).

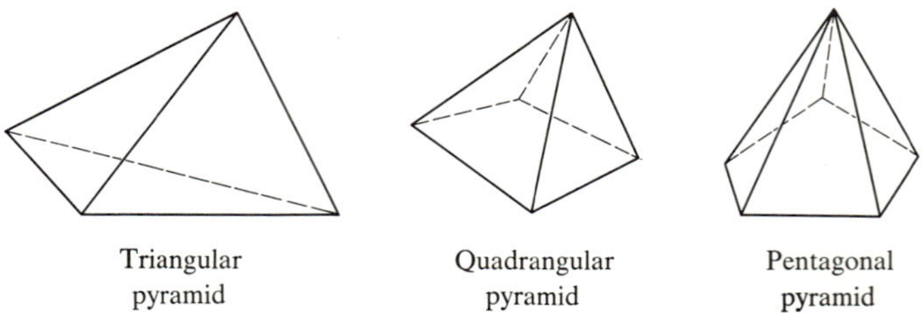

FIGURE 6-43

Theorems About Pyramids

THEOREM 6-32 *The lateral edges of a regular pyramid are congruent.*

THEOREM 6-33 *The lateral faces of a regular pyramid are congruent isosceles triangles.*

THEOREM 6-34 *The altitudes of the lateral faces of a regular pyramid are congruent.*

Cylinders

Before defining a *cylinder* we shall define a *cylindrical surface.*

Definitions

1. If C is a curve lying in a plane P and if l is a line not in P but intersecting P in a point of C, then the union of l and all lines parallel to l passing through C is a *cylindrical surface* (FIG. 6-44). (Note that C may be *any* curve.)

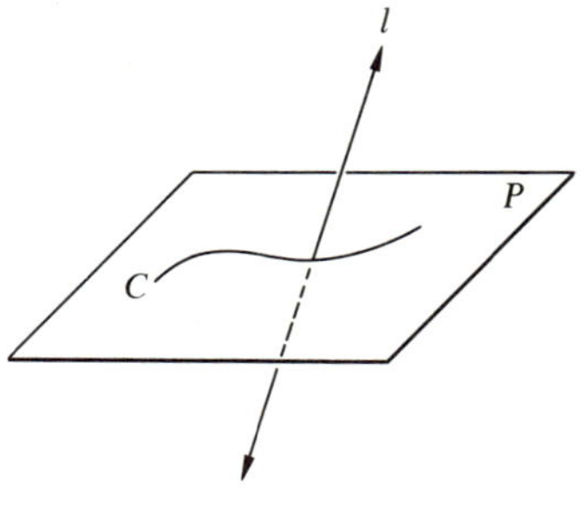

FIGURE 6-44

2. If the curve C is a simple closed curve such as a circle or a rectangle, the cylindrical surface is called a *closed cylindrical surface* (FIG. 6-45).

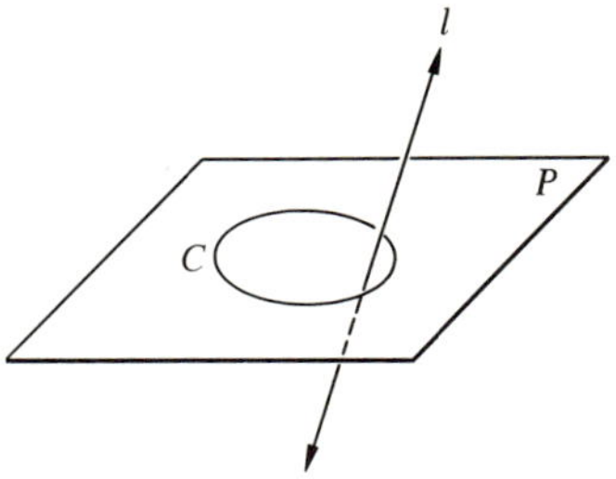

FIGURE 6-45

3. Each of the lines parallel to l and passing through a point in C is called an *element* of the cylindrical surface.
4. A *cylinder* is a solid bounded by a closed cylindrical surface and two parallel planes which intersect all the elements of the cylindrical surface. That portion of the surface of a cylinder which lies in one of the parallel planes is called a *base* of the cylinder (B_1, B_2) (FIG. 6-46). That portion of the surface which lies in the closed cylindrical surface is called the *lateral surface* of the cylinder.

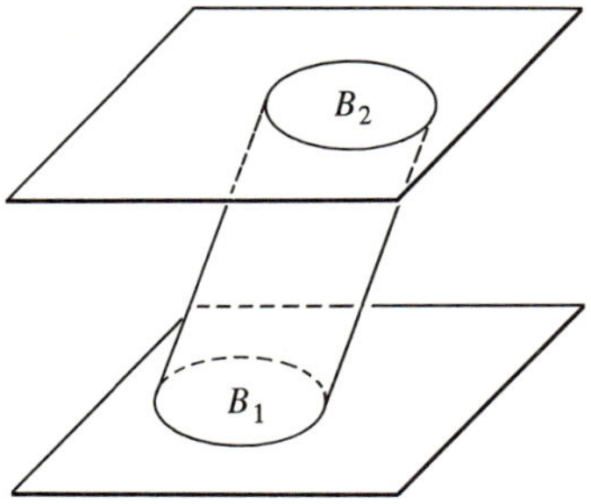

FIGURE 6-46

5. An *altitude* of the cylinder is any segment that has endpoints in the planes of the bases and is perpendicular to them.
6. A *circular cylinder* is one whose bases are circles.
7. If the line segment joining the centers of the bases of a

circular cylinder is perpendicular to the bases, then the cylinder is a *right circular cylinder* (FIG. 6-47).

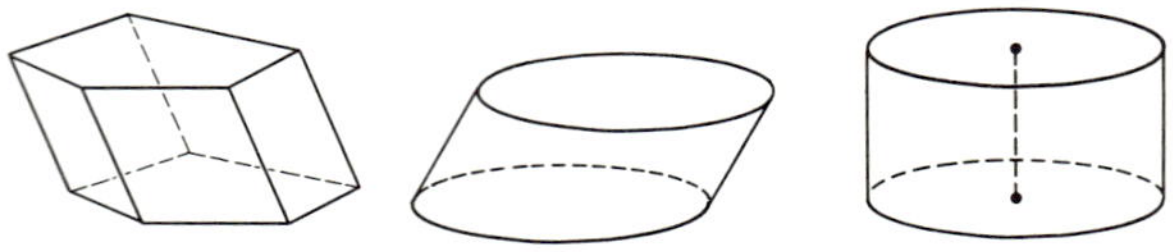

FIGURE 6-47

Cones

Before defining a *cone* we shall define a *conical surface.*

Definitions

1. If C is a curve lying in a plane P and V is a point not in P, then C and V determine a *conical surface* that is the union of all lines each of which passes through V and a point of C (FIG. 6-48). The point V is called the *vertex* of the conical surface. If C is a closed curve, the surface is called a *closed conical surface.* Any of the lines through the vertex and contained in the surface is called an *element* of the surface.

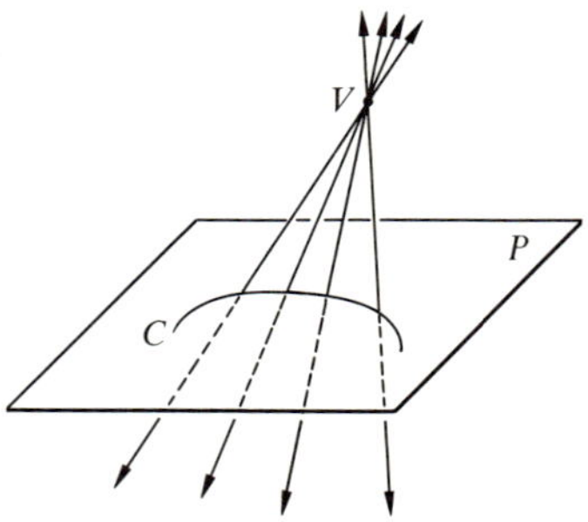

FIGURE 6-48

2. The vertex separates the conical surface into two parts called the *nappes* of the conical surface: all the points that lie on the same side of V as the points of C comprise one nappe; all the points that lie on the opposite side of V from the points of C comprise the other nappe.

3. A *cone* is a solid bounded by a closed conical surface and a plane that intersects all the elements of the surface (FIG. 6-49).

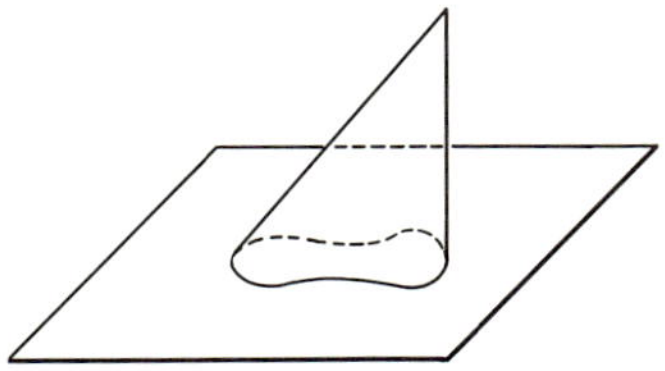

FIGURE 6-49

4. If the curve C in the plane is a circle, the cone is called a *circular cone* (FIG. 6-50).

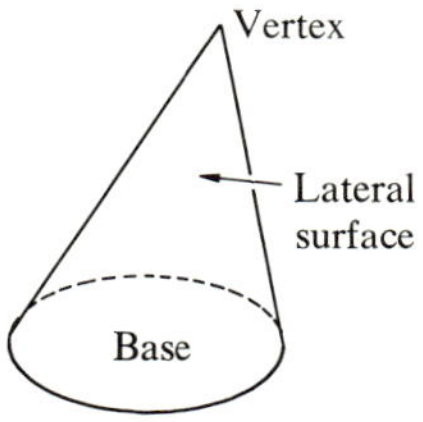

FIGURE 6-50

5. The surface of a cone consists of two parts:

   a. The *lateral surface*, which consists of those points of the cone which lie in the elements of the conical surface.
   b. The *base*.

6. The perpendicular segment from the vertex to the plane of the base is called the *altitude* of the cone, and its length is called the *height* of the cone (FIG. 6-51).

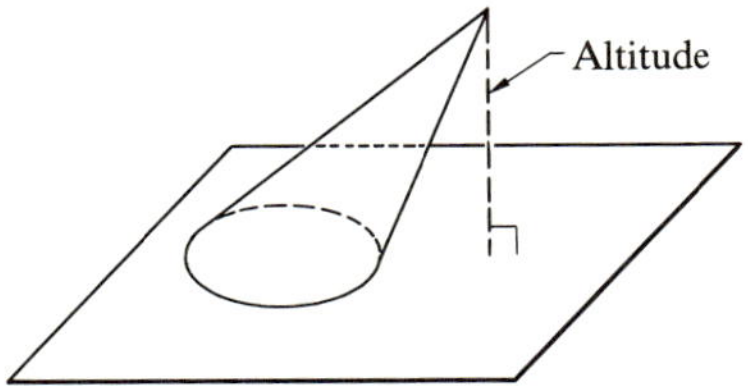

FIGURE 6-51

7. If the segment from the vertex to the center of the circle C is perpendicular to the plane of C, we call the cone a *right circular cone* (FIG. 6-52).

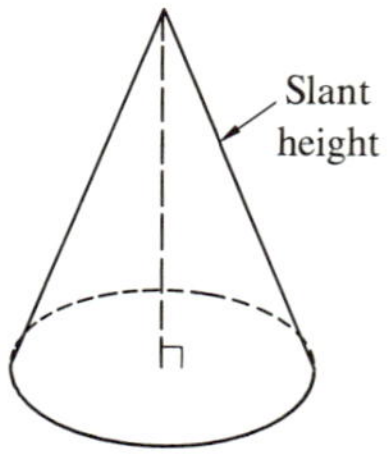

FIGURE 6-52

8. The segment that joins the vertex to any point on the base circle of a right circular cone is called the *slant height* of the cone.

9. If we cut a circular cone by a plane parallel to the base, we get a circle. If we cut a circular cone by other planes, we get other interesting curves known as *conic sections* (see pages 182–183).

Spheres

Definitions

1. A *sphere* is the set of all points at a given distance from a given point in space. The given distance r is called the *radius* of the sphere; the given point P is called the *center* of the sphere (FIG. 6-53). If A is any point of the sphere, then the segment $\overline{PA}$ also is called a *radius* of the sphere. Any segment through the center and having endpoints in the sphere is a *diameter* of the sphere. The two endpoints of the diameter are called *antipodal points* of the sphere.

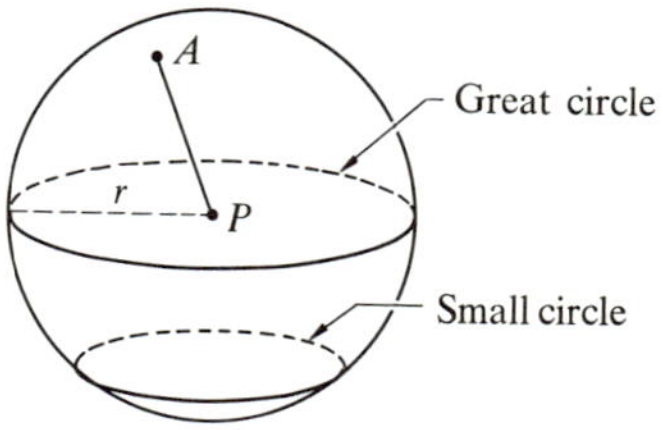

FIGURE 6-53

2. A *great circle* on a sphere is any intersection of the sphere with a plane through the center of the sphere. All circles on a sphere which are not great circles are called *small circles.*

3. A plane is *tangent* to a sphere if it intersects the sphere in exactly one point (FIG. 6-54).

4. A point is in the *interior of a sphere* if its distance from the center of the sphere is less than the radius of the sphere. A point is in the *exterior of a sphere* if its distance from the center of the sphere is greater than the radius of the sphere.[13]

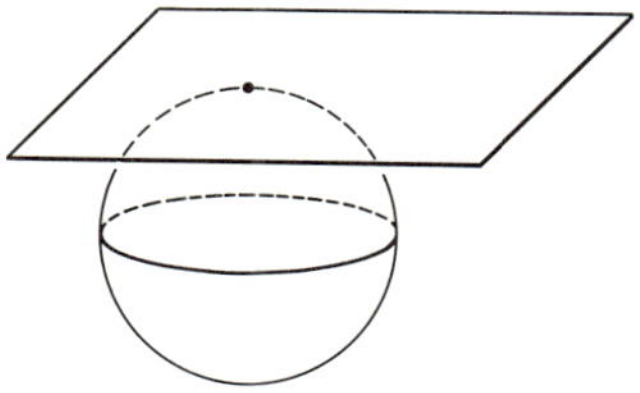

FIGURE 6-54

Theorems About Spheres

THEOREM 6-35 *The diameter of a sphere is twice its radius.*

THEOREM 6-36 *All radii of the same sphere are congruent.*

THEOREM 6-37 *All diameters of the same sphere are congruent.*

THEOREM 6-38 *Every section of a sphere made by a plane that intersects the sphere but is not tangent to it is a circle.*

THEOREM 6-39 *Circles of a sphere made by planes at equal distances from the center of the sphere have equal radii.*

THEOREM 6-40 *The planes of circles with equal radii on a sphere are equally distant from the center of the sphere.*

THEOREM 6-41 *A plane that is perpendicular to a radius of a sphere at its endpoint on the sphere is tangent to the sphere.*

THEOREM 6-42 *If a plane intersects a sphere in more than one point, then the intersection of the plane and the sphere is a circle.*

THEOREM 6-43 *Through any two points on a sphere that are not antipodal points there is one and only one great circle on the sphere.*

THEOREM 6-44 *Through any two antipodal points there are infinitely many great circles on the sphere.*

[13] 1. The sphere as just defined is a *surface*, not a solid. There are occasions, however, when it is convenient to regard a sphere as a solid. Then the sphere consists of the set of points described above and its interior. We call this set of points a *closed spherical region*.

2. Great circles are so named because they are the circles of greatest radius that lie on the sphere.

3. Note the similarity between the definition and terminology of a circle and a sphere. The difference between them is that in the case of a circle we are concerned only with points on a plane, whereas with a sphere we are concerned with points in space.

4. The shape of the earth is *almost* a sphere. The earth is slightly flattened at the poles. The length of the equator is 24,902 miles; the length of a "great circle" (really an ellipse) through the poles is 24,860 miles.

TEACHING SUGGESTIONS AND ACTIVITIES

VISUAL AIDS

Visual 1

PURPOSE

To provide visual representations of common plane and space figures.

CONSISTS OF

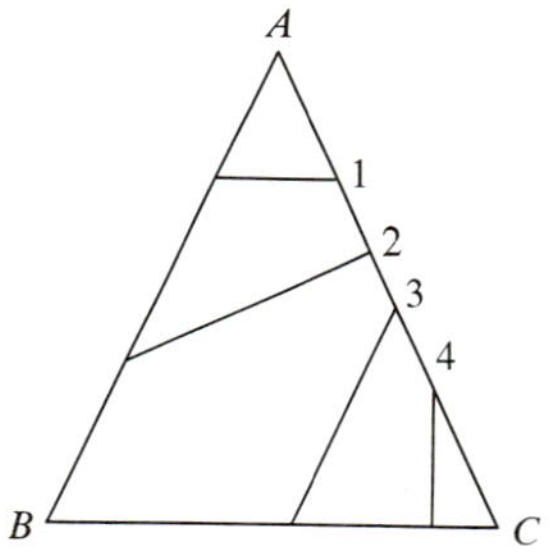

FIGURE 6-55

1. Plane-figure cutouts (with heavy outline of boundaries): five triangles (right, acute, obtuse, isosceles, equilateral); quadrilateral; pentagon; hexagon; octagon; decagon; parallelogram; rectangle; square; trapezoid; rhombus; circle.
2. Space figures: rectangular prism; triangular prism; pentagonal prism; hexagonal prism; oblique prism; cube; triangular pyramid; square pyramid; pentagonal pyramid; oblique pyramid; cylinder (noncircular); circular cylinder; oblique cylinder; cone; sphere.

HOW TO USE

1. Use with Introducing the Chapter to Students and as needed with Content Review.
2. With plane figures stress the difference between the figure and its interior, between the curve and the surface; with space figures stress the difference between the surface and the solid.

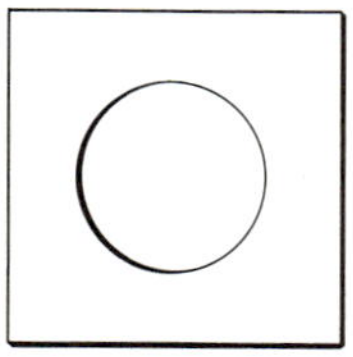

FIGURE 6-56

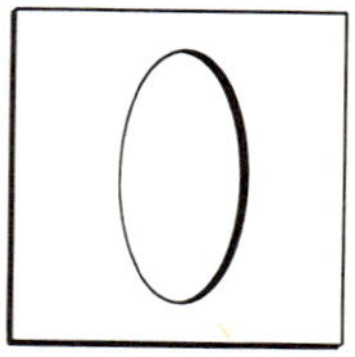

FIGURE 6-57

Visual 2

PURPOSE

To show that when a cone is cut by a variety of planes, the result is a group of curves called conic sections, that is, the circle, ellipse, parabola, and hyperbola.

CONSISTS OF

1. Cone (from Visual 1).
2. Four planes, each with a different cutout, as shown in FIG. 6-55.

   a. Plane 1: makes a cut parallel to the base of the cone. Result: *circle* (FIG. 6-56).
   b. Plane 2: makes a cut *not* perpendicular to the base of the cone and not cutting the base of the cone. Result: *ellipse* (FIG. 6-57).
   c. Plane 3: makes a cut parallel to just $\overline{AB}$ or just $\overline{AC}$. Result: part of *parabola* (FIG. 6-58).
   d. Plane 4: makes a cut perpendicular to the base of the cone. Result: part of *hyperbola* (FIG. 6-59).

Note: A section through the vertex of the cone will form a pair of lines, one line, or just the vertex.

HOW TO USE

Slip each plane over the cone. Let the class see the resulting curve.

Visual 3 (For Student Use)

PURPOSE

To provide enrichment material in the form of "jigsaw" puzzles with geometric figures. Such puzzles are called *dissection problems*.

CONSISTS OF

1. A square cut up into four pieces as shown in FIG. 6-60 (painted blue).
2. A square cut up into five pieces as shown in FIG. 6-61 (painted green).
3. A square cut up into six pieces as shown in FIG. 6-62 (painted silver).

HOW TO USE

1. Scatter pieces of each puzzle.
2. Ask the students to arrange pieces of each puzzle into the shape of a square.
3. Solutions are shown above.

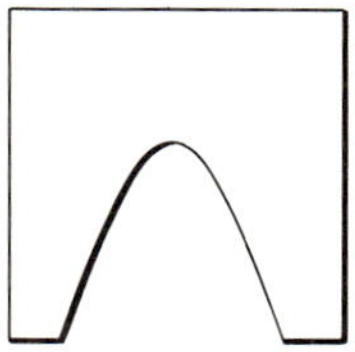

FIGURE 6-58

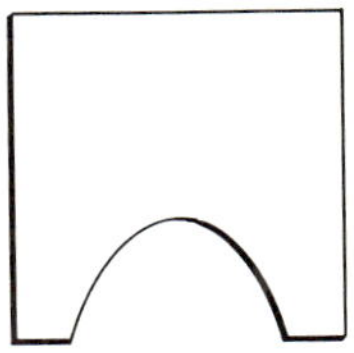

FIGURE 6-59

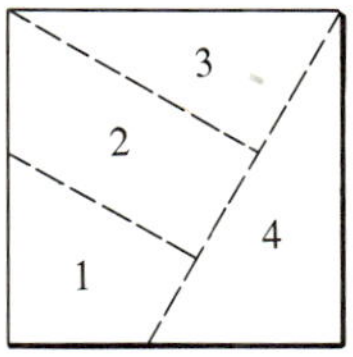

FIGURE 6-60

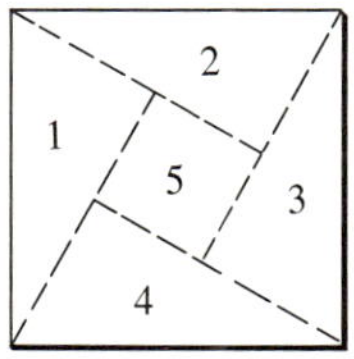

FIGURE 6-61

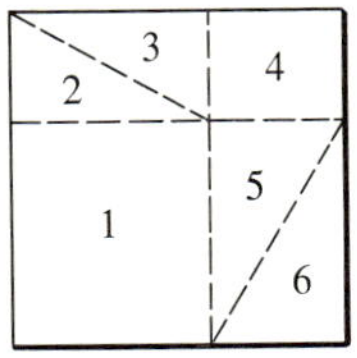

FIGURE 6-62

Visual 4

Note: This visual consists of two commercially available sets of manipulative materials and devices that can enrich the study of geometric figures.

PURPOSE

1. To provide means for building a large variety of plane and space figures.
2. To provide means for drawing a great variety of geometric shapes, designs, and curves.

CONSISTS OF

1. Geo D Stix: 350 color-coded vinyl sticks, in various lengths, that fit into flexible plastic connectors.
2. Spirograph: transparent plastic wheels, rings, and racks; different-colored pens; baseboard; pack of paper; illustrated pattern booklet.

HOW TO USE

1. Use the Geo D Stix to build plane and space figures.
2. Use the Spirograph to draw a variety of geometric shapes, designs, and curves.
3. Explore some of the mathematics behind the curves in 2 and behind the method used in their construction.

USING THE COMPASS

In work with geometric figures it is essential that students learn how to use the compass correctly, first when drawing circles and, later, in constructions. In their early experience with the compass, stress

1. The need for a sharp pencil in the compass.
2. Not to press the point of the compass too hard into the paper.
3. Not to alter the distance between the point and the pencil tip while drawing a circle.
4. That a circle is drawn in *one* direction without pause, not first in one and then in the other direction.
5. To grasp the compass at the top.

QUESTIONS

1. Here is a list of geometric figures: rectangle, parallelogram, rhombus, square, trapezoid. Below is a list of properties. Next to each property write the names of *all* the figures (listed above) about which the property is true.

   a. The opposite sides are congruent.
   b. The diagonals bisect each other.
   c. The diagonals are congruent.
   d. The angles are all right angles.
   e. The diagonals are perpendicular to each other.
   f. At least one pair of opposite sides is parallel.
   g. All the sides are congruent.
   h. At least one angle is a right angle.
   i. Any two consecutive angles are supplementary.

2. a. The intersection of a sphere with a plane passing through the center of the sphere is called____________________.

   b. The intersection of a sphere with a plane not passing through the center of the sphere is called ________________.

3. If the midpoints of the sides of a rectangle are joined in order, will the figure formed be (a) another rectangle? (b) a square? (c) a rhombus? (d) a trapezoid?

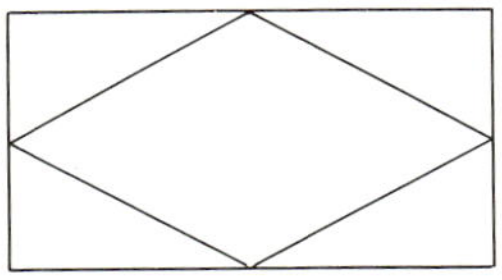

4. Why do the diagonals of a square form four isosceles triangles?

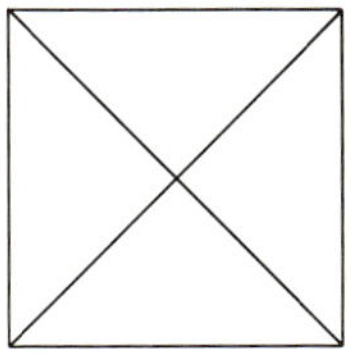

5. Prove that the diagonals of a rhombus are perpendicular bisectors of each other.

6. a. What is the relationship between the number of diagonals and the number of vertices of a polygon?

b. Does this relationship hold for a convex as well as a concave polygon?

7. In trapezoid $ABCD$, $AB = CD$ and $\overline{BC} \parallel \overline{AD}$. Prove that $\angle A \cong \angle D$.

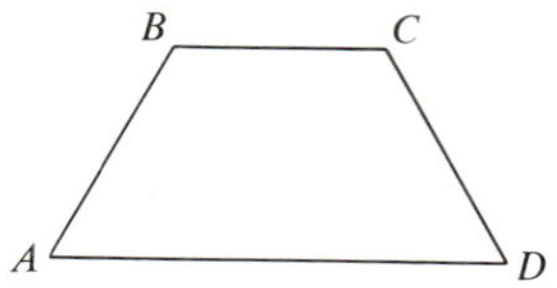

8. Imagine a 3-inch cube painted blue.

   a. How many cuts will be required to divide the cube into 1-inch cubes?
   b. How many 1-inch cubes would there be altogether?
   c. How many of these cubes would have *four* blue faces?
   d. How many would have *three* blue faces?
   e. How many would have *two* blue faces?
   f. How many would have *one* blue face?
   g. How many would have *no* blue faces?
   h. Into how many cubes, other than 1-inch cubes, can the 3-inch cube be cut? (1) How many of these cubes would have four blue faces? (2) How many would have three blue faces? Two? One? None?

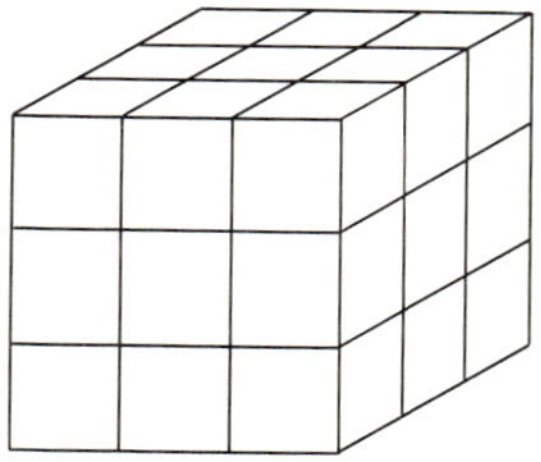

9. Show that the sum of the measures of the angles of a convex polygon of n sides is $S = (n - 2) \cdot 180$. *Hint*: Divide the polygon into a set of triangles.

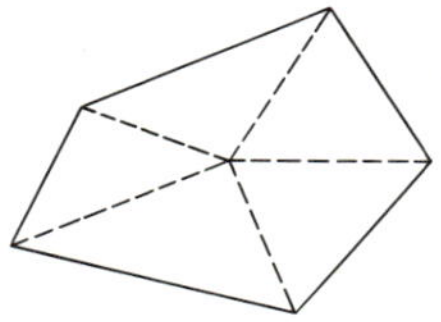

10. Prove that medians $\overline{CD}$ and $\overline{AE}$ of $\triangle ABC$ do not bisect each other.

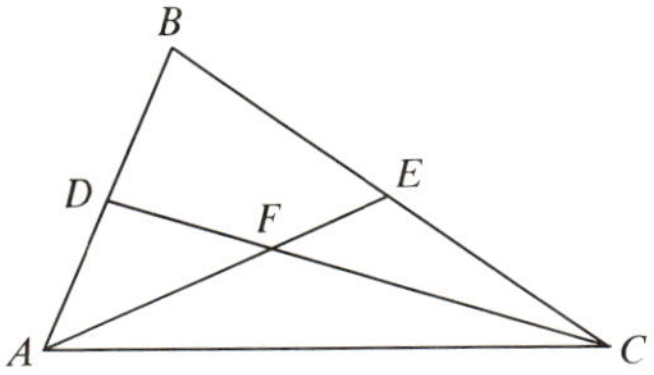

11. Consider a block of wood whose surface is an octahedron. If each of the corners is sawed off, how many vertices, faces, and edges are there in the new polyhedron?

12. a. Define a prism in terms of a cylinder.

    b. Define a pyramid in terms of a cone.

13. If $m\widehat{AC} = m\widehat{BC}$, prove that $AB < 2(AC)$.

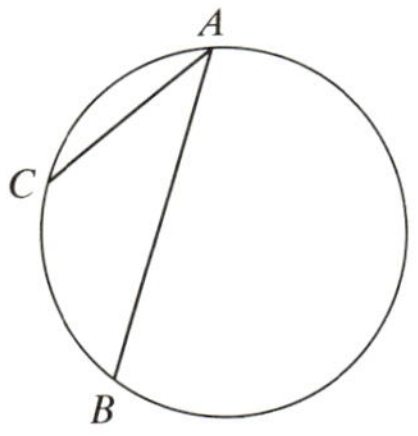

14. True–false. Give a reason for your answer.

    a. If a line intersects a circle at one point, then it intersects the circle at two points.
    b. All great circles of a sphere intersect.
    c. A cylindrical surface can have two skew lines as two of its elements.
    d. A right prism with square lateral faces must be a cube.
    e. All tetrahedra are also pyramids.
    f. Every square is a rhombus.
    g. If the vertices of a quadrilateral are on a circle, then any two opposite angles are supplementary.

SOLUTIONS

1.

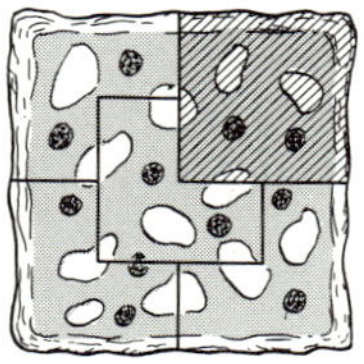

2.

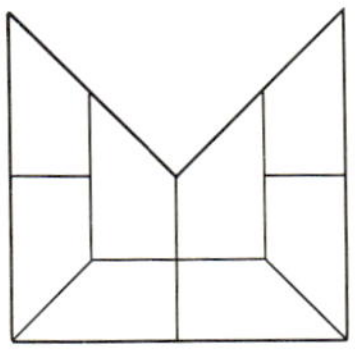

3. The coins cannot be so arranged.

4. The solution for each dissection is on page 183.

PUZZLES

1. A pizza pie in the shape of a square has a quarter of the pie missing. How can you divide what is left of the pie into four pieces that have the same size and shape?

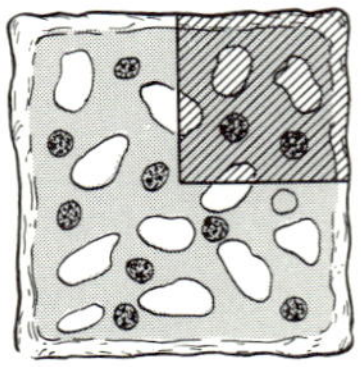

2. Divide the figure shown into eight parts such that each part has the same size and shape.

3. Can you arrange four coins, all the same size, on a table so that they are all the same distance apart?

4. Use Visual 3 as a set of dissection puzzles. The object of each of these puzzles is to arrange the pieces so as to form a square.

ENRICHMENT

ANOTHER VIEW OF CONIC SECTIONS

Although we have defined conic sections by the way in which a plane intersects a cone, there are more useful ways of defining these curves —definitions that make no reference to the cone.

For instance, the Greeks found one property that is common to the parabola, ellipse, and hyperbola. They found that for each of these curves there exists a certain *line* on the plane, and a certain *point* on the plane not on the line, such that for every point on the curve the ratio of its distance from the fixed point to its distance from the fixed line remains constant.

The fixed point is called a *focus*.

The fixed line is called a *directrix*.

The fixed ratio is called the *eccentricity* (FIG. 6-63):

$$\text{eccentricity} = \frac{m}{n}.$$

A *parabola* is defined as a curve where $m/n = 1$.
An *ellipse* is defined as a curve where $m/n < 1$.
A *hyperbola* is defined as a curve where $m/n > 1$.

More conventional definitions of the conic sections, which also make no reference to the cone, are:

1. A *parabola* is the set of points in a plane such that every point is equally distant from a fixed point F and a fixed line l (FIG. 6-64). The fixed point is called the *focus*, and the fixed line is called the *directrix*.
2. An *ellipse* is the set of all points in a plane such that for each point the *sum* of its distances from two fixed points F and F' in the plane is constant (FIG. 6-65). F and F' are called the *foci* of the ellipse. If the two foci coincide, the figure is a circle.
3. A *hyperbola* is the set of all points in a plane such that for each point the *difference* of its distances from two fixed points F and F' is constant (FIG. 6-66). F and F' are called the *foci* of the hyperbola.

These definitions enable us to represent each of the curves by an algebraic equation. By this approach we learn much more about the properties and applications of the conic sections than when we merely think of them in terms of a cone. (See Chapter 13, pp. 377–382.)

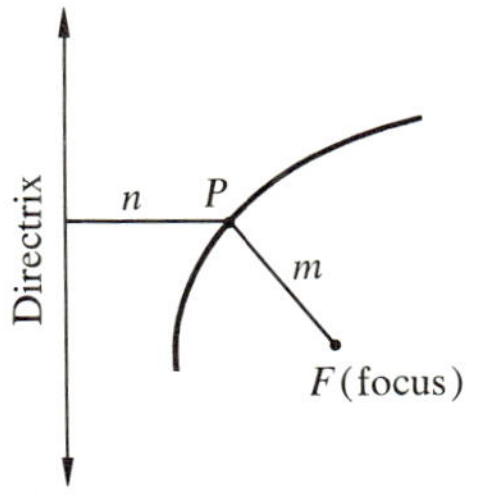

FIGURE 6-63

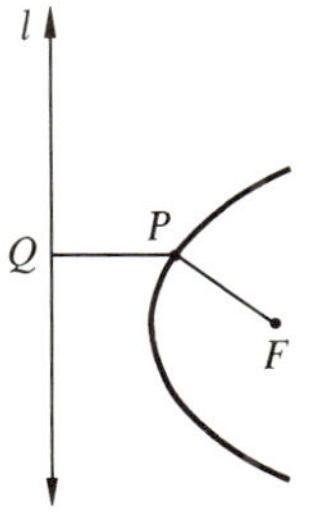

$FP = QP$

FIGURE 6-64

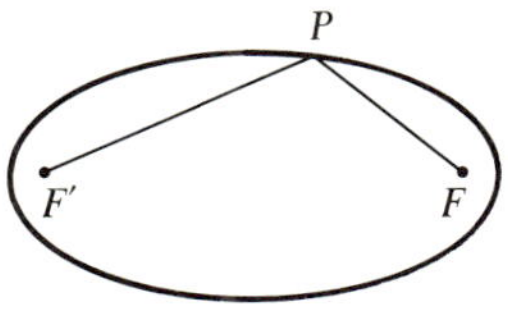

$F'P + PF =$ a constant

FIGURE 6-65

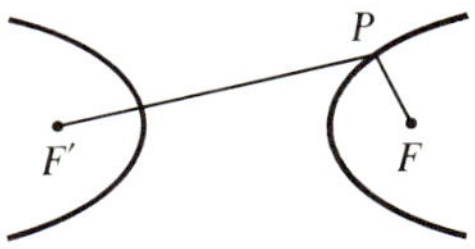

$F'P - FP =$ a constant

FIGURE 6-66

APPLICATIONS OF CONIC SECTIONS

The conic sections are not only interesting in themselves, but they have some very important applications.

Parabola

A projectile, fired into space, follows a path that is approximately a parabola. For the path to be parabolic we assume (1) that the earth is planar, (2) no resistance, and (3) gravitational pull independent of height of projectile from earth.

Cables of a suspension bridge hang in parabolic arcs if the weight of the bridge is uniformly distributed. A reflecting surface in the shape of a parabola of revolution reflects light from the focus in the direction of the axis. This idea is put to use in automobile headlights, in telescopes, and in radar. The same principle, applied to sound waves, is used in listening and broadcasting devices. Parabolic reflectors with a microphone at the focus catch distant sounds and "focus" them on the microphone.

The Greeks discovered how the parabola can be used to control light. Archimedes was said to have put to use this discovery by constructing a huge parabolic mirror which focused the sun's rays on the Roman ships besieging Syracuse and so set them on fire.

Ellipse

The planets have elliptic paths with the sun at a focus. Gears are sometimes made in the shape of an ellipse rather than a circle. This is done when a slow, powerful stroke followed by a short, quick return is needed.

Hyperbola

A very interesting and useful application of the hyperbola is that of locating the place from which a sound, as of a gun G, originates.

If we know the difference in the times at which the sound reaches two listening posts, P_1 and P_2, we can determine the difference between the distances of the posts from the gun G (FIG. 6-67). We then know that the gun is located on a branch of a hyperbola of which the posts are foci.

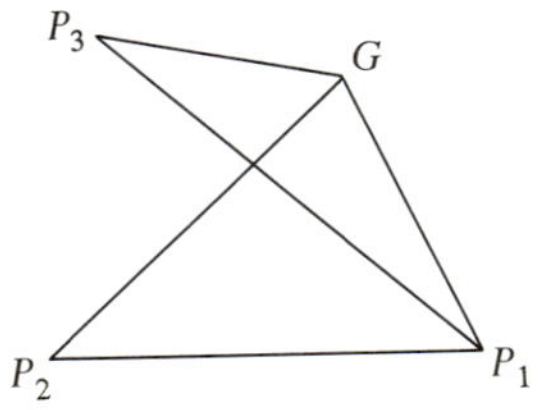

FIGURE 6-67

By using a third listening post, P_3, either of the first two posts and the third are now foci of a branch of another hyperbola on which the gun is located. The gun is at the intersection of the two branches.

The conic sections were discovered by the Greeks in their quest for solutions to the three famous construction problems of antiquity: how, by the use of compass and straightedge only,

1. To trisect any angle.
2. To construct a square equal in area to a given circle.
3. To construct the side of a cube whose volume is twice that of a cube of given side.

MORE ON THE SPHERE

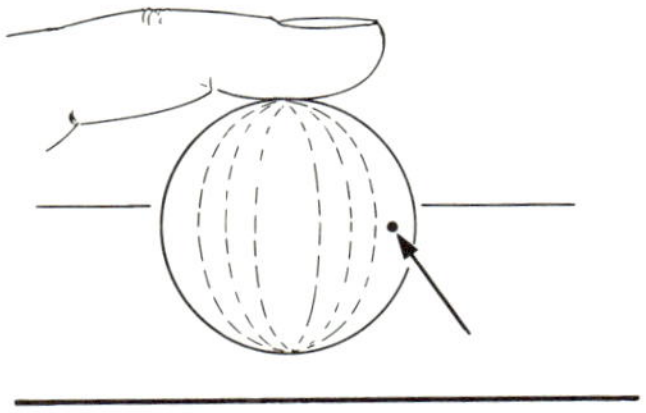

FIGURE 6-68

An interesting way for students to see how a sphere can be generated is to stand a coin in an upright position by pressing it lightly at the top with a finger (FIG. 6-68). Then strike one edge of the coin sharply so as to make the coin rotate. This will generate a sphere.

A right circular cylinder can be generated in a similar way if we use a rectangular disc instead of a circular disc or coin.

In the design of spacecraft that could survive the radiation of the earth-encircling Van Allen belts, the geometry of the sphere has become important. The Van Allen belts are composed of high-energy atomic particles that are trapped by the earth's magnetic field. These particles travel at speeds close to the speed of light from about 500 miles to 40,000 miles above the earth. The survival of high-flying communications and surveillance satellites is dependent primarily on the extent of the damage inflicted on their delicate electronic systems by the absorption of this intense radiation. Recent tests have shown that the life of such a spacecraft can be greatly prolonged if it were built in the shape of a large sphere about 10 feet in diameter. A spherical surface deflects and scatters these radiation particles because they strike such a surface at wide angles, thereby decreasing the amount of absorption.

PLATONIC SOLIDS

We have seen that a *polygon* is a planar simple closed curve which is a union of three or more line segments. If all the sides of the polygon are congruent and all its angles are congruent, the polygon is called a *regular polygon.*

The space figures that correspond to polygons are called *polyhedra.* Whereas the boundary of a polygon is a union of line segments, the surface of a polyhedron is a union of polygons, called faces. And whereas a polygon contains only plane angles, a polyhedron also contains dihedral and polyhedral angles. Prisms and pyramids are examples of polyhedra. The triangle is the simplest polygon, and the tetrahedron is the simplest polyhedron.

A *regular polyhedron* is a polyhedron all of whose faces are congruent regular polygons and all of whose polyhedral angles are congruent. An example of a regular polyhedron is a cube.

A *simple polyhedron* is one that is topologically equivalent to a sphere; that is, a polyhedron is simple if by stretching each face we can deform the polyhedron into a sphere.

One of the most interesting theorems related to polyhedra is the famous Euler–Descartes formula, which establishes a simple relationship among the number of vertices (V), edges (E), and faces (F) of any simple polyhedron: $V - E + F = 2$; that is, the sum of the number of vertices and faces of any simple polyhedron is two more than the number of edges of the polyhedron.

Only Five Possible Regular Polyhedra

Although there are infinitely many regular polygons, there are only five different regular polyhedra. Because Plato recognized this in his *Dialogues*, the five regular polyhedra have been called the *Platonic solids*.

We can use Euler's formula to prove that the only possible regular polyhedra are those with 4, 6, 8, 12, or 20 faces.

Proof

Assume that a regular polyhedron has F faces, each face being a regular n-sided polygon, and that r of the E edges of the polyhedron meet at each vertex.

1. Since each face has n edges and each edge belongs to two faces,

   $$nF = 2E$$

   or

   $$F = \frac{2E}{n}. \qquad (6\text{-}1)$$

2. Since each vertex is on r edges and each edge joins two vertices,

   $$rV = 2E$$

or

$$V = \frac{2E}{r}. \tag{6-2}$$

3. Substitnting these values for F and V in Euler's formula, we get

$$V - E + F = 2,$$

$$\frac{2E}{r} - E + \frac{2R}{n} = 2,$$

or

$$\frac{1}{r} + \frac{1}{n} = \frac{1}{2} + \frac{1}{E}. \tag{6-3}$$

4. Since a polygon must have at least three sides, and at least three edges meet at each polyhedral angle, we need an integral solution to equation (6-3) with $n \geq 3$ and $r \geq 3$.

5. When $n = 3$, equation (6-3) becomes

$$\frac{1}{r} - \frac{1}{6} = \frac{1}{E}. \tag{6-4}$$

Since E is positive, r can only be 3, 4, or 5. Therefore, when $n = 3$, we have the values in Table 1.

Table 1

r	3	4	5
E	6	12	30
F	4	8	20

6. For $n > 3$, r must be 3, since E in equation (6-3) must be positive. Equation (6-3) then becomes

$$\frac{1}{n} - \frac{1}{6} = \frac{1}{E},$$

and n can only be 4 or 5. When $n > 3$, and $r = 3$, we have the values in Table 2.

Table 2

n	4	5
E	12	30
F	6	12

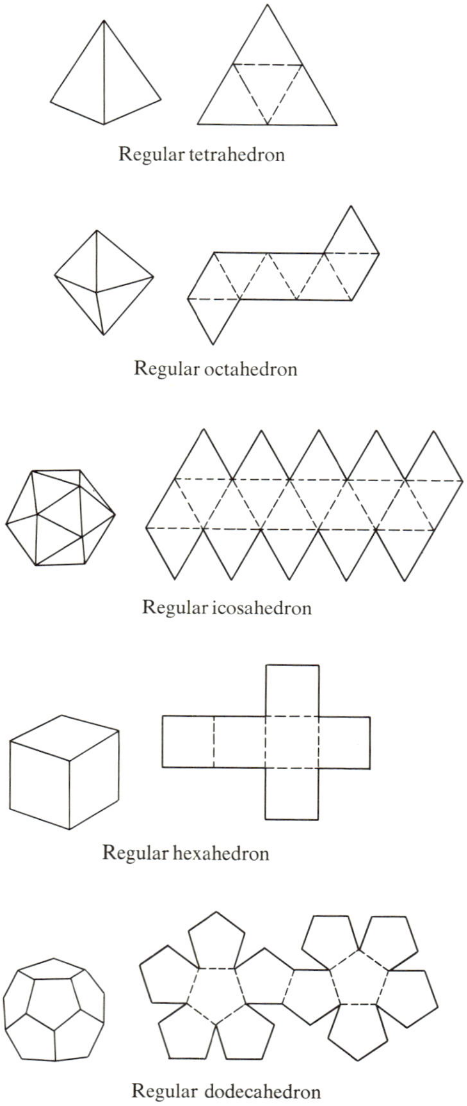

FIGURE 6-69

We have now shown that any regular polyhedron can be bounded only by triangles, squares, or pentagons, and that there are only the following five possible polyhedra:

1. The *tetrahedron*, with 4 triangular faces.
2. The *hexahedron* (cube), with 6 square faces.
3. The *octahedron*, with 8 triangular faces.
4. The *dodecahedron*, with 12 pentagonal faces.
5. The *icosahedron*, with 20 triangular faces.

Construction of the Regular Polyhedra

The regular polyhedra can be constructed by drawing on cardboard the patterns shown in FIG. 6-69 next to each pictured polyhedron. Cut out the patterns along the heavy lines and then fold them along the dashed lines, using adhesive tape to join the faces which intersect. Each pattern will fold into the corresponding polyhedron.

Physical representations of the tetrahedron, hexahedron, and octahedron can be found in crystals. Skeletons of microscopic sea animals called Radiolaria have been found in the shape of dodecahedra and icosahedra.

Surprise

It may be surprising to note that if a cube and a regular octahedron are inscribed in the same sphere, the cube has the greater volume; and if a regular dodecahedron and a regular icosahedron are inscribed in the same sphere, the dodecahedron has the greater volume.

Crime and Punishment

The history of the Platonic solids is steeped in mysticism and legend. One legend has it that the Pythagorean Hippasus was drowned in a shipwreck as punishment for claiming that *he* discovered the regular dodecahedron, instead of attributing the discovery, out of reverence, to his master Pythagoras.

Kepler associated the tetrahedron with fire, the hexahedron with the earth, the octahedron with air, the icosahedron with water, and the dodecahedron with the signs of the zodiac. He also believed that the Platonic solids accounted for the number of planets (only five were known at the time) and their location with respect to the sun.

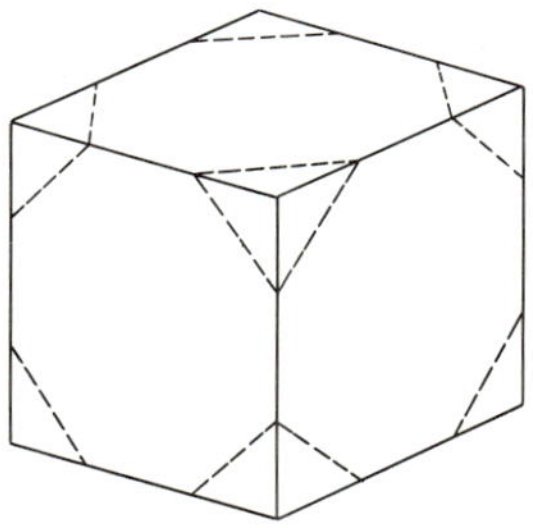

FIGURE 6-70

Archimedean Solids

We defined a regular polyhedron as one that satisfies three conditions: (1) Its polyhedral angles are congruent, (2) its faces are regular polygons, and (3) its faces are congruent.

In his study of polyhedra, Archimedes (287–212 B.C.) concerned himself with semiregular solids now known as *Archimedean solids*. An Archimedean solid is a polyhedron that satisfies only conditions (1) and (2) but not condition (3), that is, a polyhedron whose polyhedral angles are congruent, and whose faces are bounded by regular but not congruent polygons.

For example, if we properly cut off the eight corners of a cube, the resulting figure is a semiregular or Archimedean solid with faces made up of eight equilateral triangles and six regular octagons (FIG. 6-70).

Archimedes discovered 13 semiregular solids, and about 1800 years later Kepler showed how to construct them.

7

THE PYTHAGOREAN THEOREM

INTRODUCING THE CHAPTER TO STUDENTS

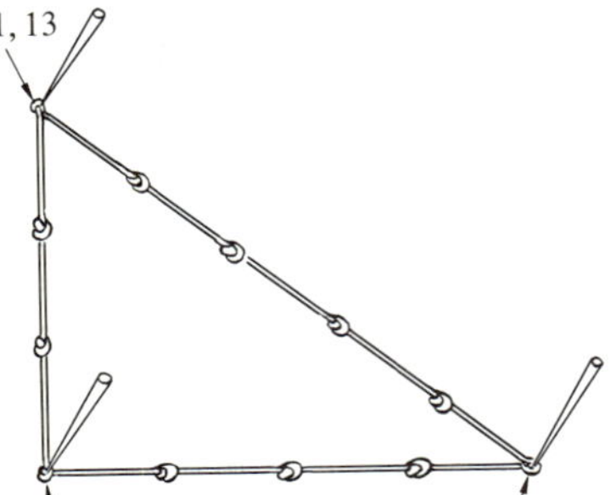

FIGURE 7-1

In ancient Egypt, every spring, the waters of the Nile River overflowed their banks and washed out the boundaries that separated one man's property from his neighbor's. So every year, after the floods, the Egyptians were faced with the problem of reestablishing their boundaries. But to do this they had to construct right angles, and the construction of right angles was not well understood in those days. So whenever the washed-out boundaries had to be redrawn, special consultants known as "rope stretchers" had to be called in.

(See Visual 1.) These specialists took a rope and tied 13 knots in it in such a way as to get 12 equal spaces (FIG. 7-1). They then drove stakes into the ground in this manner: A stake was driven through the first and thirteenth knots. Another stake was then driven through the fourth knot. A third stake was driven through the eighth knot. In each case the rope was stretched as tightly as possible before the stake was driven into the ground. They found that the angle opposite the side containing the 5 spaces turned out to be a right angle. The rope stretchers did not understand why stretching a rope in this manner resulted in a right angle, but they knew it worked and so were perfectly happy.

At about the same time in India, the Hindus also needed to construct right angles. They went a step further than the Egyptians. They discovered that in addition to the number combination (3, 4, 5) forming a right angle, there were other combinations that also resulted in a right angle: (5, 12, 13), (15, 20, 25), (12, 35, 37), (8, 15, 17), (12, 16, 20), (15, 36, 39). But like the Egyptians, the Hindus did not understand why these number combinations yielded a right angle.

According to tradition, the answer to this question was finally found about 2500 years ago by the Greek mathematician Pythagoras. His important discovery has been called ever since the Pythagorean Theorem.

CONTENT REVIEW

PREREQUISITES FOR LEARNING THIS CHAPTER

1. Knowing that
   a. The area of a square region can be found by multiplying the length of a side by itself; that is, $A = s \times s$, or $A = s^2$.
   b. The area of a triangular region is half the product of the length of any base and the altitude to that base; that is, $A = \frac{1}{2}ab$.
2. Being able to find the square root of a number.

THE PYTHAGOREAN THEOREM AND ITS CONVERSE[1]

[1] The Pythagorean theorem gives us a way of finding the length of any side of a right triangle if we know the lengths of the other two sides. We use the converse of the theorem to test whether or not a triangle is, in fact, a right triangle.

Before stating the theorem, we shall define the parts of a right triangle.

Definitions
A *right triangle* is a triangle that has one right angle. The side opposite the right angle is called the *hypotenuse* (FIG. 7-2).

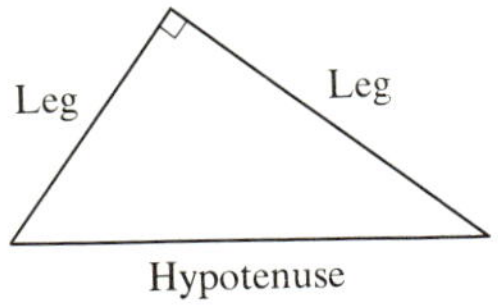

FIGURE 7-2

The sides that include the right angle are called the *legs*.

THEOREM 7-1 (Pythagorean Theorem) *In any right triangle, the square of the length of the hypotenuse is equal to the sum of the squares of the lengths of the legs* (FIG. 7-3).

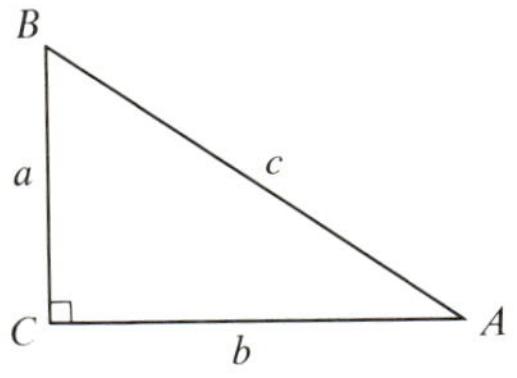

FIGURE 7-3

That is, if $\triangle ABC$ is a right triangle with right angle at C, then $c^2 = a^2 + b^2$.[2]

Proof

Given a right triangle $\triangle ABC$, with legs of lengths a, b and hypotenuse of length c (FIG. 7-4).

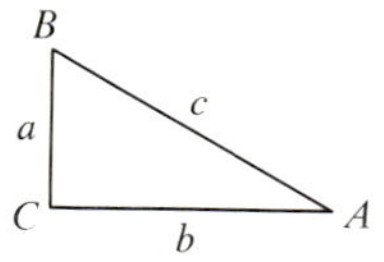

FIGURE 7-4

1. Let S be a square with side of length $a + b$, containing four right triangles, each congruent to $\triangle ABC$ (FIG. 7-5). Then

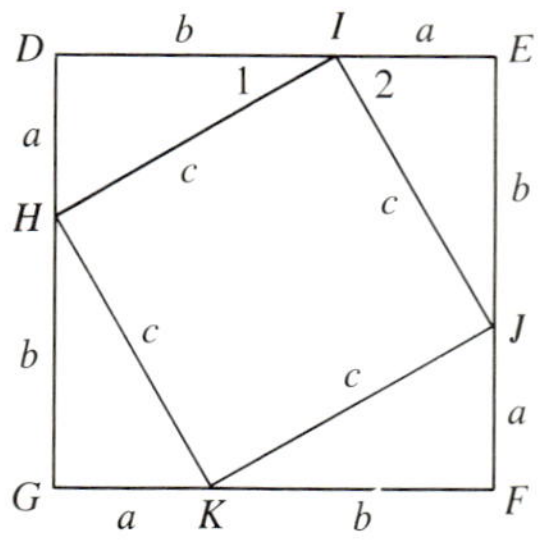

FIGURE 7-5

2. $\angle HIJ$ is a right angle because $\angle 1$ and $\angle 2$ are complementary.
3. For the same reason $\angle IJK$, $\angle JKH$, and $\angle KHI$ are right angles.
4. Therefore, quadrilateral $HIJK$ is a square.
5. By the Addition Postulate,

   area of $DEFG$ = area of $HIJK$ + $4 \times$ area of $\triangle DIH$.

6. That is,

$$(a + b)^2 = c^2 + 4 \times \tfrac{1}{2}ab,$$
$$a^2 + 2ab + b^2 = c^2 + 2ab,$$
$$a^2 + b^2 = c^2.$$

[2] The following is one of hundreds of proofs that have been given for this theorem. The proof given here uses the concept of area (which will be discussed in Chapter 11) and the Addition Postulate for area; that is, if two polygonal regions intersect only in edges and vertices (or do not intersect at all), then the area of their sum is the sum of their areas. In FIG. 7A,

$$\text{area of } ABCD\ (R_1 + R_2 + R_3 + R_4 + R_5)$$
$$= \text{area of } R_1 + \text{area of } R_2 + \text{area of } R_3$$
$$+ \text{area of } R_4 + \text{area of } R_5.$$

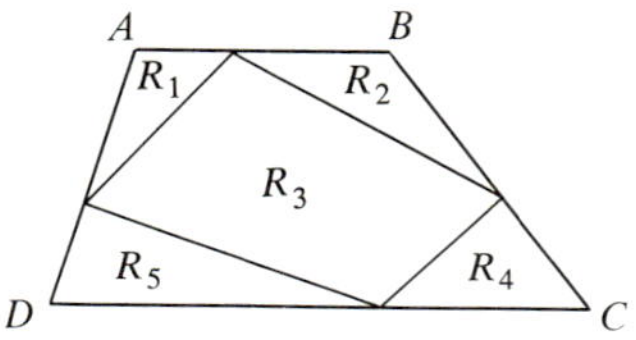

FIGURE 7A

THEOREM 7-2 (Converse of the Pythagorean Theorem) If the square of the length of one side of a triangle is equal to the sum of the squares of the lengths of the other two sides, then the triangle is a right triangle whose right angle is opposite the longest side (FIG. 7-6).

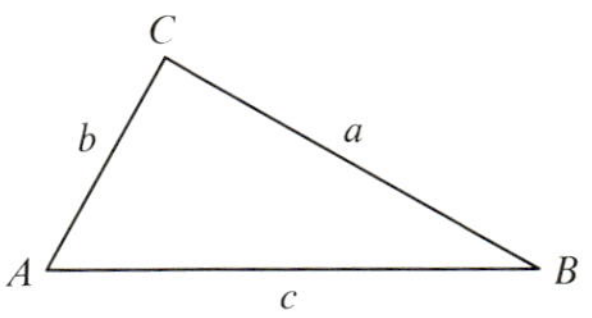

FIGURE 7-6

That is, if in $\triangle ABC$, $c^2 = a^2 + b^2$, then $\triangle ABC$ is a right triangle with its right angle at C.

Proof

1. Let $\angle F$ be a right angle, and let D and E be the points on the sides of $\angle F$ such that $FE = a$ and $FD = b$ (FIG. 7-7).

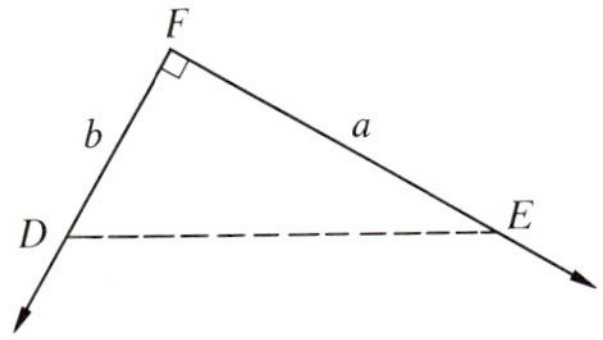

FIGURE 7-7

2. By the Pythagorean Theorem,

$$(DE)^2 = a^2 + b^2 = c^2,$$

$$DE = c.$$

3. $\triangle ABC \cong \triangle DEF$ by SSS.
4. Therefore, $\triangle ABC$ is a right triangle with its right angle at C.[3]

TWO SPECIAL RIGHT TRIANGLES[4]

THEOREM 7-3 (30–60 Triangle Theorem) In a right triangle whose acute angles have measures of 30 *and* 60, *the hypotenuse*

[3] 1. As a consequence of Theorem 7-2 we can test whether a triangle is a right triangle if we know the lengths of its three sides. We substitute the values of the lengths in $c^2 = a^2 + b^2$, where c is the largest of the three sides. If the three lengths satisfy this relationship, then they form a right triangle.

2. Students should learn all forms of the Pythagorean Theorem:

$c^2 = a^2 + b^2$, $a^2 = c^2 - b^2$,

$b^2 = c^2 - a^2$,

and remember that the theorem applies only to a right triangle.

[4] There are two types of right triangles that are particularly useful (FIG. 7B):

1. The 30-60-90 triangle.
2. The 45-45-90 triangle (isosceles right triangle).

In each of these triangles, if we know the length of any *one* side we can find the lengths of the other two sides. Theorems 7-3 and 7-4 can both be proved by using the Pythagorean Theorem.

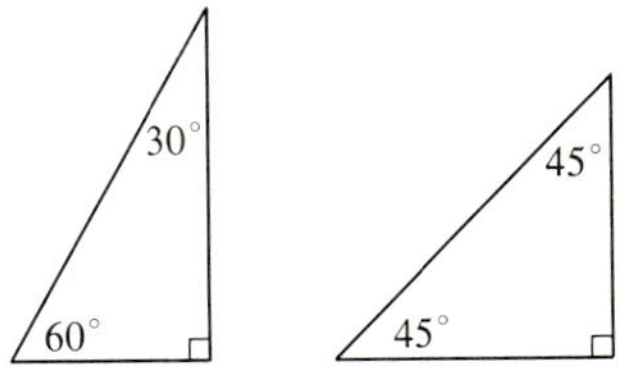

FIGURE 7B

is twice as long as the shorter leg, and the longer leg is $\sqrt{3}$ *times the shorter leg* (FIG. 7-8).

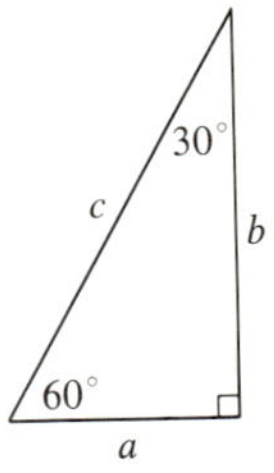

FIGURE 7-8

That is, $c = 2a$ and $b = a\sqrt{3}$.

THEOREM 7-4 (Isosceles Right Triangle Theorem) *In a right triangle whose acute angles have measures of* 45 *and* 45, *the hypotenuse is* $\sqrt{2}$ *times the length of a leg* (FIG. 7-9).

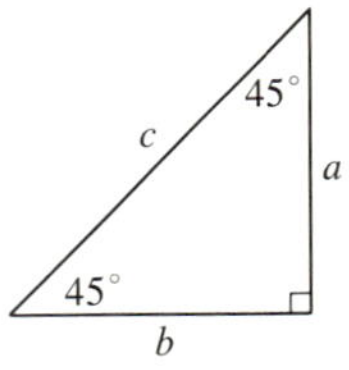

FIGURE 7-9

That is, $a = b$ and $c = a\sqrt{2}$.

THE HYPOTENUSE-LEG THEOREM

For right triangles, we have a special congruence theorem.

THEOREM 7-5 (The Hypotenuse-Leg Theorem) *If the hypotenuse and a leg of one right triangle are congruent to the hypotenuse and a leg of another, then the triangles are congruent.*

That is, given $\triangle ABC$ with a right angle at C, and $\triangle DEF$ with a right angle at F (FIG. 7-10). If $a = d$ and $c = f$, then $\triangle ABC \cong \triangle DEF$.

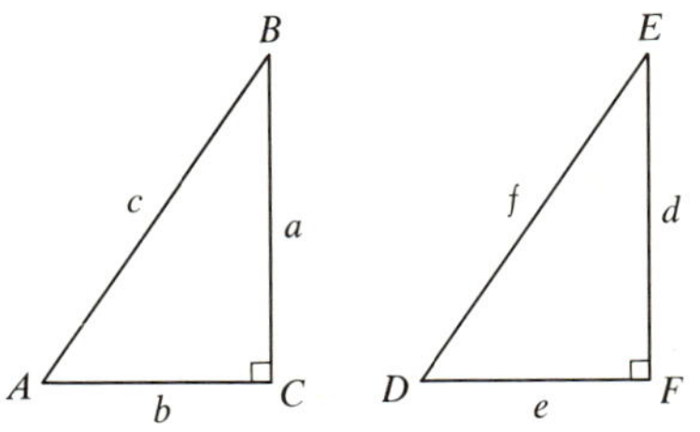

FIGURE 7-10

Proof

1. By the Pythagorean Theorem, $b = \sqrt{c^2 - a^2}$ and $e = \sqrt{f^2 - d^2}$.
2. Since $a = d$ and $c = f$, we have $b = e$.
3. Therefore, $\triangle ABC \cong \triangle DEF$ by SSS.

TEACHING SUGGESTIONS AND ACTIVITIES

VISUAL AIDS

Visual 1

PURPOSE

To show how ancient Egyptians used a knotted rope to construct right angles.

CONSISTS OF

1. Knotted rope.
2. Board.
3. "Stakes."

HOW TO USE

See Introducing the Chapter to Students.

Visual 2

PURPOSE

To provide a proof of the Pythagorean Theorem.

CONSISTS OF

1. Three squares.
2. Eight congruent triangles.

HOW TO USE

1. Arrange the parts as shown in FIG. 7-11, forming two congruent squares with side $a + b$.
2. Remove the four congruent triangles from square 1, and the four congruent triangles from square 2. This leaves FIG. 7-12.
3. Since equal areas were removed from equal areas, the resulting areas are equal; that is,

$$c^2 = a^2 + b^2.$$

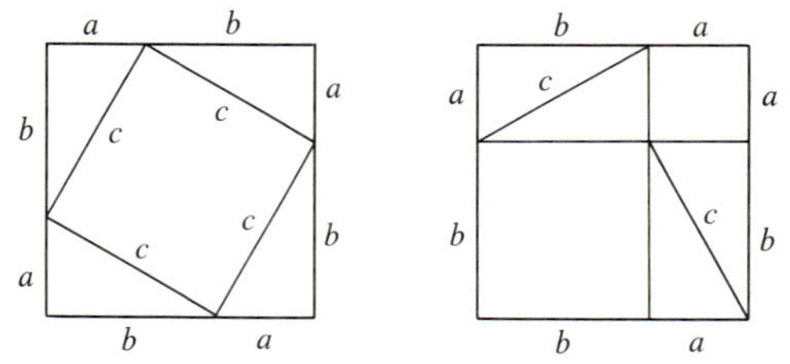

Square 1 Square 2

FIGURE 7-11

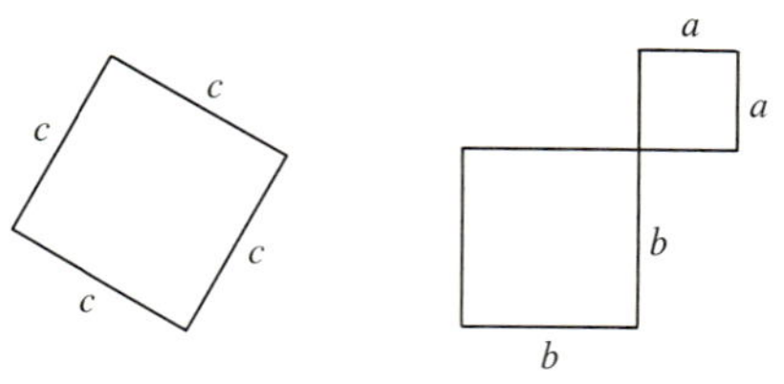

Square 1 Square 2

FIGURE 7-12

Visual 3

PURPOSE

To exhibit application of the Pythagorean Theorem to sets of similar figures other than squares.

CONSISTS OF

A right triangle (of contrasting color with other pieces in this visual); three squares; three equilateral triangles; three similar rectangles; three regular hexagons.

HOW TO USE

1. Arrange pieces as shown in FIG. 7-13.
2. In each instance indicate that the Pythagorean area relationship holds.
3. For proof, see page 210.

Visual 4

PURPOSE

A device for computing a side of a right triangle.

CONSISTS OF

A "slide rule" constructed as follows: Take two strips of stiff material and mark off each strip in tenths of an inch (FIG. 7-14). Start the scale by marking a 0 at the left. Then mark a 1, one unit to the right from 0, to represent 1^2; mark 2, four units to the right from 0, to represent 2^2; mark 3, nine units to the right from 0, to represent 3^2; etc.

HOW TO USE

1. Use like a slide rule. For example, to find the hypotenuse of a right triangle whose legs are 3 and 4,
   a. Set the 0 reading of Scale 1 over the 3 reading of Scale 2.
   b. Locate the 4 reading on Scale 1; under it on Scale 2 appears the answer 5.
2. Also use the "rule" for finding one leg, given the hypotenuse and other leg.
3. Ask the students to explain why the device "works."

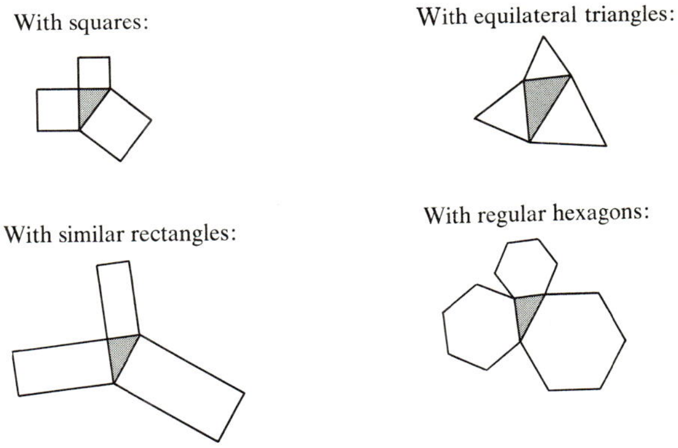

FIGURE 7-13

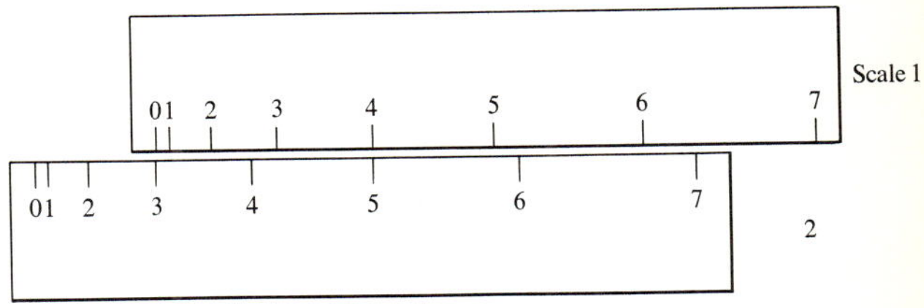

FIGURE 7-14

AREA RELATIONSHIPS FOR ANY TRIANGLE

The Pythagorean Theorem states that if squares are constructed on the three sides of any right triangle, there will exist a certain area relationship among the three squares: The area of the square on the hypotenuse will equal the sum of the areas of the squares on the legs. But what is the area relationship among three squares constructed on the sides of *any* triangle, not just a right triangle?

Students can discover an intuitive answer to this question through experimentation with squares of different sizes.

Give each student a set of squares, cut out of graph paper, with areas of 4, 9, 16, 25, 36, 49, 64, 81, 100, 121, 144, and 169 square units. Ask students to form a triangle with the sides of any three squares (FIG. 7-15).

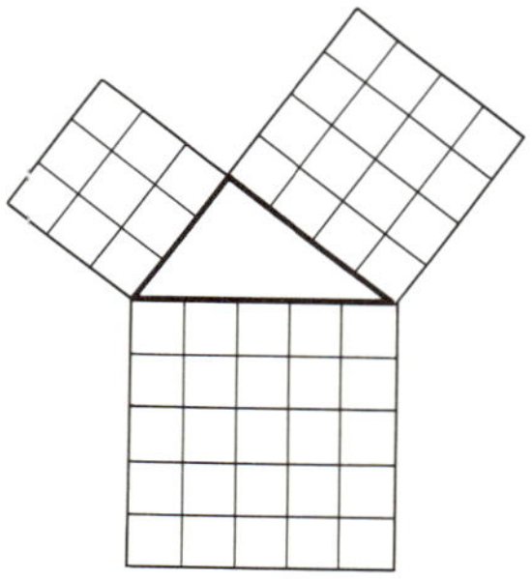

FIGURE 7-15

Each student then measures with a protractor the angles of the triangle so formed to see whether it is right, acute, or obtuse.

Class continues with this procedure, using various combinations of squares, three at a time, and forming in this way many different triangles.

Ask students to tabulate their results (FIG. 7-16).

	Areas of squares used			Type of triangle formed
①	25	144	169	Right
②	16	36	49	Acute
③	16	25	49	Obtuse
④	4	9	36	No triangle formed
⑤				
⑥				

FIGURE 7-16

Guide students to discover these relationships:

1. In an *acute* triangle the sum of the areas of the two smaller squares is *greater* than the area of the largest square; or when $a^2 + b^2 > c^2$, the triangle is acute.

2. In a *right* triangle the sum of the areas of the two smaller squares is *equal* to the area of the largest square; or when $a^2 + b^2 = c^2$, the triangle is right.

3. In an *obtuse* triangle the sum of the area of the two smaller squares is *less* than the area of the largest square; or when $a^2 + b^2 < c^2$, the triangle is obtuse.

4. When the sum of the lengths of the two smaller sides is *less than or equal to* the length of the largest side, *no triangle* is formed; or a triangle is formed only when the sum of the two sides of a triangle is greater than the third side.

QUESTIONS

1. Combine the Pythagorean Theorem and its converse into one statement
   a. Using the phrase "if and only if."
   b. Using the phrase "necessary and sufficient condition."

2. Prove that each leg of a right triangle is shorter than the hypotenuse.

3. True–false. Give a reason for your answer.

   a. If you know the length of the diagonal of a square, you can find the length of the side of the square.
   b. If you know the length of the diagonal of a rectangle, you can find the lengths of the sides of the rectangle.
   c. The sides of a right triangle can have lengths of 3, 5, and 7.
   d. Tripling the lengths of the sides of a right triangle changes the triangle to an obtuse triangle.
   e. The Pythagorean Theorem was known to the Hindus by a different name.

4. Why can the sides of three squares of equal area never form a right triangle?

5. One angle of a rhombus has a measure of 60 and one side has a length of 12. Find the length of each diagonal of the rhombus.

6. Find the length of the altitude to the hypotenuse of a right triangle in terms of the lengths of the legs.

7. Find the length of $\overline{AD}$ in the 3-inch cube shown.

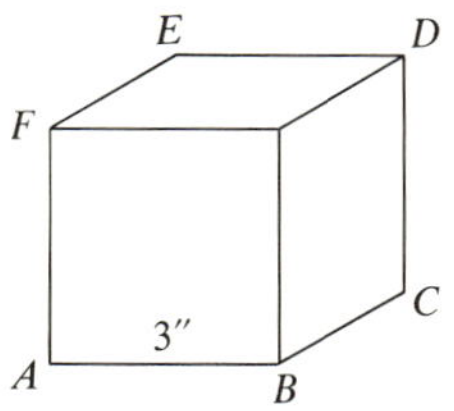

8. Find the length of $\overline{BD}$ in the triangle shown.

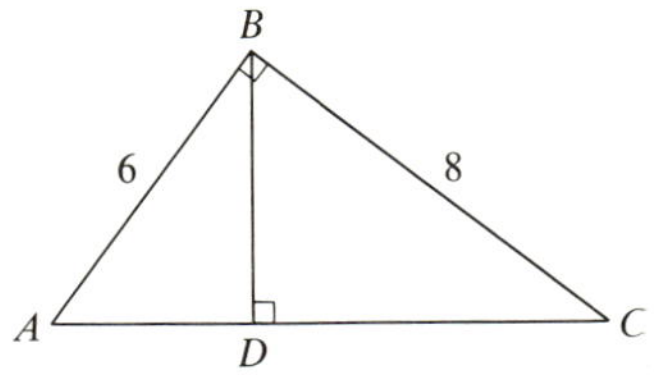

9. What is the longest pipe that can be placed inside a tool box 12 inches wide, 15 inches long, and 9 inches high? What is the diameter of this pipe?

10. The distance around a square is 800. Find the diagonal.

11. The distance around a rectangle is 68 and a diagonal is 26. Find the dimensions of the rectangle.

12. The diagonal of a square is 8 inches. Find the side of the square.

13. The area of a square is 25 square inches. Find its perimeter.

14. Use the Pythagorean Theorem to construct a segment of length (a) $\sqrt{2}$, (b) $\sqrt{5}$, (c) $\sqrt{3}$.

15. Two cars leave from your house at the same time. One travels north at a rate of 36 miles per hour, and the other travels east at a rate of 48 miles per hour. How far apart are they at the end of 30 minutes?

16. On a square baseball field measuring 90 feet on each side, how far is a ball thrown from the shortstop to home plate if the shortstop is on the baseline midway between second and third base?

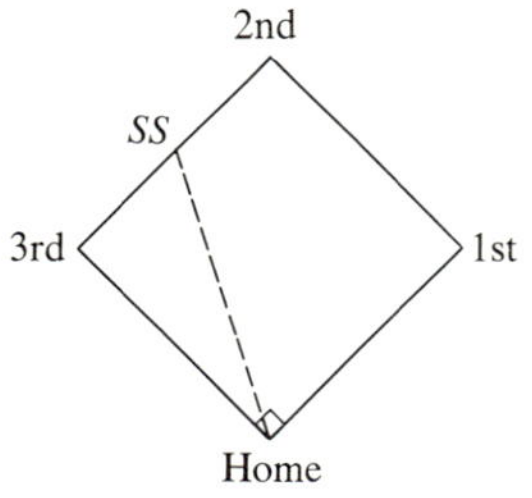

17. In the figure shown, find the length of (a) $\overline{PB}$, (b) $\overline{PC}$, (c) $\overline{PD}$, (d) $\overline{PE}$.

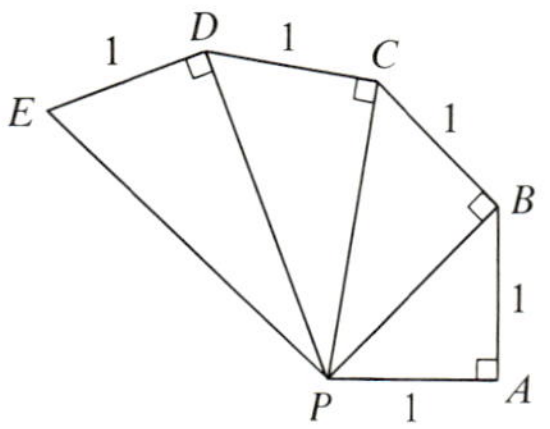

18. If we continue the pattern shown in Question 17, and if we think of $\overline{PA}$ as the first segment, $\overline{PB}$ as the second segment, $\overline{PC}$ as the third segment, etc.,

a. What would be the length of the sixth segment? Of the seventh segment? Of the thirteenth segment?
b. What would be the length of the kth segment?

19. Find the area of the region bounded by an equilateral triangle whose sides are of length (a) 6, (b) 10, (c) a.

20. Consider the right $\triangle ABC$. Why is each of the following statements true?

a. $AC = 25$.
b. Area of triangle $ABC = 150$.
c. Area of triangle $ABC = \frac{1}{2}BD \times AC$.
d. Area of triangle $ABC = \frac{1}{2}BD \times 25$.
e. $BD = 12$.

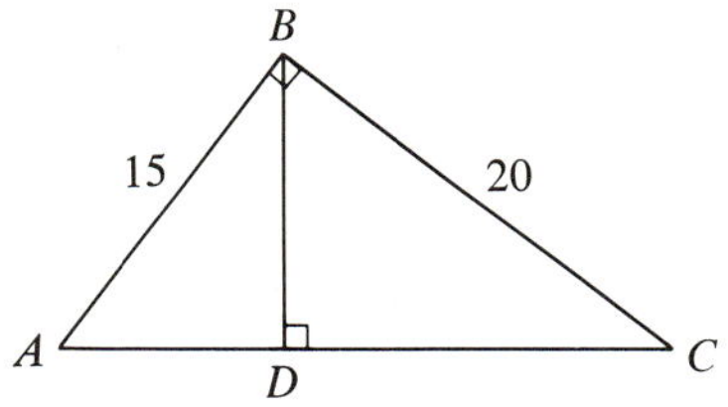

21. What method does Question 20 suggest for finding the altitude to the hypotenuse of a right triangle? Use this method to find the altitude to the hypotenuse in $\triangle LMN$.

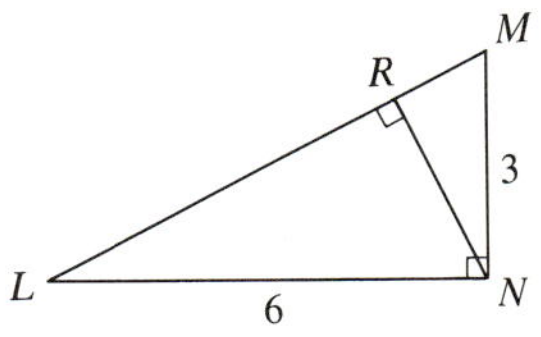

PUZZLES

1. A carpenter has a piece of wood in the shape of a right triangle with legs 21 and 28 inches. From it he cuts a square piece as shown. Find the length of a side of the square piece.

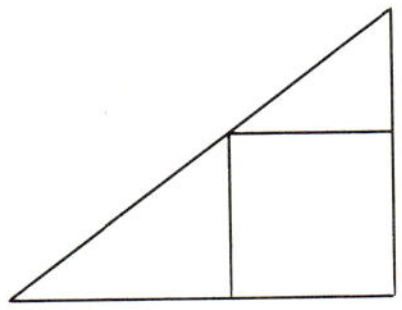

2. How can you divide a square into six smaller squares, all the squares not being necessarily equal in area?

SOLUTIONS

1. 12 inches. [Note that this length is half the *harmonic* mean of the legs, where the harmonic mean is $H = 2ab/(a + b)$.]

2.

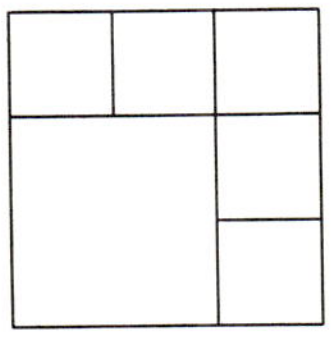

ENRICHMENT

THE PYTHAGOREAN THEOREM AND A PRESIDENT

Even a man who later became President of the United States tried his hand at devising an original proof of the Pythagorean Theorem. Several years before he became President (about 1875), James A. Garfield submitted this proof.

1. Prove that $\angle ABE$ is a right angle (FIG. 7-17).

2. Then prove that $a^2 + b^2 = c^2$ from the fact that the area bounded by trapezoid *ACDE* equals the sum of the areas of the three triangular regions.

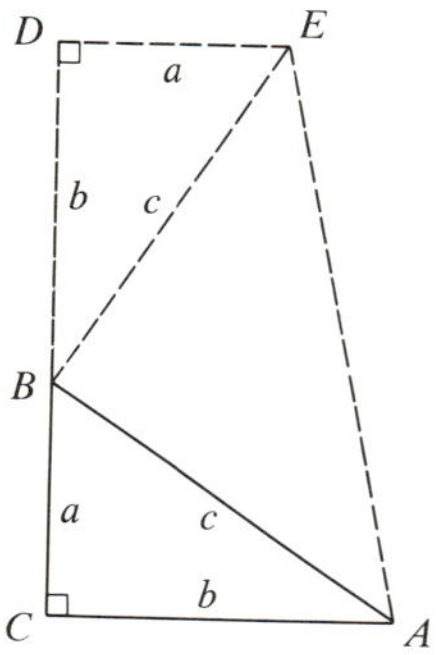

FIGURE 7-17

AN EXTENSION OF THE PYTHAGOREAN THEOREM

Visual 3 suggests that *any* three figures that are similar to each other, when similarly constructed on the sides of a right triangle, have the same area relationship as squares that are constructed on the sides of a right triangle: The area of the largest region equals the sum of the areas of the other two regions.

The reason for this relationship is that the areas of any two similar figures are proportional to the squares of a corresponding dimension.

For instance, if *A*, *B*, *C* are three similar figures similarly constructed on the sides of a right triangle (FIG. 7-18), then

$$\frac{\text{area } A}{\text{area } C} = \frac{a^2}{c^2} \quad \text{and} \quad \frac{\text{area } B}{\text{area } C} = \frac{b^2}{c^2},$$

$$\frac{\text{area } A}{\text{area } C} + \frac{\text{area } B}{\text{area } C} = \frac{a^2}{c^2} + \frac{b^2}{c^2},$$

$$\frac{\text{area } A + \text{area } B}{\text{area } C} = \frac{a^2 + b^2}{c^2}.$$

But

$$a^2 + b^2 = c^2.$$

Therefore,

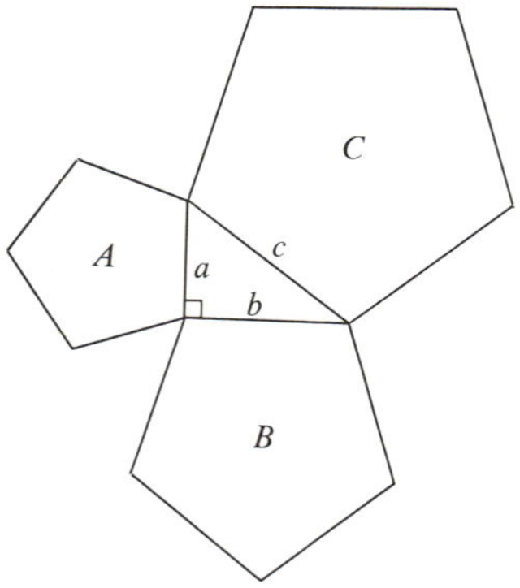

FIGURE 7-18

$$\frac{\text{area } A + \text{area } B}{\text{area } C} = \frac{c^2}{c^2} = 1$$

and

$$\text{area } A + \text{area } B = \text{area } C.$$

GENERATING PYTHAGOREAN TRIPLES

A set of three positive integers, such as (3, 4, 5), which can be the lengths of the sides of a right triangle, is called a *Pythagorean triple.* If the triple contains no common positive integral factor other than 1, it is called a *primitive Pythagorean triple.* For example, (5, 12, 13) is a primitive triple, but (9, 12, 15) is not.

One of the achievements of the Greeks was to show that all primitive Pythagorean triples (a, b, c) satisfy these relations:

$$a = m^2 - n^2, \qquad b = 2mn, \qquad c = m^2 + n^2,$$

where (1) m and n are relatively prime, (2) $m > n$, and (3) m and n are not both odd or both even.

When (a, b, c) satisfy these relations, they will always have the property that $c^2 = a^2 + b^2$; and all triples of primitive sets of integers having this property can be determined from these relations.

For example, if we choose, say, $m = 3$ and $n = 2$, then

$$a = m^2 - n^2 = 9 - 4 = 5,$$

$$b = 2mn = 2 \times 3 \times 2 = 12,$$

$$c = m^2 + n^2 = 9 + 4 = 13.$$

So the two relatively prime numbers (3, 2) generate the primitive Pythagorean triple (5, 12, 13).

From an examination of the Babylonian tablet *Plimpton* 322, it appears that the ancient Babylonians were aware of this method of generating primitive triples, but it is unlikely they had any proof of it.

Pythagorean Triples and Area

It can be shown that in any right triangle whose sides have positive integral measures,

1. The measure of one leg is always divisible by 3.
2. The measure of one leg is always divisible by 4.
3. The measure of one side is always divisible by 5.

An immediate consequence of 1 and 2 is that the area of any right triangle whose sides have positive integral measures is divisible by 6. This follows because (a) by 1 and 2 the product of the lengths of the legs of such a right triangle is divisible by 12, and (b) the area of a right triangle is $\frac{1}{2}$ the product of the lengths of the legs.

A Famous Conjecture

Related to Pythagorean triples is perhaps the most famous conjecture in all number-theory history, known as *Fermat's Last Theorem.*

We have seen that for all Pythagorean triples, the sum of the squares of two of the integers is equal to the square of the third; that is, $a^2 + b^2 = c^2$. But are there three integers where the sum of the *cubes* of two of them is equal to the *cube* of the third; that is, is there an integral solution to the equation $a^3 + b^3 = c^3$?

More generally, is there an integral solution to the equation $a^n + b^n = c^n$ if n is an integer greater than 2?

The generalization occurred to the great number theorist Pierre de Fermat (1601?–1665) when considering a problem in Diophantus' (250 A.D.?) *Arithmetica*: "To divide a given square into two squares." In the margin of the book he was using, Fermat wrote this note: "It is impossible to write a cube as the sum of two cubes, a fourth power as the sum of two fourth powers; and, in general, any power beyond the second as the sum of two similar powers. For this I have discovered a truly wonderful proof but the margin is too small to contain it."

This is Fermat's famous last theorem, which most mathematicians believe to be true but which no one has been able either to prove or disprove. The theorem has been shown to be true for values of $n < 4003$. But what needs to be proved is that the theorem holds (or does not hold) for *every* n greater than 2.

LINKING NUMBERS WITH GEOMETRY

Pythagoras was a Greek mathematician and philosopher who is believed to have lived between 569 and 500 B.C. After much travel, he settled down in 509 B.C. in Crotona, a Greek city of southern Italy, where he organized the Pythagorean Society. Membership in the society was restricted, and the members, who were called Pythagoreans, were pledged to secrecy. Although the greatest interest of the Pythagoreans was in moral reform, they were also interested in science and in mathematics and are credited with many important ideas in mathematics, astronomy, and the science of music.

Pythagoras and his followers were intensely interested in numbers and were the first to think of arithmetic as a science rather than as a tool in computation. They were the first to study the properties of whole numbers and to take the first steps in the development of number theory. One of their most notable achievements was the discovery and proof that the real number $\sqrt{2}$ is irrational.

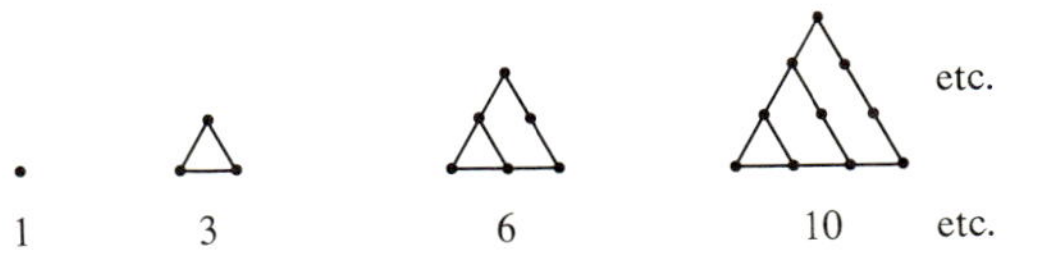

FIGURE 7-19

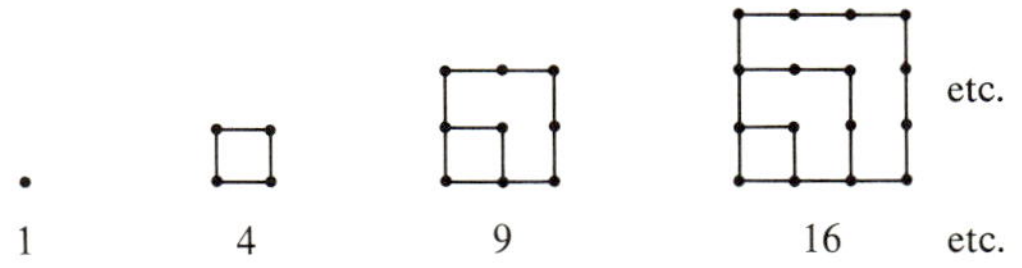

FIGURE 7-20

Figurate Numbers

The Pythagoreans linked numbers with geometric patterns by originating the *figurate numbers*, also called *polygonal numbers*. These numbers can be thought of as the number of dots in certain geometrical configurations. For example, the numbers 1, 3, 6, 10, 15, . . . , or, in general, numbers given by the formula $[n(n+1)]/2$, are called *triangular numbers*, because they can form the triangular configurations shown in FIG. 7-19.

The numbers 1, 4, 9, 16, . . . , given by the formula n^2, are called *square numbers*, because they can form square configurations (FIG. 7-20).

The numbers 1, 5, 12, 22, . . . , given by the formula $[n(3n-1)]/2$ are called *pentagonal numbers*, because they can form pentagonal configurations (FIG. 7-21).

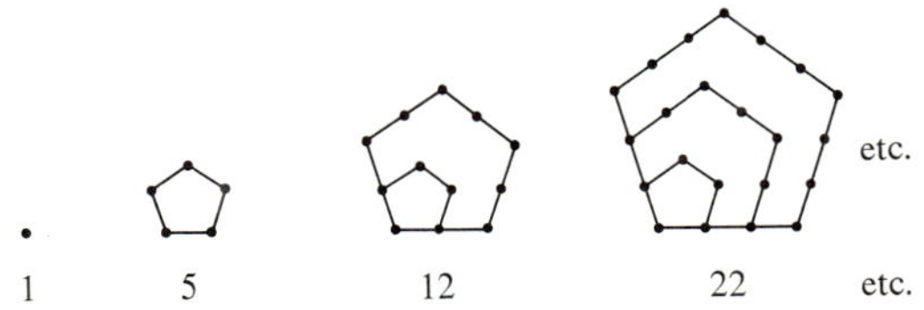

FIGURE 7-21

The *hexagonal numbers* 1, 6, 15, 28, . . . , derived from the formula $2n^2 - n$, can form hexagonal configurations (FIG. 7-22).

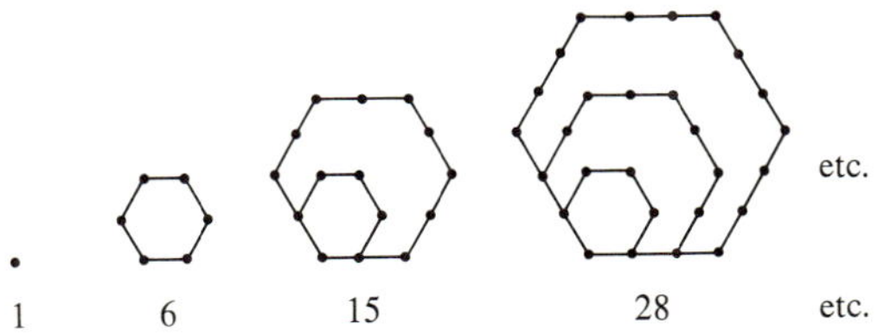

FIGURE 7-22

Polygonal numbers of all orders are designated in a similar way.

By means of these geometric configurations we can show some interesting properties of figurate numbers. For example,

1. Any triangular number added to the next higher triangular number is a square number (FIG. 7-23). Since any geometric configuration representing a square number can be divided as shown above, we can conclude that any square number is the sum of two successive triangular numbers.

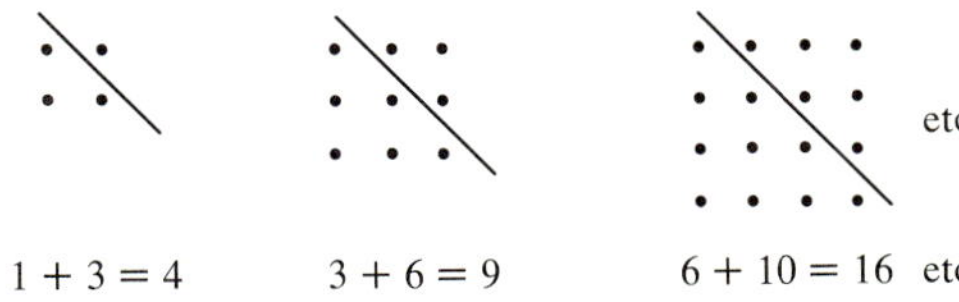

FIGURE 7-23

2. The sum of any number of consecutive odd integers, starting with 1, is always a square number (FIG. 7-24).

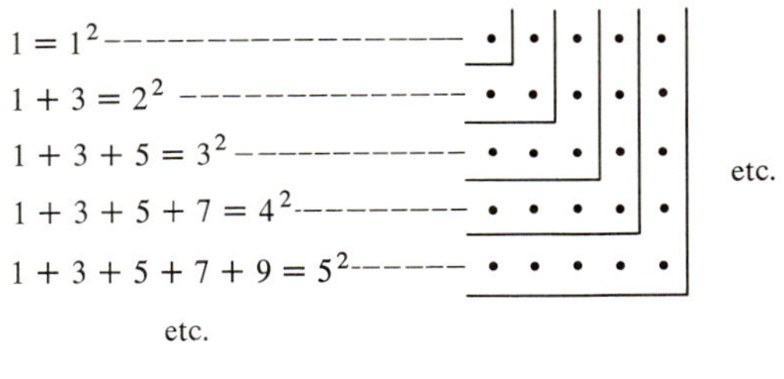

FIGURE 7-24

[*Note:* For an interesting discussion of polygonal numbers see *The Arithmetic Teacher*, pp. 33–38 (Jan. 1970).]

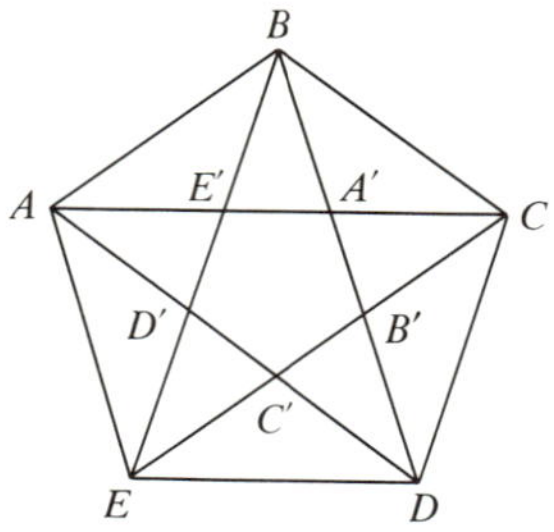

FIGURE 7-25

Symbol of the Pythagoreans

The figure of a five-pointed star is said to have been the special symbol of the Pythagorean Society. A five-pointed star is formed by drawing the five diagonals of a regular pentagon. In FIG. 7-25 the points A', B', C', D', E' divide the diagonals into two segments such that the ratio of the length of the whole diagonal is to the length of the larger segment as the length of the larger segment is to the length of the smaller segment; for example,

$$\frac{BD}{A'D} = \frac{A'D}{A'B}.$$

We recognize this ratio to be the "golden section" of a line segment which the Greeks designated as "the division of a segment in mean and extreme ratio."

8 SIMILARITY

INTRODUCING THE CHAPTER TO STUDENTS

CONTENT REVIEW

Prerequisites for Learning This Chapter
Informal Remarks
Proportionalities
Similarities Between Triangles
The Basic Proportionality Theorem and Its Converse
The Basic Similarity Theorems
Similarities in Right Triangles
Other Similarity Theorems
General Concept of Similarity

TEACHING SUGGESTIONS AND ACTIVITIES

Visual Aids
An Experimental Approach to Similarity
A Guide to Similarity Proofs
Applications of Similar Triangles
Questions
A Student Project

ENRICHMENT

Using Similarity to Simplify Solutions
Homothetic Figures
Tiling and Similarity

INTRODUCING THE CHAPTER TO STUDENTS

(Use Visual 1.) Here is a pair of rectangles. Both have the same size and the same shape. Here is a pair of triangles. They, too, have the same size and shape. The same is true of these pentagons. You know, of course, that figures that have the same size and the same shape are called congruent figures.

Now look at these two figures (rectangle and its enlargement). These rectangles obviously do not have the same size, but they do have the same shape. We call such figures *similar* figures. Here are other examples of similar figures (from Visual 1).

Similarity is the mathematical concept behind blueprints, maps, and picture enlargements. Scale models of houses, ships, planes, spacecraft, and other large structures are often built to assist in their planning and construction. The mathematics of similarity guarantees the similarity of the model and the actual structure.

What is there about similar figures that gives them the same shape but not necessarily the same size? What conditions are needed to make any two figures similar? This question was answered about 2500 years ago by a Greek mathematician named Thales (600 B.C.), who may have been the teacher of Pythagoras.

Thales was a successful businessman who retired from business at an early age to spend his time studying mathematics and philosophy. Thales made so many important contributions to mathematics and science that he was named the "Father of Greek Mathematics" and later became known as the first of the Seven Wise Men of Greece, about whom there are many legends.

Thales showed that two conditions must be true for figures to be similar. If both conditions are true, then the figures are similar; and if two figures are similar, then both of these conditions will be true.

In this chapter we shall state precisely what is meant by *similarity*, and the exact conditions needed to achieve it.

CONTENT REVIEW

PREREQUISITES FOR LEARNING THIS CHAPTER

1. Understanding the meaning of: congruence; correspondence between vertices of two triangles, for example $ABC \leftrightarrow A'B'C'$.

2. The ability to work with and solve simple fractional equations such as $a/b = c/x$.

INFORMAL REMARKS

1. Two figures are *similar* if they have the same shape but not necessarily the same size. Since congruent figures have the same shape, we may regard congruence as a special case of similarity. FIGURE 8-1 gives several examples of similar figures.

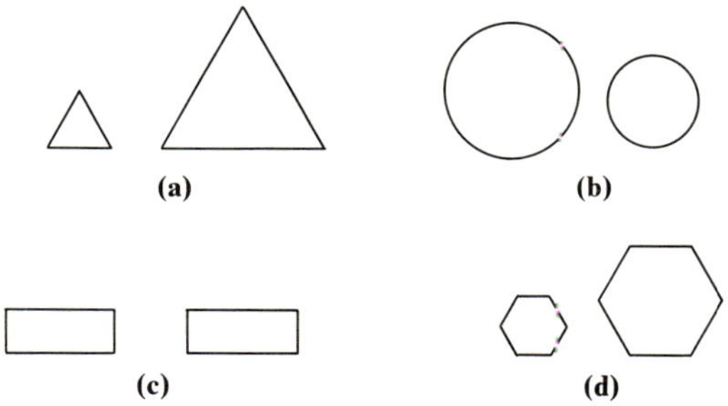

FIGURE 8-1

2. Another way to describe similarity is to say that two figures are similar if one of them is a photographic enlargement or reduction of the other (FIG. 8-2).

FIGURE 8-2

3. Consider the two similar triangles shown in FIG. 8-3 with the correspondence $ABC \leftrightarrow A'B'C'$. Note that the lengths of the

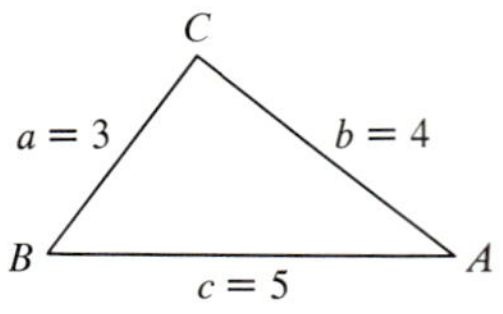

FIGURE 8-3

sides of the two triangles form two sequences of positive numbers: 3, 4, 5 and 6, 8, 10, and that each number in the second sequence is twice the corresponding number in the first sequence; that is, $a' = 2a$, $b' = 2b$, $c' = 2c$ (FIG. 8-4). We

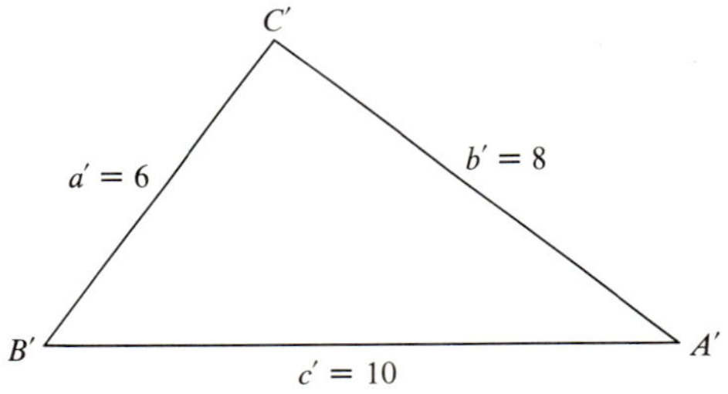

FIGURE 8-4

can also say that each member in the first sequence is half of the corresponding number in the second sequence; that is, $a = \frac{1}{2}a'$, $b = \frac{1}{2}b'$, $c = \frac{1}{2}c'$. In terms of a photograph, we might say that the sides of $\triangle ABC$ would have to be enlarged to twice their lengths to produce $\triangle A'B'C'$, whereas the sides of $\triangle A'B'C'$ would have to be reduced to half their lengths to produce $\triangle ABC$. We call such a correspondence between the triangles a *similarity*.

4. Another way of expressing the above ideas is to write

$$\frac{a'}{a} = \frac{b'}{b} = \frac{c'}{c} = 2 \qquad \text{and} \qquad \frac{a}{a'} = \frac{b}{b'} = \frac{c}{c'} = \frac{1}{2}.$$

Sequences that are related in this way are called *proportional*.[1]

[1] 1. Remark 1 is not, of course, a definition of similarity. But this and the other remarks will lead to such a definition.

2. It should be noted with students that the word "similar" in mathematics, unlike its use in everyday speech, has a very precise meaning.

3. Similar figures cannot exist in non-Euclidean geometry except when the figures are also congruent. (See pages 147, 151.)

PROPORTIONALITIES[2]

Definitions

1. Given two sequences of numbers $a, b, c, \ldots, a', b', c', \ldots,$ if

$$\frac{a}{a'} = \frac{b}{b'} = \frac{c}{c'} = \cdots,$$

then we say that the two sequences are *proportional.* When a proportionality involves only four numbers, such as

$$\frac{a}{b} = \frac{c}{d},$$

we refer to such a proportionality as a *proportion.*

2. If a, b, and c are positive numbers and

$$\frac{a}{b} = \frac{b}{c},$$

then b is called the *geometric mean* between a and c. It follows from this definition that the geometric mean between a and c is $\sqrt{ac}$. Another name for geometric mean is *mean proportional.*

We regard proportionalities as fractional equations. By the application of the ordinary rules of algebra, we can derive the following properties:

If $\frac{a}{b} = \frac{c}{d}$,

then

(a) $ad = bc$, (b) $\frac{a}{c} = \frac{b}{d}$, (c) $\frac{b}{a} = \frac{d}{c}$,

(d) $\frac{a+b}{b} = \frac{c+d}{d}$, (e) $\frac{a-b}{b} = \frac{c-d}{d}$.

SIMILARITIES BETWEEN TRIANGLES

Definition

Given a correspondence between two triangles, if corresponding angles are congruent and corresponding sides are propor-

[2] We are about to make precise what we mean when we say that two figures "have the same shape." We shall also investigate the conditions which guarantee that two figures will have the same shape. Because the concept of similarity is closely related to the concept of proportionality, we shall begin with the definition of *proportion*.

tional, then the correspondence is called a *similarity*, and the triangles are said to be similar (FIG. 8-5).

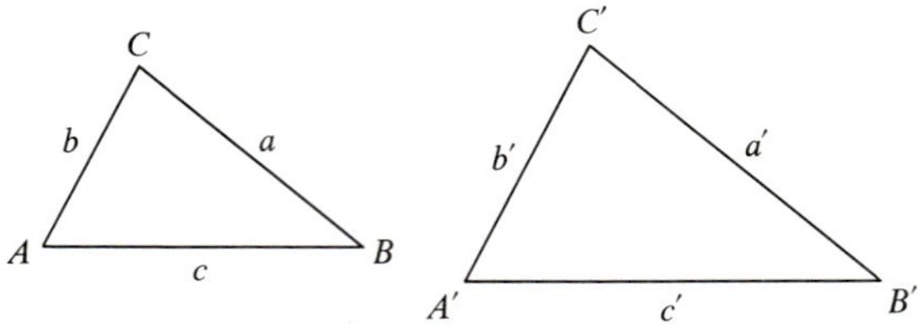

FIGURE 8-5

That is, given a correspondence $ABC \leftrightarrow A'B'C'$ between $\triangle ABC$ and $\triangle A'B'C'$, if (1) $\angle A \cong \angle A'$, $\angle B \cong \angle B'$, $\angle C \cong \angle C'$, and (2)

$$\frac{a}{a'} = \frac{b}{b'} = \frac{c}{c'},$$

then the correspondence $ABC \leftrightarrow A'B'C'$ is a similarity, and we write $\triangle ABC \sim \triangle A'B'C'$.[3]

[3] From the expression "$\triangle ABC \sim \triangle A'B'C'$" we can conclude, without reference to any figure, these facts:

$$\angle A \cong \angle A', \quad \angle B \cong \angle B', \quad \angle C \cong \angle C',$$

$$\frac{a}{a'} = \frac{b}{b'} = \frac{c}{c'}.$$

The proportionality can, of course, be written as

$$\frac{BC}{B'C'} = \frac{AC}{A'C'} = \frac{AB}{A'B'}.$$

THE BASIC PROPORTIONALITY THEOREM AND ITS CONVERSE

THEOREM 8-1 (Basic Proportionality Theorem) If a line parallel to one side of a triangle intersects the other two sides in distinct points, then it cuts off segments that are proportional to these sides (FIG. 8-6).

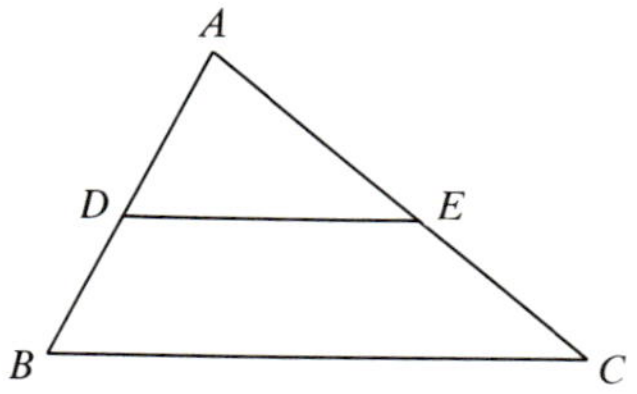

FIGURE 8-6

That is, if in $\triangle ABC$, D and E are points of $\overline{AB}$ and $\overline{AC}$ such that $\overleftrightarrow{DE} \parallel \overline{BC}$, then

$$\frac{AB}{AD} = \frac{AC}{AE}.$$ [4]

[4] As a consequence of Theorem 8-1 and of the definition of similarity of two triangles, we can prove line segments proportional by showing that

1. They are corresponding sides of similar triangles (or other polygons), or
2. They are segments on two sides of a triangle formed by a line parallel to the third side.

THEOREM 8-2 (Converse of Theorem 8-1) *If a line intersects two sides of a triangle, and cuts off segments that are proportional to these sides, then it is parallel to the third side* (FIG. 8-7).

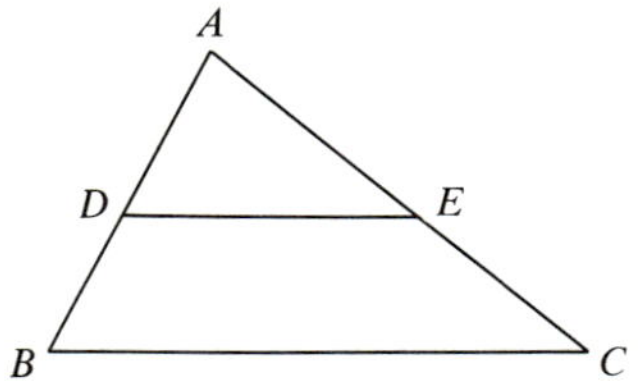

FIGURE 8-7

That is,

$$\text{if } \frac{AB}{AD} = \frac{AC}{AE},$$

then $\overleftrightarrow{DE} \parallel \overline{BC}$.

THE BASIC SIMILARITY THEOREMS[5]

THEOREM 8-3 (AAA Similarity Theorem) *Given a correspondence between two triangles, if corresponding angles are congruent, then the correspondence is a similarity* (FIG. 8-8).

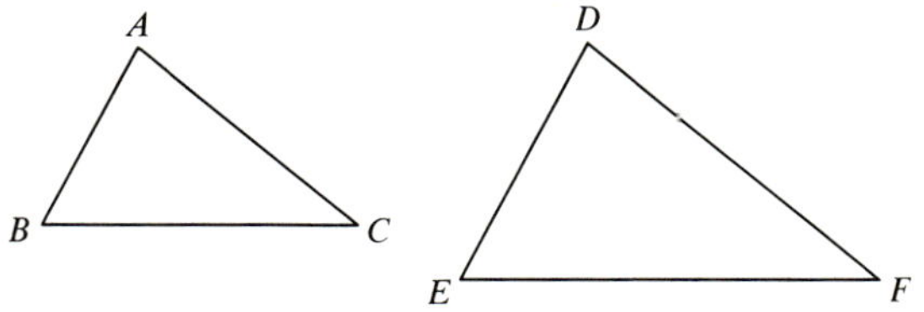

FIGURE 8-8

That is, given the correspondence $ABC \leftrightarrow DEF$ between $\triangle ABC$ and $\triangle DEF$, if $\angle A \cong \angle D$, $\angle B \cong \angle E$, $\angle C \cong \angle F$, then $\triangle ABC \sim \triangle DEF$.

The following is an important corollary of Theorem 8-3.

The AA Corollary *Given a correspondence between two triangles, if two pairs of corresponding angles are congruent, then the correspondence is a similarity.*

[5] By the definition of similarity, two triangles are similar if (1) corresponding angles are congruent, and (2) corresponding sides are proportional.

Theorems 8-3 and 8-5 prove that for triangles, both conditions for similarity are met when either condition is met.

[6] 1. Theorems 8-3 and 8-5 imply that we could have defined similarity for triangles by *either* of the two conditions given in the definition, and we then could have proved the other. But such a definition would not have held for other polygons. For example, consider a square and a rectangle (FIG. 8A). Under the

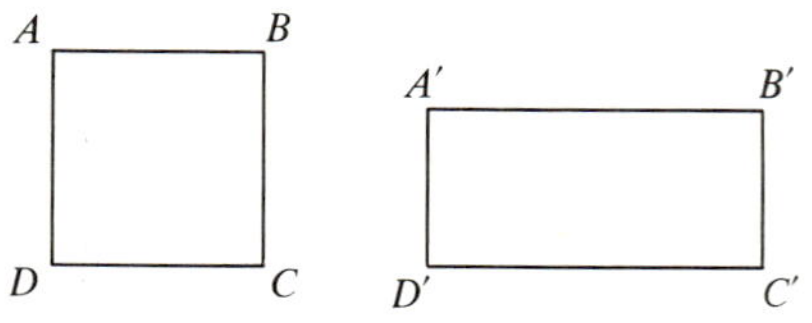

FIGURE 8A

correspondence $ABCD \leftrightarrow A'B'C'D'$, the corresponding angles are congruent, but the two figures do not have the same shape. In this instance, fulfilling only the first condition (congruent angles) is not sufficient to produce a similarity.

Now consider a square and a rhombus (FIG. 8B). Under the correspondence $ABCD \leftrightarrow$

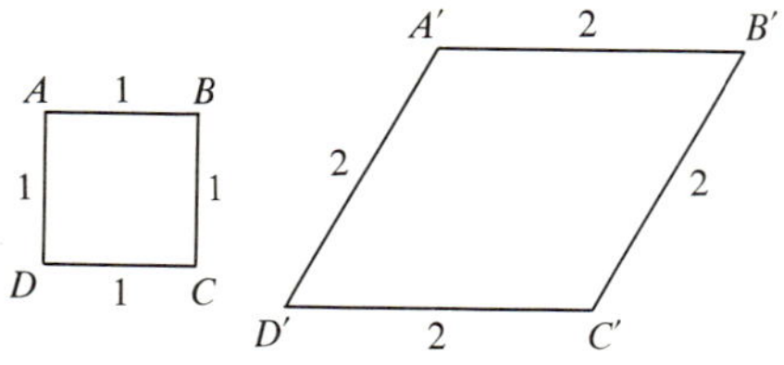

FIGURE 8B

$A'B'C'D'$, corresponding sides are proportional, yet the shapes are different. In this instance, fulfilling only the second condition (proportional sides) is insufficient to produce a similarity.

The fact that congruence of angles is sufficient to determine similarity for triangles but not for quadrilaterals is due to the rigidity of the triangle and to the deformability of the quadrilateral.

2. It is apparent from the fact that congruence is a special kind of similarity that the concepts of congruence and similarity are very much alike. The differences between them must

THEOREM 8-4 (SAS Similarity Theorem) Given a correspondence between two triangles, if two pairs of corresponding sides are proportional and the included angles are congruent, then the correspondence is a similarity (FIG. 8-9).

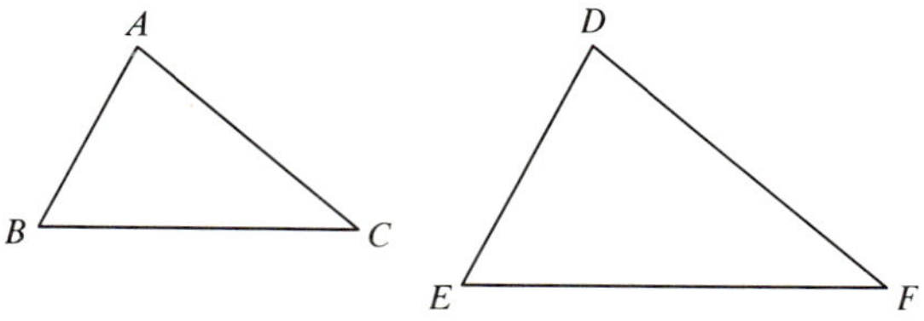

FIGURE 8-9

That is, given the correspondence $ABC \leftrightarrow DEF$ between $\triangle ABC$ and $\triangle DEF$,

$$\text{if } \frac{AB}{DE} = \frac{AC}{DF}$$

and $\angle A \cong \angle D$, then $\triangle ABC \sim \triangle DEF$.

THEOREM 8-5 (SSS Similarity Theorem) Given a correspondence between two triangles, if corresponding sides are proportional, then the correspondence is a similarity (FIG. 8-10).

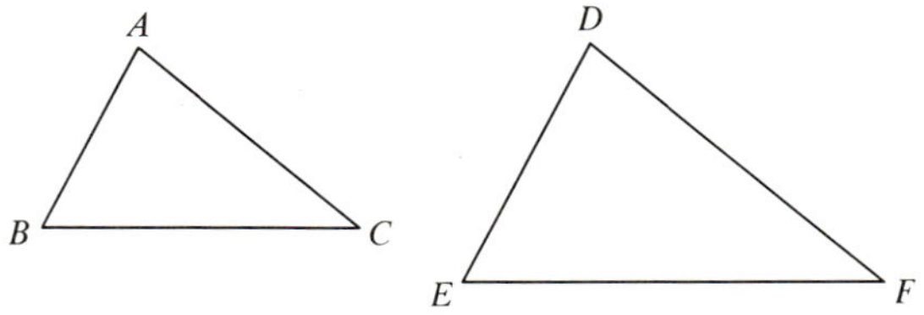

FIGURE 8-10

That is, given the correspondence $ABC \leftrightarrow DEF$ between $\triangle ABC$ and $\triangle DEF$,

$$\text{if } \frac{AB}{DE} = \frac{AC}{DF} = \frac{BC}{EF},$$

then $\triangle ABC \sim \triangle DEF$.[6]

SIMILARITIES IN RIGHT TRIANGLES

THEOREM 8-6 *In any right triangle, the altitude to the hypotenuse separates the triangle into two triangles which are similar to each other and to the original triangle* (FIG. 8-11).

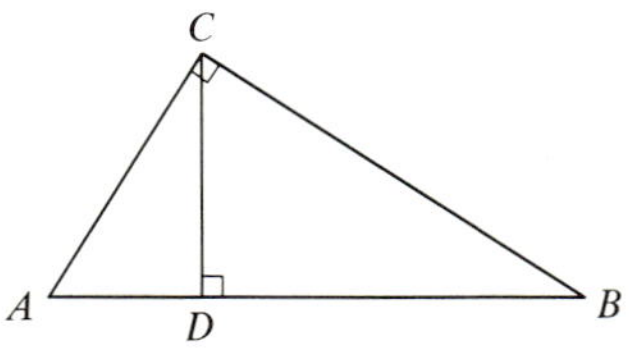

FIGURE 8-11

That is, if $\overline{CD}$ is the altitude to the hypotenuse $\overline{AB}$ in $\triangle ABC$, then $\triangle ACD \sim \triangle ABC \sim \triangle CBD$.

THEOREM 8-7 *Given a right triangle and the altitude to the hypotenuse* (FIG. 8-12),

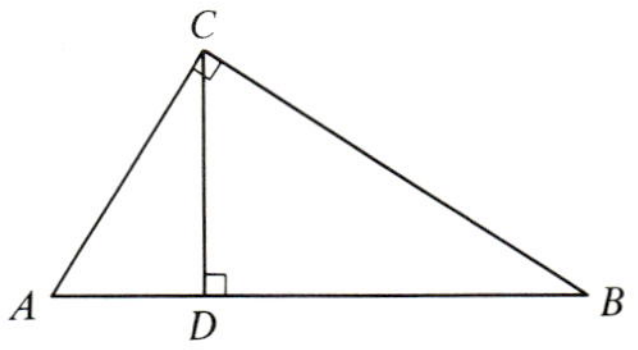

FIGURE 8-12

1. *The altitude is the geometric mean of the segments into which it separates the hypotenuse.*
2. *Each leg is the geometric mean of the hypotenuse and the segment of the hypotenuse adjacent to the leg.*

That is, if $\overline{CD}$ is the altitude to the hypotenuse $\overline{AB}$ in $\triangle ABC$, then

(1) $\frac{AD}{CD} = \frac{CD}{BD}$, (2) $\frac{AD}{AC} = \frac{AC}{AB}$, and $\frac{BD}{BC} = \frac{BC}{BA}$.

OTHER SIMILARITY THEOREMS

THEOREM 8-8 *Similarity between triangles is an equivalence relation.*

be related to the measures of the sides rather than the angles, because there are no differences in the measures of the corresponding angles both in congruent and similar figures.

It is interesting to note, therefore, that if we change "congruent sides" to "proportional sides" and "congruent triangles" to "similar triangles" in the SAS and SSS properties for congruence (pages 111 and 112), we come out with Theorems 8-4 and 8-5.

THEOREM 8-9 *If two triangles are similar, then the altitudes drawn to corresponding sides are proportional.*

THEOREM 8-10 *If three or more parallel lines are cut by two transversals, then the lengths of the segments intercepted on one transversal are proportional to the lengths of the corresponding segments intercepted on the other transversal.*

THEOREM 8-11 *The perimeters of two similar polygons have the same ratio as any two corresponding sides.*

GENERAL CONCEPT OF SIMILARITY

Just as we saw a need to define congruence for *any set of points*, so is there a need to define similarity not just for triangles or other polygons, but for any two figures.

Definition

Two *figures F and F' are similar* if a one-to-one correspondence exists between the points of F and those of F' such that for any pair of points A, B of F and the corresponding pair of points A', B' of F', we have $AB/A'B' =$ a constant k.

TEACHING SUGGESTIONS AND ACTIVITIES

VISUAL AIDS

Visual 1

PURPOSE

To provide models of similar figures.

CONSISTS OF

1. Pairs of congruent figures.
   a. Rectangles.
   b. Triangles.
   c. Pentagons.
2. Enlargements of each of the above figures, that is,
   a. One enlarged rectangle.
   b. One enlarged triangle.
   c. One enlarged pentagon.
3. Photograph and an enlargement of
   a. The letter "A."
   b. A house.
   c. A face.
4. U.S. map.
5. Scale drawing of a car.

HOW TO USE

See Introducing the Chapter to Students.

AN EXPERIMENTAL APPROACH TO SIMILARITY

Initially, develop notions of similarity through experimentation. For instance, ask students to construct two triangles whose corresponding angles are congruent. Let them then measure the corresponding sides to see whether their ratios are equal.

Use the same experimental approach with Theorem 8-1, Theorem 8-2, the corollary to Theorem 8-3, Theorem 8-4, and Theorem 8-5.

A GUIDE TO SIMILARITY PROOFS

A summary of methods that can be used to prove similarity problems is most helpful to students. Let them extract from the definitions and theorems related to similarity the following guide.

1. To prove two triangles similar show that one of the following is true:
   a. Corresponding angles are congruent.
   b. Two angles of one triangle are congruent to two angles of the other.
   c. Two pairs of corresponding sides are proportional and the included angles are congruent.
   d. Corresponding sides are proportional.
   e. Corresponding sides are parallel.
   f. The triangles are right triangles with an acute angle of the first congruent to an acute angle of the second.
   g. Both triangles are similar to a third triangle.
2. To prove line segments proportional show that one of the following is true:
   a. They are corresponding segments on two sides of a triangle formed by a line parallel to the third side.
   b. They are corresponding segments formed by parallel lines cut by two transversals.
   c. They are corresponding sides of similar triangles.
   d. They are corresponding altitudes, medians, or angle bisectors of similar triangles.

APPLICATIONS OF SIMILAR TRIANGLES

Illustrate how similar triangles are used in indirect measurement with these examples:

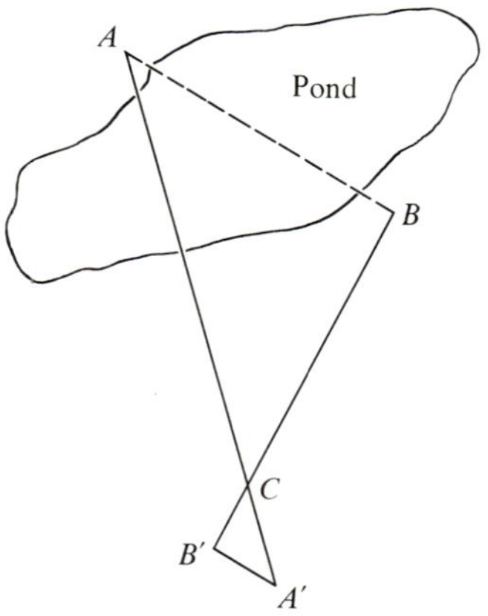

FIGURE 8-13

1. They can be used for measuring an inaccessible distance, such as the distance across a pond.
   a. To find the distance across the pond, drive stakes into the ground at points A and B (FIG. 8-13). Drive a third stake at any convenient point C.
   b. In a line with $\overleftrightarrow{BC}$, drive another stake into the ground at a point B' such that CB' is a convenient fraction of the distance BC, say $\frac{1}{5}BC$.
   c. In a line with $\overleftrightarrow{AC}$, drive another stake at point A' such that CA' is the same fraction of AC as CB' is of BC (in this case, $\frac{1}{5}$).
   d. Measure $\overline{A'B'}$. $A'B'$ will then be $\frac{1}{5}$ as large as AB; or AB will be 5 times as large as $A'B'$.

2. Similar right triangles can be used to measure indirectly the height of vertical objects, such as a tree, by a method known as the "shadow method." Let us find the height of the tree, AB, if it casts a shadow (BC) 20 feet long, while a 3-foot stick (DE) casts a shadow (EF) 4 feet long (FIG. 8-14).

   a. Since the tree and stick are both standing vertically, each forms a right angle with the ground: $m\angle B = m\angle E$.
   b. Because the tree and stick are near each other, $m\angle C = m\angle F$.
   c. Since $\triangle ABC \sim \triangle DEF$, the ratios of the corresponding sides are equal; that is,

$$\frac{AB}{DE} = \frac{BC}{EF}.$$

   d. Therefore,

$$\frac{h}{3} = \frac{20}{4},$$

   or $h = 15$ feet.

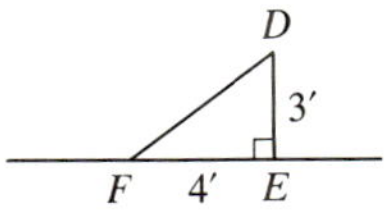

FIGURE 8-14

QUESTIONS

1. Tell whether the figures mentioned in each statement below are similar, congruent, or neither similar nor congruent.

   a. If the three angles of one triangle are congruent to the three corresponding angles of another triangle.
   b. If the three sides of one triangle are congruent to the three corresponding sides of another triangle.
   c. If all the angles of a figure are congruent to the corresponding angles of another figure.
   d. If one angle of a triangle is congruent to an angle of another triangle.
   e. If two angles of a triangle are congruent to two angles of another triangle.

2. Which of the following will always be similar? A pair of

   a. Line segments.
   b. Circles.
   c. Triangles.
   d. Squares.
   e. Rectangles.
   f. Equilateral triangles.
   g. Right triangles.
   h. Isosceles triangles.
   i. Spheres.
   j. Regular hexagons.
   k. Isosceles right triangles.

3. True–false. Give a reason for your answer.

   a. If an acute angle of one right triangle is congruent to an acute angle of another right triangle, then the triangles are similar.
   b. If the corresponding angles of two quadrilaterals are congruent, then the ratios of their corresponding sides are equal.
   c. If the ratios of the corresponding sides of two quadrilaterals are equal, then the corresponding angles are congruent.
   d. If the ratios of the corresponding sides of two triangles are equal, then the corresponding angles are congruent.
   e. Two congruent figures are always similar.
   f. Congruence is a special kind of similarity.
   g. It is possible to construct two triangles whose corresponding angles are congruent but whose corresponding sides are not proportional.

4. In an architect's drawing, the scale is given as $\frac{1}{4}$ inch to 1 foot. What should be the measurements of the drawing of a room measuring 18 by 25 feet?

5. The scale of a map is $\frac{1}{4}$ inch to 10 miles. What is the distance from city X to city Y if the map shows a distance of $4\frac{3}{8}$ inches?

6. From the measurements shown, determine the distance AB across a stream.

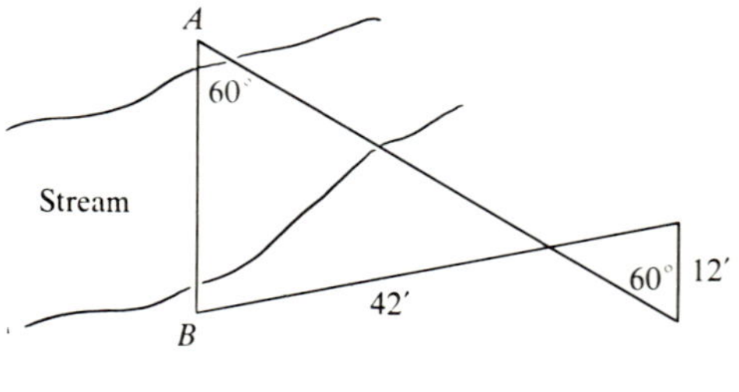

7. Prove that proportionality of sequences is an equivalence relation.

8. Prove that lengths of corresponding medians of similar triangles have the same ratio as lengths of corresponding sides.

9. Prove the Pythagorean Theorem by using the concept of

similar triangles.
Suggestion:

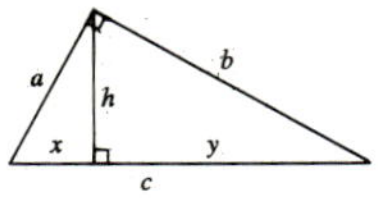

10. How tall is a flagpole that casts an 18-foot shadow at the same time a yardstick casts a 2-foot shadow?

11. Two rectangular fences are situated as shown. If the width of the larger rectangle (w') is twice the width of the smaller rectangle (w), and the length of the smaller fence is 60 feet, what is the length of the larger fence?

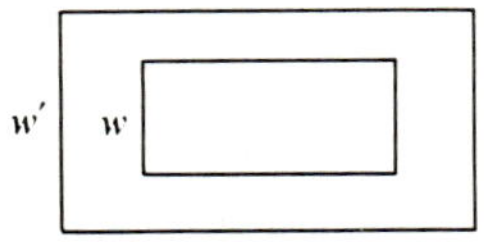

12. If the midpoints of the sides of a triangle are joined, a second triangle is formed. Prove that the second triangle is similar to the original triangle.

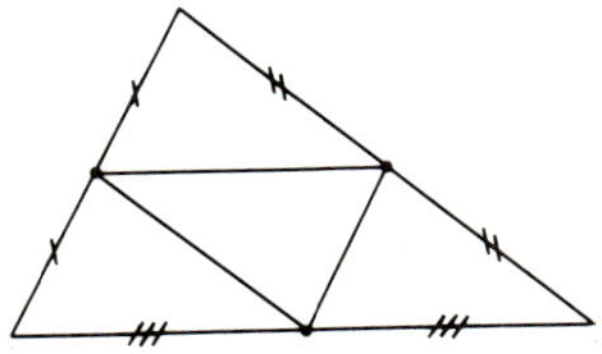

13. Prove that two regular polygons with the same number of sides are similar.

14. Prove that if two triangles are similar, then the ratio of their perimeters is the same as the ratio of two corresponding sides.

15. Prove that in similar triangles corresponding angle bisectors have the same ratio as corresponding sides.

16. Prove that the diagonals of a trapezoid intersect each other at a point that makes the lengths of the segments of one diagonal proportional to the lengths of the segments of the other diagonal.

17. Square $DEFG$ is inscribed in $\triangle ABC$ with $\angle C$ a right angle. Prove that $AG \cdot CE = CD \cdot FG$.

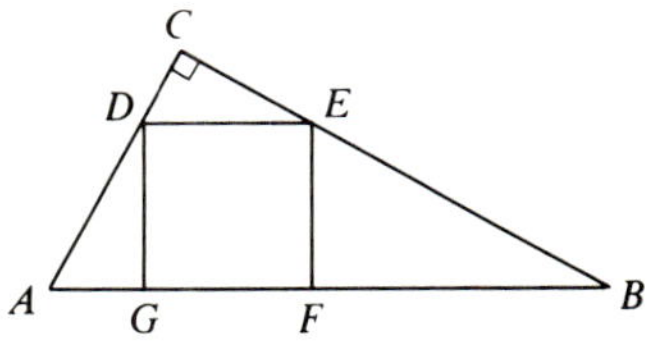

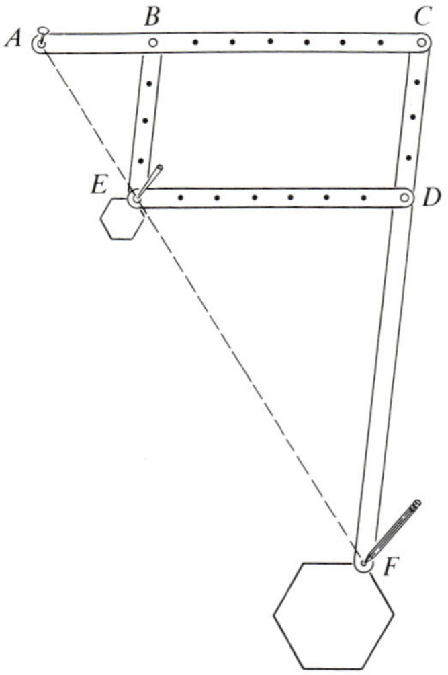

FIGURE 8-15

A STUDENT PROJECT

An interesting and useful application of similar polygons is the *pantograph*, an instrument used for drawing similar plane figures. The pantograph can be used to reduce or enlarge maps and drawings. Building a pantograph can make an interesting project for students.

The instrument consists of four bars hinged at points B, C, D, and E so as to form parallelogram $BCDE$, with points A, E, and F collinear (FIG. 8-15). Pencils are inserted into holes at E and F.

Point A is kept fixed, and as the pencil at E traces the given figure it is enlarged by the pencil at F.

Additional holes are punched in the four bars to provide for various settings to give a desired ratio of similitude CD/CF. If BC is 5 units and AB is 2 units, the ratio of the length of the sides of the two hexagons traced at E and F would be $\frac{2}{7}$.

ENRICHMENT

USING SIMILARITY TO SIMPLIFY SOLUTIONS

The ideas of similarity are not only useful in indirect measurement and in the trigonometric functions, but they often help in the solution of geometric problems, such as constructions, by reducing the original problem to a simpler problem.

For example, consider the problem of constructing a square if we know the sum of the lengths of its side (a) and its diagonal (d); that is, we are given the length $(a + d)$.

By using the fact that all squares are similar, we can proceed as follows:

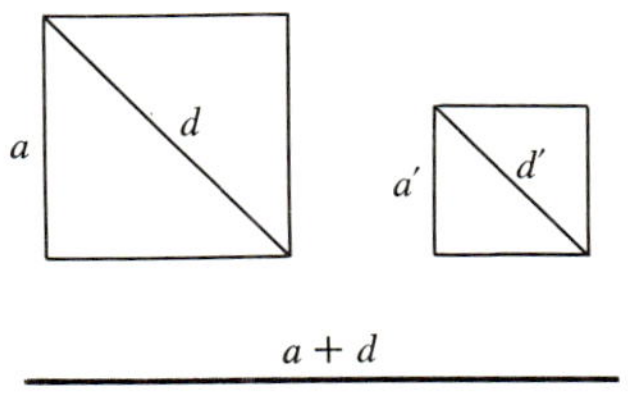

FIGURE 8-16

1. Construct an arbitrary square with side a' and diagonal d' (FIG. 8-16).
2. Since all squares are similar, we have

$$\frac{a}{a'} = \frac{d}{d'}$$

or

$$\frac{a+d}{a'+d'} = \frac{a}{a'}.$$

3. Since we know $(a + d)$, $(a' + d')$, and a', we can find the length a by constructing the fourth proportional to the other three lengths. (See page 252.)
4. We have now reduced the original problem to the problem of constructing a square with side of known length a.

HOMOTHETIC FIGURES

Similar geometric figures give rise to *homothetic figures* whose properties are useful in solving a wide variety of problems.

Two similar polygons are *homothetic* if their corresponding sides are parallel (FIG. 8-17). A basic property of homothetic polygons is that the lines joining the corresponding vertices of two homothetic polygons are concurrent or parallel. We can prove this property in the following way:

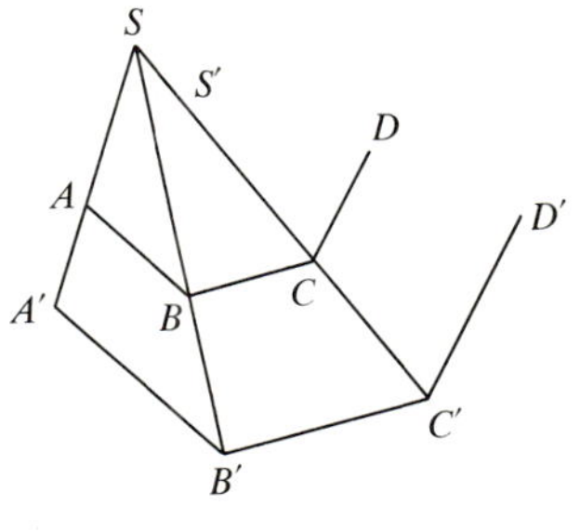

FIGURE 8-17

1. Let $ABCD\cdots$, $A'B'C'D'\cdots$ be two homothetic polygons and let S be the point of intersection of $\overleftrightarrow{AA'}$ with $\overleftrightarrow{BB'}$. We must now show that the line $\overleftrightarrow{CC'}$ joining any other pair of corresponding vertices C and C' will also pass through S.

2. Assume that $\overleftrightarrow{CC'}$ does not pass through S, and that it intersects $\overleftrightarrow{BB'}$ in S'.

3. From the similar triangles $\triangle SAB$ and $\triangle SA'B'$ we get

$$\frac{SB'}{SB} = \frac{A'B'}{AB},$$

and from the similar triangles $\triangle S'B'C'$ and $\triangle S'BC$ we get

$$\frac{S'B'}{S'B} = \frac{B'C'}{BC}.$$

4. By hypothesis,

$$\frac{B'C'}{BC} = \frac{A'B'}{AB}.$$

5. Therefore,

$$\frac{SB'}{SB} = \frac{S'B'}{S'B},$$

or

$$\frac{SB'}{SB - SB'} = \frac{S'B'}{S'B - S'B'},$$

or

$$\frac{SB'}{BB'} = \frac{S'B'}{BB'}.$$

6. Since the denominator of both fractions is BB', the numerators SB' and $S'B'$ must be the same, and point S' must coincide with point S.

The point S is called the *homothetic center* or the *center of similitude* of the two homothetic polygons. If k is the ratio of the corresponding sides, that is, if

$$\frac{SA'}{SA} = \frac{SB'}{SB} = \cdots \frac{A'B'}{AB} = \frac{B'C'}{BC} = \cdots = k,$$

then k is called the *homothetic ratio* or the *ratio of similitude*.

In the case where $\overleftrightarrow{AA'} \parallel \overleftrightarrow{BB'}$, the lines joining the corresponding vertices of the two homothetic polygons are parallel.

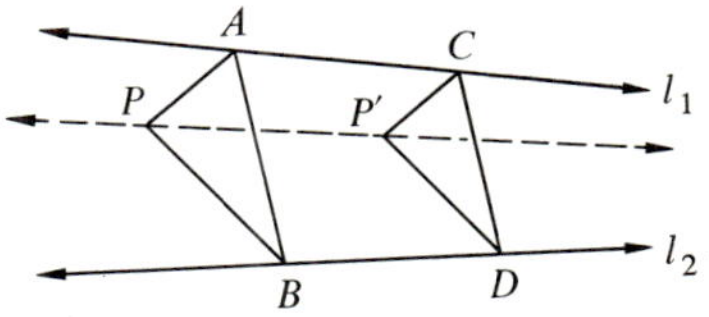

FIGURE 8-18

The theorem just proved provides an easy solution to the following problem.

Given two nonparallel lines l_1 and l_2 and point P between the two lines (FIG. 8-18), the problem is to draw a line that passes through point P and through the inaccessible point of intersection of l_1 and l_2.

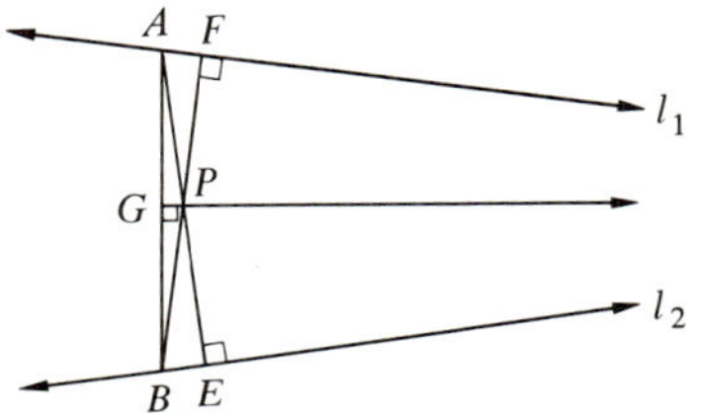

FIGURE 8-19

Solution

1. Draw any two parallel segments $\overline{AB}$ and $\overline{CD}$ intersecting l_1 at A and C, and l_2 at B and D.
2. Join the points A and B to P. Through C and D draw parallels to $\overline{AP}$ and $\overline{BP}$, respectively, and let P' be the point of intersection of these parallels.
3. Since $\triangle APB$ is homothetic to $\triangle CP'D$, the lines $\overleftrightarrow{AC}$, $\overleftrightarrow{BD}$, and $\overleftrightarrow{PP'}$ are concurrent. Therefore, $\overleftrightarrow{PP'}$ is the required line.

Another interesting solution of this problem, using the concurrence of the altitudes of a triangle rather than concepts of similar figures, is as follows. Draw $\overline{APE} \perp l_2$, $\overline{BPF} \perp l_1$, and $\overline{PG} \perp \overline{AB}$ (FIG. 8-19). $\overleftrightarrow{GP}$ is the required line.

TILING AND SIMILARITY

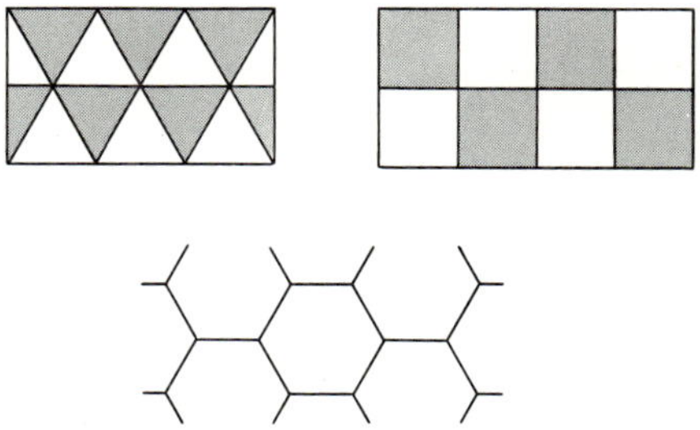

FIGURE 8-20

There are three regular polygons—the equilateral triangle, the square, and the regular hexagon—that are used often as floor tiles because they fit together so well. In fact, these three are the only regular polygons that can be arranged in such a way that the floor is covered completely with nonoverlapping congruent figures (FIG. 8-20).

There are, however, infinitely many *irregular* polygons that can be used for the same purpose. We can, for example, cut out of a piece of stiff material a sufficient number of, say, congruent triangles of any shape and fit them together to cover the floor the way we would fit the pieces of a jigsaw puzzle.

There is a less familiar but more interesting way of tiling a floor by making use of similar figures. The method takes advantage of the property of polygons whereby they can be replicated into enlarged and reduced copies of themselves. Tiling a floor by this method depends, therefore, upon the dissection of an irregular polygon into smaller, congruent polygons that are similar to the original polygon.

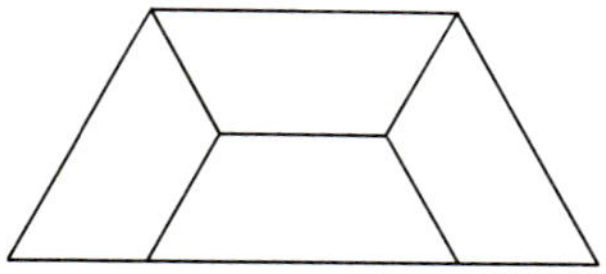
FIGURE 8-21

For example, let us use the trapezoid at the right as our irregular polygon.

This trapezoid can be dissected into four smaller congruent trapezoids that are similar to the original trapezoid (FIG. 8-21).

The four smaller trapezoids can, in turn, be subdivided in the same way into four still smaller trapezoids. This process can be continued indefinitely.

To utilize this idea in tiling, we only need to reverse the process: we put together four trapezoids to form a still larger model; and four of these, in turn, will form a still larger model. We continue this process until the floor is covered completely.

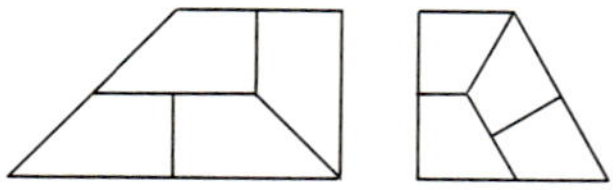
FIGURE 8-22

A trapezoid that can be divided into four replicas congruent to one another and similar to the original trapezoid is called a *replicating trapezoid of order* 4. FIGURE 8-22 shows two other replicating trapezoids of order 4. The only two known replicating polygons of order 2 are the isosceles right triangle and the parallelogram with sides in the ratio of $1:\sqrt{2}$ (FIG. 8-23). Every triangle and parallelogram is a replicating polygon of order 4 (FIG. 8-24).

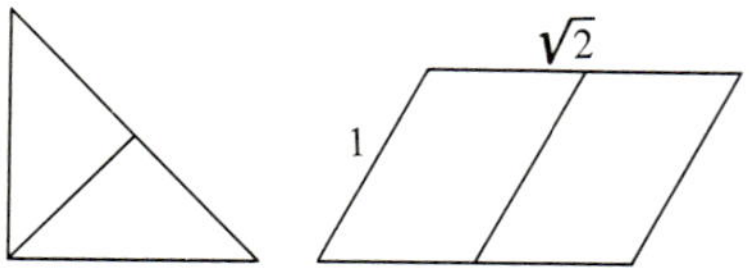

FIGURE 8-23

We can extend this replication process to three dimensions. One obvious case is the cube that can be subdivided into 8 smaller cubes, all congruent to each other and similar to the original cube. The cube can likewise be subdivided into 27 or 64 smaller cubes—or to any cubical number of replicas.

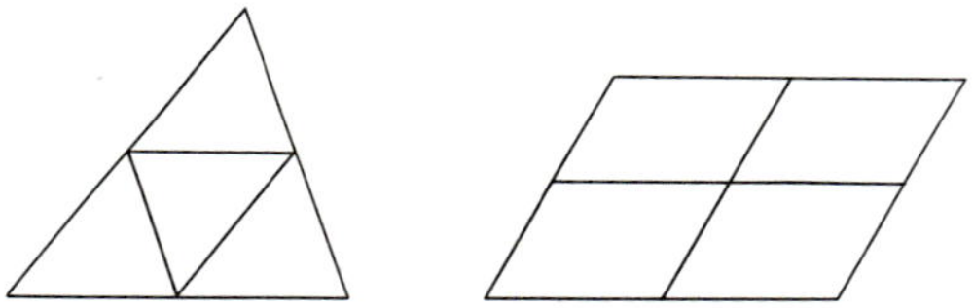
FIGURE 8-24

9 CONSTRUCTIONS

INTRODUCING THE CHAPTER TO STUDENTS

CONTENT REVIEW

Prerequisites for Learning This Chapter
Introduction
Constructions

TEACHING SUGGESTIONS AND ACTIVITIES

Visual Aids
Geometry Through Paper-Folding
Questions
Puzzles
Solutions

ENRICHMENT

The Famous Construction Problems of Antiquity
The Problem of Apollonius

INTRODUCING THE CHAPTER TO STUDENTS

In making drawings, a modern draftsman or artist uses such instruments as a ruler, compass, protractor, T-square, draftman's triangle, French curve, parallel ruler, proportional divider, and linkages.

In this chapter, however, we shall see how to construct a variety of figures using only a compass and straightedge. A straightedge is an unmarked ruler. We shall use the straightedge only to draw lines; we shall use the compass only to draw a circle with a given center and radius.

Geometric drawings that are made only by compass and straightedge are called *constructions*. We shall first learn how to make several basic constructions which we shall later apply to more complicated constructions.

To make these constructions will be like playing a game in which we agree in advance to impose certain restrictions. In this game, we are restricted to using only a compass and straightedge in certain ways. If we wish, we can impose one more restriction. We can agree to use *only a compass*! In fact, about 175 years ago an Italian mathematician named L. Mascheroni (1750–1800) discovered that all geometrical constructions possible by compass and straightedge, insofar as the given and required elements are points, can be made by compass *alone*! About 125 years before that a Danish mathematician, G. Mohr (1640–1697), made the same discovery.

We will, however, find the use of both straightedge and compass more interesting. These are the instruments the Greeks used many years ago, and in their efforts to find solutions to certain problems discovered many important and unforeseen ideas in mathematics.

When we speak of a straightedge and compass as they are related to geometric constructions, we mean an "ideal straightedge" and an "ideal compass" with which straight lines and circles can be drawn precisely. So, in performing geometric constructions our main concern is not whether we can draw figures with a certain degree of accuracy, but to determine whether, by the use of straightedge and compass alone, the solution is theoretically possible.

In solving construction problems we shall be required to do two things:

1. To discover a method for constructing a certain figure, under given conditions, with the allowable instruments.
2. To prove that the method used actually yields a figure with the required properties.

CONTENT REVIEW

PREREQUISITES FOR LEARNING THIS CHAPTER

Understanding the following

1. Terms and concepts: congruence of segments, angles and triangles; bisector of an angle; bisector, midpoint of a segment; perpendicular, perpendicular bisector of a segment; parallel lines; arc; tangent; geometric mean.

2. Postulates and theorems.

   a. Postulates 9-1, 9-2, and 9-3.
   b. The SAS, ASA, SSS properties of congruent triangles.
   c. If two angles are congruent and supplementary, then they are right angles.
   d. If two lines are cut by a transversal so that a pair of alternate interior angles is congruent, then the two lines are parallel.
   e. If three or more parallel lines intercept congruent segments on one transversal, then they intercept congruent segments on any other transversal.
   f. The perpendicular bisectors of the sides of a triangle meet in a point equidistant from the three vertices of the triangle.
   g. The angle bisectors of a triangle meet in a point equidistant from the three sides of the triangle.
   h. Any line perpendicular to a radius at its outer end is tangent to the circle.
   i. If a line parallel to one side of a triangle intersects the other two sides in distinct points, then it cuts off segments that are proportional to these sides.
   j. If an angle is inscribed in a semicircle, then it is a right angle.
   k. In any right triangle, the altitude to the hypotenuse is the geometric mean of the segments into which it separates the hypotenuse.
   l. If three parallel lines are cut by two transversals, then the lengths of the segments intercepted on one transversal are proportional to the lengths of the corresponding segments intercepted on the other transversal.

Postulate 9-1:
2 Nonparallel lines Lying in a plane intersect in a point.

9-2:
A LINE AND A CIRCLE INTERSECT IN 2 POINTS IF THE LINE CONTAINS A POINT INSIDE THE CIRCLE

9-3:

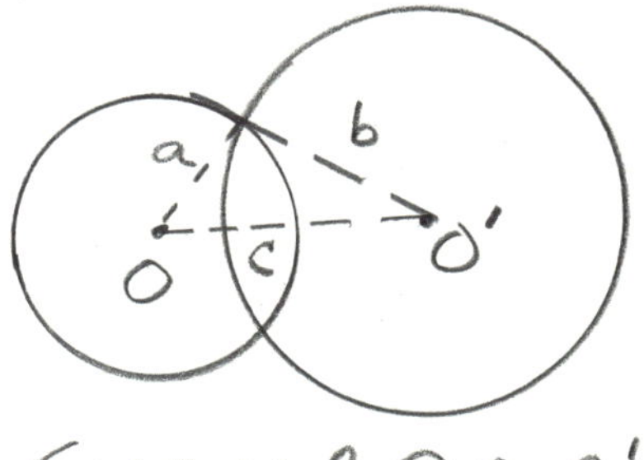

GIVEN: 2 ⓢ O, O'
a → radius O
b → radius O'
c → M $\overline{OO'}$

$a+b>c$ $b+c>a$
$a+c>b$

Then: ⓢ O, O' intersect in 2 places.

INTRODUCTION

This may be the first time that students will be using a compass for something other than drawing circles. Now they will be using it also as an instrument for measuring lengths. Also, they will be using a ruler in a new way. They will ignore the marks on it and use it only to draw the line through two given points. But they may not use it to measure distances between points, to tell whether two segments are congruent, or to tell whether three points are collinear.[1]

To prove the validity of our constructions, we must establish

1. Proof of the existence of certain points and lines.
2. Proof that the construction accomplishes its objective.

We shall establish (1) by assuming the three postulates below. We shall establish (2) by proceeding in the usual way of proving a geometric problem, relying on previous definitions, postulates, and theorems.

Since all our constructions will depend on (1) finding the intersection of two straight lines, (2) finding an intersection of a straight line and a circle, or (3) finding an intersection of two circles, we shall assume the following postulates:

POSTULATE 9-1 *Two nonparallel lines lying in a plane intersect in a point* (FIG. 9-1).

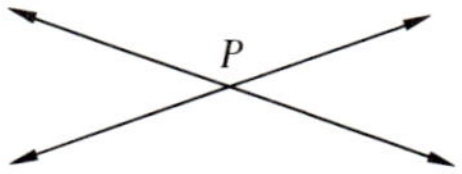

FIGURE 9-1

POSTULATE 9-2 *A line and a circle intersect in two points if the line contains a point inside the circle* (FIG. 9-2).

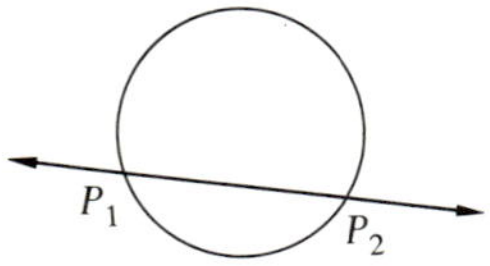

FIGURE 9-2

[1] We shall use the compass a little differently from the way the Greeks used it. The Euclidean compass was used only in the following way: Given a point *P* and a point *A* in the plane, we can draw the circle with center at *P* and radius *PA*. If we are given a third point *P′*, it is improper to move the spike of the compass to *P′* and then draw the circle with center at *P′* and radius *PA*. What we would be doing, in effect, would be to transfer a line segment (the radius) from one point to another. We are not permitted to move the spike of the compass because "when you lift the spike off the paper, the compass collapses." This is why the Euclidean compass is sometimes called "collapsible."

We will, however, assume that segments may be transferred by lifting and moving the compass.

POSTULATE 9-3 *Let C and C′ be circles of radius a and b, and let c be the distance between their centers* (FIG. 9-3). *If each of the numbers a, b, and c is less than the sum of the other two, then C and C′ intersect in two points. The two points of intersection lie on opposite sides of the line of centers.*

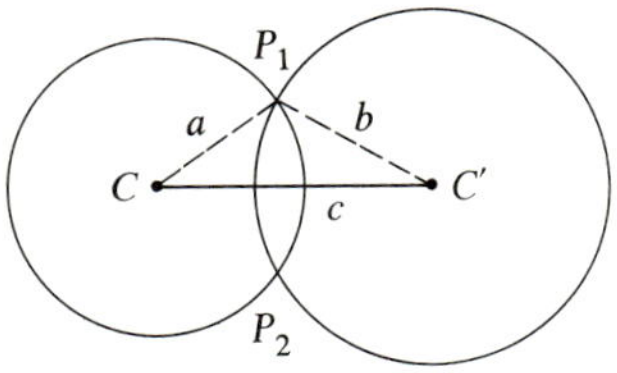

FIGURE 9-3

(Note that Postulate 9-3 is the Two-Circle Theorem on pages 171–172. We use the theorem here as a postulate for convenience.)

We shall now begin with a few basic constructions and then use these in more complicated constructions. All constructions will relate to figures in a given plane.

CONSTRUCTIONS

CONSTRUCTION 1 To copy a given line segment on a given line.

Problem

Given line segment $\overline{AB}$ and line l (FIG. 9-4), how can you construct on l a segment $\overline{CD}$ that is congruent to $\overline{AB}$?

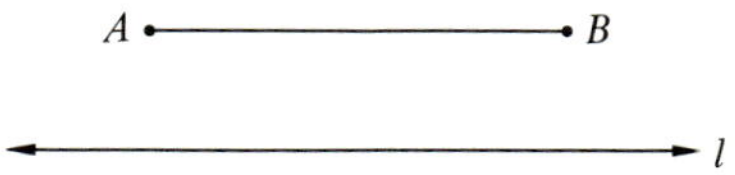

FIGURE 9-4

Solution

1. Place the spike of the compass at point A and the pencil tip at point B (FIG. 9-5). The distance between the spike and the pencil tip is called the *radius of the compass.*

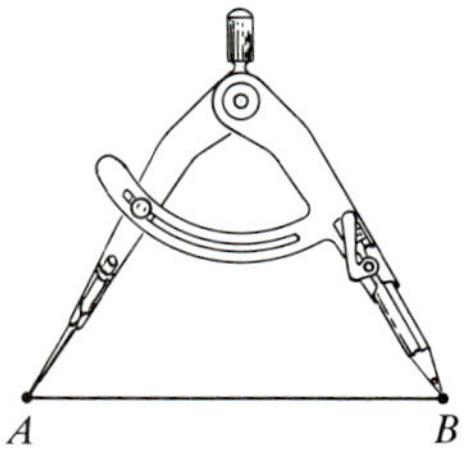

FIGURE 9-5

2. Keeping the radius fixed, place the spike at any point C on l and draw a short arc intersecting l at D (FIG. 9-6). Then $CD = AB$.

FIGURE 9-6

Proof
$\overline{CD}$ and $\overline{AB}$ have the same measure since the measure of each is the radius of the compass. Therefore, by the definition of congruent line segments, $\overline{CD} \cong \overline{AB}$.

Note: When we construct $\overline{CD}$ congruent to $\overline{AB}$, we say that we have "copied" segment $\overline{AB}$.

CONSTRUCTION 2 To bisect a line segment.

Problem
Given line segment $\overline{AB}$ (FIG. 9-7), how can you locate a point C on $\overline{AB}$ that will *bisect* $\overline{AB}$, that is, divide $\overline{AB}$ into two congruent parts such that $\overline{AC} \cong \overline{CB}$?

FIGURE 9-7

Solution
1. Place the spike of the compass at either endpoint A or B with the radius set at more than half the length of $\overline{AB}$. Draw arcs above and below $\overline{AB}$ (FIG. 9-8).

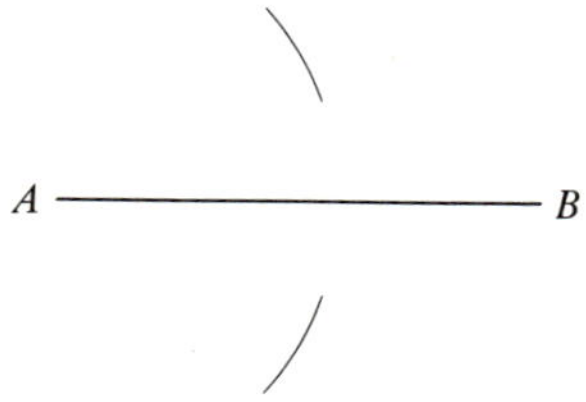

FIGURE 9-8

2. Without changing the radius of the compass, place the spike at the other endpoint of $\overline{AB}$, and draw arcs that intersect the arcs in 1 at points R and S (FIG. 9-9).

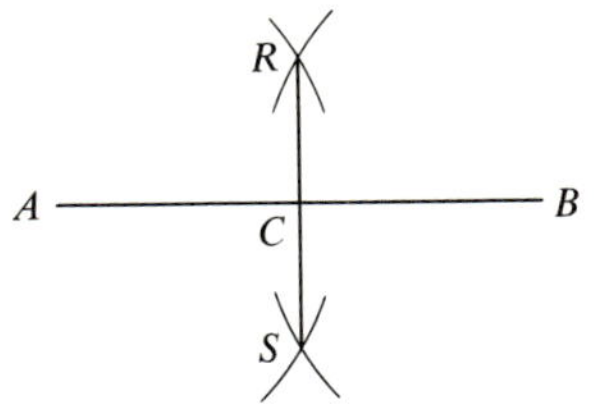

FIGURE 9-9

3. Draw a line segment through points R and S, intersecting $\overline{AB}$ at C. Point C bisects $\overline{AB}$ and is called the *midpoint* of $\overline{AB}$.

Proof

1. $AR = RB = BS = SA$ since the measure of each segment is the same radius (FIG. 9-10).

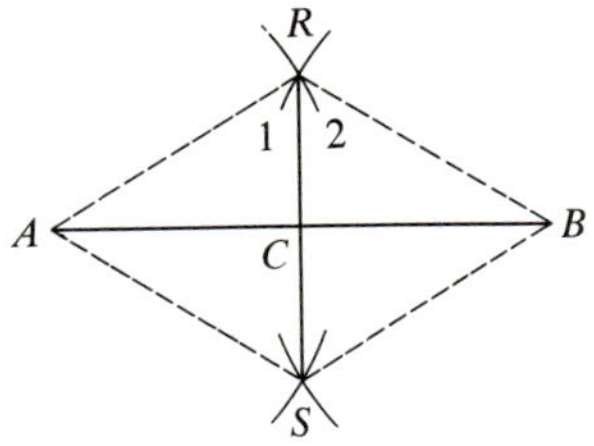

FIGURE 9-10

2. $\triangle ARS \cong \triangle BRS$ (SSS) and the corresponding angles $\angle 1$ and $\angle 2$ are congruent.
3. $\triangle ARC \cong \triangle BRC$ (SAS) and $\overline{AC} \cong \overline{CB}$ since they are corresponding sides of congruent triangles.

CONSTRUCTION 3 To construct the perpendicular bisector of a given segment.

Problem

Given line segment $\overline{AB}$ (FIG. 9-11), how do you construct $\overleftrightarrow{RS}$ (1) to be perpendicular to $\overline{AB}$, and (2) to bisect $\overline{AB}$?

FIGURE 9-11

Solution

Bisect $\overline{AB}$ as in Construction 2. $\overleftrightarrow{RS}$ is perpendicular to $\overline{AB}$ at point C and also bisects $\overline{AB}$ at point C (FIG. 9-12).

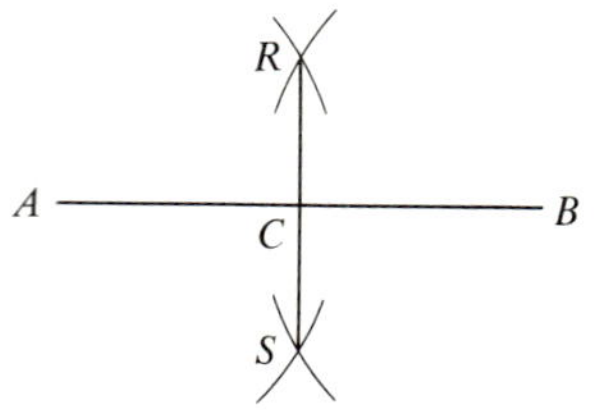

FIGURE 9-12

Proof

1. $\triangle ARC \cong \triangle BRC$ by the proof in Construction 2; $\overline{AC} \cong \overline{CB}$.
2. $\angle 3 \cong \angle 4$ since they are corresponding angles of congruent triangles (FIG. 9-13).

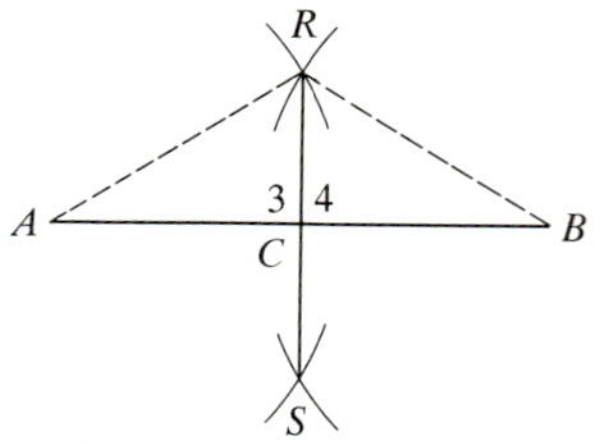

FIGURE 9-13

3. $\angle 3$ and $\angle 4$ are right angles since they are congruent and supplementary.
4. Therefore, $\overleftrightarrow{RS} \perp \overline{AB}$ at C, and also bisects $\overline{AB}$ at C.[2]

[2] 1. Note that Constructions 2 and 3 depend upon Postulate 9-3. This postulate assures us that points R and S exist.

2. It is easy to prove that every point on $\overleftrightarrow{RS}$, the perpendicular bisector of $\overline{AB}$, is equidistant from points A and B.

3. The converse is also true; that is, every point that is equidistant from points A and B lies on $\overleftrightarrow{RS}$, the perpendicular bisector of $\overline{AB}$.

CONSTRUCTION 4 To copy a given angle on a given side of a given ray.

Problem

Given $\angle A$ and a ray with endpoint at R (FIG. 9-14), how do we construct an angle with vertex R and having the given ray as a side, congruent to $\angle A$?

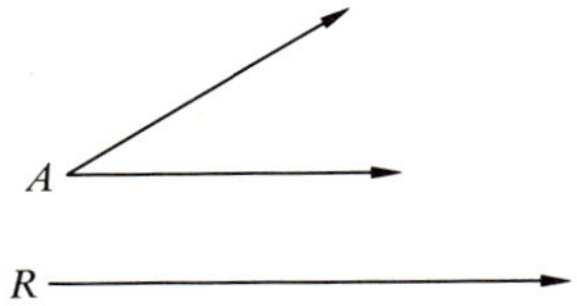

FIGURE 9-14

Solution

1. With A as center, draw an arc intersecting the sides of the given angle in points B and C (FIG. 9-15).

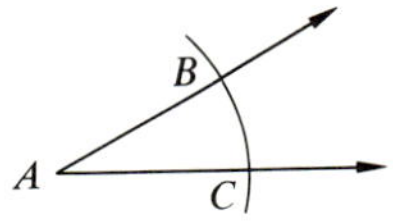

FIGURE 9-15

2. Without changing the radius, draw an arc with R as center, intersecting the given ray in T (FIG. 9-16).

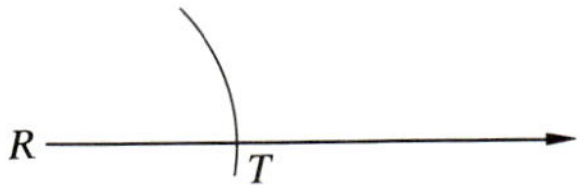

FIGURE 9-16

3. With T as center and BC as radius, draw an arc intersecting the arc in step 2 in S (FIG. 9-17).

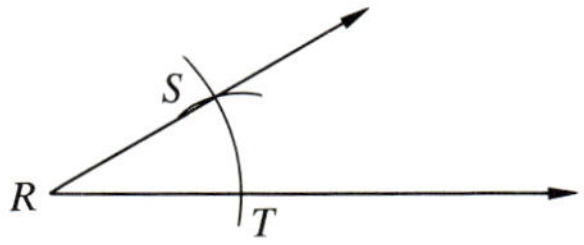

FIGURE 9-17

4. Draw ray $\overrightarrow{RS}$. $\angle SRT \cong \angle BAC$.

Proof

1. $\triangle ABC \cong \triangle RST$ (SSS).

2. $\angle SRT \cong \angle BAC$ since they are corresponding angles of congruent triangles (FIG. 9-18).[3]

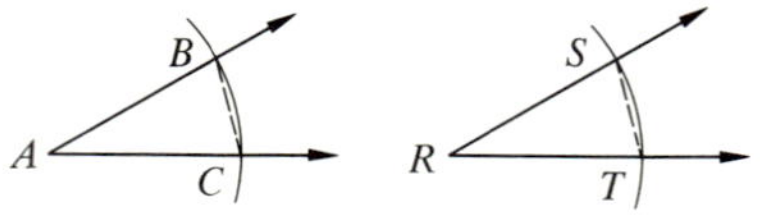

FIGURE 9-18

CONSTRUCTION 5 To bisect an angle.

Problem

Given $\angle A$ (FIG. 9-19), how can you construct a ray with endpoint A that will bisect $\angle A$?

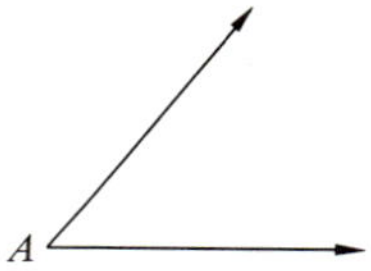

FIGURE 9-19

Solution

1. With A as center, draw an arc intersecting the sides of $\angle A$ at B and C (FIG. 9-20).

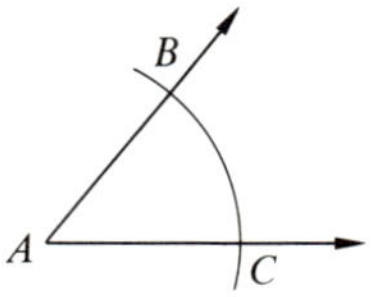

FIGURE 9-20

2. With B as center, and a radius greater than half of BC, draw an arc in the interior of $\angle A$ (FIG. 9-21). Keeping the same radius and with C as center, draw an arc that intersects the first arc in point P.

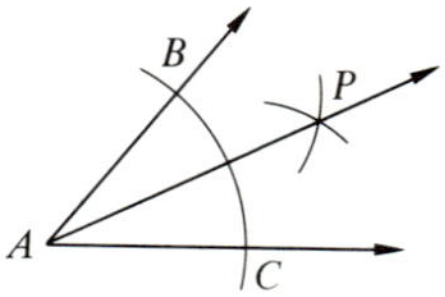

FIGURE 9-21

[3] 1. There are *two* angles with vertex R and the given ray $\overrightarrow{RT}$ as a side that can be constructed congruent to $\angle A$, as shown in FIG. 9A. The proof that $\angle S'RT \cong \angle BAC$ is similar to the proof that $\angle SRT \cong \angle BAC$.

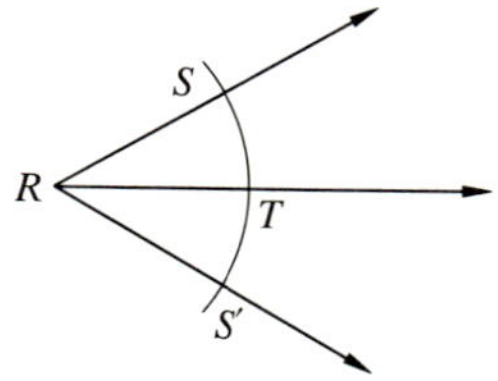

FIGURE 9A

2. In this construction we are usually given a ray with an endpoint as was done here. If the ray is not given, we start by drawing the ray that is to be one of the sides of the angle.

3. When we construct the angle at R congruent to $\angle A$, we say that we have "copied" $\angle A$.

3. Draw the ray $\overrightarrow{AP}$. $\angle BAP \cong \angle PAC$. $\overrightarrow{AP}$ is called the *bisector* of $\angle BAC$.

Proof

$\triangle ABP \cong \triangle ACP$ (SSS). Therefore, $\angle BAP \cong \angle PAC$, since they are corresponding angles of congruent triangles (FIG. 9-22).[4]

[4] 1. The arcs in step 2 will

a. Not intersect at all if the radius chosen is less than $\frac{1}{2}BC$.
b. Intersect in one point if the radius chosen equals $\frac{1}{2}BC$.
c. Intersect in two distinct points if the radius chosen is greater than $\frac{1}{2}BC$.

2. Every point on $\overrightarrow{AP}$ is equidistant from $\overrightarrow{AB}$ and $\overrightarrow{AC}$.

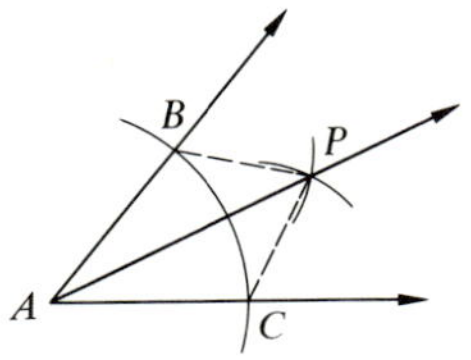

FIGURE 9-22

CONSTRUCTION 6 To copy a given triangle on a given side of a given ray.

Problem

Given $\triangle ABC$ and a ray with endpoint at D (FIG. 9-23), how can you construct a $\triangle DEF$ congruent to $\triangle ABC$, with the side $\overline{DF}$ lying on the given ray?

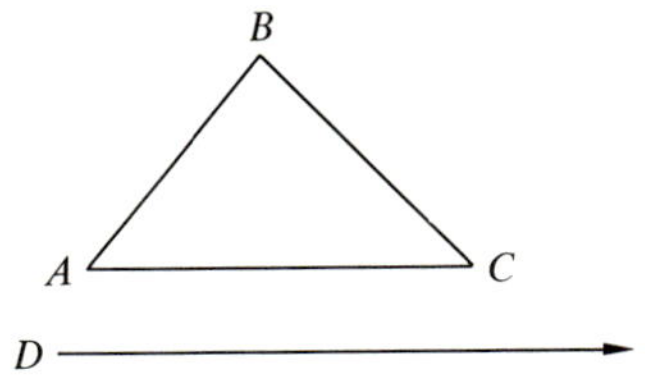

FIGURE 9-23

Solution

See FIG. 9-24.

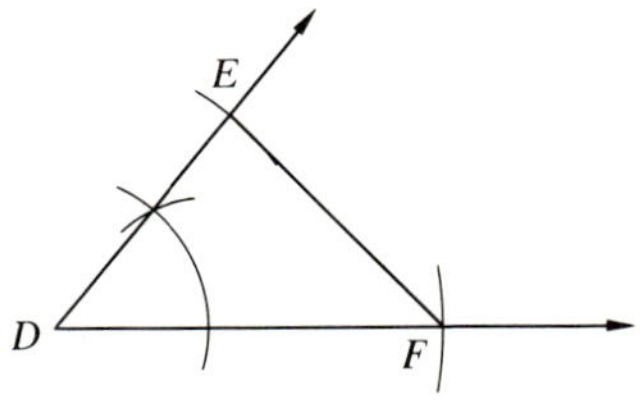

FIGURE 9-24

[5] We can construct $\triangle DEF \cong \triangle ABC$ in two other ways: by using either of the two other congruence properties, SSS or ASA.

1. Using the SSS property, the construction would look like this:
   a. With D as center and AC as radius, draw an arc intersecting the given ray in F (FIG. 9B).

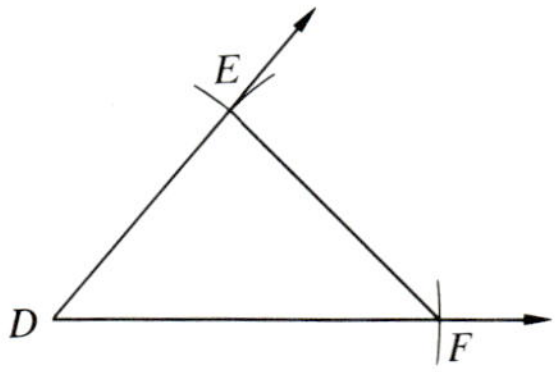

FIGURE 9B

   b. With D as center and AB as radius, draw an arc; with F as center and BC as radius, draw an arc intersecting the first arc in E.
   c. $\triangle DEF \cong \triangle ABC$ (SSS).
2. Using the ASA property, the construction would look like this:
   a. With D as center and AC as radius, draw an arc intersecting the given ray in F (FIG. 9C).

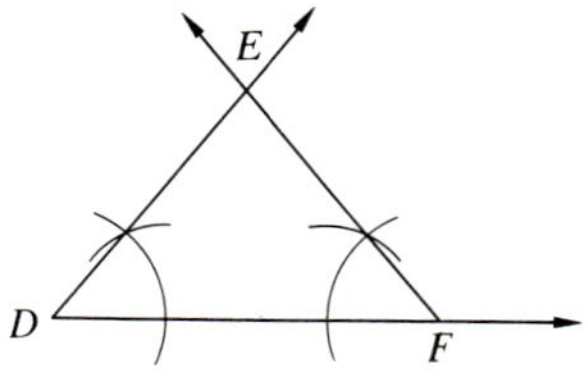

FIGURE 9C

   b. At D construct $\angle A$; at F construct $\angle C$. Extend both rays until they intersect in E.
   c. $\triangle DEF \cong \triangle ABC$ (ASA).

1. With D as center and AC as radius, draw an arc intersecting the given ray in F. $DF = AC$.
2. With D as vertex and $\overrightarrow{DF}$ as a side, construct $\angle D \cong \angle A$.
3. With D as center and AB as radius, draw an arc intersecting the other side of $\angle D$ at E. $DE = AB$. $\triangle DEF$ is the required triangle.

Proof
$\triangle DEF \cong \triangle ABC$ (SAS).[5]

CONSTRUCTION 7 To construct a perpendicular to a given line through a given point on the line.

Problem
Given line $\overleftrightarrow{AB}$ with P on the line (FIG. 9-25), how can you construct a line $\overleftrightarrow{PE} \perp \overleftrightarrow{AB}$ at point P?

FIGURE 9-25

Solution
1. With center at P, draw an arc intersecting $\overleftrightarrow{AB}$ at C and D (FIG. 9-26).

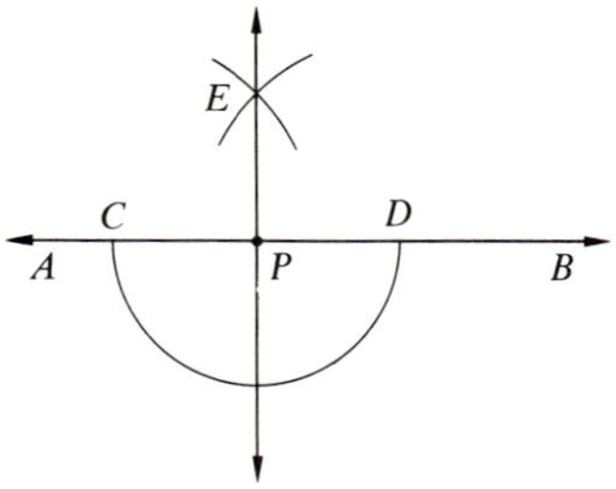

FIGURE 9-26

2. With C and D as centers and a radius greater than $\frac{1}{2}CD$, draw two arcs intersecting in E.
3. The line $\overleftrightarrow{PE}$ is perpendicular to the given line $\overleftrightarrow{AB}$ at P.

Proof

1. $\triangle CPE \cong \triangle DPE$ (SSS) (FIG. 9-27).

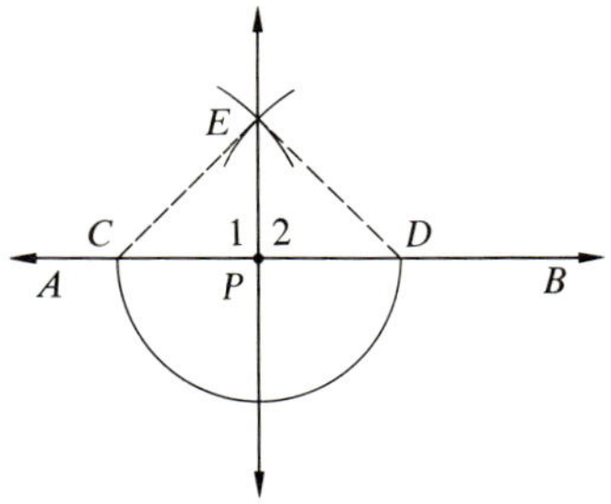

FIGURE 9-27

2. $\angle 1 \cong \angle 2$ (corresponding angles of congruent triangles).
3. $\angle 1$ and $\angle 2$ are right angles since they are congruent and supplementary.
4. Therefore, $\overleftrightarrow{EP} \perp \overleftrightarrow{AB}$, by definition.[6]

[6] Our construction, in effect, was to make P the center of a segment $\overline{CD}$ of the given line $\overleftrightarrow{AB}$, and then construct the perpendicular bisector of $\overline{CD}$.

CONSTRUCTION 8 To construct a perpendicular to a given line through a given point not on the line.

Problem

Given line $\overleftrightarrow{AB}$ and point P not on the line (FIG. 9-28), how can you construct a perpendicular to $\overleftrightarrow{AB}$ through P?

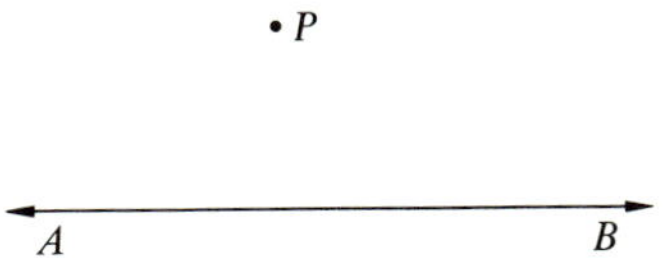

FIGURE 9-28

Solution

1. Choose any point C on $\overleftrightarrow{AB}$. With center at P and radius PC, draw an arc intersecting $\overleftrightarrow{AB}$ in C and D (FIG. 9-29).
2. With C and D as centers and a radius $> \frac{1}{2}CD$, draw two arcs intersecting at E.
3. Draw $\overleftrightarrow{PE}$. $\overleftrightarrow{PE}$ is perpendicular to $\overleftrightarrow{AB}$.

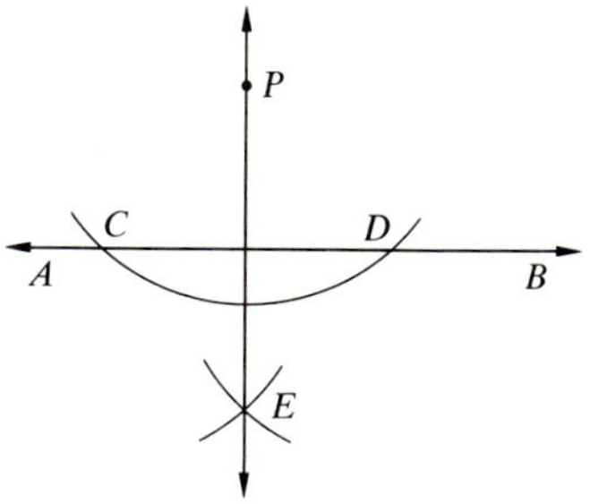

FIGURE 9-29

Proof

1. P and E are each equidistant from C and D by construction (FIG. 9-30).

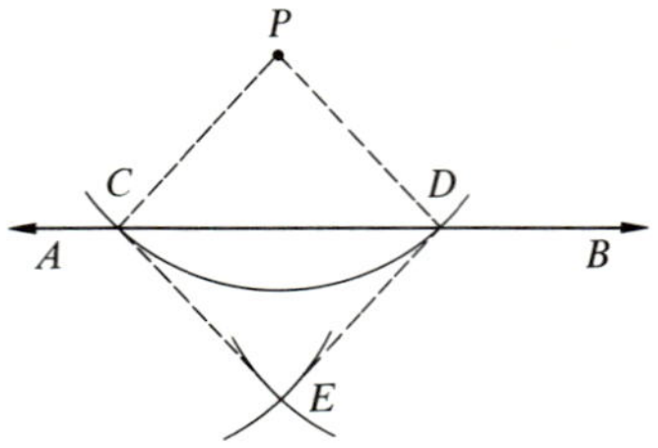

FIGURE 9-30

2. P and E must lie on the perpendicular bisector of $\overline{CD}$, by Construction 3.
3. Therefore, $\overleftrightarrow{PE} \perp \overleftrightarrow{AB}$ by definition.

CONSTRUCTION 9 To construct a parallel to a given line, through a given point not on the line.

Problem

Given line $\overleftrightarrow{AB}$ and point P not on the line (FIG. 9-31), how can you construct a line that contains the point P and is parallel to $\overleftrightarrow{AB}$?

FIGURE 9-31

Solution

1. Choose any point C of the line $\overleftrightarrow{AB}$ and join P to C (FIG. 9-32).

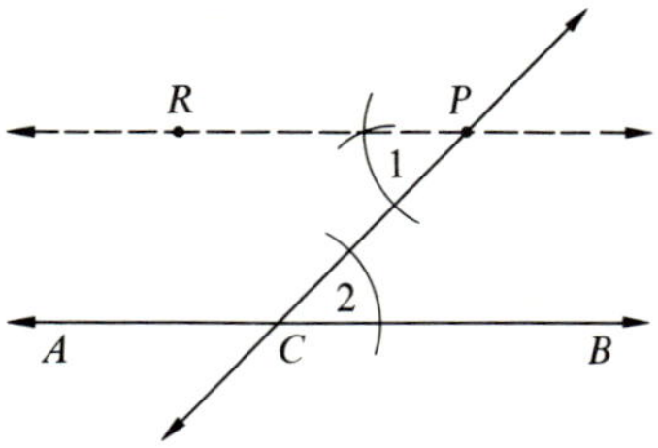

FIGURE 9-32

2. Construct $\angle CPR$ congruent to $\angle PCB$.
3. $\overleftrightarrow{RP}$ is parallel to $\overleftrightarrow{AB}$.

Proof

1. $\angle 1 \cong \angle 2$ by construction.
2. $\overleftrightarrow{RP} \parallel \overleftrightarrow{AB}$, because if two lines are cut by a transversal so that a pair of alternate interior angles is congruent, then the two lines are parallel.

CONSTRUCTION 10 To divide a segment into a given number of congruent segments.

Problem

Given segment $\overline{AB}$ (FIG. 9-33), how can you divide this segment into, say, three congruent segments?

FIGURE 9-33

Solution

1. Draw any ray $\overrightarrow{AT}$, not on the line $\overleftrightarrow{AB}$.
2. Choose any radius on your compass and lay off on $\overrightarrow{AT}$ three congruent segments $\overline{AT_1}$, $\overline{T_1T_2}$, $\overline{T_2T_3}$, end to end.
3. Join T_3 to B. Through the other points T_1, T_2, construct lines parallel to $\overline{T_3B}$. These lines intersect $\overline{AB}$ in B_1, B_2 (FIG. 9-34).

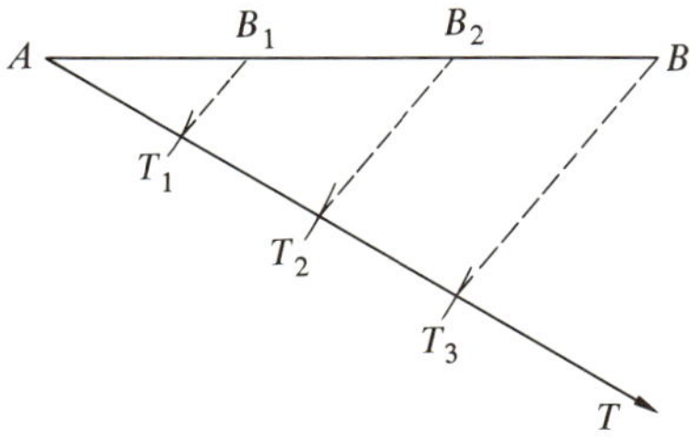

FIGURE 9-34

4. The points B_1, B_2 divide $\overline{AB}$ into three congruent segments.

Proof

1. $\overleftrightarrow{T_1B_1} \parallel \overleftrightarrow{T_2B_2} \parallel \overleftrightarrow{T_3B}$ by construction.
2. $\overline{AT_1} \cong \overline{T_1T_2} \cong \overline{T_2T_3}$ by construction. That is, the three parallel lines are intercepting congruent segments on the transversal $\overrightarrow{AT}$.
3. Therefore, $\overline{AB_1} \cong \overline{B_1B_2} \cong \overline{B_2B}$, because if three or more parallel lines intercept congruent segments on one transversal, then they intercept congruent segments on any other transversal.[7]

[7] 1. In this construction we showed the case where the given segment is divided into three congruent segments. If we wish to divide the given segment into *n* congruent segments, we follow a similar procedure by drawing *n* parallel lines.

2. This construction suggests an easy way to divide a segment $\overline{AB}$ into, say, five congruent segments by using a sheet of ruled paper (assuming the lines are equally spaced and sufficiently close together), FIG. 9D.

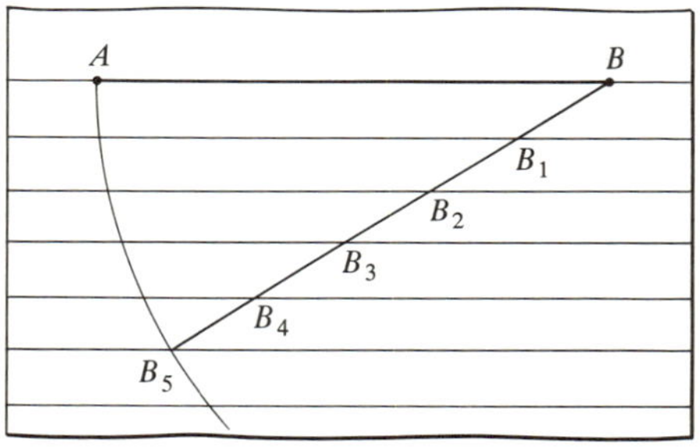

FIGURE 9D

CONSTRUCTION 11 To circumscribe a circle about a given triangle.

Definition

A circle is *circumscribed* about a triangle if each vertex of the triangle lies on the cricle. We say that circle O is *circumscribed* about $\triangle ABC$, or that $\triangle ABC$ is *inscribed* in circle O (FIG. 9-35).

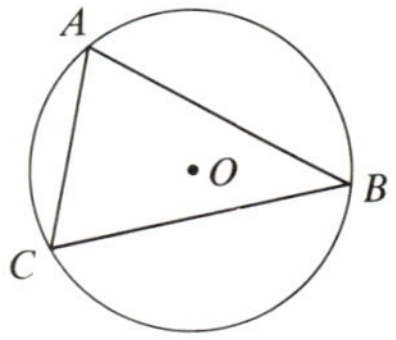

FIGURE 9-35

Problem

Given $\triangle ABC$ (FIG. 9-36), how can you circumscribe a circle around the given triangle?

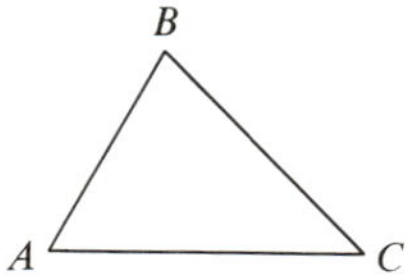

FIGURE 9-36

Solution

1. Construct the perpendicular bisectors of two sides of the triangle, intersecting in point P (FIG. 9-37).
2. With P as center and PA as radius, construct a circle. This circle will also pass through B and C and is the desired circle.

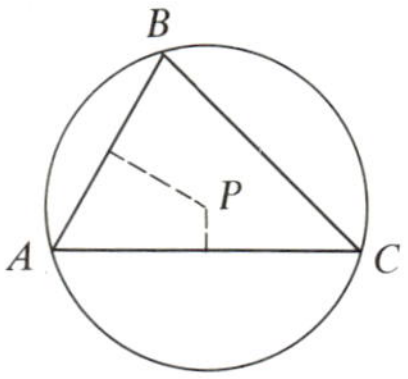

FIGURE 9-37

Proof

The proof of this construction rests on Theorem 6-5 which states that: *The perpendicular bisectors of the sides of a triangle meet in a point equidistant from the three vertices of the triangle.*

1. Since the three bisectors meet in a point, P is this point.
2. Since P is equidistant from the three vertices, $AP = BP = CP$, and the circle passes through A, B, and C.

CONSTRUCTION 12 To inscribe a circle in a given triangle.

Definition

A circle is *inscribed* in a triangle if each side of the triangle is tangent to the circle (FIG. 9-38). We say that circle O is *inscribed* in $\triangle ABC$, or that $\triangle ABC$ is *circumscribed* about the circle.

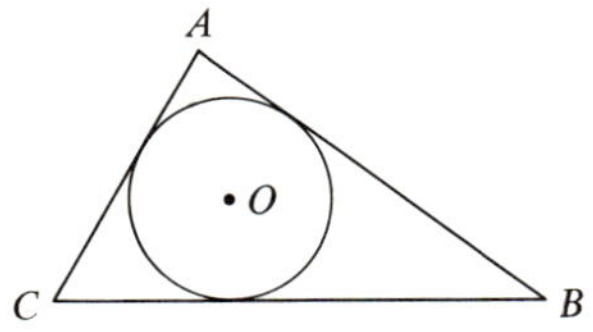

FIGURE 9-38

Problem

Given $\triangle ABC$ (FIG. 9-39), how can you inscribe a circle in the given triangle?

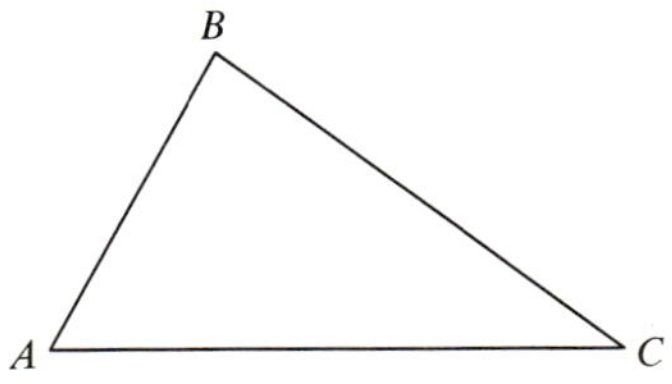

FIGURE 9-39

Solution

1. Bisect $\angle A$ and $\angle B$, and let P be the point where the bisectors intersect (FIG. 9-40).
2. Construct a perpendicular $\overline{PD}$ from P to $\overline{AC}$.
3. With P as center and PD as radius, construct a circle. This circle is tangent to all three sides of $\triangle ABC$ and is therefore the desired circle.

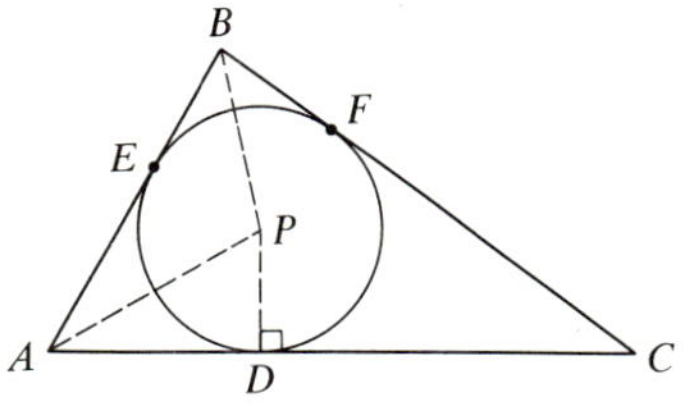

FIGURE 9-40

Proof

The proof of this construction rests on Theorem 6.4, which states that: *The angle bisectors of a triangle meet in a point equidistant from the three sides.*

1. Since the three bisectors meet in a point, P is this point.
2. Since the point P is equidistant from the three sides, $PD = PE = PF$, where E and F are the feet of the perpendiculars from P to $\overline{AB}$ and $\overline{BC}$, respectively, and the circle passes through D, E, and F.
3. The circle is tangent to $\overline{AC}$ because $\overline{AC}$ is perpendicular to radius $\overline{PD}$, and any line perpendicular to a radius at its outer end is tangent to the circle. For the same reason, the circle is also tangent to sides $\overline{AB}$ and $\overline{BC}$.

CONSTRUCTION 13 To construct the fourth proportional to three given line segments.

Problem

Given the segments a, b, and c (that is, the lengths of the segments) (FIG. 9-41), how can you construct a fourth segment

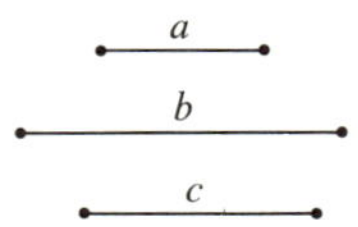

FIGURE 9-41

x such that

$$\frac{a}{b} = \frac{c}{x} \quad ?$$

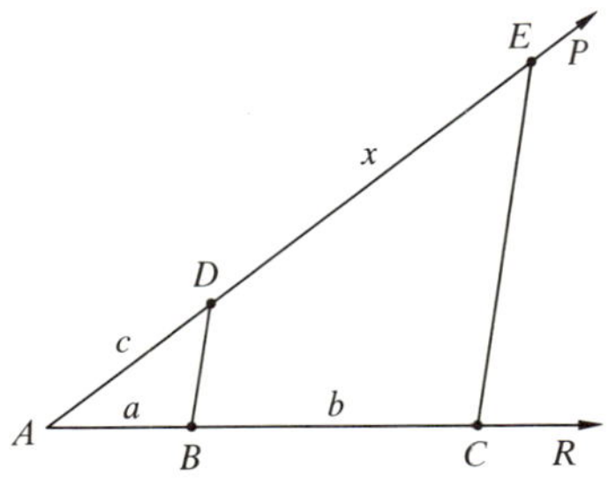

FIGURE 9-42

Solution

1. Draw any convenient angle $\angle PAR$ (FIG. 9-42).
2. On ray $\overrightarrow{AR}$ of $\angle PAR$ construct $AB = a$, $BC = b$.
3. On ray $\overrightarrow{AP}$ construct $AD = c$.
4. Draw $\overline{BD}$. Construct $\overline{CE} \parallel \overline{BD}$.
5. $DE = x$.

Proof

1. $\overline{BD}$ is parallel to side $\overline{CE}$ of $\triangle ACE$ by construction.

2. Therefore,

$$\frac{a}{b} = \frac{c}{x},$$

because if a line parallel to one side of a triangle intersects the other two sides in distinct points, then it cuts off segments that are proportional to these sides.

CONSTRUCTION 14 To construct the geometric mean (or mean proportional) between two given line segments.

Problem

Given line segments a and b (FIG. 9-43), how can you construct a third segment x such that

$$\frac{a}{x} = \frac{x}{b} \quad ?$$

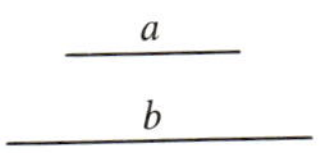

FIGURE 9-43

Solution

1. On ray $\overrightarrow{AR}$ construct $AB = a$, $BC = b$ (FIG. 9-44).

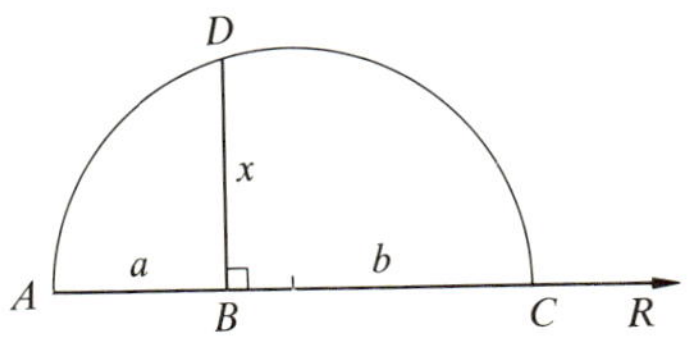

FIGURE 9-44

2. On $\overline{AC}$ as a diameter, construct semicircle $\widehat{ADC}$.

3. At point B construct $\overline{DB} \perp \overline{AC}$.

4. $DB = x$.

Proof

1. $\angle ADC$ is a right angle, since it is inscribed in a semicircle (FIG. 9-45).

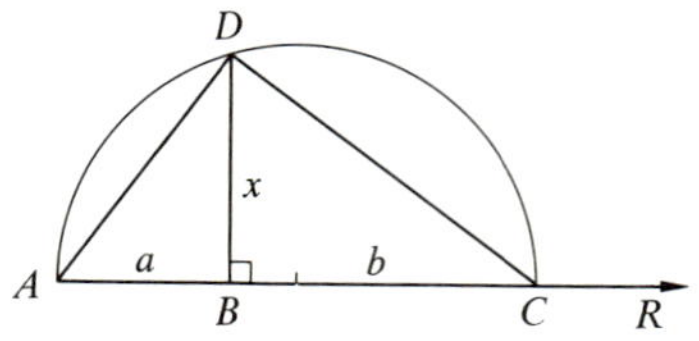

FIGURE 9-45

2. $\overline{DB}$ is the altitude to the hypotenuse $\overline{AC}$ of right triangle $\triangle ADC$ by construction.
3. Therefore,

$$\frac{AB}{DB} = \frac{DB}{BC};$$

that is,

$$\frac{a}{x} = \frac{x}{b},$$

because in any right triangle, the altitude to the hypotenuse is the geometric mean of the segments into which it separates the hypotenuse.

CONSTRUCTION 15 To divide a line segment into segments proportional to given segments.

Problem

Given line segments $a, b, c, \overline{AB}$ (FIG. 9-46), how can you divide $\overline{AB}$ into segments proportional to a, b, and c?

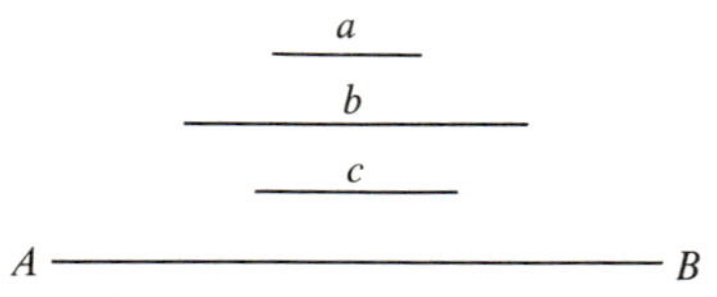

FIGURE 9-46

Solution

1. Draw any ray $\overrightarrow{AR}$ forming a convenient angle $\angle BAR$ (FIG. 9-47).

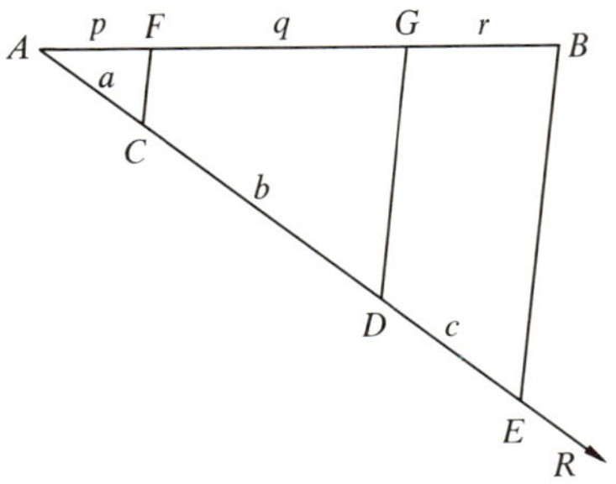

FIGURE 9-47

2. On $\overrightarrow{AR}$ construct $AC = a$, $CD = b$, $DE = c$.
3. Draw $\overline{EB}$. Construct $\overline{CF}$ and $\overline{DG}$ parallel to $\overline{EB}$.
4. F and G are the required points of division.

Proof

1. By construction, parallel lines $\overleftrightarrow{CF}$, $\overleftrightarrow{DG}$, $\overleftrightarrow{EB}$ are cut by transversals $\overrightarrow{AB}$ and $\overrightarrow{AR}$.
2. $\dfrac{p}{a} = \dfrac{q}{b} = \dfrac{r}{c}$,

   because if three parallel lines are cut by two transversals, then the lengths of the segments intercepted on one transversal are proportional to the lengths of the corresponding segments intercepted on the other transversal.

TEACHING SUGGESTIONS AND ACTIVITIES

VISUAL AIDS

By creating devices to draw curves other than the circle and the straight line, we can obviously enlarge the number of constructible figures. We shall now describe several such devices that can be made easily.

Visual 1

PURPOSE

A device with which to trisect any angle.

CONSISTS OF

A "geometer's square," an instrument that can be made of a piece of cardboard having the shape shown in FIG. 9-48. Note that

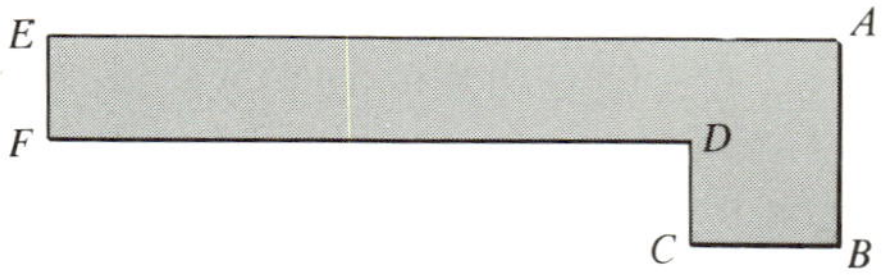

FIGURE 9-48

1. All the angles are right angles.
2. $EF = CD = \frac{1}{2}AB$.

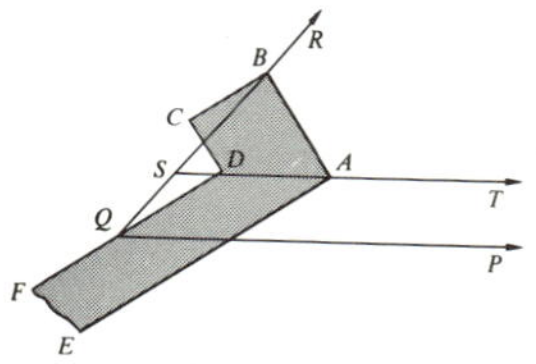

FIGURE 9-49

HOW TO USE

1. To trisect $\angle PQR$ use the long side to construct $\overrightarrow{ST} \parallel \overrightarrow{QP}$ at distance EF (FIG. 9-49).
2. Then place the geometer's square so that $\overleftrightarrow{DF}$ passes through Q, A lies on $\overrightarrow{ST}$, and B lies on $\overrightarrow{QR}$.
3. Then $m\angle PQA = \frac{1}{3}(m\angle PQR)$.

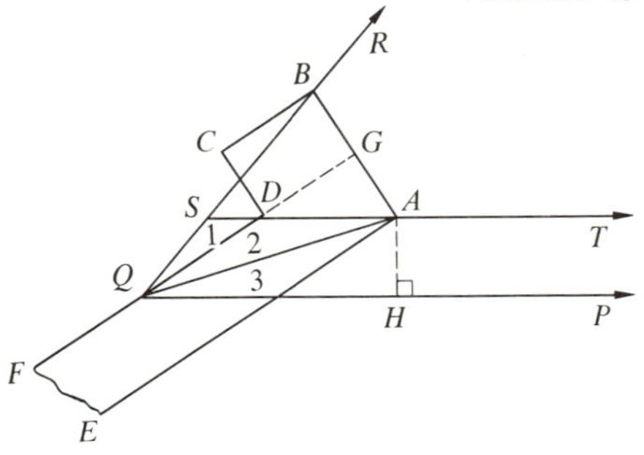

FIGURE 9-50

PROOF

1. Draw $\overline{AH} \perp \overrightarrow{QP}$, and let $\overleftrightarrow{QD}$ intersect $\overline{AB}$ at G (FIG. 9-50).
2. $\triangle QHA \cong \triangle QGA \cong \triangle QGB$.
3. $\angle 1 \cong \angle 2 \cong \angle 3$.
4. Therefore, $\overrightarrow{QA}$ and $\overrightarrow{QG}$ are trisectors of $\angle PQR$.

Visual 2

PURPOSE

Another device to trisect an angle, attributed to Archimedes.

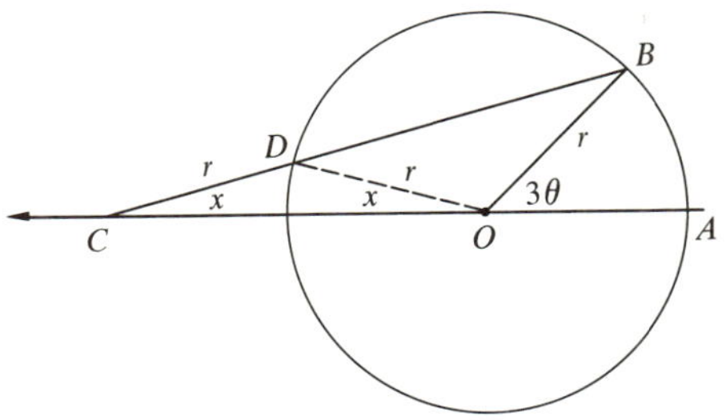

FIGURE 9-51

CONSISTS OF

A ruler with two points marked on it at distance r apart.

HOW TO USE

1. Let $\angle AOB$ be the given angle, and let $m\angle AOB = 3\theta$ (FIG. 9-51).
2. Describe circle (O) with arbitrary radius $OA = r$.
3. Mark the points C, D on a ruler at distance r apart.
4. Then, sliding point C along line $\overleftrightarrow{AO}$ produced, align the points C, D, B along the edge of the ruler so that C lies on $\overleftrightarrow{AO}$ and D lies on the circle.
5. $\angle OCB$ is the required angle whose measure is equal to $\frac{1}{3}$ of the given $\angle AOB$.

Note: For other methods of trisecting an angle see R. C. Yates, *The Trisection Problem*, Baton Rouge, La.; The Franklin Press, 1942.

PROOF

1. $m\angle B = 3\theta - x$, since $m\angle B + x = 3\theta$.
2. $m\angle BDO = 3\theta - x$, since $\triangle OBD$ is isosceles.
3. $3\theta - x = x + x$, since $m\angle BDO = x + x$.
4. $3x = 3\theta$.
5. $x = \theta$.

Visual 3

PURPOSE

A device with which to draw a *cycloid*. [These curves were used by C. Ptolemy (ca. 85–ca. 165) to describe the movements of the planets.]

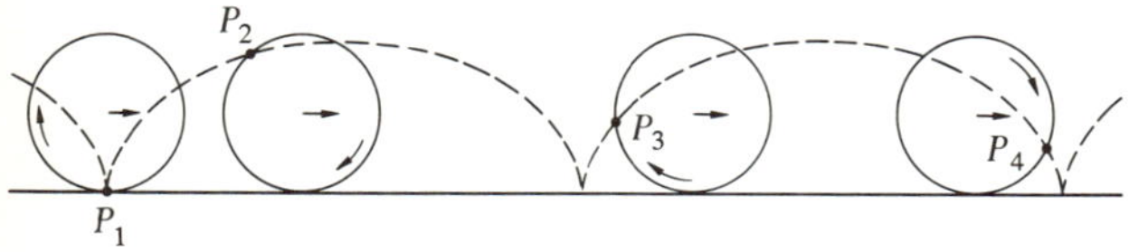

FIGURE 9-52

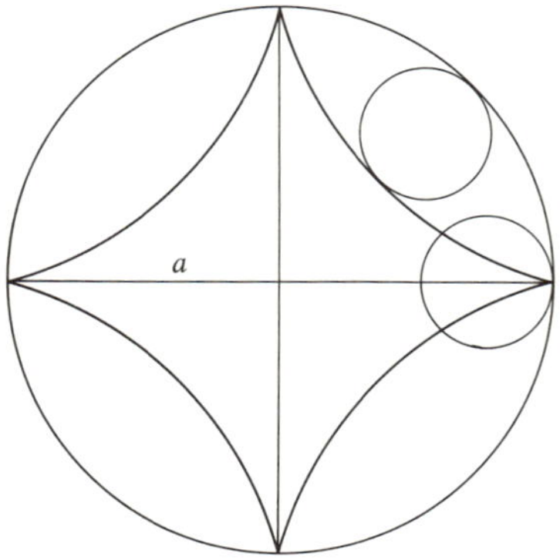

FIGURE 9-53

CONSISTS OF

1. A circular disc with a pencil inserted through a fixed point P near the circumference of the disc.
2. A grooved track on which to roll the disc. A stiff piece of writing paper is tacked to the back of the track.

HOW TO USE

1. A cycloid will be described on the paper as the disc is rolled along the track without slipping. The cycloid will look like a series of arches resting on the line (FIG. 9-52).
2. Variations of the cycloid may be obtained by choosing the point P elsewhere on the disc or by allowing the disc to roll not along a straight line, but on another circle. For example, the path traced by a point on the circumference of a circle of radius $a/4$ which rolls around the inside of a circle of radius a is a *hypocycloid of four cusps* (FIG. 9-53).

Visual 4

PURPOSE

To exhibit modern instruments used in drawing geometric figures.

CONSISTS OF

Ruler, compass, protractor, T-square, parallel ruler, proportional divider, draftsman's triangle, French curve, linkages.

HOW TO USE

Exhibit to class in connection with Introducing the Chapter to Students.

GEOMETRY THROUGH PAPER-FOLDING

Paper-folding can provide a different and interesting way of performing geometric constructions and verifying properties of geometric figures. The only "tools" required are paper, a pencil, and a pair of scissors. Waxed paper is preferable because it shows the creases better.

With a few assumptions, it can be shown that all Euclidean plane constructions can be accomplished by folding and creasing. In some instances, a construction can be accomplished more easily by paper-folding than with the use of a straightedge and compass. Examples of this are the division of segments and angles into two or more congruent parts, and the construction of perpendiculars and parallels to straight lines.

We assume that a *crease* represents a straight line. We can justify this assumption on two grounds: (1) The intersection of two planes is a line; and (2) if A and B are two points on the paper that are brought into coincidence by the folding process, then any point C on the crease must be equidistant from A and B, since the lines $\overleftrightarrow{AC}$ and $\overleftrightarrow{BC}$ are folded into coincidence. Therefore, the crease, being the locus of all such points C, is the perpendicular bisector of $\overline{AB}$.

We also make the following assumptions:

1. One point may be placed upon another, thus creating a crease on the sheet of paper.
2. This crease may be created through two distinct points.
3. A given point may be placed upon a given line in such a way that the resulting crease passes through a second given point. This implies the ability to fold a crease over upon itself or upon another crease.

It is evident from what has already been said that the basic idea on which paper-folding rests is that of congruence by superposition. By folding a piece of paper in a certain way we form congruent figures. In folding an edge of the paper over itself in such a way that the endpoints coincide, we are forming two congruent segments, the intersection of the crease with the edge of the paper being the midpoint of the edge. By folding a corner of a piece of paper in such a way that the sides of the angle coincide, we are bisecting the angle by dividing it into two congruent angles through superposition.

Note: We shall use the word "construction" in connection with the results accomplished through paper-folding, although the word is generally reserved for drawings made only with straightedge and compass.

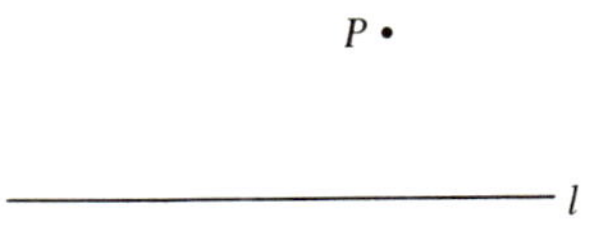

FIGURE 9-54

Elementary Constructions by Paper-Folding

PROBLEM 1 Given line l and point P not on l (FIG. 9-54), how can you construct a perpendicular to l and passing through P?

Solution
Fold l over upon itself so that the crease passes through P (FIG. 9-55).

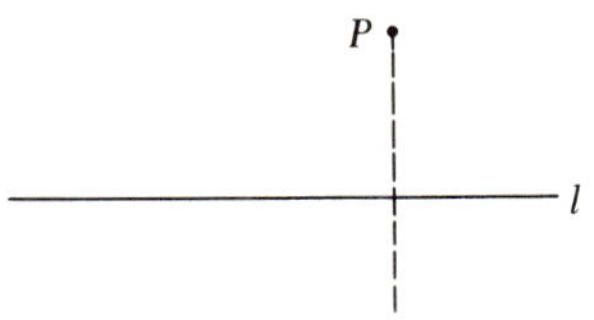

FIGURE 9-55

FIGURE 9-56

PROBLEM 2 Given line l and point P not on l (FIG. 9-56), how can you construct a line parallel to l and passing through P?

Solution

1. Construct $l_1 \perp l$, passing through P (FIG. 9-57).
2. Construct $l_2 \perp l_1$, passing through P. l_2 is the required line.

PROBLEM 3 Given $\triangle ABC$ (FIG. 9-58), how can you construct the altitude from B to $\overline{AC}$?

Solution

Fold A onto line $\overleftrightarrow{AC}$ so that the crease passes through B (by assumption 3) (FIG. 9-59). The crease $\overline{BD}$ is the sought altitude.[8]

[8] This model verifies two important theorems:

1. The sum of the measures of the angles of a triangle is 180.

2. The segment joining the midpoints of two sides of a triangle is parallel to the third side, and the length of the segment is half the length of the third side.

PROBLEM 4 Given a triangle, how can you construct its angle bisectors?

Solution

1. Fold each side over upon the other two sides (FIG. 9-60).
2. The creases represent the angle bisectors and the intersection of the creases is the incenter.

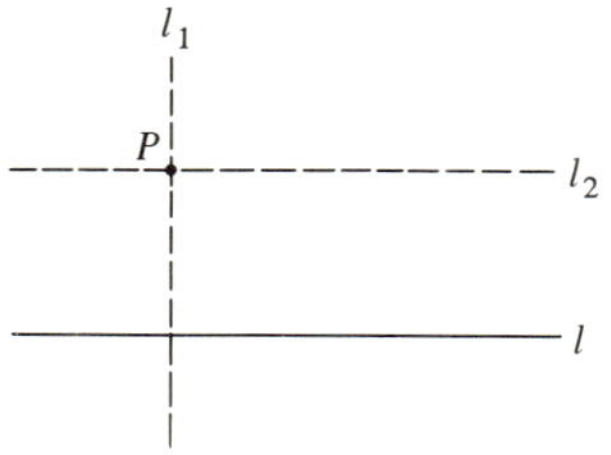

FIGURE 9-57

PROBLEM 5 Given a triangle, how can you construct the perpendicular bisectors of its sides?

Solution

1. Fold each vertex over upon the other two (FIG. 9-61).
2. Each crease represents a perpendicular bisector, and the intersection of the creases is the circumcenter.

PROBLEM 6 Given a triangle, how can you construct its altitudes?

Solution

Fold each side over upon itself so that the crease passes through the opposite vertex (FIG. 9-62). These creases represent the altitudes, and the intersection of the creases is the orthocenter.

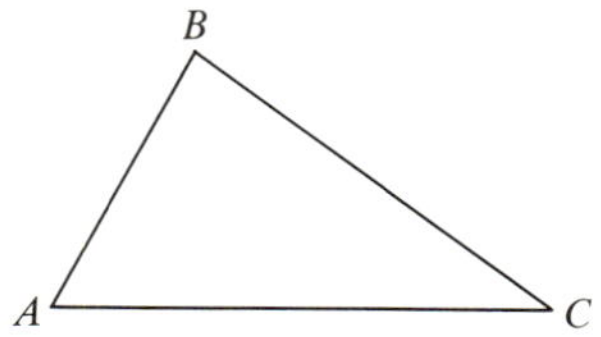

FIGURE 9-58

PROBLEM 7 Given a triangle, how can you construct its medians?

Solution

Form a crease through each vertex and midpoint of the opposite side (FIG. 9-63). These creases represent the medians, and their intersection is the centroid.

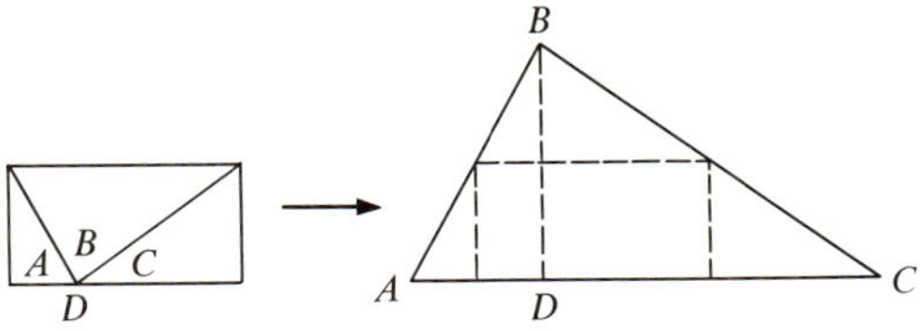

FIGURE 9-59

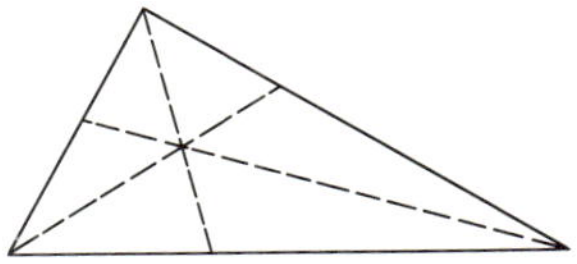

FIGURE 9-60

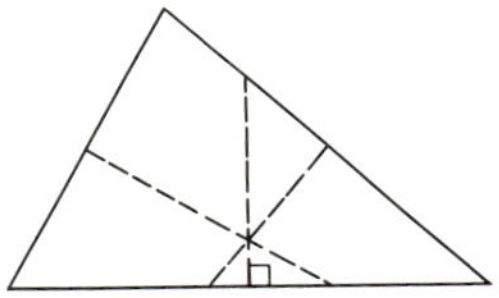

FIGURE 9-61

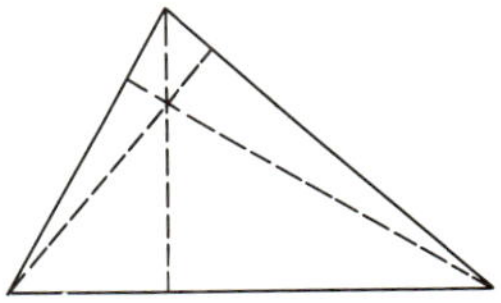

FIGURE 9-62

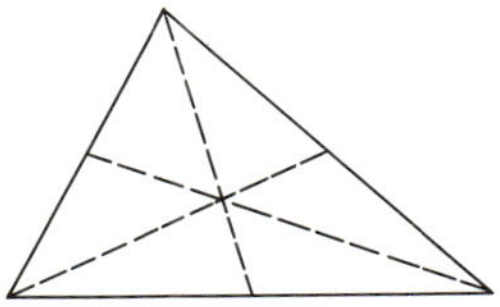

FIGURE 9-63

Other Constructions and Uses

These elementary constructions can be used to construct other geometric figures such as regular polygons and even the conic sections. Paper-folding can be used to prove the Pythagorean theorem and to divide a line segment by the "golden section" into extreme and mean ratio. That is, we can divide the segment in such a way that the shorter section is to the longer section as the longer section is to the whole line segment: $a/b = b/(a + b)$. We can also construct regular polygons and regular polyhedra through paper-folding.

Paper-folding can be used to test properties relating to concurrence of lines and collinearity of points. For instance, students can test the following theorems:

1. The medians of a triangle are concurrent.
2. The altitudes of a triangle are concurrent.
3. The perpendicular bisectors of the sides of a triangle are concurrent.
4. The angle bisectors of a triangle are concurrent.
5. In any triangle, the orthocenter, circumcenter, and centroid are collinear.
6. The points of intersection of the external bisectors of the angles of a triangle and the opposite sides are collinear.

The teacher can extend the list of paper-folding activities indefinitely.

QUESTIONS

1. Our construction of an angle bisector was based on the SSS congruence property. Show a method of constructing an angle bisector that is based on the SAS congruence property.
2. Construct an angle of (a) 45°, (b) 60°, (c) 30°, (d) 15°, (e) 75°.
3. Construct a square, given the length of its diagonal.
4. Construct a square, given the length of its perimeter.
5. Given length a. Construct an isosceles triangle in which a is the

perimeter and in which the length of one of the congruent sides is twice the length of the base.

6. Circumscribe a circle about an obtuse triangle.

7. Inscribe a circle in an obtuse triangle.

8. Circumscribe a circle about (1) a square, (b) a rectangle.

9. Given three noncollinear points, locate the center of the circle which passes through the three points.

10. Draw any $\triangle ABC$. Construct parallelogram $ABCD$.

11. Construct an equilateral triangle the length of whose median is 1 inch.

12. Given $\angle ABC$ with point P in its interior, through P construct a line that intersects $\overrightarrow{BA}$ and $\overrightarrow{BC}$ at points D and E such that $DB = EB$.

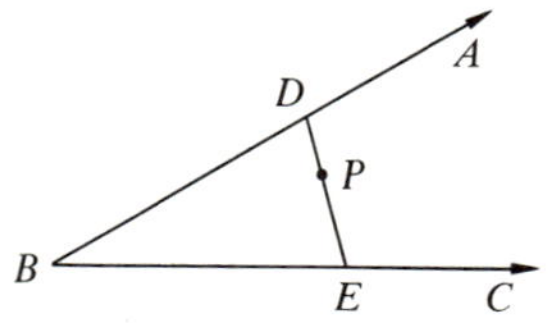

13. We wish to construct a triangle with sides of lengths a, b, and c. What condition must be true of a, b, and c for the construction to be possible?

14. Construct an equilateral triangle, given the radius of the inscribed circle.

15. Given one acute angle of a right triangle and the sum of the lengths of its legs, construct the triangle.

16. Construct a parallelogram if given one side, one acute angle, and the longer diagonal.

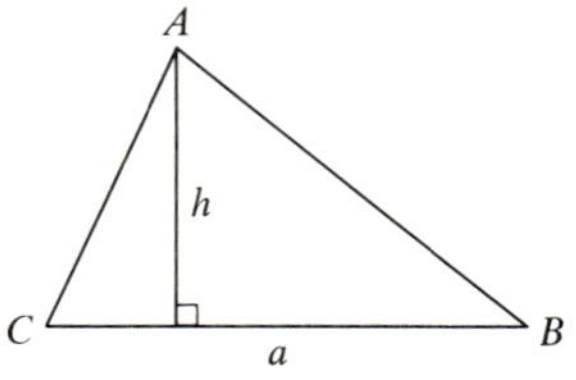

17. Given a, h, and B, construct the triangle.

18. Given point A on side $\overrightarrow{BC}$ of $\angle CBD$, find a point P on $\overrightarrow{BC}$ such that the distance from P to A is the same as the distance from P to $\overrightarrow{BE}$; that is, $AP = PD$.

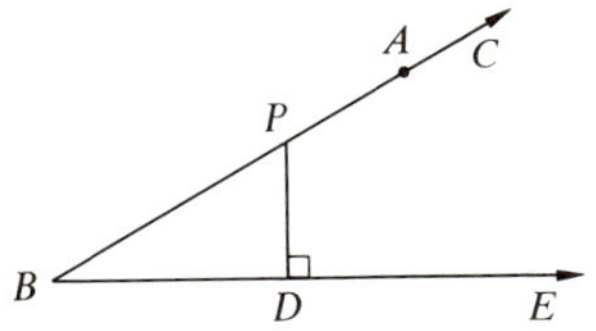

19. Given points A, B, C, and D, construct a square so that each side passes through one of these points.

PUZZLES

1. Given a circle but not the location of its center, how can you locate the center of the circle?

2. Inscribe a square in a given triangle.

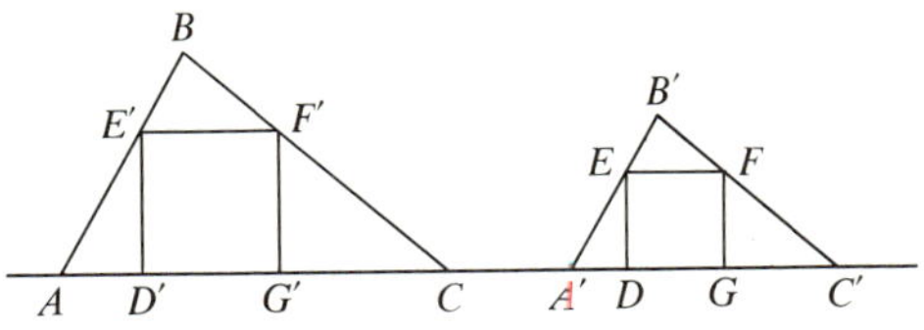

SOLUTIONS

1. Inscribe any triangle in the circle. The intersection of the perpendicular bisectors of any two sides is the center of the circle.

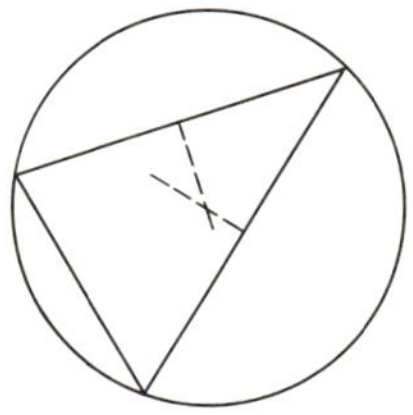

2. Our solution is based on the ideas of similarity.

   a. Let $\triangle ABC$ be the given triangle, and let $\overline{DG}$ be any arbitrary segment on the line $\overleftrightarrow{AC}$.
   b. Construct the square $DEFG$ on the same side of AC as $\triangle ABC$.
   c. Through E and F construct parallels to $\overline{AB}$ and $\overline{BC}$, determining with the line AC the $\triangle A'B'C'$.
   d. Construct the point E' such that $BE'/E'A = B'E/EA'$.
   e. Through E' construct lines perpendicular and parallel to AC meeting AC in D' and $\overline{BC}$ in F'.
   f. Construct $\overline{F'G'} \perp \overline{AC}$ and meeting it in G'. $D'E'F'G'$ is the required square.

3. Construct a square whose side is the length of the diagonal of the given square.

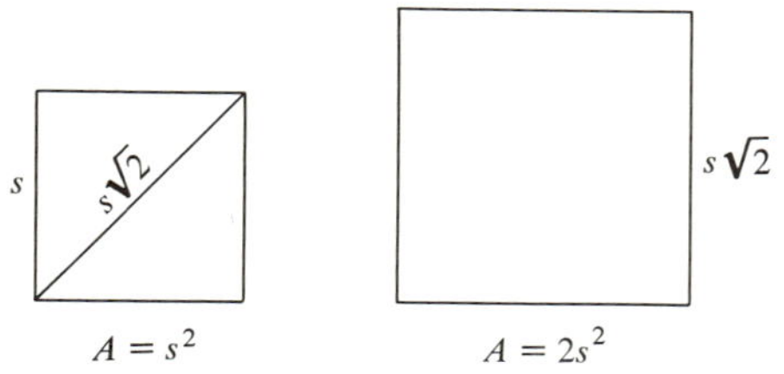

4. Let the given segment be $\overline{AB}$.

   a. Draw a circle with B as center and AB as radius.

   b. Starting with A as center and with radius AB, mark off on the circle the congruent arcs $\widehat{AC}$, $\widehat{CD}$, and $\widehat{DE}$.

   c. Points A, B, and E will be collinear and $AE = 2 \times AB$.

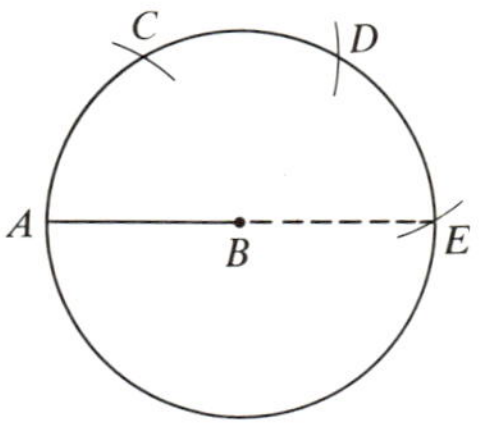

3. Construct a square whose area is double the area of a given square.

4. Assume that a straight line is considered as constructed when two of its points are located. How can you, *with only a compass*, double a given line segment?

ENRICHMENT

THE FAMOUS CONSTRUCTION PROBLEMS OF ANTIQUITY

During their study of geometry, the Greeks tried to solve all construction problems by the use of only two tools—the straightedge and compass. This was a kind of game with self-imposed restrictions. You could use these two tools only in a special way: With the straightedge you were permitted to draw a line segment of indefinite length through any two given distinct points; with the compass you were permitted to draw a circle with any given point as center and passing through any given second point.

But the Greeks soon learned that even though they could construct many complicated curves with these two tools alone, not all problems could be solved this way. Three of the most famous of these "failures" were the problems of how (1) to trisect any angle, (2) to construct a square whose area is equal to that of a given circle, and (3) to construct a cube whose volume is twice that of a given cube.

In their fruitless efforts to find solutions to these problems, the Greeks discovered the conic sections, cubic curves, and even some transcendental curves.

It was not until more than 2000 years later that the *reason* for the failure of the Greeks to solve these seemingly easy construction problems was established. The reason turned out to be that these problems are *impossible* of solution by straightedge and compass alone. This impossibility was established by algebraic means.

Trisection of an Angle

Given *any* angle $\angle BAC$ (FIG. 9-64), the problem is to construct, by straightedge and compass, two rays $\overrightarrow{AD}$ and $\overrightarrow{AE}$ so that $\angle BAD \cong \angle DAE \cong \angle EAC$.

The impossibility of this construction rests on a theorem which states: From *a given unit length*, it is impossible to construct with Euclidean tools a segment the magnitude of whose length is a root of a cubic equation with rational coefficients but no rational roots.

If we can prove that any *one* angle, say an angle of 60°, cannot be trisected, then we will have established the nonexistence of a *general* method of trisecting an angle.

The problem of trisecting an angle of 60° can be shown to be

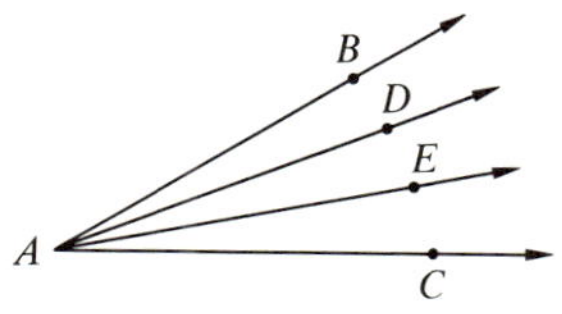

FIGURE 9-64

equivalent to the problem of constructing a segment the magnitude of whose length is a root of the cubic equation $8x^3 - 6x - 1 = 0$ that has no rational root.

Therefore, by the theorem just stated, this construction is impossible. Despite this proof, a new supply of "solutions" are offered each year, the latest publicized one in March 1969 (*New York Times*).

To say that the trisection of an angle is impossible is not to say that *no* angle can be trisected by straightedge and compass. Special angles such as 90°, 45°, or 135° can be so trisected.

If we remove the restriction of having to use *only* a straightedge and compass, the problem becomes solvable. A number of instruments have been devised for trisecting an angle. (See Visuals 1 and 2 on pages 256–257.)

Construction of Regular Polygons

Related to the problem of trisecting an angle is the series of problems of how to construct regular polygons of n sides (or n-gons) with straightedge and compass only. The early Greeks knew how to construct regular n-gons for $n = a \cdot 2^x\,(a = 2, 3, 5, 15; x = 1, 2, 3, \ldots)$. They were also able to construct regular polygons of 3 and 5 sides, but not those with 7, 11, 13, or 17 sides. It was not until about 200 years ago that a 19-year-old boy named Gauss found a way of constructing a regular polygon of 17 sides. In fact, he later proved that a necessary and sufficient condition that a regular n-gon be constructible by straightedge and compass alone is that n be a number of the form $n = 2^x \cdot p_1 \cdot p_2 \cdots p_k$, where $x = 0, 1, 2, \ldots$ and the $p_i (i = 1, \ldots, k)$ are primes of the form $2^{2^r} + 1$.

Polygons with a *prime number* of sides that may be constructed are as follows: From $p = 2^{2^r} + 1$, where p is the number of sides and r is any integer up to 4, we obtain

r	0	1	2	3	4
p	3	5	17	257	65,537

When $r > 4$, there are no known primes of the form $2^{2^r} + 1$. Therefore, the only regular polygons with a prime number of sides known to be constructible by straightedge and compass are those with 3, 5, 17, 257, and 65,537 number of sides.

When Gauss made this discovery he was so overwhelmed by it that he gave up his plans to become a philologist and decided to become a mathematician. After his death, a bronze statue of him was erected in Göttingen and a regular 17-gon was engraved on the base of the monument.

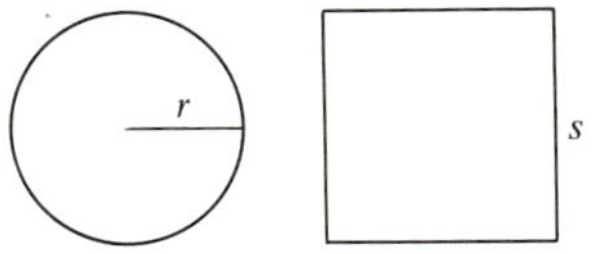

FIGURE 9-65

Squaring the Circle

Given a circle, the problem is to construct, by straightedge and compass, a square whose area is the same as that of the circle (FIG. 9-65). That is, we wish $s^2 = \pi r^2$, or $s = r\sqrt{\pi}$. We could construct s if we could construct a segment whose length is π units.

The impossibility of this construction rests on a theorem which states that the magnitude of any length constructible with Euclidean tools from *a given unit length* is an algebraic number.

A number r is an *algebraic* number if it satisfies an equation of the form $a_0x^n + a_1x^{n-1} + \cdots + a_{n-1}x + a_n = 0$, where the a's are all rational numbers and n is a positive integer. Otherwise, r is *transcendental*. In other words, a number that cannot be the root of an algebraic equation with rational coefficients is called a *transcendental number*.

About a hundred years ago (1882), F. Lindemann (1852–1939) showed that the number π is transcendental. Therefore, a segment of length π cannot be constructed from a given unit segment by straightedge and compass alone, and our problem becomes inconstructible.

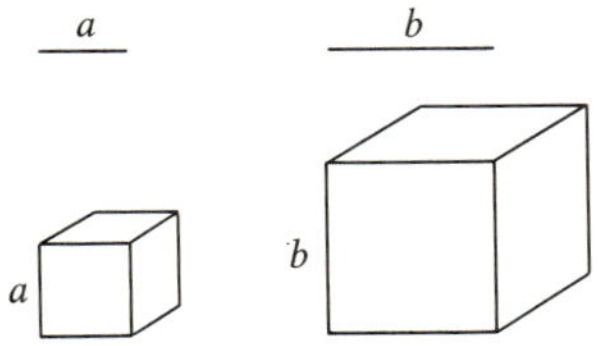

FIGURE 9-66

Duplication of the Cube

Given a segment of length a, the problem is to construct, by straightedge and compass, a segment of length b such that a cube of edge b has twice the volume of a cube of edge a (FIG. 9-66). That is, we wish $b^3 = 2a^3$. If a is unit length, then $b^3 = 2$ or $b^3 - 2 = 0$.

Since this equation has no rational root, b cannot be constructed from segment a with straightedge and compass.

By removing the restriction of having to use only a straightedge and compass, the Greeks solved the problem of doubling the cube by essentially proposing this instrument:

1. Start with right $\angle MZN$, with a movable right-angled cross B, VW, PQ (FIG. 9-67).
2. Two more edges RS and TU are allowed to slide perpendicularly to the arms of $\angle MZN$.
3. Choose any two fixed points E and G on the cross such that $GB = a$ and $BE = f$, a and f being prescribed lengths.

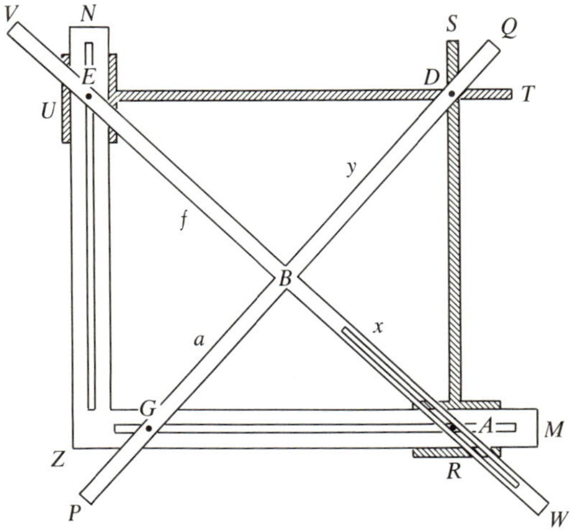

FIGURE 9-67

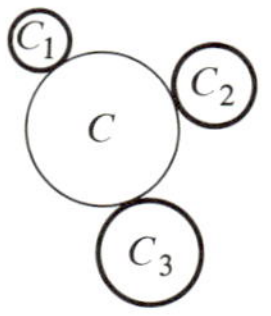

FIGURE 9-68

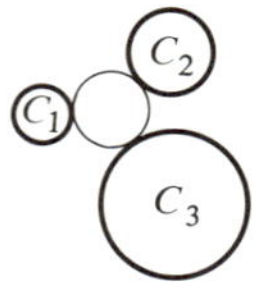

FIGURE 9-69

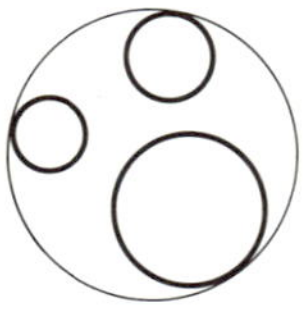

FIGURE 9-70

FIGURE 9-71

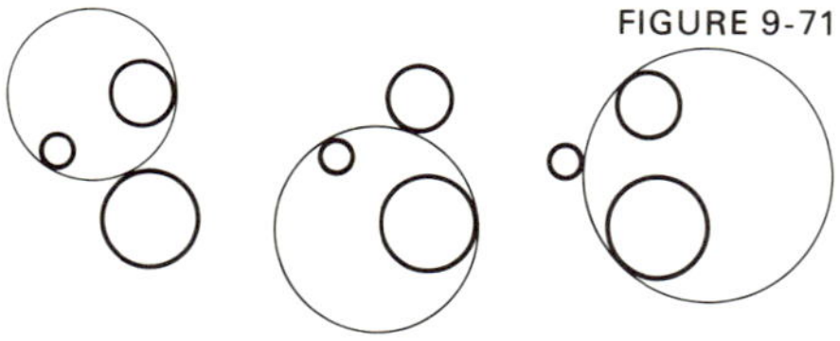

FIGURE 9-72

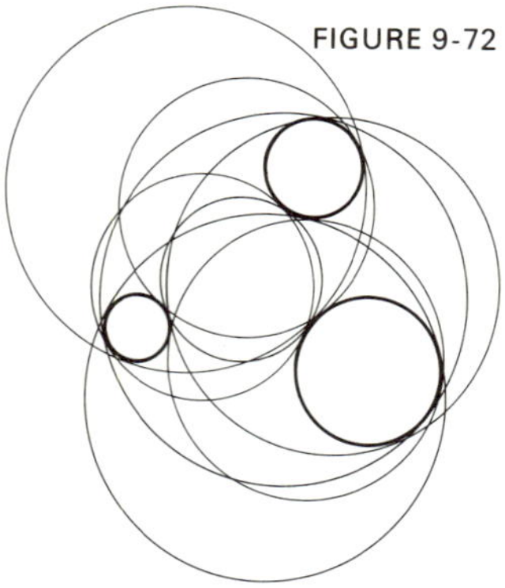

FIGURE 9-73

4. By placing the cross so that the points E and G lie on NZ and MZ, respectively, and sliding the edges TU and RS, we can form rectangle $ADEZ$. This will always be possible if $f > a$.

5. $a/x = x/y = y/f$, because the altitude to the hypotenuse of a right triangle is the mean proportional between the segments of the hypotenuse. Therefore, $x^2 = ay$, and $xy = af$.

6. If we make $f = 2a$, we get (from step 5) $x^3 = 2a^3$. That is, x is the edge of a cube whose volume is double that of a cube with edge a.

THE PROBLEM OF APOLLONIUS

One of the most famous of the ancient construction problems (circa 200 B.C.) is the Problem of Apollonius. In this problem we are asked to construct the circle (or circles) tangent to three given circles in the plane. The contact of the required circle with each of the given circles may be external or internal.

For instance, given three circles C_1, C_2, and C_3 in the plane (FIG. 9-68), the problem requires us to construct, with straightedge and compass only, all possible circles C that are tangent to all three circles. One such C is shown.

Since the contact of the required circle with each of the given circles may be external or internal, eight distinct solutions are sometimes possible:

1. All three given circles tangent externally to the required circle (one solution) (FIG. 9-69).
2. All three given circles tangent internally to the required circle (one solution) (FIG. 9-70).
3. One tangent internally and two tangent externally (three solutions) (FIG. 9-71).
4. One tangent externally and two tangent internally (three solutions) (FIG. 9-72).

The eight solutions merged into one diagram look as shown in FIG. 9-73.

The Problem of Apollonius may be regarded as a generalized version of a set of problems that require us to determine a circle that must meet three conditions taken from among the following:

1. The circle is to pass through one or more points.
2. The circle is to be tangent to one or more lines.
3. The circle is to be tangent to one or more circles.

If we let P, L, C represent a point, a line, and a circle, respectively, then we have these 10 problems:

(1) *PPP* (2) *PPL* (3) *PLL* (4) *LLL* (5) *PPC*
(6) *PLC* (7) *PCC* (8) *LLC* (9) *LCC* (10) *CCC*

If we permit one or more of the given circles to degenerate into a point or a straight line by regarding a circle with radius zero as a point, and a circle with radius "infinity" as a straight line, then problems (1) through (9) are really special cases of the general problem (10).

Some of these constructions, such as *PPP*, are obvious. It is, in essence, Construction 11 on page 250. Others, such as *PCC* and *CCC*, are more difficult. The solution in some of these cases is obtained by reducing the problem to a previous problem (from among the 10 listed) to which the solution is already known.

For example, the problem of constructing a circle through two given points and tangent to a given circle (*PPC*) is reduced to the problem of constructing a circle through three points (*PPP*). This is done as follows: Let P_1 and P_2 be the given points and C_1 the given circle (FIG. 9-74). Draw any circle S through P_1 and P_2, intersecting circle C_1 at A and B. Let $\overleftrightarrow{AB}$ meet $\overline{P_1P_2}$ at O. Draw tangent $\overline{OP_3}$ from O to C_1. The point P_3 is also the point of tangency for the required circle C. Therefore, to construct a circle through points P_1 and P_2 and tangent to circle C_1, we construct circle C through the three points P_1, P_2, and P_3. C is the required circle. Since there are two tangents from O to circle C_1, yielding two points P_3, there are two circles C.

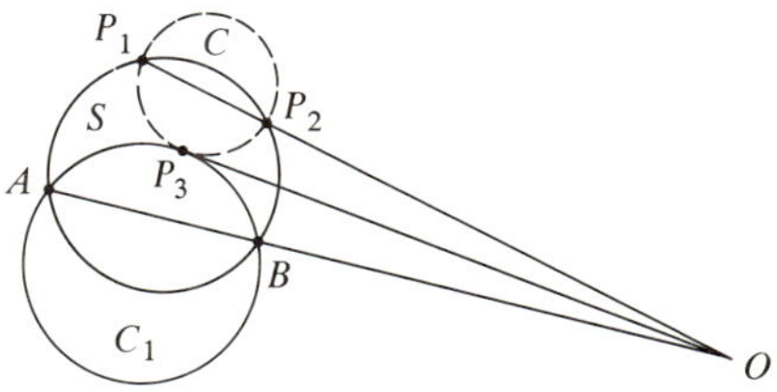

FIGURE 9-74

In a similar way, we can show that the problem of constructing a circle tangent to three given circles is equivalent to the problem of constructing a circle through a given point and tangent to two given circles. In other words, *CCC* can be reduced to *PCC*. We can do this in the following way: Let O_1, O_2, and O_3 be the centers of the given circles, and let r_1, r_2, and r_3 be their respective radii, r_1 being the smallest (FIG. 9-75). Let O be the center of the required circle. The circle with O as center and $\overline{OO_1}$ as radius will pass through O_1 and be tangent to the two circles with centers O_2 and O_3 and radii $(r_2 - r_1)$ and $(r_3 - r_1)$, respectively. The center, O, of this circle is also the center of the required circle. But how do we construct *PCC*? *PCC*, in turn, can be reduced to *PPC*! What about *PPC*? *PPC* is reduced to *PPP*!

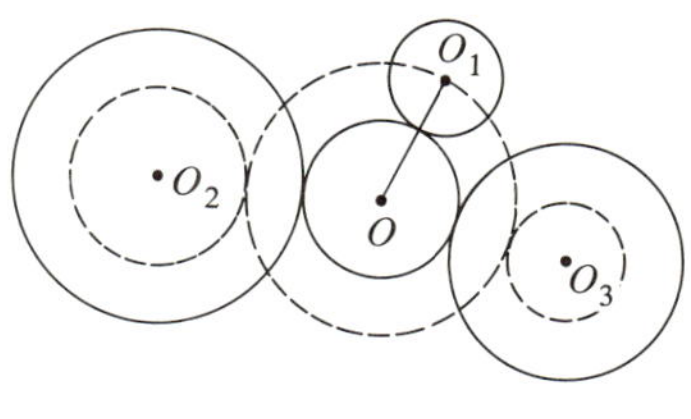

FIGURE 9-75

Other solutions of the Problem of Apollonius, such as one using the concepts of inversion in a circle or the theory of cycles, have been developed over the years.

10 LINEAR MEASUREMENT

INTRODUCING THE CHAPTER TO STUDENTS

CONTENT REVIEW

Prerequisites for Learning This Chapter
Introduction
Counting and Measuring
The Measuring Process
The Approximate Nature of Measurement
The Ruler

TEACHING SUGGESTIONS AND ACTIVITIES

Visual Aids
Making Finer Distinctions Between Lengths
Another Use for the Ruler
Measurement Exhibit
Questions
Puzzles
Solutions

ENRICHMENT

Other Standard Units of Measurement
The Greatest Possible Error
Significant Figures
Computing with Numbers Arising from Approximations
Archimedean Postulate

INTRODUCING THE CHAPTER TO STUDENTS

Although we measure things almost every day, how many of us really understand what we mean by "measurement"? When we lay a ruler against the edge of a table and say that "it measures 36 inches," what does this statement mean?

Look at these different measuring devices (describe the devices in Visual 1). What do all these devices have in common?

In this chapter we shall analyze what it is that we are doing when we measure the length of a room or the height of a pole. We will see that when we measure the length of a desk, the diameter of a pipe, the surface of a wall, the volume of a can, or when we measure heat, power, sound, light, or anything at all, we are in every instance engaged in the same process.

Although, today, we have many different kinds of measurements, the measurement of *length* was among the first measurements man needed. In this chapter we shall deal mostly with measuring lengths.

The early measures of length were based on parts of the human body. For instance, the width of the index finger was called a *digit*. The distance from the elbow to the end of the middle finger was called a *cubit*.

Since these units of length differed from one person to another, it became necessary to define the units more precisely. For instance, King Henry I of England decreed 850 years ago that a *yard* should be the distance from the end of *his* nose to the end of *his* thumb. In the sixteenth century a *rod* was the length of the left foot of 16 men lined up, as they left church on Sunday morning.

It was soon realized that using parts of the body for measuring may be convenient but is not reliable. A way had to be found by which a yard in England represented the same length as a yard in the United States. For these reasons certain units were agreed upon by large numbers of people. Such units are called *standard units*. The most common standard units of length are the inch, foot, yard, mile, and meter.

The only actual standard unit of length is the *meter*. The other units, such as the inch, foot, and yard, are specified by law in relation to the meter. The length of the meter was chosen to be one ten-millionth (0.0000001) of the meridional distance from the North Pole to the Equator.

Various methods have been used by the United States Bureau of Standards for maintaining a precise model of the standard meter.

For many years the model was a platinum bar kept under carefully controlled atmospheric conditions. In 1960, a new definition of the meter was adopted. The meter is now defined as having a length that is 1,650,763.73 times the wavelength of orange light from krypton 86. This definition was adopted because it is easier to reproduce in scientific laboratories in different parts of the world, and because it provides a more precise model than the platinum bar.

In 1966, North America's most accurate ruler was built by Ohio State University scientists. This ruler is a 500-meter line measured so accurately that its margin of error is less than the thickness of a sheet of paper. This remarkably accurate ruler, which is more than 1600 feet long, is used to calibrate instruments and other smaller measuring devices.

In the previous chapters we considered properties of line segments, including the relative sizes of two segments (i.e., congruent, less than, greater than). But we never asked the question : How long is this segment? In this chapter we shall concern ourselves with such questions. We shall be interested not only in knowing the lengths of segments, but also in the process for finding these lengths. Measurement of line segments is called *linear measurement*.

CONTENT REVIEW

PREREQUISITES FOR LEARNING THIS CHAPTER

1. Familiarity with the table of common linear measurements:

   12 inches = 1 foot,

   3 feet = 1 yard,

   5280 feet = 1 mile,

   1 meter = 39.37 inches (approximately).

2. Familiarity with a ruler marked in sixteenths.

3. Ability to measure to the nearest inch, half-inch, . . . , sixteenth-inch.

4. An understanding of the Distance Postulate and the Ruler Postulate:

   a. Distance Postulate (Postulate 2-7): To each pair of distinct points there corresponds a unique positive number called the *distance* between the two points.

   b. Ruler Postulate (Postulate 2-8): The points of a line can be placed in correspondence with the real numbers in such a way that (1) to every point on the line there corresponds exactly one real number, (2) to every real number there corresponds exactly one point of the line, and (3) the distance between two points equals the absolute value of the difference between the corresponding numbers.

INTRODUCTION

Measurement as a Quality

A measurement is a quality of an object. It describes a characteristic of the object in the same sense that a color describes a characteristic of a picture. If we know that a picture is green, we know more about the picture than if we knew nothing about its color. Similarly, if we know that the picture is 26 inches high, we know more about the picture than if we knew nothing about its height. A green picture differentiates it from a yellow picture or a brown picture. A height of 26 inches differentiates it from some other height.

Measurement Postulates

The concept of measurement rests upon several basic assumptions:

> *POSTULATE 10-1 (Comparison Postulate)* *We assume that two continuous quantities of the same kind (for example, two segments, or two surfaces) may be* compared *to determine whether they have the same size, or which one is the larger.*

Implied in this assumption is still another assumption: that when we compare two such quantities, a and b, one and only one of the following relations is true: $a = b$, $a > b$, or $a < b$.

> *POSTULATE 10-2 (Motion Postulate)* *We assume that an object may be* moved *from one position to another without changing its size or shape.*

We use this postulate to determine which of the three possible relations in the Comparison Postulate is true for any two quantities.

> *POSTULATE 10-3 (Matching Postulate)* *We assume that two continuous quantities have the same size if every part of one can be* matched *to a part of the second in such a way that no part of either is left out.*

This postulate is needed if we want differently shaped polygons, for example, to be able to have equal perimeters.

> *POSTULATE 10-4 (Subdivision Postulate)* *We assume that a quantity may be subdivided without producing a change in the size of the whole.*

This postulate makes it possible to measure geometric figures by subdividing them into unit parts.

In dealing with length, or the distance between two points, we also rely on two other assumptions: the Distance Postulate (Postulate 2-7) and the Ruler Postulate (Postulate 2-8).

COUNTING AND MEASURING

Although *counting* and *measuring* are related activities, there are important differences between them:

1. By counting, we answer the question, "How many?" By measuring, we answer such questions as "How much?", "How long?", "How fast?", or "How hot?"

2. Counting yields one, exact answer. Measuring yields varying, approximate answers.

3. To count, we only need the *counting numbers*. To measure, the counting numbers are not sufficient; we need the *rational numbers*.

4. Counting matches the elements of a set against the set of counting numbers $\{1, 2, 3, \ldots\}$. Measuring involves a comparison between a unit and the object being measured.

5. We count separate, distinct, discrete elements. We measure a *continuous* quantity such as height, time, weight.[1]

[1] To answer the question "How many people are in the room?", we *count* the people. Each person is a separate, whole object. There are no fractions of a person in the room. Our answer will, therefore, be a counting number.

To answer the question "How high is the fence?" we cannot find the answer by "counting the height" because height is a *continuous* quantity, not a collection of separate elements which we can count.

We find the size of such a continuous quantity by *measuring*. The result will not be described by a counting number, but by a number *and* a unit of measure with which the height of the fence is compared.

THE MEASURING PROCESS

When we *measure* a line segment $\overline{AB}$ (FIG. 10-1) we do three things:

FIGURE 10-1

1. We first decide upon some segment, say $\overline{XY}$, to serve as a *unit* with which $\overline{AB}$ will be compared (FIG. 10-2). By selecting $\overline{XY}$ as our unit, we are agreeing to consider its measure to be exactly 1 (FIG. 10-3).

FIGURE 10-2

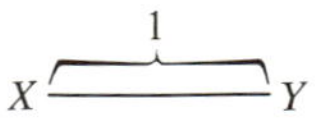

FIGURE 10-3

2. Then, starting at one end of $\overline{AB}$, we lay off a series of units $\overline{XY}$, end to end along $\overline{AB}$, until we reach (approximately) the other end of the segment $\overline{AB}$ (FIG. 10-4).

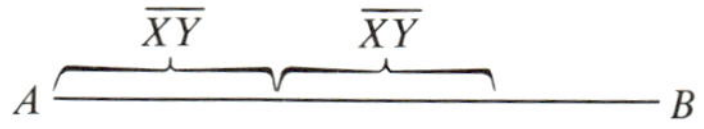

FIGURE 10-4

3. We then count the number of units $\overline{XY}$ needed to approximate the length of $\overline{AB}$. This number is the approximate measure of the length of $\overline{AB}$, expressed in terms of $\overline{XY}$.

a. If $\overline{AB}$ is "covered" by laying off the unit $\overline{XY}$ exactly twice, we say that the measure of $\overline{AB}$ is the number 2, and that the length of $\overline{AB}$ is exactly two $\overline{XY}$'s.
b. It is more likely, however, that $\overline{XY}$ will not fit into $\overline{AB}$ exactly a *whole* number of times (FIG. 10-5). In this case we can determine visually that the length of $\overline{AB}$ is, say, *nearer* to 3 units than to 2 units. So we say that, *to the nearest unit*, the length of $\overline{AB}$ is 3 units.[2]

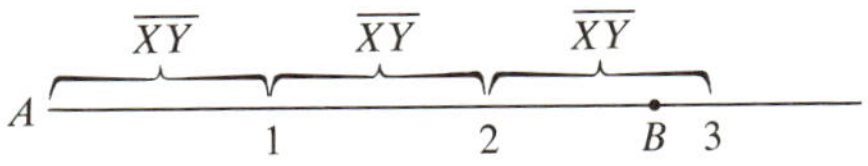

FIGURE 10-5

Although the size of the unit of length we use is arbitrary, in actual practice it has been found convenient to use certain units upon which large groups of people have agreed. Such units are called *standard units*. The most common standard units used for measuring length are the inch, foot, yard, mile, and meter.[3]

[3] The use of standard units makes it possible for people in different places to compare their results, to communicate with each other, to engage in mass production, to manufacture interchangeable parts, and to be sure that a 6-inch rod in California will have the same length as a 6-inch rod in Pennsylvania.

THE APPROXIMATE NATURE OF MEASUREMENT

All measurements are approximate because unavoidable sources of error are always present.

Weather conditions affect the accuracy of the measuring instrument. The marks on the instrument are not exact and are not designated exactly. Neither are, say, the endpoints of the segment we wish to measure.

Carelessness, poor eyesight, and inexperience with the instrument are other sources of error that may creep into our measurements.[4]

We can make finer distinctions between lengths by using a *smaller unit* of measurement. For instance, let us compare the lengths

[2] 1. Note that the *measure* of a segment is a number, while the *measurement* of a segment is a number *and* the unit used. For example, in the illustration above the *measure* of $\overline{AB}$ is approximately 3; the *measurement* of AB is approximately 3 $\overline{XY}$'s.

2. To help estimate whether the measure of $\overline{AB}$ is nearer to 3 than to 2, we locate the midpoint of the unit (FIG. 10A).

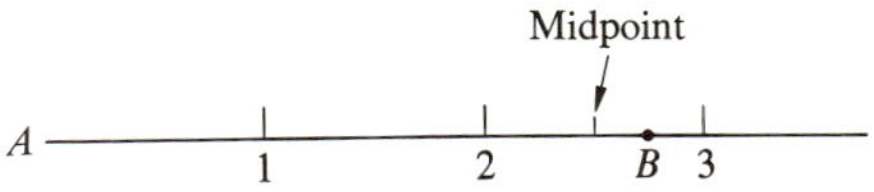

FIGURE 10A

3. Stress that the size of the unit $\overline{XY}$ is completely arbitrary.

4. Let students measure their desks using as the unit a pencil, eraser, or book. A desk may measure "about 4 pencils long."

5. After using objects as units of measure, students will be ready for the more abstract notion of measuring a line segment by using another line segment as a unit of measurement.

[4] 1. Comparing measurements of the same object made by different students, and also measures determined by instruments marked

with varying degrees of precision, will help students see the approximate nature of measurement.

2. We have been talking only about *direct* measurement, where we apply a measuring instrument directly to the object whose measurement we wish to determine. But not all measurements can be obtained by such direct procedures. Some are obtained by *indirect measurement.* In indirect measurement we *calculate* certain quantities from other directly measured quantities.

5 1. It is the *line* at the edge of the ruler that is important. The width of the ruler only provides space for writing the numerals and for giving the instrument more "body."

2. A ruler is a convenient and clever instrument for measuring lengths because it combines all the requirements for measuring into one simple instrument:

1. The unit of measurement is already indicated.
2. The units have already been "laid off," end to end.
3. The counting is done for us, since the units have been numbered.
4. The ruler contains a series of smaller subdivisions ($\frac{1}{2}$ inch, $\frac{1}{4}$ inch, $\frac{1}{8}$ inch, etc.) for finer distinctions between lengths that are close together.

of $\overline{AB}$ and $\overline{CD}$ using different units of measurement (FIG. 10-6). If the *inch* is used as the unit of measurement, then $\overline{AB}$ and $\overline{CD}$, although they apparently do not have the same length, do have the same measure *to the nearest inch.* But $\overline{AB}$ and $\overline{CD}$ have different measures *to the nearest half-inch.*

Unit of measurement: *Inch*

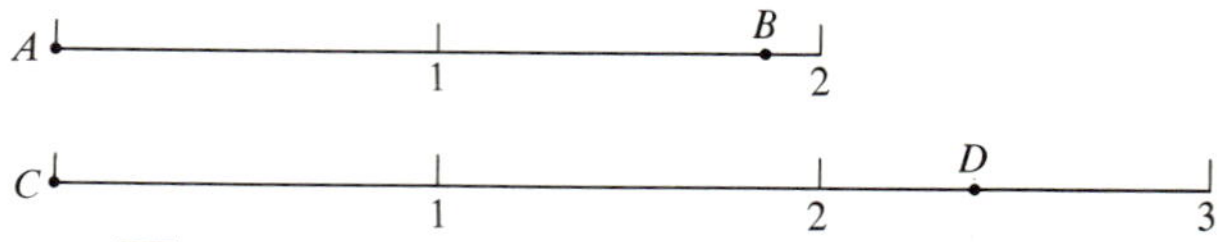

(a) $\overline{AB}$ is closer to 2″ than to 1″. Therefore, the length of $\overline{AB}$ is 2″ *to the nearest inch.*

(b) $\overline{CD}$ is closer to 2″ than to 3″. Therefore, the length of $\overline{CD}$ is 2″ *to the nearest inch.*

Unit of measurement: *Half-inch*

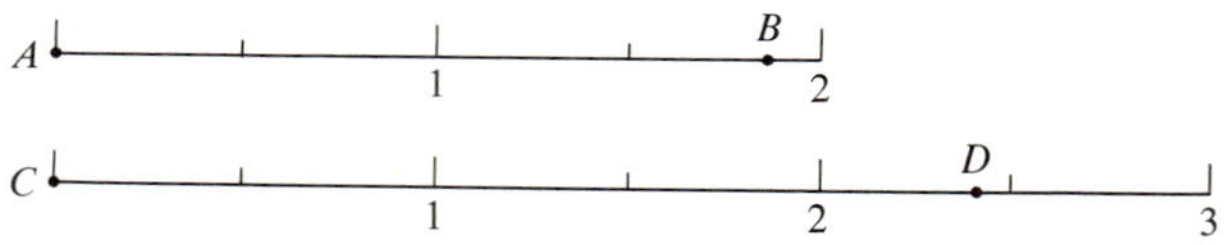

(a) $\overline{AB}$ is closer to 2″ than to $1\frac{1}{2}$″. Therefore, the length of $\overline{AB}$ is 2″ *to the nearest half-inch.*

(b) $\overline{CD}$ is closer to $2\frac{1}{2}$″ than to 2″. Therefore, the length of $\overline{CD}$ is $2\frac{1}{2}$″ *to the nearest half-inch.*

FIGURE 10-6

THE RULER

A *ruler* is an instrument with a number line that has been subdivided into standard units, usually inches. When we measure the edge of a desk with a ruler, we compare the length of the edge with the number line marked on the ruler.[5]

TEACHING SUGGESTIONS AND ACTIVITIES

VISUAL AIDS

Visual 1

PURPOSE

To exhibit a variety of measuring devices.

CONSISTS OF

Photos of a variety of measuring devices.

HOW TO USE

See Introducing the Chapter to Students.

Visual 2

PURPOSE

To illustrate early history of measurement.

CONSISTS OF

Photos of early measures based on parts of human body.

HOW TO USE

See Introducing the Chapter to Students.

Visual 3

PURPOSE

1. To illustrate the difference between counting and measuring.
2. To show what we do when we "measure."
3. To show why all measurements are approximate.
4. To show why smaller units result in greater precision.
5. To show how a ruler combines into one instrument all the requirements for measuring.

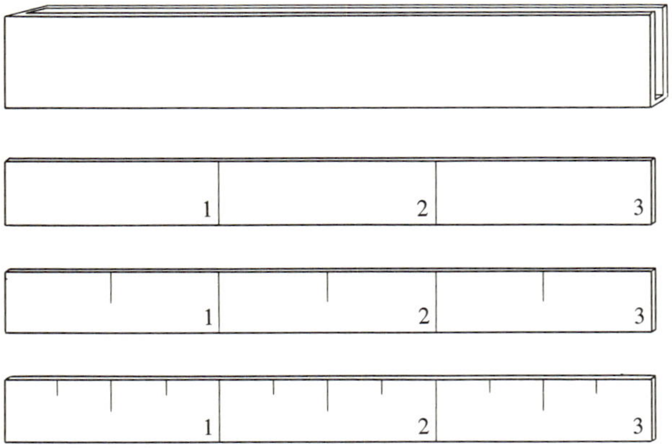

FIGURE 10-7

CONSISTS OF

1. A set of discs (about six).
2. An arbitrary linear unit.
3. Two "lines" with several movable "points."
4. An enlarged ruler consisting of a base and three removable scales (FIG. 10-7):
   a. First scale subdivided into *inches.*
   b. Second scale subdivided into *half-inches.*
   c. Third scale subdivided into *quarter-inches.*

HOW TO USE

1. To illustrate difference between counting and measuring.
   a. Place set of discs on magnetic board and ask five students to count them and tell how many there are.
   b. Then place a line segment on board and ask five students to measure the segment and say how long it is.
   c. There will be agreement on the number of discs but differences in the reported length of the segment.
2. To show why measurements are approximate.
   a. Show that endpoints of segment measured, as well as ruler marks used in determining the measure of the segment, are inexact locations.
   b. Show difficulty of lining up precisely the endpoints with the ruler marks.
   c. Show difficulty of determining visually whether a segment measures, say, $1\frac{1}{4}$ inches or $1\frac{11}{40}$ inches.
3. Use as per Content Review.

MAKING FINER DISTINCTIONS BETWEEN LENGTHS

Use Visual 3 as follows to demonstrate that we can make finer distinctions between lengths by using a smaller unit of measurement.

1. First set "points" on a line so that segment measures a *whole* number of units.
2. Then set "points" so that segment does *not* measure a whole number of units. Here develop notion of "to the nearest unit."

3. By shifting the "points" to various locations on the line, guide students into seeing that the unit selected cannot always make a fine enough distinction between two different lengths.

4. Introduce a *smaller* unit. By setting the "points" at certain (preplanned!) locations, show that by using the *smaller* unit we can get a more precise measure of length than by using the larger unit.

Duplicate with the Visual the settings illustrated on page 278 with $\overline{AB}$ and $\overline{CD}$.

An error in measurement is, at most, *half the unit* being used. If a segment is measured in inches, the error cannot be more than half an inch. If it is measured in half-inches, the error cannot be more than half of a half-inch, or a quarter-inch.

A measurement with an error of a *quarter*-inch is closer to the true measurement than one with an error of *half* an inch. Therefore, a measurement made to the nearest half-inch is more precise than one made to the nearest inch.

ANOTHER USE FOR THE RULER

By thinking of a ruler as a fractional number line, we can use it for adding and subtracting fractions.

EXAMPLE 1 $1\frac{3}{4} + \frac{3}{16}$.

1. On a straight line mark off $1\frac{3}{4}$ inches (FIG. 10-8).
2. Then mark off another $\frac{3}{16}$ inches.
3. Measure the combined lengths with ruler.

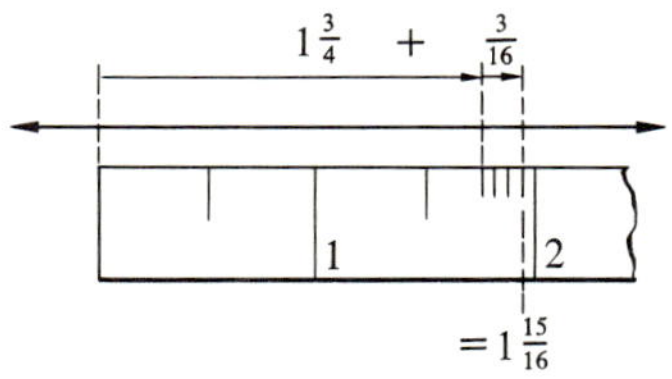

FIGURE 10-8

EXAMPLE 2 $3\frac{1}{8} - 1\frac{7}{16}$.

1. On a straight line mark off $3\frac{1}{8}$ inches (FIG. 10-9).
2. Then mark of $1\frac{7}{16}$ inches in the opposite direction.
3. Measure the segment that is left.

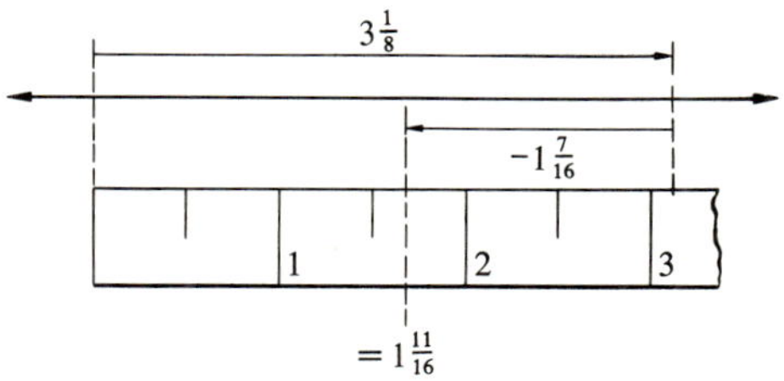

FIGURE 10-9

MEASUREMENT EXHIBIT

Immediately before this chapter is taught, a classroom exhibit on measurements, including a display of old and new measuring instruments and units of measurement used in different parts of the world, would serve as excellent motivation. Little-known units of length such as

1. The Indian "guz" (a yard in Bengal),
2. The Spanish "pulgada" (0.914 inch in Spain and Chile),
3. The "French foot" (12.19 inches long)

might be displayed.

The exhibit might bring out the fact that not only were parts of the human body used in the early history of measurement as standards of length, but the occupations of people were also used sometimes for this purpose. For example, a "furlong" (220 yards) was the length of a furrow along which an ox could pull the plow without stopping to rest. Sailors use "cable length" (120 fathoms); weavers use the "skein" (120 yards); and surveyors use the "link" (7.92 inches).

QUESTIONS

1. Which of the following questions would you answer by counting, and which by measuring?

   a. How far is it to his house?
   b. How many people saw the game?
   c. How much sand is in the box?
   d. How old is your brother?
   e. How many questions did you answer?
   f. How warm is the water?

2. Which of the following are discrete sets and which are continuous quantities?

   a. The whole numbers between 1 and 10.
   b. The distance you traveled.
   c. The school enrollment.
   d. The weight of a package.
   e. The length of your driveway.
   f. The age of your dog.
   g. The number of sweaters you own.

3. Find the measure of each of the following segments to the nearest unit, using $\overline{AB}$ as your unit: A ________________ B

a. ____________________
b. ___________________________
c. _________
d. ____

4. Find the measure of segments in Question 3 to the nearest unit, using $\overline{CD}$ as your unit: C ______ D

5. In Questions 3 and 4, which unit gave a more precise answer? Why?

6. Here are five segments:

(1) __________________ (2) _______
(3) ______________ (4) ________________________
(5) ____

Record the following data in the chart shown.

a. Estimate the length of each segment to the nearest inch.
b. Check your estimates with a ruler.
c. Measure each segment to the nearest $\frac{1}{4}$ inch; to the nearest $\frac{1}{8}$ inch; to the nearest $\frac{1}{16}$ inch.

	To Nearest Inch		*To Nearest $\frac{1}{4}$ Inch*	*To Nearest $\frac{1}{8}$ Inch*	*To Nearest $\frac{1}{16}$ Inch*
	Estimate	*Ruler*			
1					
2					
3					
4					
5					

7. a. $\overline{AB}$ contains ______ units, to the nearest whole unit.
b. $\overline{AC}$ contains ______ units, to the nearest $\frac{1}{2}$ unit.
c. $\overline{AD}$ contains ______ units, to the nearest $\frac{1}{4}$ unit.
d. $\overline{BC}$ contains ______ units, to the nearest $\frac{1}{4}$ unit.
e. If the length of a segment measured by this ruler is given as $1\frac{3}{4}$ units, its actual length lies between ______ and ______.

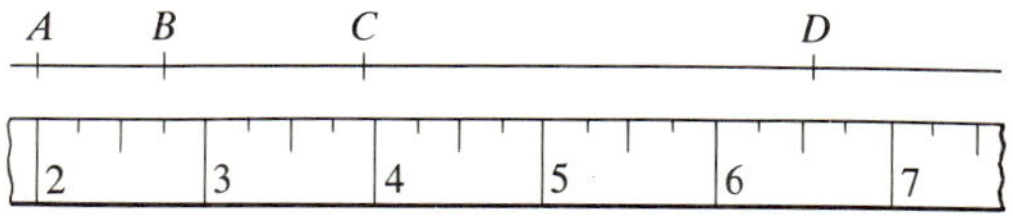

8. a. Measure to the nearest $\frac{1}{16}$ inch: (1) $\overline{AB}$, (2) $\overline{BC}$, (3) $\overline{CD}$, (4) $\overline{DE}$.
   b. Add the four measurements.
   c. Check your answer by measuring $\overline{AE}$.

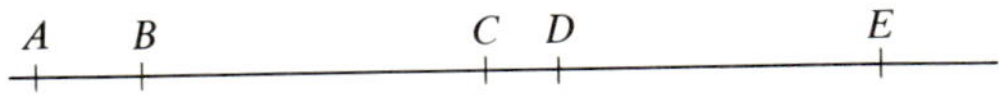

9. Draw a line segment across your paper and on it mark these segments next to each other: $2\frac{1}{2}$ inches; $\frac{7}{8}$ inch; $\frac{1}{4}$ inch.
   a. Using your ruler, read the total length of the three segments.
   b. Check your reading by adding the three lengths.

10. a. Use your ruler to find the sum of (1) $\frac{1}{4}$ and $1\frac{3}{8}$; (2) $2\frac{1}{4}$, $1\frac{3}{8}$, and $\frac{9}{16}$.
    b. Check each result by addition.

11. a. Use your ruler to find the difference between (1) $4\frac{1}{2}$ and $1\frac{3}{8}$, (2) $3\frac{1}{4}$ and $1\frac{5}{16}$.
    b. Check each result by subtraction.

12. Change to inches:

a. 12 feet	b. $2\frac{1}{2}$ feet
c. $\frac{2}{3}$ foot	d. 3.25 feet
e. 1 foot 5 inches	f. 6 meters

13. Change to feet:

a. 36 inches	b. 60 inches
c. 6 inches	d. 9 inches
e. 1 inch	f. 10 meters

14. If the length of a segment is stated to be $8\frac{3}{4}$ inches, its actual length might be as short as ______ or as long as ______.

15. Do you get a smaller or larger number in your answer when
    a. You change a larger unit of measurement to a smaller unit of measurement?
    b. You change a smaller unit of measurement to a larger unit of measurement?

16. True–false. Give a reason for your answer.

a. The counting numbers are all that are needed for both counting and measuring.
b. A ruler is a type of number line.
c. We may choose any size for a measuring unit.
d. The smaller the unit, the more precise the measurement.
e. The more precise the measurement, the greater the possible error.
f. The measurement of 7 miles has the same greatest possible error as a measurement of 18 miles.
g. The term "greatest possible error" of measurement refers to a mistake made in the measurement.

PUZZLES

1. A rope ladder with the rungs 1 foot apart is hanging over the side of a ship. The bottom rung is resting on the surface of the ocean. If the tide rises at the rate of 9 inches per hour, how long will it take for the first four rungs to be covered with water?

2. Two bicycle riders are 50 miles apart. Rider *A* pedals toward rider *B* at 10 miles per hour; rider *B* pedals toward rider *A* at 15 miles per hour. At the same time, a bee, starting on the nose of *A*, flies towards *B*'s nose at the rate of 25 miles per hour. The bee continues to fly back and forth from nose to nose until *A* and *B* meet. How far does the bee fly?

3. You are given nine boxes of candy, identical in every way except that one is slightly lighter than any of the others. Using only a balance scale, how can you identify the underweight box by making only *two* weighings?

SOLUTIONS

1. The first four rungs will never be covered with water because the ladder and ship will rise with the tide.

2. Since the bicyclists will meet in 2 hours, the bee will fly for 2 hours. At 25 miles per hour, the bee will fly 50 miles in 2 hours.

3. Divide the boxes into three groups of three boxes each. Call these groups *A*, *B*, and *C*.

   a. First weighing: Weigh *A* against *B*. If they balance, then the underweight box is in group *C*. If one side of the scale goes up, then the group on that side of the scale contains the underweight box.
   b. Second weighing: Now that the group containing the underweight box has been identified, take two boxes from this group and weigh them against each other. The underweight box will be on the side of the scale that goes up. If neither side goes up, the underweight box is the remaining box.

ENRICHMENT

OTHER STANDARD UNITS OF MEASUREMENT

Scientific developments in recent years have brought about an extension of some of our already familiar standard units of measurement. For example, to facilitate the measurement of supersonic speeds, *Mach numbers* have been invented. These numbers have been defined in terms of the speed of sound. A plane traveling at, say, Mach 2, would be flying at twice the speed of sound, or about 1500 miles per hour.

The *light-year* was defined for better measurement of astronomic distances. One light-year is defined to be the distance light will travel in 1 year's time—a distance of about 5,870,000,000,000 miles. The North Star is a distance of about 47 light-years from the earth.

The *atomic mass unit* was invented for use with atomic weights. This unit is defined to be 0.00000000000000000000000000366 pound. Atomic nuclei are made up of *protons* and *neutrons*. The proton has a mass of 1.00758 atomic mass units, and the neutron has a mass of 1.00897 atomic mass units. The nucleus of an atom is surrounded by *electrons*, each of which has a mass of 0.00055 atomic mass unit.

If we should attempt to express these measurements in terms of miles and pounds, we shall quickly appreciate the convenience of having defined these new standard units of measurement.

THE GREATEST POSSIBLE ERROR

We have seen that if a measurement is given to the nearest inch, the greatest possible difference between the true length and the given length is half of an inch, or 0.5 inch: If the measurement given is to the nearest half-inch, then the greatest possible difference between the true length and the given length is half of a half-inch, or 0.25 inch.

The greatest possible difference between the true length of a segment and the reported length is called the *greatest possible error*.

If we report a measurement as $1\frac{3}{8}$ inches, the greatest possible error is half of $\frac{1}{8}$ inch, or $\frac{1}{16}$ inch. Therefore, the true length of the segment lies in the interval between $(1\frac{3}{8} - \frac{1}{16})$ and $(1\frac{3}{8} + \frac{1}{16})$, or between $1\frac{5}{16}$ inches and $1\frac{7}{16}$ inches. The smaller the unit size used in measurement, the smaller the greatest possible error.

By stating the greatest possible error of a measurement we are indicating the *precision* of the measurement. The smaller the greatest possible error, the more precise is our measurement. A measurement made with a smaller unit will, therefore, be more precise than one made with a larger unit.

If we measure the length of our lawn and find it to be 50 feet to the nearest foot, and the length of our bookshelf to be 5 feet to the nearest foot, both measurements are equally precise, since both have the same greatest possible error—0.5 foot. But an error of 0.5 foot in measuring the lawn is not nearly as serious as an error of 0.5 foot in measuring the bookshelf. Obviously, an error of 0.5 is a larger part of 5 than it is of 50.

The ratio of the greatest possible error to the stated measurement of the lawn is $0.5/50 = 1$ percent. The ratio of the greatest possible error to the stated measurement of the bookshelf is $0.5/5 = 10$ percent.

The ratio of the greatest possible error to a given measurement is called the *relative error* of the measurement. The *relative error* of the measurement determines the *accuracy* of the measurement. The smaller the relative error, the greater the accuracy of the measurement.

The difference between the *precision* and *accuracy* of a measurement can be stated this way: The *precision* of a measurement is the *difference* between the true value and the measured value. The *accuracy* of a measurement is the *ratio* of this difference to the measured value. Precision reflects the size of the greatest possible error; accuracy reflects the importance of the error.

Given any two measurements, the more precise measurement may or may not be the more accurate measurement, and the more accurate measurement may or may not be the more precise measurement.

In industry, the permissible error in measurement is of great importance. This permissible error is known as *tolerance*. The specified tolerance of, say, 0.005 on a shaft whose dimension is specified as 2.000 inches means that any shaft measuring anywhere between 2.000 ± 0.005 inches, or between 1.995 and 2.005 inches, will be acceptable. The interval 1.995 to 2.005 inches is called the *tolerance interval*.

SIGNIFICANT FIGURES

When a number expresses a measurement, the *significant figures* in that number are of prime importance because they affect the accuracy of the measurement.

The significant figures in any number may be determined as follows:

1. Any nonzero digit is always significant.

   35—2 significant figures

   7.94—3 significant figures

2. Zeros occurring between nonzero digits are always significant.

   205—3 significant figures

   4002—4 significant figures

   3.08—3 significant figures

3. Terminal zeros following the decimal point are always significant.

   2.0—2 significant figures

   105.00—5 significant figures

4. Zeros used merely for the purpose of placing the decimal point are not significant; or, in a number less than 1, zeros immediately following the decimal point are not significant.

   0.002—1 significant figure

   0.015—2 significant figures

5. Terminal zeros in an integer should not be considered significant unless so indicated.

   7500—2 significant figures (unless otherwise indicated)

Let us now see how the number of significant figures in a measurement affects the *accuracy* of the measurement. (See the accompanying table.)

Measurement (feet)	*Greatest Possible Error (feet)*	*Percent of Error*
50	0.5	1
50.0	0.05	0.1
50.00	0.005	0.01

As the number of significant figures in the measurement increases, the percent of error decreases and, therefore, the accuracy of the measurement increases.

COMPUTING WITH NUMBERS ARISING FROM APPROXIMATIONS

Addition and Subtraction

When approximations are added or subtracted, the precision of the answer is no greater than that of the least precise number involved. Therefore, to add or subtract numbers arising from approximations, first round each number to the unit of the least precise number and then perform the operation.

For example, to add 3.5, 0.4, and 6.79 inches we first round 6.79 to 6.8 inches and then add: $3.5 + 0.4 + 6.8 = 10.7$ inches.

Multiplication and Division

Our rule for adding or subtracting approximations depends upon the idea of precision; that is, a sum or difference is not more precise than the least precise number used in the operation. Our rule for multiplication or division depends upon the idea of accuracy, or the number of significant figures. In multiplying or dividing approximations, keep as many significant figures in the answer as there are in the number with the fewer significant figures.

For example, when we multiply 15.2 by 4.8 we get

$$\begin{array}{r} 15.2 \\ \times 4.8 \\ \hline 1216 \\ 608 \\ \hline 72.96 \end{array} \qquad \text{or } 73$$

Since 4.8 has the fewer significant figures (2), the answer is rounded to 73.

Square Root

A similar rule applies to square root. To find the square root of an approximation, retain as many significant figures in the answer as the number contains. For example, $\sqrt{5.42} = 2.33$.

ARCHIMEDEAN POSTULATE

On page 276 we said that to measure the length of a segment we choose a unit of length, lay it out end to end on the segment to be measured, and then count the number of units required to "cover" the segment. If the unit fits exactly k times, then the measure of the

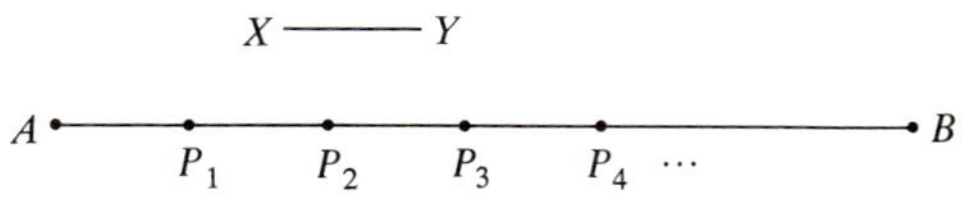

FIGURE 10-10

segment is k. If the segment is not "covered" by an exact number of units, then we use subunits to determine its measure.

This process raises a basic question that needs to be answered. Assume that we wish to measure segment $\overline{AB}$ with the unit $\overline{XY}$ (FIG. 10-10). We begin by laying off $AP_1 = P_1P_2 = P_2P_3 = P_3P_4 = \cdots = XY$, until we reach point B or a point beyond it.

Now the question is: How do we know that this process will eventually lead us to point B or a point beyond it? Although, intuitively, this appears to be the case, can we prove this from any of our previous postulates or theorems?

Our question can be formulated in a slightly different way: Given two segments of unequal lengths, can we prove that there exists a finite multiple of the shorter segment which is equal to or longer than the other segment?

It has been demonstrated that what we have assumed to be the answer to this question cannot be proved from our previous assumptions or theorems. Therefore, to assure us of an affirmative answer we must introduce another postulate, called the *Archimedean Postulate*, named after the man who first recognized the need for it.

> *Postulate 10-5 (Archimedean Postulate)* *If* $\overline{AB}$ *and* $\overline{XY}$ *are any segments* (FIG. 10-10), *and on* $\overline{AB}$ *we choose points* P_1, $P_2, P_3, P_4, \ldots$ *such that* $\overline{AP_1}, \overline{P_1P_2}, \overline{P_2P_3}, \overline{P_3P_4}, \ldots$ *are all congruent to* XY, *then eventually we shall find points* P_n *and* P_{n+1} *for which* $B = P_n$ *or* B *is between* P_n *and* P_{n+1}.

This postulate assures us that no matter how small a unit segment $\overline{XY}$ we may choose, there is always an integral multiple of that unit, $n \times (\overline{XY})$, which is greater than any other segment $\overline{AB}$ no matter how long.

It is interesting to note that in the late nineteenth and early twentieth centuries geometric systems that reject the Archimedean Postulate were constructed. These geometries are referred to as "non-Archimedean geometries."

11

SQUARE AND CUBIC MEASUREMENT

INTRODUCING THE CHAPTER TO STUDENTS

CONTENT REVIEW

Prerequisites for Learning This Chapter
Area: Informal Remarks
Area Postulates
Area Theorems
Volume: Informal Remarks
Volume Postulates
Volume Theorems

TEACHING SUGGESTIONS AND ACTIVITIES

Visual Aids
Milestones in the History of Measurement
Area Under Certain Distortions
Questions
Puzzles
Solutions

ENRICHMENT

Another Formula for the Area of a Triangle
Another View of Area
Maximum and Minimum Areas and Perimeters

INTRODUCING THE CHAPTER TO STUDENTS

In Chapter 10 we saw how to compare the lengths of two segments. In this chapter we shall develop ways of comparing the "sizes" of two polygonal regions, and then the "sizes" of two solids.

The "sizes" of segments can be compared easily by placing one over the other and seeing which "sticks out" or whether both match. But if we wish to compare the "size" of, say, a triangular region with the "size" of, say, a trapezoidal region, we cannot do what we did with the segments. We cannot place the triangle over the trapezoid or the trapezoid over the triangle and obtain a meaningful comparison because neither region necessarily fits onto or into the other. If we decide to snip off pieces of one figure in such a way as to make all the pieces fit into the other figure, we *might* then be able to compare their "sizes." But such a procedure is not always possible and surely not efficient.

In this chapter we shall see what is meant by the "size" of a region or of a solid, and then develop a way of finding these "sizes" by a process very similar to the one we followed when we found the "size" of a segment. We shall then be able to compare "sizes" of regions and "sizes" of solids without having to superimpose one figure over the other.

CONTENT REVIEW

PREREQUISITES FOR LEARNING THIS CHAPTER

1. Familiarity with the tables of square and cubic measure.
2. Knowing the meaning of: boundary; perimeter; region; closed region; square; rectangle; trapezoid; parallelogram; solid; rectangular parallelepiped; prism; cube; pyramid.

AREA: INFORMAL REMARKS

1. By the *area* of a closed region we mean a *number* that measures the "amount of surface" in the region. This number is called the *measure of the area* of the region.
2. We must be careful to distinguish between
   a. A *region*, which is a set of points, and its *area*, which is a number.
   b. *Area*, which is a property of a region, and *perimeter*, which is a property of a boundary.
   c. *Length*, which answers such questions as "How far?", and *area*, which answers such questions as "How much floor space?"
3. To determine the area of a region we follow the same process as when we want to determine the length of a segment.

To Determine Length of Segment	*To Determine Area of Region*
1. We chose a unit of length.	1. We choose a unit of area.
2. We estimated the number of units needed to "cover" the segment.	2. We estimate the number of units needed to "cover" the region.
3. This number of units is the *length* of the segment.	3. This number of units is the *area* of the region.
4. More precise estimates of length can be gotten by using a smaller unit.	4. More precise estimates of area can be gotten by using a smaller unit.

4. Although our choice of shape and size for our *unit of area* is arbitrary, using a *square* unit has advantages over other shapes. For example, when circular units are used to cover a rectangular region there are parts of the region left uncovered. Square units, however, cover the *entire* region (FIG. 11-1).[1]

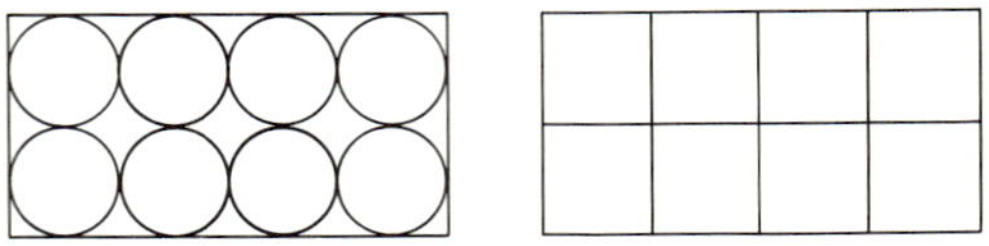

FIGURE 11-1

[1] Using a unit in the shape of a right isosceles triangle would, like using squares, also completely cover a region. But squares are more convenient.

5. A *square inch* is the area of a region bounded by a square whose side is 1 inch (FIG. 11-2). A standard unit for measuring area is a *square inch*.[2]

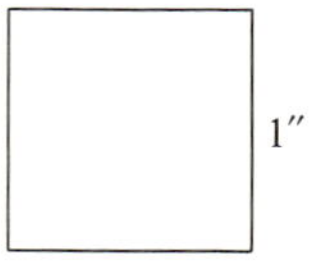

FIGURE 11-2

[2] 1. By using the square inch as a standard unit for measuring area, we are relating our unit of area to our unit of length.

2. Ask students to define a *square foot;* a *square yard;* a *square mile*.

3. Students should know these relationships:

144 square inches = 1 square foot,

9 square feet = 1 square yard.

4. We always select the unit most appropriate to the problem: for example, square inches to measure a sheet of paper; square feet to measure a living room floor; square miles for the area of a city.

AREA POSTULATES[3]

[3] 1. Formally, we shall begin our study of area by postulating its properties. We shall then use these properties to develop the usual formulas for the areas of various polygonal regions.

2. One of the postulates defines the area of a square. From this definition we shall derive the formulas for the area of a rectangle, triangle, trapezoid, and parallelogram, in that order.

3. Let us recall these definitions:

a. A *triangular region* is the union of a triangle and its interior (Fig. 11A).

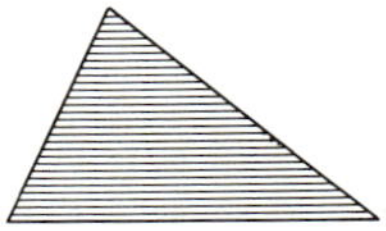

FIGURE 11A

b. A *polygonal region* is the union of a finite number of triangular regions in a plane such that if two of these intersect, their intersection is either a point or a segment (Fig. 11B).

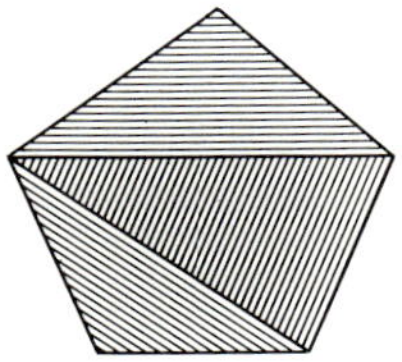

FIGURE 11B

POSTULATE 11-1 (Area Postulate) *To each polygonal region there corresponds a unique positive number which is called the* area *of the region.*

POSTULATE 11-2 (Congruence Postulate for Area) *If two triangles are congruent, then their triangular regions have the same area.*[4]

POSTULATE 11-3 (Area Addition Postulate) *If a polygonal region R is the union of two nonoverlapping polygonal regions R_1 and R_2, then the area of R = the area of R_1 + the area of R_2* (FIG. 11-3).[5]

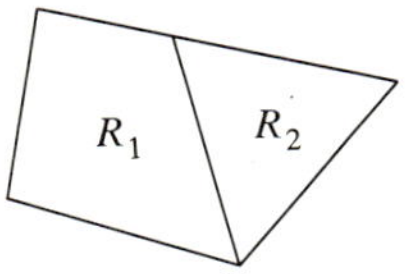

FIGURE 11-3

POSTULATE 11-4 (Unit Area Postulate) *The area of a square region is the square of the length of its side* (FIG. 11-4).

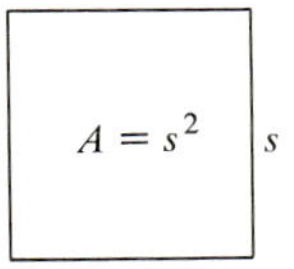

FIGURE 11-4

That is, the area of a square with side of length s is $A = s^2$.[6]

[4] 1. By defining *area*, Postulate 11-1 tells us what is meant by the "size" of a polygonal region.
2. Postulate 11-2 satisfies our intuitive notion that two polygonal regions of the same size and same shape will have the same area.

[5] 1. If we cut a region into two pieces, we would expect the area of the region to be the sum of the areas of the two pieces. Postulate 11-3 makes our geometry conform to this intuitive expectation.
2. Postulate 11-3 is also the basis for subtraction of areas.

[6] 1. Postulate 11-4 relates our concept of area to our concept of length and describes a way of assigning a measure to a square region.
2. Hereafter, for the sake of simpler communication, we shall refer to "the area of a triangular region" as "the area of a triangle"; by "the area of a rectangle" we shall mean "the area of a rectangular region"; etc. By "the base" we shall mean "the length of the base," by "the altitude" we shall mean "the length of the altitude," etc.

AREA THEOREMS

THEOREM 11-1 (Area of a Rectangle) *The area of a rectangle is the product of its base and its altitude* ($A = bh$).

Proof

1. Consider a square with side $(b + h)$ and with subregions as shown in FIG. 11-5.

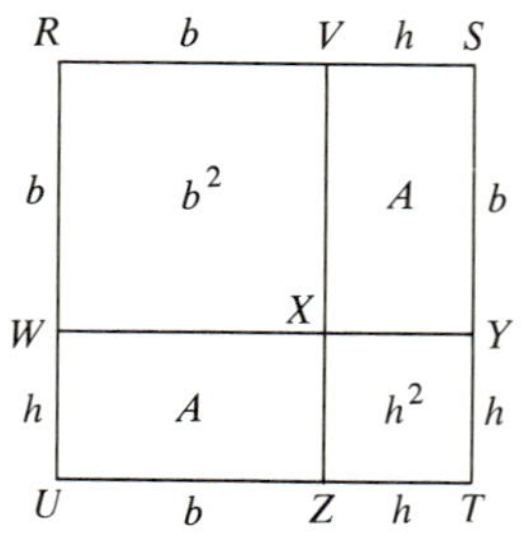

FIGURE 11-5

2. Let A represent the area of rectangle $WXZU$ as well as the area of rectangle $VSYX$. Both rectangles have the same area A because
   a. The four triangles in FIG. 11-6 are congruent.
   b. The area of each rectangle is the sum of the areas of two of these triangles.

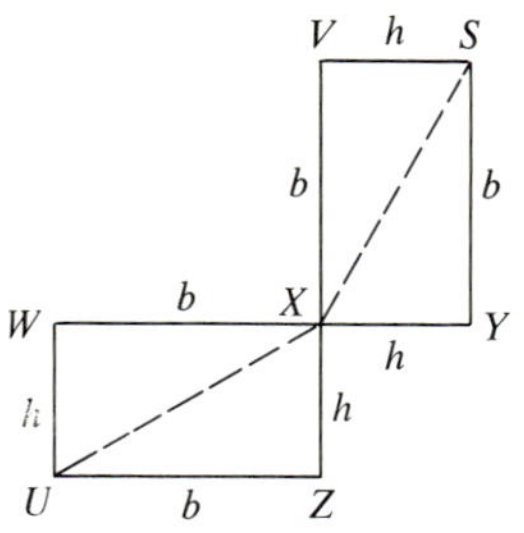

FIGURE 11-6

3. By Postulate 11-4, the areas of the two squares $RVXW$ and $XYTZ$ are b^2 and h^2; the area of the given square $RSTU$ is $(b + h)^2$.

4. By Postulate 11-3, the area of square $RSTU$ equals the sum of the areas of the subregions into which it is cut; that is,

$$b^2 + A + A + h^2 = (b + h)^2$$
$$b^2 + 2A + h^2 = b^2 + 2bh + h^2$$

or

$$A = bh.$$

THEOREM 11-2 (Area of a Right Triangle) *The area of a right triangle is half the product of its legs* ($A = \frac{1}{2}bh$).

Proof

1. Form rectangle $BDAC$ (FIG. 11-7).

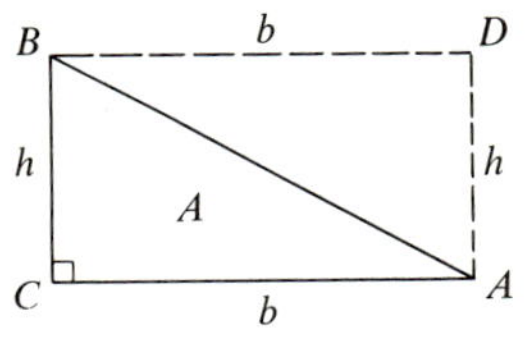

FIGURE 11-7

2. Then $\triangle BCA \cong \triangle ADB$, and both triangles have the same area, A.
3. By the Area Addition Postulate and Theorem 11-1, $A + A = bh$; or $A = \frac{1}{2}bh$.

THEOREM 11-3 (Area of a Triangle) *The area of a triangle is half the product of any base and the altitude to that base* ($A = \frac{1}{2}bh$).

In proving this theorem there are three cases to consider:

1. If the foot of the altitude is between the endpoints of the base [FIG. 11-8(a)].

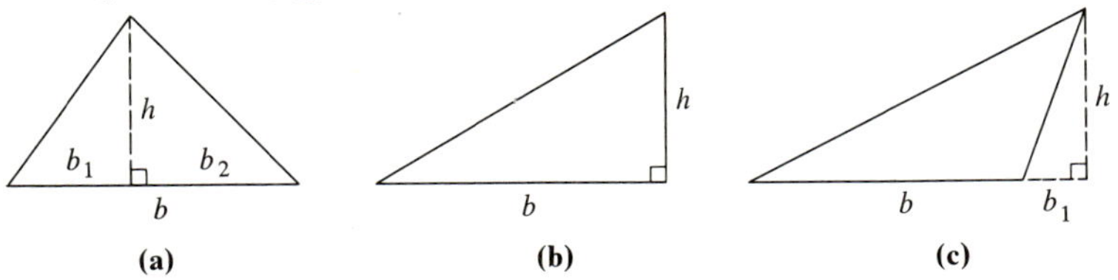

FIGURE 11-8

2. If the foot of the altitude is an endpoint of the base [FIG. 11-8(b)].
3. If the foot of the altitude is not on the base [FIG. 11-8(c)].

Proof
Case 1 [FIG. 11-8(a)]

1. The altitude cuts the triangle into two triangles with bases b_1 and b_2, and $b_1 + b_2 = b$.
2. By Theorem 11-2, the areas of these triangles are $\frac{1}{2}b_1h$ and $\frac{1}{2}b_2h$.
3. By the Area Addition Postulate,

$$A = \tfrac{1}{2}b_1h + \tfrac{1}{2}b_2h$$
$$= \tfrac{1}{2}(b_1 + b_2)h$$

or

$$A = \tfrac{1}{2}bh.$$

Case 2 [FIG. 11-8(b)]
By Theorem 11-2, $A = \frac{1}{2}bh$.

Case 3 [FIG. 11-8(c)]
By the Area Addition Postulate,

$$\tfrac{1}{2}b_1h + A = \tfrac{1}{2}(b_1 + b)h$$

or

$$A = \tfrac{1}{2}bh.$$[7]

THEOREM 11-4 *If two triangles have the same base and the same altitude, then they have the same area* (since each area is $\frac{1}{2}bh$) (FIG. 11-9).

[7] 1. Note that Theorem 11-2 is a special case of Theorem 11-3.

2. Theorem 11-3 can be applied to any triangle in three ways because any side may be chosen as the base.

Since the area of the triangle will remain the same regardless of which formula is used to compute it, the three formulas

$$\tfrac{1}{2}b_1h_1, \quad \tfrac{1}{2}b_2h_2, \quad \tfrac{1}{2}b_3h_3$$

will yield the same number (FIG. 11C).

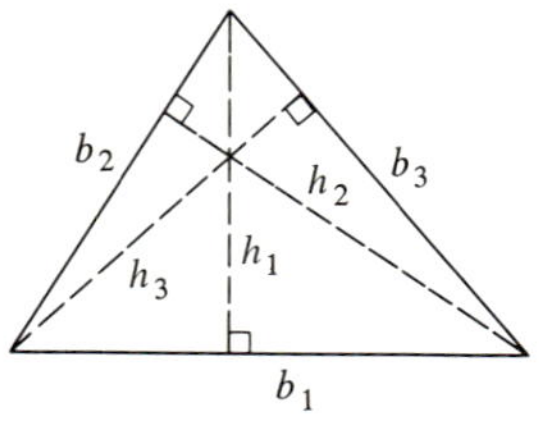

FIGURE 11C

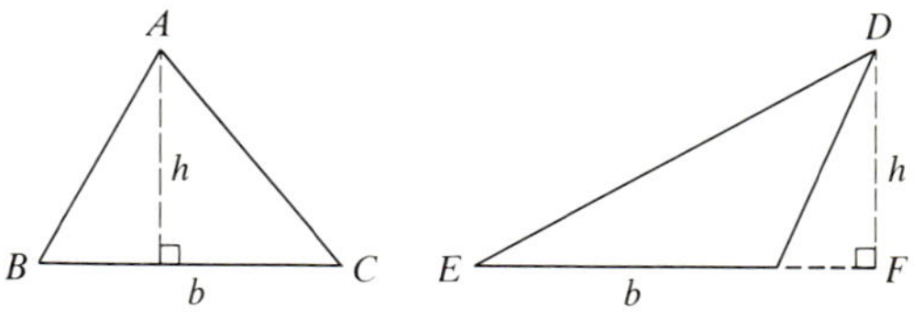

FIGURE 11-9

THEOREM 11-5 *If two triangles have the same altitude, then the ratio of their areas is equal to the ratio of their bases* (FIG. 11-10).

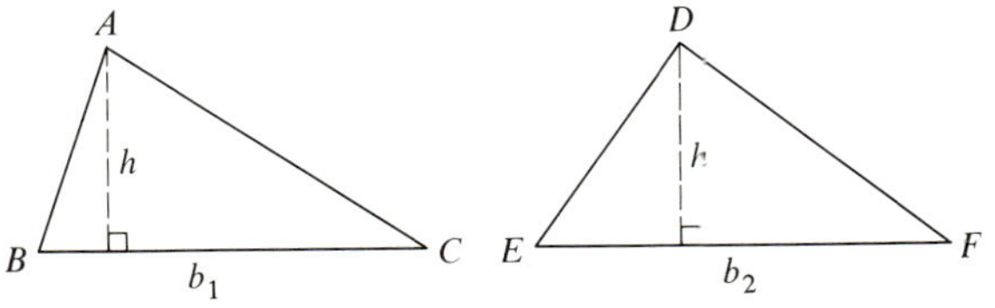

FIGURE 11-10

That is, if $\triangle ABC$ and $\triangle DEF$ have the same altitude h, then

$$\frac{\text{area of } \triangle ABC}{\text{area of } \triangle DEF} = \frac{b_1}{b_2}.$$

Proof

1. Area of $\triangle ABC = \frac{1}{2}b_1h$,

   Area of $\triangle DEF = \frac{1}{2}b_2h$.

2. $$\frac{\text{Area of } \triangle ABC}{\text{Area of } \triangle DEF} = \frac{\frac{1}{2}b_1h}{\frac{1}{2}b_2h} = \frac{b_1}{b_2}.$$

THEOREM 11-6 *If two triangles have the same base, then the ratio of their areas is equal to the ratio of their corresponding altitudes* (FIG. 11-11).

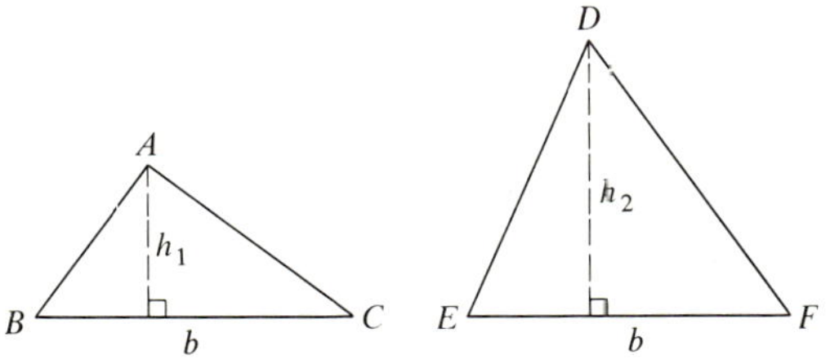

FIGURE 11-11

That is, if $\triangle ABC$ and $\triangle DEF$ have the same base b, then

$$\frac{\text{area of } \triangle ABC}{\text{area of } \triangle DEF} = \frac{h_1}{h_2}.$$

THEOREM 11-7 *If two triangles are similar, then the ratio of their areas is the square of the ratio of any two corresponding sides* (FIG. 11-12).

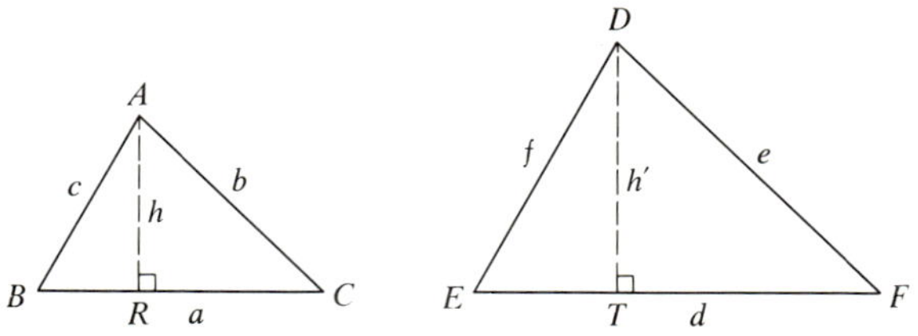

FIGURE 11-12

That is, if $\triangle ABC \sim \triangle DEF$, then

$$\frac{\text{area of } \triangle ABC}{\text{area of } \triangle DEF} = \left(\frac{a}{d}\right)^2 = \left(\frac{b}{e}\right)^2 = \left(\frac{c}{f}\right)^2.$$

Proof

1. Since $\triangle ABR \sim \triangle DET$ and $\triangle ABC \sim \triangle DEF$,

$$\frac{h}{h'} = \frac{c}{f} = \frac{a}{d} = \frac{b}{e}.$$

2. $$\frac{\text{Area of } \triangle ABC}{\text{Area of } \triangle DEF} = \frac{\frac{1}{2}ah}{\frac{1}{2}dh'} = \frac{a}{d} \cdot \frac{h}{h'}.$$

3. $$\frac{\text{Area of } \triangle ABC}{\text{Area of } \triangle DEF} = \left(\frac{a}{d}\right)^2 = \left(\frac{b}{e}\right)^2 = \left(\frac{c}{f}\right)^2.$$

THEOREM 11-8 (Area of a Trapezoid) *The area of a trapezoid is half the product of its altitude and the sum of its bases [that is,* $A = \frac{1}{2}h(b_1 + b_2)$*]* (FIG. 11-13).

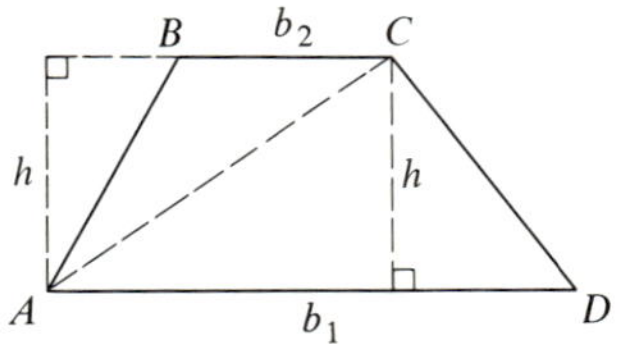

FIGURE 11-13

Proof

By the Area Addition Postulate, the area of trapezoid $ABCD$ is equal to the sum of the areas of $\triangle ACD$ and $\triangle ABC$. That is,

$$\text{area of trapezoid } ABCD = \tfrac{1}{2}b_1h + \tfrac{1}{2}b_2h = \tfrac{1}{2}h(b_1 + b_2).$$

THEOREM 11-9 (Area of a Parallelogram) *The area of a parallelogram is the product of any base and the corresponding altitude* ($A = bh$) (FIG. 11-14).

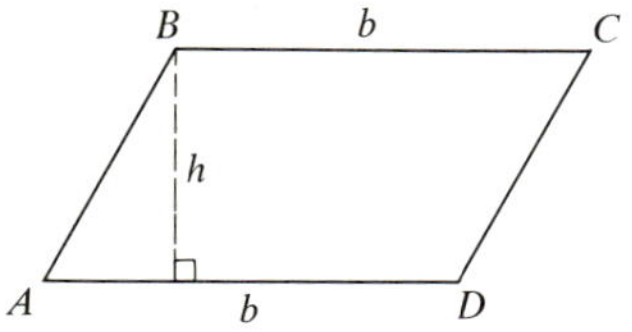

FIGURE 11-14

Proof

Since a parallelogram is a special case of a trapezoid with $b_1 = b_2 = b$, it follows from Theorem 11-8 that

$$\text{area of parallelogram } ABCD = \tfrac{1}{2}h(b + b) = bh.$$ [8]

VOLUME: INFORMAL REMARKS

1. The *volume* of a solid is a *number* that measures the "amount of space" the solid occupies. The number is called the *measure of the volume* of the solid.
2. To determine the volume of a solid we proceed the same way as when we determine length or area:
   a. We choose a unit volume, or *cubic unit* (FIG. 11-15).

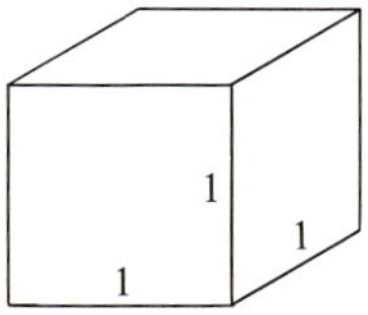

FIGURE 11-15

   b. We estimate the number of these units needed to fill the solid (FIG. 11-16).

[8] 1. To find the area of any polygonal region, we subdivide it into nonoverlapping triangular regions (FIG. 11D). Then, by the Area Addition Postulate, the area of the polygonal region will equal the sum of the areas of these triangular regions.

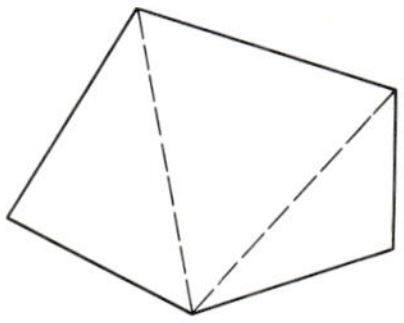

FIGURE 11D

2. The area of a polygon is independent of the way in which we divide it into triangles. For example, the area of polygon *ABCD* will be the same whether it is divided as in FIG. 11E (a), (b), or (c).

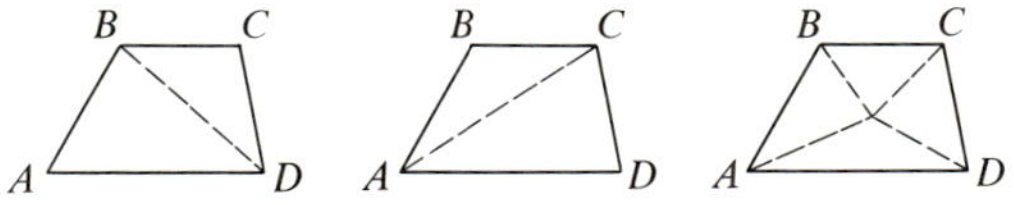

FIGURE 11E

3. The theorem states "*any* base and the *corresponding* altitude." Stress that bases do not necessarily lie on horizontal lines and that altitudes of figures do not always extend in a vertical direction (FIG. 11F).

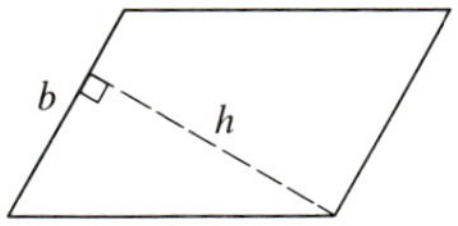

FIGURE 11F

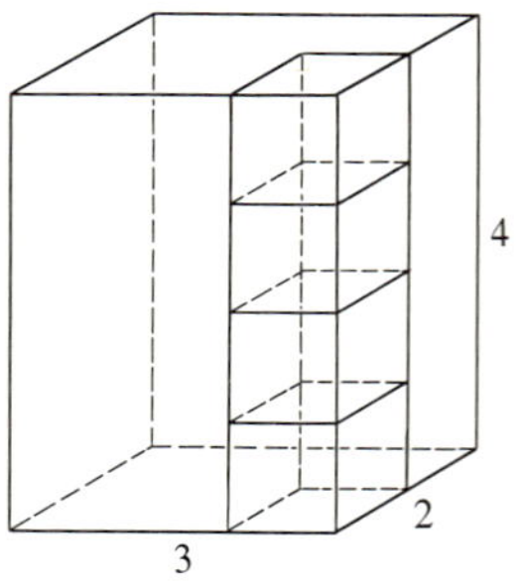

FIGURE 11-16

c. This *number* is the *volume* of the solid.
d. More precise estimates of volume can be obtained by using a smaller unit.

3. A standard unit for measuring volume is a *cubic inch*. A *cubic inch* is the volume of a cube each side of which is 1 inch (FIG. 11-17).[9]

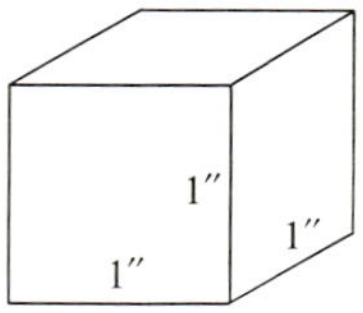

FIGURE 11-17

VOLUME POSTULATES[10]

POSTULATE 11-5 (Unit Volume Postulate) *The volume of a rectangular parallelepiped is the product of the altitude and the area of the base* (FIG. 11-18).

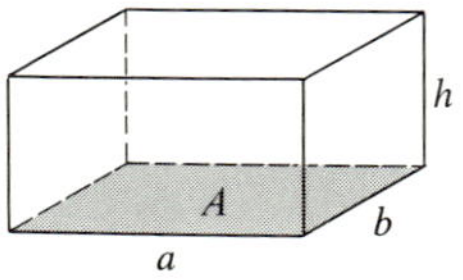

FIGURE 11-18

That is, $V = Ah = abh$.

[9] 1. Although the shape and size of the unit of volume is arbitrary, using a unit in the shape of a cube has advantages over other shapes similar to the advantages that a *square unit* has when we want to determine area.

2. Stress that a unit of *length* is a *segment* one unit long; a unit of *area* is a square one unit in each of two dimensions; a unit of *volume* is a *cube* one unit in each of three dimensions.

3. Students should know these relationships:

1728 cubic inches = 1 cubic foot,

27 cubic feet = 1 cubic yard.

[10] In our discussion of area we postulated the formula for the area of a square and then used this formula to derive the formulas for the areas of other polygonal regions. In the case of volume, we shall postulate the formula for the volume of a rectangular parallelepiped and from it derive the other volume formulas. However, before we can derive these formulas we need a definition, another postulate, and two preliminary theorems.

Definition
A *cross section* of a prism is the intersection of the prism with a plane parallel to the plane of the base (assuming that the intersection is not empty) (FIG. 11-19).

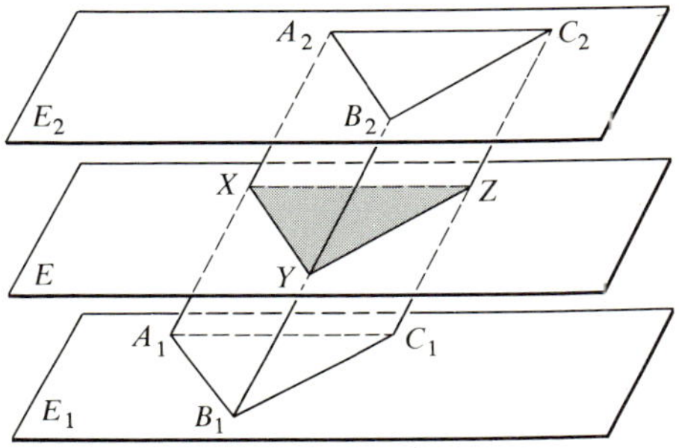

FIGURE 11-19

POSTULATE 11-6 (Cavalieri's Principle) *Given two solids and a plane, suppose that every plane parallel to the given plane, intersecting one of the two solids, also intersects the other in such a way that the two cross sections have equal areas. Then the two solids have the same volume.*

That is, if N is any plane parallel to plane M and intersects the solids A and B in such a way that the area of section a equals the area of section b, then the solids A and B have the same volume (FIG. 11-20).[11]

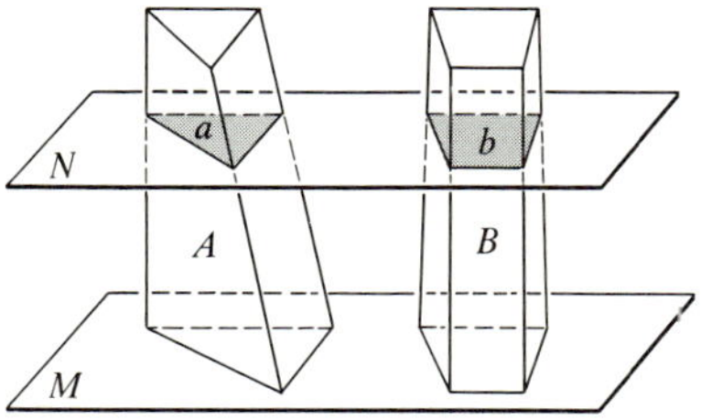

FIGURE 11-20

THEOREM 11-10 (Prism Cross-Section Theorem) *All cross sections of a prism have the same area* (FIG. 11-21).

[11] 1. Cavalieri's Principle can also be stated this way: If two solids have equal altitudes, and if sections of these solids formed by planes parallel to the bases and at equal distances from the bases have equal areas, then the solids have equal volumes.

2. Cavalieri's Principle can be illustrated with two stacks of cards, one stack consisting of, say, 100 square cards and the other stack consisting of, say, 100 triangular cards (FIG. 11G).

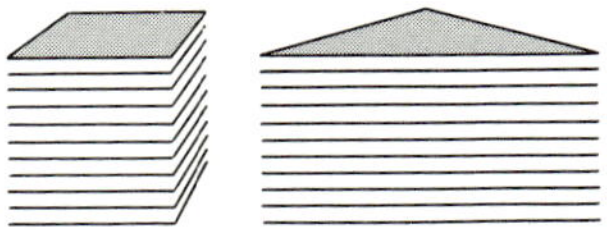

FIGURE 11G

If all the cards have the same thickness and if each card in one stack has the same area as the corresponding card in the other stack, then the volumes of the two stacks are equal.

3. We can also use an ordinary deck of playing cards to illustrate Cavalieri's Principle. The volume of the deck when it stands erect is equal to its volume when it leans to one side (FIG. 11H). Obviously, sections at equal distances from the base have equal areas.

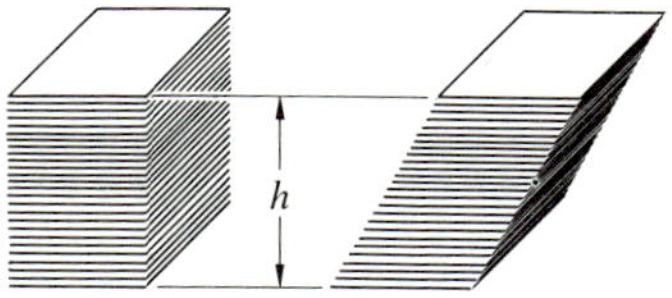

FIGURE 11H

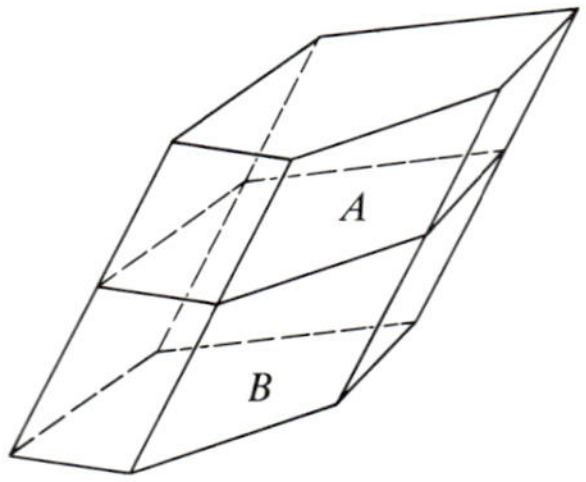

FIGURE 11-21

That is, if B is the base and A a cross section, then the area of A is the same as the area of B.

THEOREM 11-11 (Pyramid Cross-Section Theorem) If two pyramids have the same base area and the same altitude, then cross sections equidistant from the vertices have the same area (FIG. 11-22).

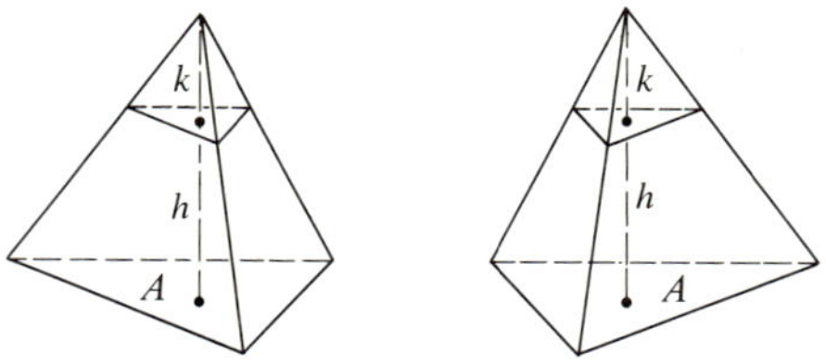

FIGURE 11-22

That is, if the base area of each pyramid is A, the altitude of each pyramid is h, and the distance from each cross section to the vertex is k, then the areas of the two cross sections are the same.

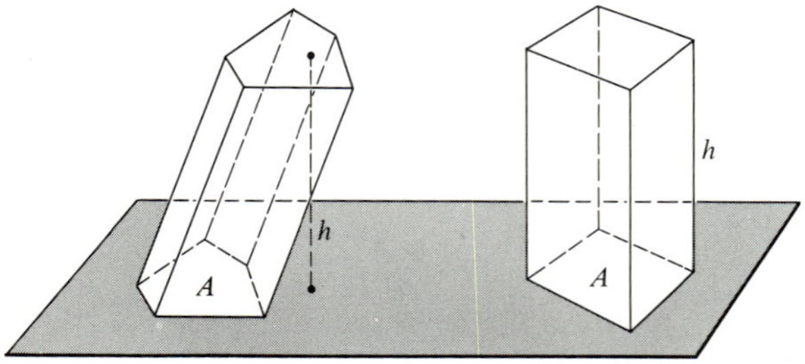

FIGURE 11-23

VOLUME THEOREMS

THEOREM 11-12 (The Volume of a Prism) The volume of any prism is the product of the altitude and the area of the base ($V = Ah$) (FIG. 11-23).

That is, if h is the altitude and A is the base area of the given prism, then its volume is $V = Ah$.

Proof

1. Consider a rectangular parallelepiped with the same altitude h and base area A and with its base in the same plane as the base of the given prism.
2. By the Prism Cross-Section Theorem, all cross sections for both prisms have the same area A.
3. By Cavalieri's Principle, both prisms have the same volume.
4. Since, by Postulate 11-5, the volume of the rectangular parallelepiped is Ah, the volume of the prism is also Ah.

THEOREM 11-13 *If two pyramids have the same base area and the same height, then they have the same volume* (FIG. 11-24).

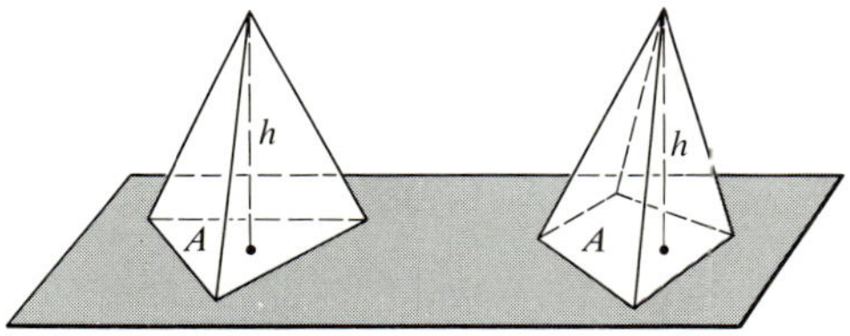

FIGURE 11-24

Proof

1. By the Pyramid Cross-Section Theorem, corresponding cross sections of the two pyramids have the same area.
2. By Cavalieri's Principle, the volumes of the two pyramids are the same.

THEOREM 11-14 (Volume of a Pyramid) *The volume of a pyramid is one third the product of its altitude and its base area* ($V = \frac{1}{3}Ah$) (FIG. 11-25).

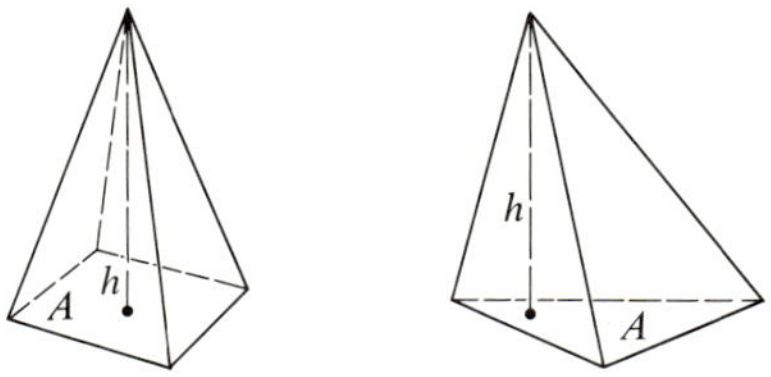

FIGURE 11-25

That is, if h is the altitude and A the base area of the given pyramid, then its volume is $V = \frac{1}{3}Ah$.

Proof

The proof of this theorem depends on another theorem, which says that *the volume of a triangular pyramid is one third the product of its altitude and its base area.*

1. Let us compare the given pyramid with a triangular pyramid of the same altitude and base area. The cross sections of the two pyramids that are equidistant from the vertices will have the same area, by the Pyramid Cross-Section Theorem.
2. By Cavalieri's Principle, the two pyramids have the same volume.
3. Since, by the theorem just cited, the volume of the triangular pyramid is $\frac{1}{3}Ah$, the volume of the given pyramid is also $V = \frac{1}{3}Ah$.[12]

[12] 1. Theorem 11-14, in effect, tells us that the volume of a pyramid is one third the volume of a prism with the same base and the same altitude as the pyramid. For instance, in FIG. 11-26, the prism *ABCDEF* can be seen as the union of three triangular pyramids. All three pyramids have the same volume because in pairs they have the same altitude and the same base area.

2. Use Visual 6 to illustrate this theorem.

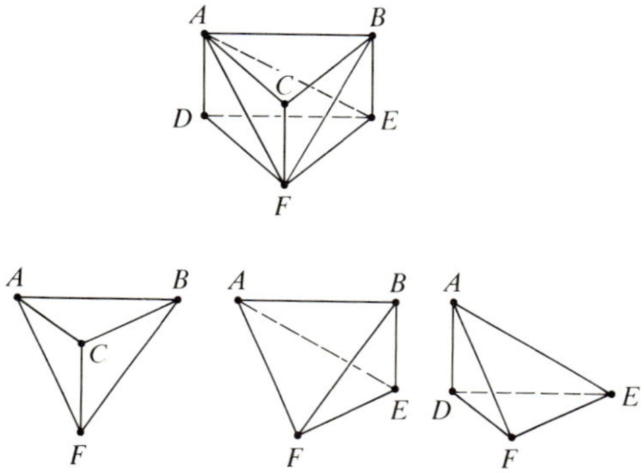

FIGURE 11-26

TEACHING SUGGESTIONS AND ACTIVITIES

VISUAL AIDS

Visual 1

PURPOSE

To show that the area of a parallelogram is the same as the area of a rectangle having the same base and the same altitude.

CONSISTS OF

A rectangle cut into two parts (Fig. 11-27).

HOW TO USE

1. Start with rectangle as in Fig. 11-27(a).
2. Shift region 1 as shown in Fig. 11-27(b), forming a parallelogram.
3. Although the shapes in Fig. 11-27(a) and (b) are different, their areas remain the same; that is, $A = bh$.

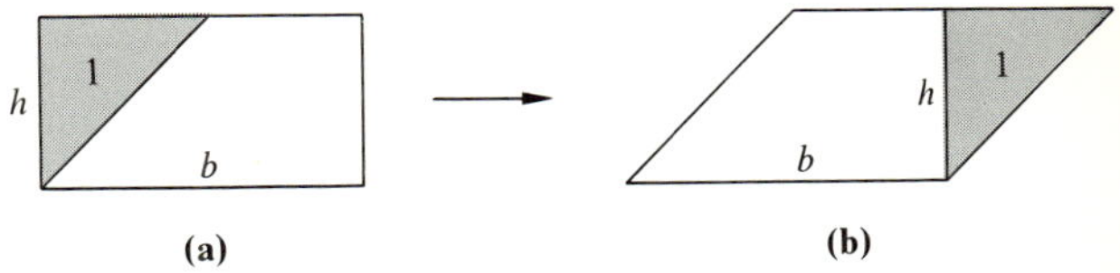

FIGURE 11-27

Visual 2

PURPOSE

To show that the area of a parallelogram is twice the area of a triangle having the same base and the same altitude.

CONSISTS OF

Two congruent triangles, $\triangle ABC$ and $\triangle A'B'C'$ (Fig. 11-28).

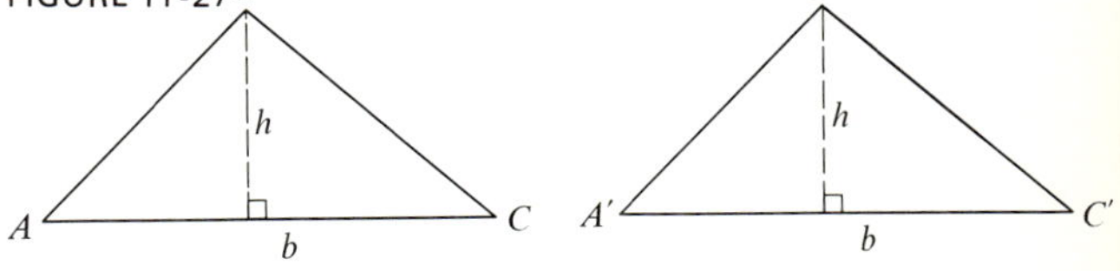

FIGURE 11-28

HOW TO USE

1. Arrange the two triangles so as to form parallelogram $ABA'C$ (Fig. 11-29).
2. The area of the parallelogram is the sum of the areas of the two congruent triangles, or twice the area of one of them.

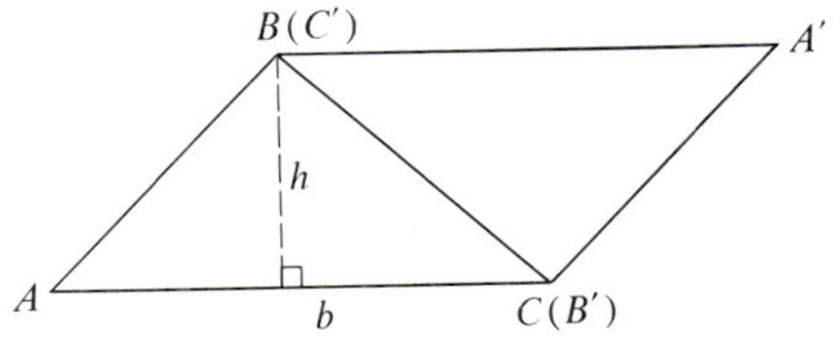

FIGURE 11-29

Visual 3

PURPOSE

To show that the area of a trapezoid is

$$A = \frac{1}{2}h(b_1 + b_2).$$

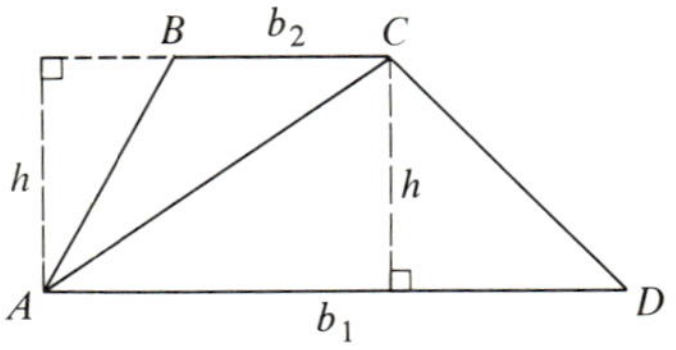

FIGURE 11-30

CONSISTS OF

A trapezoid *ABCD* cut into two triangles, $\triangle ACD$ and $\triangle ABC$ (FIG. 11-30).

HOW TO USE

1. Take $\triangle ACD$ and show that its area is $\frac{1}{2}b_1h$ (FIG. 11-31).
2. Take $\triangle ABC$ and show that its area is $\frac{1}{2}b_2h$ (FIG. 11-32).
3. The area of the trapezoid *ABCD* will equal the sum of the areas of the two triangles; that is,

   $$A = \tfrac{1}{2}b_1h + \tfrac{1}{2}b_2h$$

   or

   $$A = \tfrac{1}{2}(b_1 + b_2).$$

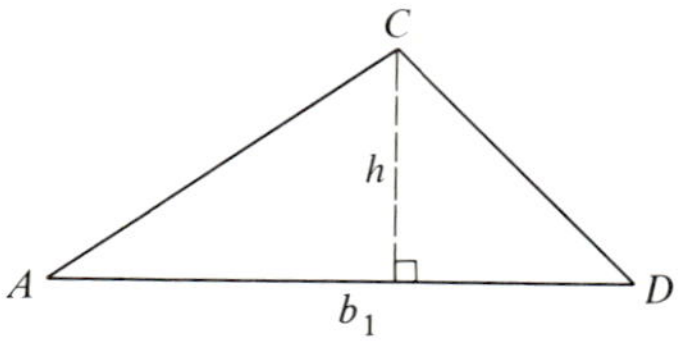

FIGURE 11-31

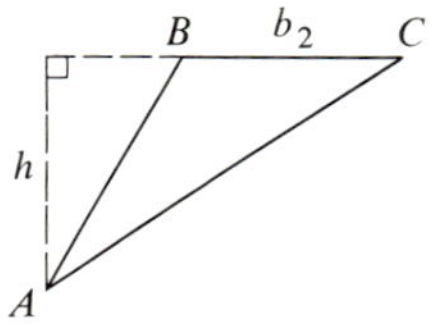

FIGURE 11-32

Visual 4

PURPOSE

To show that the volume of a prism is $V = Ah$.

CONSISTS OF

1. A hollow, transparent rectangular prism measuring 3 by 2 by 5 [FIG. 11-33(a)].
2. 5 layers of cubes, each layer measuring 3 by 2 by 1 [FIG. 11-33(b)].

HOW TO USE

1. Show that the unit cube is used in the measurement of volume the same way the unit square is used in measuring area and the unit length is used in measuring length. The number of unit cubes required to "fill up" the prism is a measure of its volume.
2. Place one layer of cubes in the prism. It should be intuitively clear that the number of cubes in this layer is the same as the number of square units in the base of the prism, which is 3×2, or 6.
3. Place a second layer on top of the first. The volume now becomes $3 \times 2 \times 2$ cubic units.
4. Place a third, then a fourth, and then a fifth layer on top of the first two, making the volume first $3 \times 2 \times 3$, then $3 \times 2 \times 4$, and then $3 \times 2 \times 5$ cubic units. Since the five layers "fill up" the prism, its volume is $3 \times 2 \times 5$, or 30 cubic units.

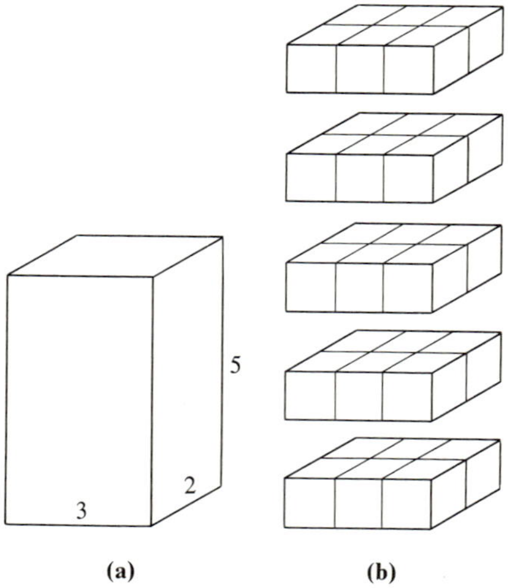

FIGURE 11-33

5. Repeat this process until it becomes clear that the volume of the prism may be found by multiplying its length by its width by its height, or by multiplying the area of the base by the height. That is, $V = Ah$.

Visual 5

PURPOSE

To illustrate Cavalieri's Principle.

CONSISTS OF

1. The following three hollow prisms *having equal altitudes and equal base areas:*
   a. A right quadrangular prism.
   b. An oblique quadrangular prism.
   c. A triangular prism.
2. A quantity of sand.

HOW TO USE

1. Fill the right prism with sand.
2. Empty the sand from this prism into the oblique prism. It will be found that the sand from the right prism will exactly fill the oblique prism, showing that both have the same volume.
3. Repeat the procedure in step 2 with the triangular prism.

Visual 6

PURPOSE

To illustrate that the volume of a pyramid is one third the product of its altitude and its base (Theorem 11-14).

CONSISTS OF

1. A hollow prism.
2. A hollow pyramid whose base area and altitude are equal to the base area and altitude of the prism.
3. A quantity of sand.

HOW TO USE

1. Fill the pyramid with sand.

2. Pour the sand from the pyramid into the prism. It will be found that it takes three pyramids of sand to fill the prism exactly.

3. Since the volume of the prism is $V = Ah$, the volume of the pyramid is $V = \frac{1}{3}Ah$.

MILESTONES IN THE HISTORY OF MEASUREMENT

Students should be left with an appreciation that during the last 500 years there have been *three* big milestones in the history of measurement:

1. The realization that measures that use parts of the human body are inaccurate and unreliable, and that *standard units* had to be established.

2. The creation of *instruments*, such as the micrometer and altimeter, which yielded more precise measurements than were possible before.

3. The creation of *more refined instruments*, such as those that use electrons, light waves, sound waves, and radio waves with which to make measurements. With these instruments we can weigh a speck of dust, measure the noise of a molecule, and measure the composition of stars.

AREA UNDER CERTAIN DISTORTIONS

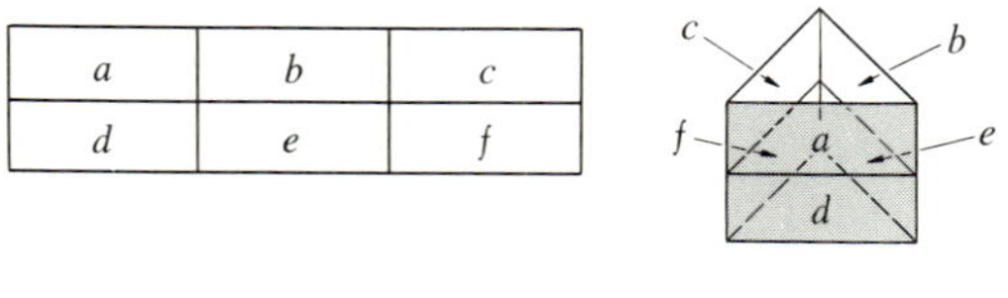

FIGURE 11-34

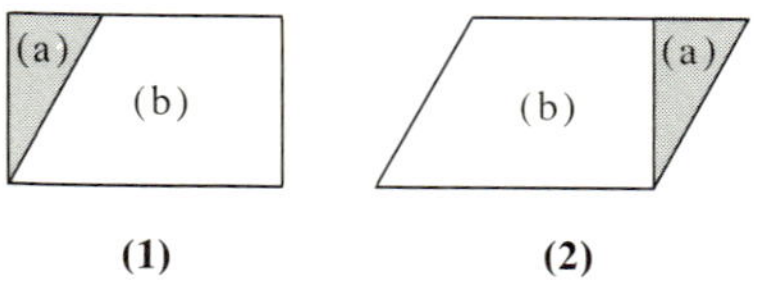

FIGURE 11-35

In mathematics we are always looking for properties that remain unchanged under changing conditions. We call such properties *invariants*.

It may be of interest to your students to see that the area of a geometric figure is invariant under certain distortions. For example,

1. In FIG. 11-34, if figure (1) is "bent" into figure (2) in such a way that there is no stretching or overlapping, the area does not change.

2. In FIG. 11-35, if the rectangle in (1) is cut into two parts, (a) and (b), and reassembled in a different way, (2) the area remains unchanged. (See Visual 1, page 307.)

QUESTIONS

1. How many bricks are needed to lay a walk 45 feet long and 12 feet wide, if seven bricks are needed to cover a square foot?

2. a. If a rectangle is twice as long as another rectangle, but both have the same width, how do their areas compare?

   b. If two rectangles have the same length, but the width of one is three times the width of the other, how do their areas compare?

   c. One rectangle is twice as wide as another and three times as long. How do their areas compare?

3. What happens to the area of a triangle if

   a. Its altitude is doubled?
   b. Its base is doubled?
   c. Both its altitude and base are doubled?

4. A cube has ______ faces, ______ vertices, and ______ edges.

5. Change the triangle shown below into a parallelogram whose area is twice that of the triangle.

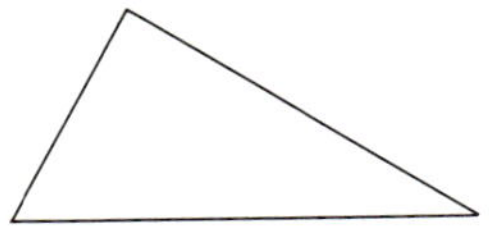

6. The volume of a box is $5\frac{1}{4}$ cubic inches. How long is the box if it is 2 inches wide and $\frac{3}{4}$ inch deep?

7. What happens to the volume of a rectangular solid

   a. If one of its measurements is doubled?
   b. If two of its measurements are doubled?
   c. If all three of its measurements are doubled?
   d. If all three of its measurements are tripled?

8. A sand box measures 54 inches by 54 inches and is 10 inches deep. How many cubic feet of sand will it hold?

9. Find the area of trapezoid *ABCD*.

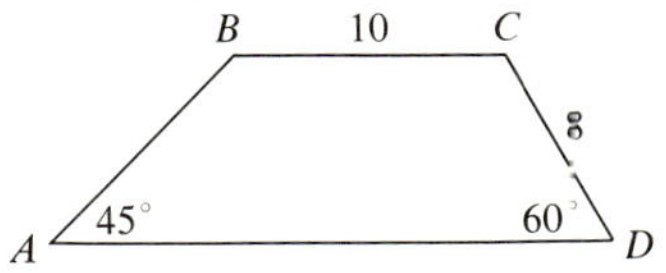

10. How is the area of a square changed if its side is doubled? Halved? Tripled?

11. The altitude of a trapezoid is 5 inches and its area is 50 square inches. If the length of one base is 7 inches, what is the length of the other base?

12. The area of a square is 225 square inches. The area of another square is 9 square inches. What is the length of the side of a third square whose area is the difference between the areas of the first two squares?

13. A triangle and a parallelogram have equal areas and equal bases. How are their altitudes related?

14. The width of a rectangle is 5 inches less than its length. If the perimeter is 50 inches, find the area of the rectangle.

15. The length of a rectangle is two times its width. If the area is 98 square feet, find the perimeter.

16. A room is 18 feet long, 15 feet wide, and 8 feet high.

    a. Find the total area of the walls and ceiling, allowing 78 square feet for windows and doorway.

    b. How many gallons of paint will be needed to cover the walls and ceiling if a gallon covers 400 square feet?

17. A classroom measures 35 by 25 by 12 feet. How many cubic feet of air space are there for each pupil if there are 30 pupils in the class?

18. Prove that the diagonals of a parallelogram subdivide the region bounded by the parallelogram into four triangular regions all having the same area.

19. $\overline{AD}$ and $\overline{BE}$ are medians of $\triangle ABC$, intersecting at P. Prove that the areas of $\triangle AEP$ and $\triangle BDP$ are equal.

20. Develop a formula for finding the area of a rhombus in terms of the lengths of its diagonals.

21. Prove that if two rectangles have the same altitude, then the ratio of their areas equals the ratio of their bases.

22. Prove that if two triangles have the same base, then the ratio of their areas is equal to the ratio of their altitudes.

23. What is the volume of a pyramid whose base is a square and whose lateral faces are equilateral triangles of side s?

24. A toy is submerged in a rectangular water tank measuring 30 inches in length and 15 inches in width. What is the volume of the toy if it raises the water level 0.45 inch?

25. The midpoints A_1, B_1, and C_1 of the sides of $\triangle ABC$ are joined to form a second triangle. The midpoints A_2, B_2, and C_2 of the sides of $\triangle A_1B_1C_1$ are then joined to form a third triangle, $\triangle A_2B_2C_2$. This process is continued until $\triangle A_4B_4C_4$ is formed. Compare the area of $\triangle A_4B_4C_4$ with the area of $\triangle ABC$.

PUZZLES

1. Construct a square whose area is the same as the area of this rectangle:

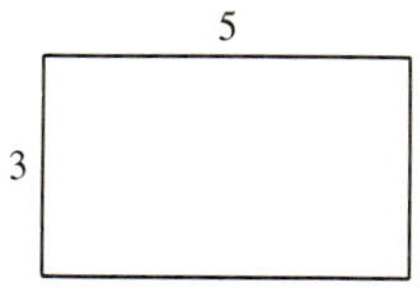

2. Divide a square into seven smaller squares, not necessarily equal in area.

3. Here are two rectangles: Draw a single line that passes through both rectangles in such a way as to separate each rectangular region into two regions of equal area.

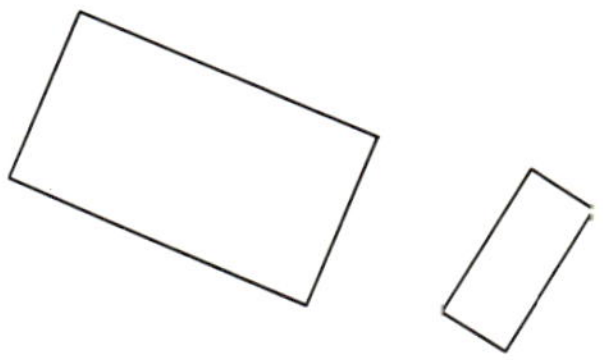

SOLUTIONS

1. *Solution (a):* Construct a right triangle with one leg measuring 1 and the hypotenuse measuring 4. The other leg, measuring $\sqrt{15}$, is the side of the required square. *Solution (b):* Construct the mean proportional between 3 and 5.

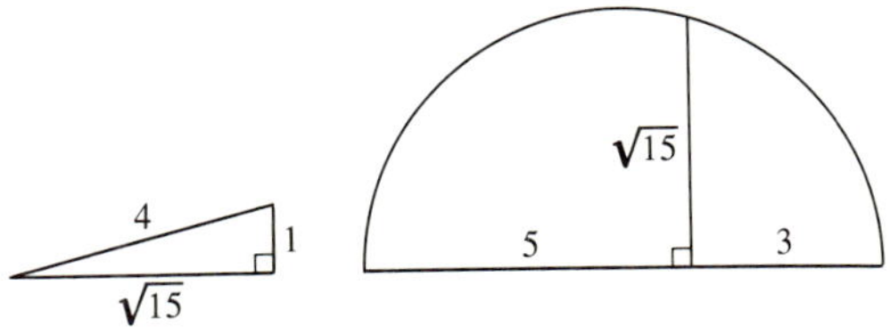

2.

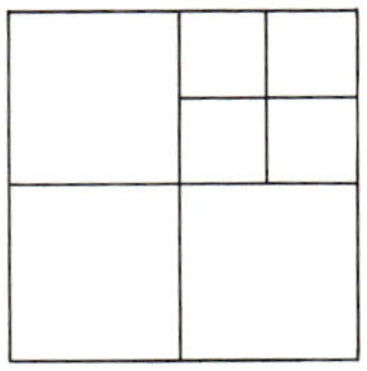

3. Find the point of intersection of the diagonals of each rectangle. The line determined by these two points of intersection separates each rectangle into two trapezoids of equal area.

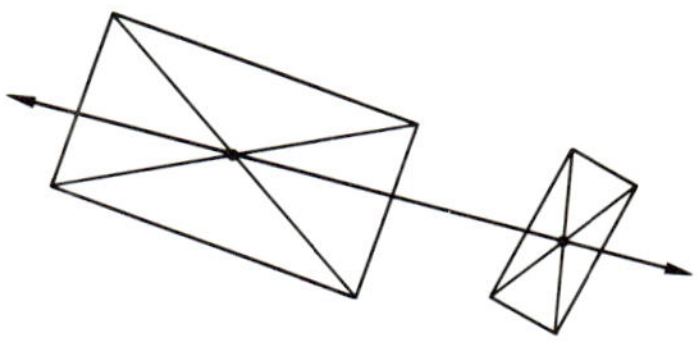

ENRICHMENT

ANOTHER FORMULA FOR THE AREA OF A TRIANGLE

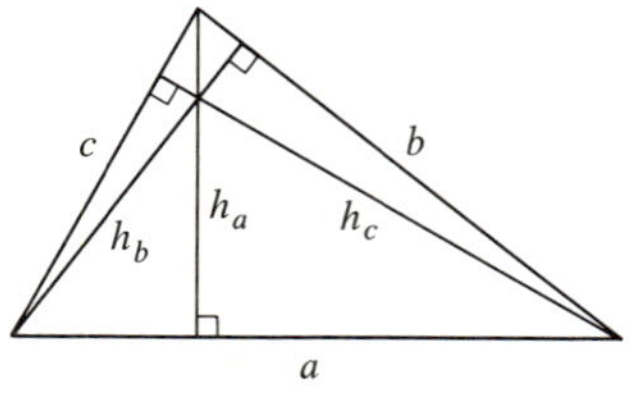

FIGURE 11-36

In this chapter we saw how to find the area of a triangle if we know a side and its altitude. The Greek mathematician Heron (ca. 75 A.D.?) found a way of finding the area of a triangle if we know the lengths of its three sides but are not given the length of an altitude. From the lengths of the sides, Heron was not only able to find the area of the triangle but also the altitude to any of the sides.

If a, b, c are the three sides; h_a, h_b, h_c are the three altitudes; and s is half the perimeter of the triangle (FIG. 11-36), then Heron's altitude formulas are

$$h_a = \frac{2}{a}\sqrt{s(s-a)(s-b)(s-c)},$$

$$h_b = \frac{2}{b}\sqrt{s(s-a)(s-b)(s-c)},$$

$$h_c = \frac{2}{c}\sqrt{s(s-a)(s-b)(s-c)}.$$

His area formula is

$$A = \sqrt{s(s-a)(s-b)(s-c)}.$$

ANOTHER VIEW OF AREA

The understanding of the concept of area can be enriched if we think of an area as being generated by a moving line segment. Just as a moving point can be considered as generating a line segment, so can a moving line segment be considered as generating a region, and a moving region can be considered as generating a solid.

From this point of view, we can think of the area of a rectangle as being generated by the moving line segment a as it moves across the distance b (FIG. 11-37). We should note that the distance the segment moves is measured in a direction perpendicular to the segment.

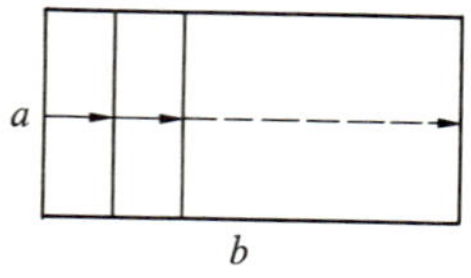

FIGURE 11-37

In the case of a triangle, the moving segment has a length of b in its initial position (FIG. 11-38). But as it moves upward its length keeps shrinking until it becomes 0, making the *average* length of the segment $(b + 0)/2$, or $b/2$.

So we can think of the area of the triangle as being generated by a moving line segment of average length $b/2$ as it moves through a distance h.

Therefore, to obtain the area of the triangle, we multiply the average length of the moving segment by the distance it sweeps along a line perpendicular to the segment; that is,

$$A = \frac{b}{2} \times h \qquad \text{or} \qquad A = \tfrac{1}{2}bh.$$

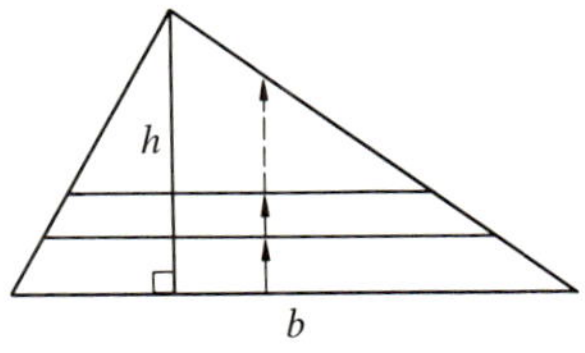

FIGURE 11-38

To find the area of a trapezoid, we note that the moving segment begins at one of the bases, say b_1, and keeps shrinking until it reaches the upper base b_2 (FIG. 11-39). Here the average length of the segment is $(b_1 + b_2)/2$, and it moves a distance h in a direction perpendicular to the segment. The area is, therefore, found by multiplying the average length of the moving segment by the distance it moves; that is,

$$A = \frac{b_1 + b_2}{2} \times h \qquad \text{or} \qquad A = \tfrac{1}{2}h(b_1 + b_2).$$

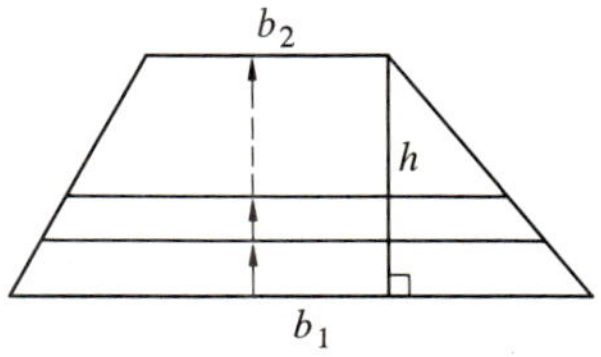

FIGURE 11-39

MAXIMUM AND MINIMUM AREAS AND PERIMETERS

The conditions under which geometric figures have maximum or minimum areas or perimeters may be of interest.

1. Of all rectangles having a given perimeter, the square has the maximum area.
2. A rectangle with a fixed area has a minimum perimeter when it is a square.
3. Given two segments a and b. The triangle of maximum area having a and b as sides will be the right triangle whose legs are a and b.
4. In a set of triangles with equal perimeters and equal bases, the triangle with equal legs has the maximum area.
5. A right triangle having a given hypotenuse has a maximum area if its legs are equal.

In general, if a and b are positive real numbers, then the following properties will hold:

1. If ab is constant, then $a + b$ is minimum when $a = b$.

2. If $a + b$ is constant, then ab is maximum when $a = b$.
3. If $a^2 + b^2$ is constant, then ab is maximum when $a = b$.

12

MEASUREMENT OF CIRCULAR FIGURES

INTRODUCING THE CHAPTER TO STUDENTS

CONTENT REVIEW

Prerequisites for Learning This Chapter
Introduction
Circumference of a Circle
Area of a Circle
Lengths of Arcs and Areas of Sectors
Area and Volume

TEACHING SUGGESTIONS AND ACTIVITIES

Visual Aids
The Wonders of π
Questions
Puzzles
Solutions

ENRICHMENT

On the History of π

INTRODUCING THE CHAPTER TO STUDENTS

To find the perimeter of a triangle we measure its three sides and add their lengths. We do the same thing to find the perimeter of a rectangle, pentagon, or any other polygon.

But how can we find the length, or circumference, of a circle? Can we measure the length of a curve the way we measure the length of a line segment? Can we use a ruler to measure the length of a circle?

We can, of course, take a tape measure and bend it into the shape of the circle. But this is not always convenient, accurate, or even possible. If we want to measure the length of a circular object, such as a disc or wheel, we might roll the object along the edge of a ruler and determine the distance covered by one complete revolution of the wheel. This distance will be the same as the circumference of the wheel. But this method, too, is not practical and seldom possible.

There is, however, a way of determining the circumference of a circle without bending the measuring instrument or rolling the object to be measured. We can determine the circumference merely by measuring a line segment! What we are saying is that it is possible to measure a curve by using a ruler! What makes this possible is the fascinating relation that exists between a particular line segment connected with a circle and the circumference of that circle.

In exploring this method of determining the circumference of a circle, we encounter a new and fascinating number—a number that has intrigued people for thousands of years. Although this number was first used by the ancient Greeks in connection with circles and is still best known in this connection, it has been found equally useful in situations that have nothing to do with circles or even with geometry. It is used, for instance, in public opinion polls to predict the winner of a forthcoming election, by insurance companies to determine the premium you pay on a life insurance policy, by a manufacturer to test the quality of his product, and by a scientist to understand the laws of heredity. We are talking, of course, about the number π.

In trying to determine the area of a circular region we again face the problem of measuring something round with something straight—in this case, measuring a circular region with a square unit. And once again we find that this new, magic number comes to the rescue.

At the end of this chapter we shall see that if we can measure a circle we can also measure a cylinder, a cone, and a sphere, which also involve circles.

More formally, we regard all these circular figures as *abstractions* and, as such, they cannot be measured by a purely physical process. We must first *define* "circumference" of a circle and "area" of a circular region and then build our theory of circular measurement on postulates and theorems. This is the course we shall follow in this chapter.

CONTENT REVIEW

PREREQUISITES FOR LEARNING THIS CHAPTER

1. Knowing these definitions: circumference (informal meaning); π (informal meaning); perimeter; regular polygon; n-gon; inscribed regular polygon; circular region; arc; slant height of a right circular cone; cylinder; cone; sphere.

2. Knowing these theorems: SAS Similarity Theorem; area of a triangle $= \frac{1}{2}bh$; area of a rectangle $= bh$; proofs of Theorem 11-13 (Volume of a Prism) and Theorem 11-15 (Volume of a Pyramid); basic theorem on limits: If two variables are always equal and each approaches a limit, then these limits are equal.

INTRODUCTION

To understand the proofs of Theorems 12-1 and 12-2, it is necessary to know how to find the perimeter and area of a regular n-gon.

To construct a regular n-gon, divide a circle into n congruent arcs. Each of the arcs has a measure of $360/n$. When we draw the corresponding chord of each arc we obtain a regular n-gon with vertices $P_1, P_2, \ldots, P_n$. (FIGURE 12-1 shows the case where $n = 6$.)

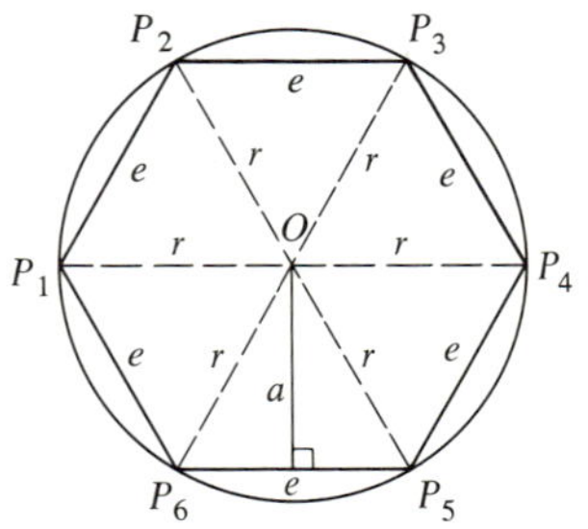

FIGURE 12-1

If we draw the radii from the center O to the vertices we obtain a set of congruent isosceles triangles with the same base e and the same altitude a. The distance a from the center to each of the sides is called the *apothem* of the polygon.

The perimeter p of the regular n-gon is $p = ne$.

Each of the isosceles triangles has an area of $\frac{1}{2}ae$. Since there are n such triangles, the area A_n of the regular n-gon is $A_n = n \cdot \frac{1}{2}ae$, or $A_n = \frac{1}{2}ap$.

CIRCUMFERENCE OF A CIRCLE

We can approximate the circumference C of a circle by inscribing in it a regular polygon with a large number of sides (FIG. 12-2). The perimeter p of such a polygon will be a good approximation of the circumference of the circle. In fact, we can make p as close as we wish to C by making the number of sides n large enough. We define the circumference as that number which the perimeter approaches as the number of sides gets larger and larger. This situation is described by the following definition.

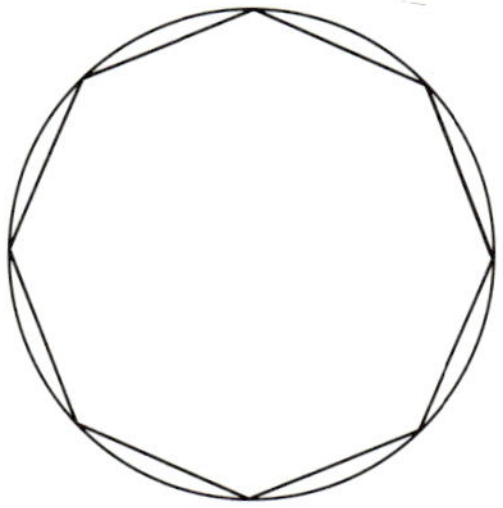

FIGURE 12-2

Definition
The *circumference* of a circle is the limit of the perimeters of the inscribed regular polygons. We denote this relationship by $p \to C$, read "p approaches C as a limit."

We shall now define the number π and show that it is the same number for all circles, regardless of their size.

THEOREM 12-1 *The ratio $C/2r$ of the circumference to the diameter is the same for all circles* (FIG. 12-3).

That is, given a circle with circumference C and radius r, and another circle with circumference C' and radius r', then $C/2r = C'/2r'$.

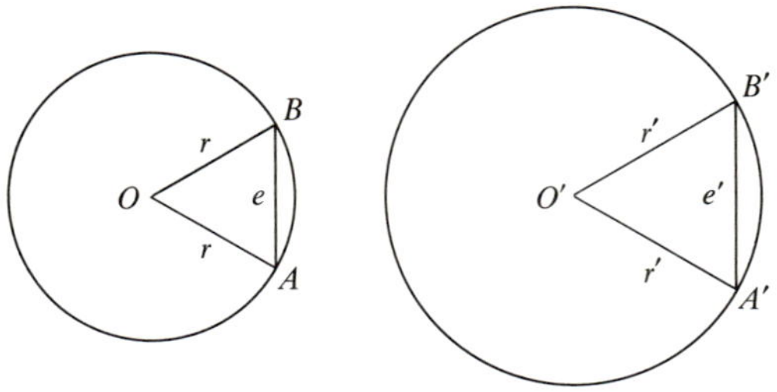

FIGURE 12-3

Proof

1. Inscribe a regular n-gon in each of the two circles, using the same value of n in each circle. Let e and e' be one side of each n-gon, and let $p = ne$ and $p' = ne'$ be their perimeters.
2. $\angle AOB \cong \angle A'O'B'$, since each has a measure of $360/n$.
3. The adjacent sides of $\angle AOB$ and $\angle A'O'B'$ are proportional, since $\dfrac{r'}{r} = \dfrac{r'}{r}$.
4. $\triangle AOB \sim \triangle A'O'B'$ by the SAS Similarity Theorem.
5. Therefore, $\dfrac{e}{r} = \dfrac{e'}{r'}$ and $\dfrac{ne}{r} = \dfrac{ne'}{r'}$, or $\dfrac{p}{r} = \dfrac{p'}{r'}$.
6. Since $p \to C$ and $p' \to C'$, by the definition of circumference, we have (by basic theorem on limits)

$$\frac{C}{r} = \frac{C'}{r'} \qquad \text{and} \qquad \frac{C}{2r} = \frac{C'}{2r'}.$$

We denote the ratio $C/2r$ by the Greek letter π. ("π" is the first letter of the Greek word for "perimeter.") This ratio is the same for all circles; that is, $C/2r = \pi$, $C = 2\pi r$, or $C = \pi d$ for every circle.

The definition of π does not immediately give its numerical value. To obtain such a value we can inscribe, successively, in a circle of radius 1, a regular 4-gon, 8-gon, 16-gon, etc., and compute in each case the ratio $p/2r$. Convenient approximations of π are 3, 3.14, $3\frac{1}{7}$, and 3.1416.

AREA OF A CIRCLE

We can approximate the area of a circle by inscribing in it a regular n-gon with a large number of sides (FIG. 12-4).

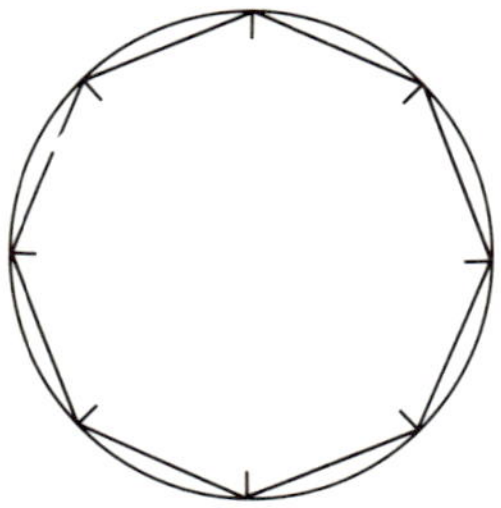

FIGURE 12-4

The area A_n of the n-gon is always less than the area A of the circle because there are always some parts that lie inside the circle but outside the n-gon. As n becomes very large, the difference between A and A_n becomes very small. Therefore, we expect that $A_n \to A$. Our definition of the area of a circle is based on this idea.

Definition′

The *area of a circle* is the limit of the areas of the inscribed regular polygons. We denote this relationship by $A_n \to A$.

THEOREM 12-2 *The area of a circle of radius r is* πr^2 (FIG. 12-5).

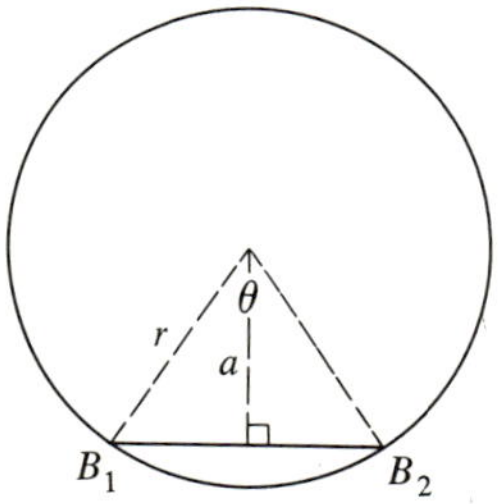

FIGURE 12-5

Proof Outline

1. Let $\overline{B_1B_2}$ be one side of a regular n-gon inscribed in a given circle of radius r, subtending a central angle θ.
2. The area A_n of the n-gon is $A_n = \frac{1}{2}ap$ and, by definition, $A_n \to A$.
3. $a < r$ because either leg of a right triangle is shorter than the hypotenuse.
4. As n becomes larger, θ becomes smaller and a approaches r; that is, $a \to r$.
5. By the definition of circumference, $p \to C$.
6. From 4 and 5, $\frac{1}{2}ap \to \frac{1}{2}rC$, and since $A_n = \frac{1}{2}ap$, $A_n \to \frac{1}{2}rC$.
7. $A = \frac{1}{2}rC$, since, by definition, $A_n \to A$.
8. Since $C = 2\pi r$, $A = \frac{1}{2}r(2\pi r)$, or $A = \pi r^2$.

LENGTHS OF ARCS AND AREAS OF SECTORS

Introduction

1. In defining the length of an arc, we use the same approach we

used in defining the circumference of a circle. If we cut up $\widehat{AB}$ into n congruent arcs and draw the corresponding chords, all having the same length e, then the sum of the lengths of the chords is $p = ne$ (FIG. 12-6).

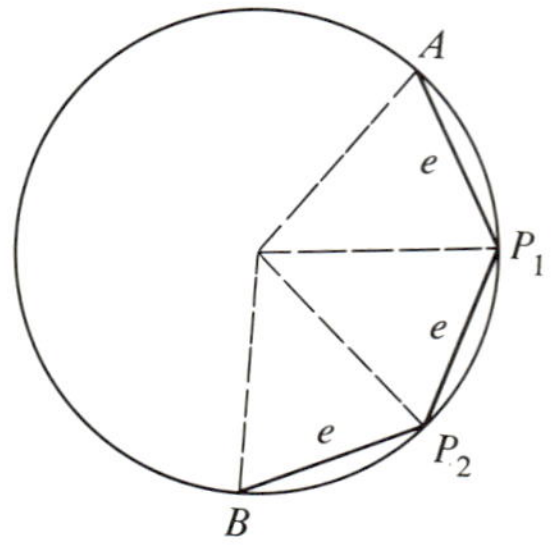

FIGURE 12-6

2. The *length* of $\widehat{AB}$ is defined to be the limit of p as n becomes larger and larger; that is, $p \to \widehat{AB}$. It is convenient to consider an entire circle as an arc of measure 360. We can then think of the circumference of a circle as the length of an arc of measure 360.

3. The basic theorem on arc length that we will state but not prove is

THEOREM 12-3 *If two arcs have equal radii, then their lengths are proportional to their measures* (FIG. 12-7).

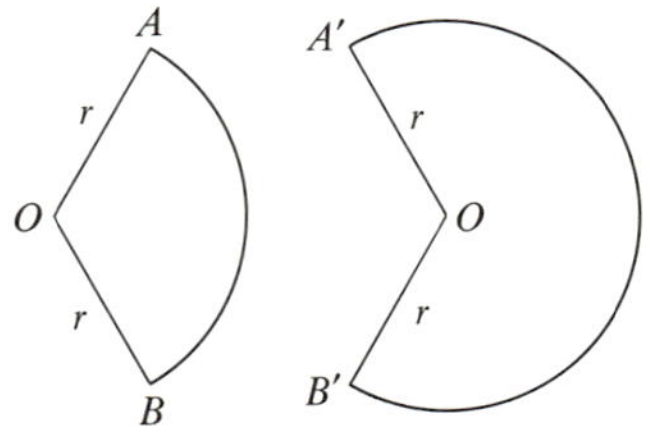

FIGURE 12-7

That is,

$$\frac{\text{length } \widehat{AB}}{m\widehat{AB}} = \frac{\text{length } \widehat{A'B'}}{m\widehat{A'B'}}.$$

Intuitively, this theorem makes sense since we would expect that doubling the measure of an arc will double its length, and multiplying the measure of an arc by any other positive number will have the effect of multiplying the length of the arc by the same number.

Length of an Arc

THEOREM 12-4 *If an arc has measure q and radius r, then its length is*

$$L = \frac{q}{180} \cdot \pi r.$$

Proof

1. If C is the circumference of a circle of radius r and L is the length of any arc, then, by Theorem 12-3,

$$\frac{L}{q} = \frac{C}{360}.$$

2. Since $C = 2\pi r$, we have

$$\frac{L}{q} = \frac{2\pi r}{360},$$

$$L = q \cdot \frac{2\pi r}{360}$$

$$= \frac{q}{180} \cdot \pi r.$$

Area of a Sector

Definition

Let $\widehat{AB}$ be an arc of a circle with center O and radius r (FIG. 12-8). The union of all segments $\overline{OP}$, where P is any point of $\widehat{AB}$, is called a *sector*. $\widehat{AB}$ is called the *arc* of the sector and r is called its *radius*.

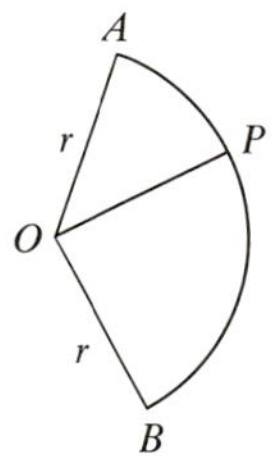

FIGURE 12-8

THEOREM 12-5 *The area of a sector is half the product of its radius and the length of its arc.*

We prove this theorem the same way we proved Theorem 12-2 for the area of a circle.

Since the length of an arc of measure q is

$$\frac{q}{180} \cdot \pi r$$

by Theorem 12-4, we get

$$A = \frac{1}{2} r \left(\frac{q}{180} \cdot \pi r \right) = \frac{q}{360} \cdot \pi r^2.$$

This proves

THEOREM 12-6 *The area of a sector of radius r and arc measure q is*

$$A = \frac{q}{360} \cdot \pi r^2.$$

AREA AND VOLUME

Cylinder

Definition

The area of the curved surface of a right circular cylinder is called its *lateral surface area.*

THEOREM 12-7 *The lateral surface area of a right circular cylinder is* $S = 2\pi rh$, *where* r *is the radius of the base and* h *is the height of the cylinder* (FIG. 12-9).

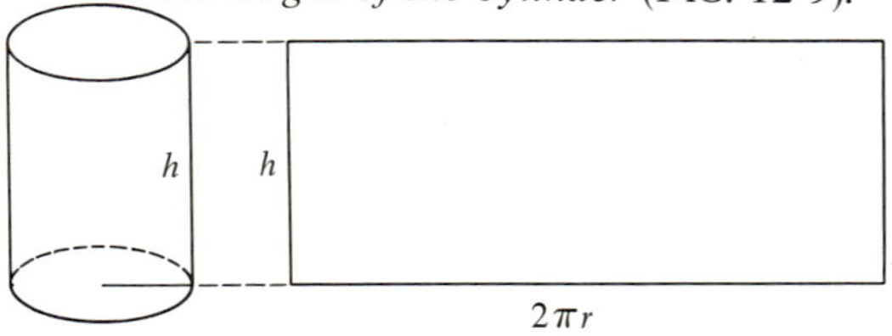

FIGURE 12-9

Think of the surface of the cylinder when it is rolled out flat. It becomes a rectangle with height h and base equal to the circumference of the cylinder.

The area of this rectangle is the product of its height h and its base $2\pi r$, that is, $2\pi rh$, or πdh.

THEOREM 12-8 *The volume of a circular cylinder is the product of its altitude and the area of its base* (FIG. 12-10).

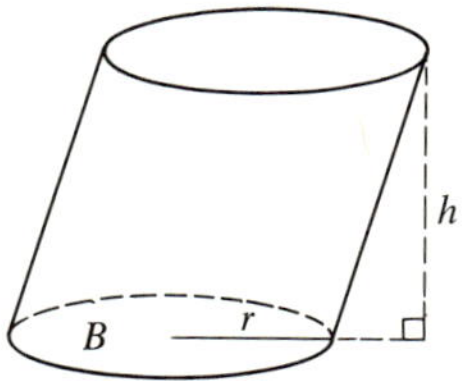

FIGURE 12-10

That is, $V = Bh$, or $V = \pi r^2 h$.

The proof of this theorem is like the proof of Theorem 11-12 for the volume of a prism.

Cone

THEOREM 12-9 *The lateral area of a right circular cone is equal to half the product of the circumference of the base and the slant height* (FIG. 12-11).

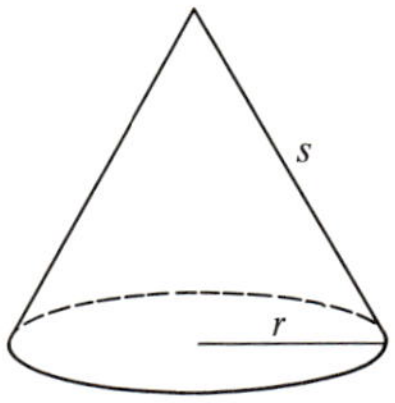

FIGURE 12-11

That is, $S = \pi rs$.

THEOREM 12-10 *The volume of a circular cone is one third the product of its altitude and the area of its base* (FIG. 12-12).

That is, $V = \frac{1}{3}Bh$, or $V = \frac{1}{3}\pi r^2 h$.

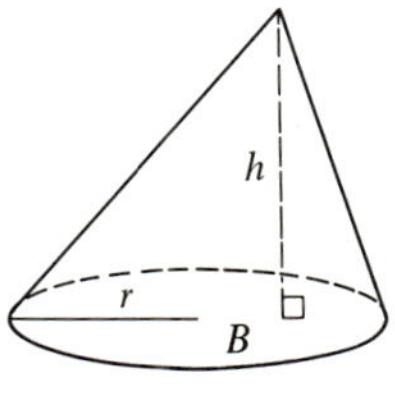

FIGURE 12-12

[1] 1. This is four times the area of the plane region inside one of its great circles. (See Visual 4.)

2. Unlike the surface of a cylinder, the surface of a sphere cannot be flattened out onto a plane without stretching.

The proof of this theorem is like the proof of Theorem 11-14 for the volume of a pyramid.

Sphere

THEOREM 12-11 *The volume of a sphere of radius r is* $V = \frac{4}{3}\pi r^3$ (FIG. 12-13).

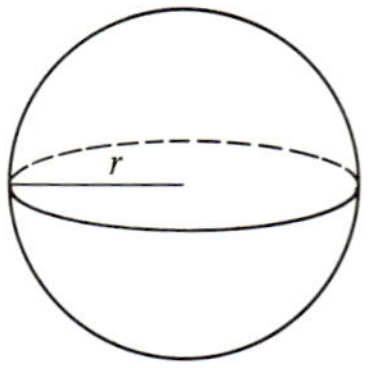

FIGURE 12-13

THEOREM 12-12 *The surface area of a sphere of radius r* is $S = 4\pi r^2$.[1]

TEACHING SUGGESTIONS AND ACTIVITIES

VISUAL AIDS

Visual 1

PURPOSE

To establish experimentally

1. The existence and value of π.
2. The formula for the circumference of a circle.

CONSISTS OF

1. Three circular objects of different diameter. On each object
   a. The center is shown.
   b. A point P is clearly marked on the circumference.
2. A transparent "track" on which the circular objects can be rolled, with front face calibrated in inches, as shown in FIG. 12-14.

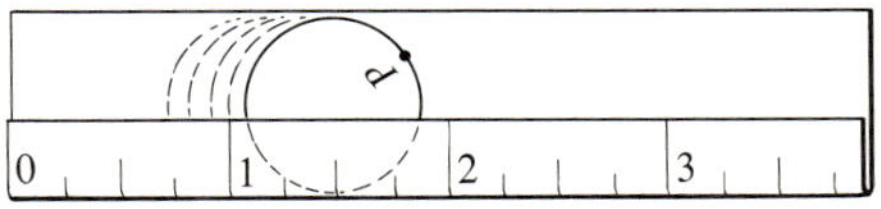

FIGURE 12-14

HOW TO USE

1. Measure the diameter of one of the circular objects.
2. Place the object in the track with point P on the zero mark of the ruler. Roll the object in the track until point P again touches the edge of the ruler. Distance rolled represents the length, or circumference, of the circle.
3. Record the data obtained:

Object	d	C	$C - d$	$\frac{C}{d}$

4. Repeat the procedure with the other circular objects, in each instance recording the data obtained.

5. The data in the chart should lead students to the relation between circumference and diameter. Indicate to class that
   a. Although the *difference* between C and d ($C - d$ column) does not show a constant relation between C and d, the *ratio* of C and d (C/d column) does establish a constant relation for *any* circle.
   b. This ratio is always the same number—about 3.14 or $3\frac{1}{7}$, called "pi" and written "π."

6. If $C/d = \pi$, then $C = \pi d$, or $C = 2\pi r$.

A variation of this experiment can be performed as follows:

1. Measure the diameters of the circular objects.
2. Instead of rolling the objects in a track, *measure each circumference with a tape.*
3. Record data and then proceed as before.

Visual 2

PURPOSE

Two ways to develop experimentally the formula for the area of a circle.

CONSISTS OF

1. Four circular discs, each disc same size, cut up as follows:
   a. One disc cut up into *four* congruent parts.
   b. One disc cut up into *six* congruent parts.
   c. One disc cut up into *eight* congruent parts.
   d. One whole disc.
2. Three circular objects, of different sizes, with centers marked.
3. Transparent grid, marked in square units.

HOW TO USE

Experiment 1

1. Measure the radius of each circular object.
2. Measure the area of each object by placing the grid over the surface of the object and estimating the number of square units it contains.

3. Record the data obtained:

Object	r	r^2	A	$\frac{A}{r^2}$

4. Guide the students to discover that $A/r^2 = \pi$, or $A = \pi r^2$.

Experiment 2

1. Take cut-up parts and arrange as follows:
   a. *Whole disc* (FIG. 12-15).
   b. *Four congruent parts* (FIG. 12-16).
   c. *Six congruent parts* (FIG. 12-17).
   d. *Eight congruent parts* (FIG. 12-18).

2. With each succeeding arrangement, it will become more apparent to the class that as you increase the number of congruent regions into which the circle is cut, the resulting figure approaches more and more a parallelogram whose base is half the circumference and whose height is the radius of the cut-up circle (FIG. 12-19).

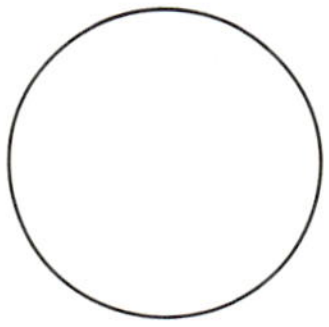

FIGURE 12-15

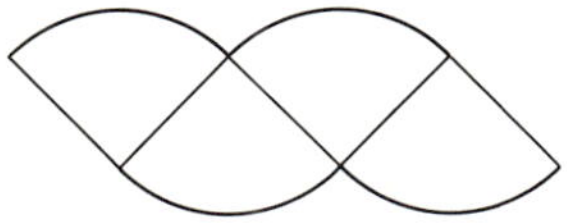

FIGURE 12-16

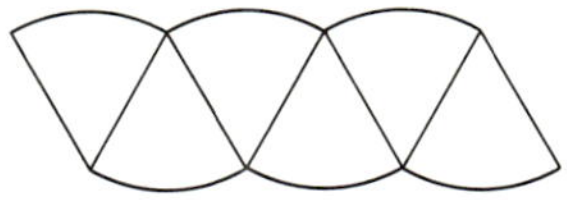

FIGURE 12-17

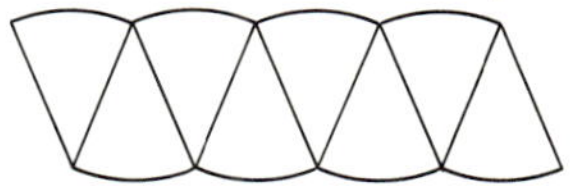

FIGURE 12-18

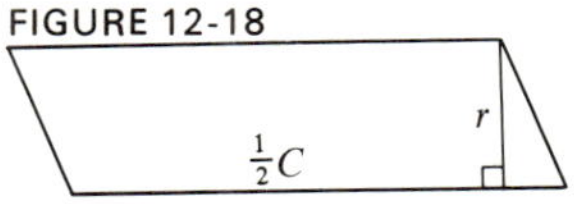

$A = r \times \frac{1}{2}C$

$A = r \times \frac{1}{2} \times 2\pi r$

$A = \pi r^2$

FIGURE 12-19

Visual 3

PURPOSE

To verify experimentally that the volume of a cone is one third the volume of a cylinder having the same base and altitude.

CONSISTS OF

1. A hollow cylinder and a hollow cone with the same base and altitude as the cylinder.

2. A quantity of sand.

HOW TO USE

1. Fill the cone with sand.
2. Empty the sand in the cone into the cylinder.
3. It will be found that three cones full of sand will exactly fill the cylinder.

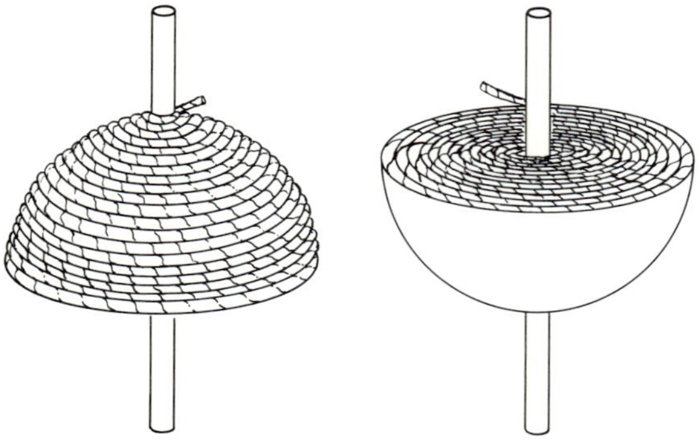

FIGURE 12-20

Visual 4

PURPOSE

To verify experimentally that the area of a sphere is exactly four times the area of a cross section through the center; that is, $A = 4\pi r^2$.

CONSISTS OF

1. A sphere cut into two hemispheres, with dowels passing through the center, as shown in FIG. 12-20.
2. Two lengths of the same heavy cord.

HOW TO USE

1. Wind one length of cord around the curved surface of a hemisphere, and wind the other length of cord so as to cover the flat circular surface as shown.
2. It will be found that the amount of cord used for the curved surface of the hemisphere is just twice that used for the flat circular surface.
3. Therefore, the amount of cord used to cover the curved surfaces of the *two* hemispheres, that is, the sphere, is four times that used to cover the flat circular surface of a cross section through the center; that is, $A = 4\pi r^2$.

THE WONDERS OF π

The student's encounter with π and the circumference formula can be used by the teacher to spark a genuine appreciation of the power and fascination of mathematics. Build upon the wonders that lie behind π and the circumference formula. For instance,

1. No matter what size of circle we are talking about—no matter how large or how small—the ratio of its circumference to its diameter will always be π.

2. The number π is a most unusual number, with an unusual history, that has fascinated people from the earliest times up to the present. One reason for this interest is the fact that π, although a geometric ratio, turns out to be useful in non-geometric situations.

3. The value of π can never be expressed exactly because it is not a rational number. Any rational number can be expressed as a terminating or as a repeating decimal, but mathematicians have proved that π cannot be expressed this way.

QUESTIONS

1. Find the missing information for each circle.

	Radius	*Diameter*	*Circumference*	*Area*
a.	5 in.			
b.		8 ft		
c.				π sq ft
d.			44 yd	
e.	$1\frac{1}{4}$ in.			
f.			15 ft	
g.		2.6 in.		
h.			150 yd	

2. A circular rug, 12 feet in diameter, costs $52. How much should a similar rug with a 6-foot diameter cost?

3. How much metal is needed to make a fruit juice can measuring 4 inches in diameter and 8 inches in height?

4. Compare the volumes of two cans of dessert, each 10 inches high, one can having a 3-inch radius and the other a 6-inch radius.

5. What is the effect on the surface of a sphere if you double its radius?

6. What is the radius of a circle whose circumference is π?

7. The side of a square is 8 inches long. Find the circumference of its inscribed circle. Of its circumscribed circle.

8. The length of a side of an equilateral triangle is 15. What is the circumference of its inscribed circle? Of its circumscribed circle?

9. Prove that the ratio of the areas of two circles is equal to the square of the ratio of their radii.

10. The diameter of a wagon wheel is 20 inches. If the wheel makes 250 revolutions as it moves from point A to point B, what is the distance from A to B?

11. How much farther will a 30-inch bike travel in two revolutions of the wheel than a 22-inch bike?

12. a. How many circular pieces of metal, each having a 3-inch diameter, can be cut from a rectangular sheet measuring 36 by 42 inches?

    b. How many square inches of metal are wasted?

13. If the area of the square shown below is 49 square inches, find the total area of the four shaded surfaces.

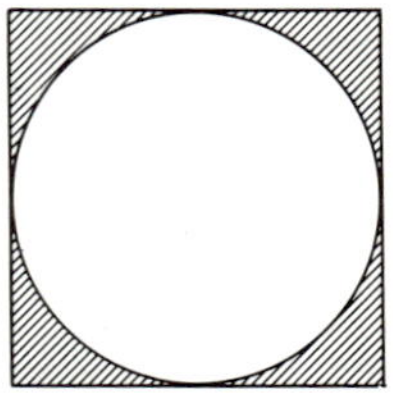

14. a. How many gallons of paint will be required for painting a cylindrical tank measuring 12 feet in diameter and 25 feet in height, if a gallon of paint covers 400 square feet?

    b. At \$6.50 a gallon, what will be the cost of applying two coats of paint?

15. Find the weight of a steel ball with a 9-inch diameter if a cubic foot of steel weighs 490 pounds.

16. A tire on a car has a diameter of 26 inches. If it revolves at 10 revolutions per second, what is the approximate speed of the car in miles per hour?

17. If the radius of a circle is 15 inches, find

    a. the length of an arc of 30°; 75°.
    b. the area of a sector with an arc of 60°; 2°.

18. If in a circle of radius 5 a sector has an area of π, what is the measure of its arc?

19. What distance does a wheel with a 15-inch-diameter roll if it turns 150°?

20. The minute hand of a wrist watch is $\frac{1}{2}$ inch long. How many inches does the tip of the hand travel in 25 minutes?

21. A *segment of a circle* is a region bounded by an arc of the circle and the chord of the arc. Derive a formula for finding the area of a segment in terms of the radius r of the circle and the measure of the arc $\widehat{AB}$.

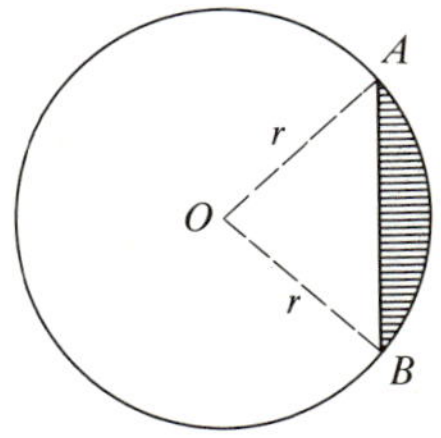

22. A regular pentagon is inscribed in a circle of radius 15. Find the area of that part of the circular region that is outside the pentagon.

23. Find the altitude of a right circular cone whose volume is 36π, if the radius of its base is 4.

24. Find the diameter of a sphere whose surface area is equal to its volume.

25. What is the ratio of the radii of two spheres if the volume of one sphere is three times the volume of the other sphere.

Not correct

26. Show that each unshaded lune is equal in area to each unshaded circular segment, and that the eight shaded regions all have equal areas.

SOLUTIONS

1. Here is one possible solution:

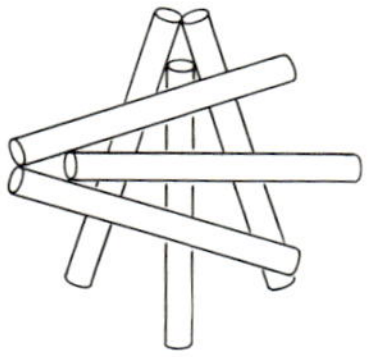

2. a. Fill the 5-quart jug and then empty it into the 3-quart jug. This leaves 2 quarts in the 5-quart jug.
   b. Empty the 3-quart jug into the 8-quart jug and pour the 2 quarts left in the 5-quart jug into the 3-quart jug.
   c. Again fill the 5-quart jug and finish filling the 3-quart jug from the 5-quart jug.
   d. Since the 3-quart jug can take only one quart more, the 5-quart jug will be left with 4 quarts and the 8-quart jug will be left with the other 4 quarts.

PUZZLES

1. Four baseballs can be placed so that each ball touches the other three. Can you place six dowel sticks so that each touches the other five?

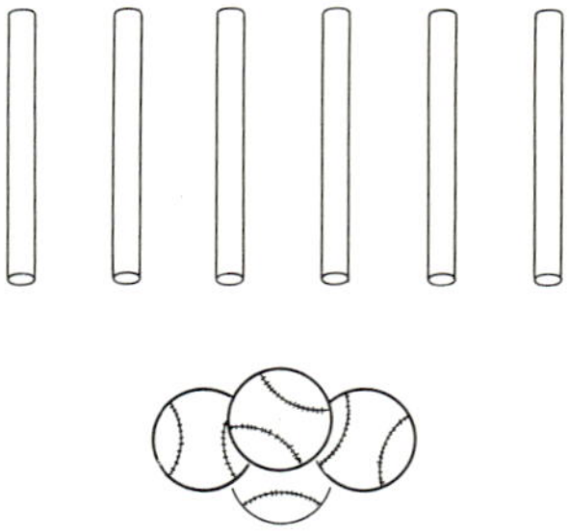

2. Two friends have an 8-quart jug of wine that they wish to share equally. They also have two empty jugs, one holding 5 quarts and the other holding 3 quarts. By using just these three jugs, how can they divide the wine into two parts, each part containing 4 quarts?

ENRICHMENT

ON THE HISTORY OF π

Interest in computing the value of π goes back 2200 years when Archimedes (287–212 B.C.), by using regular inscribed and circumscribed polygons, computed the value of π to be between $\frac{223}{71}$ and $\frac{22}{7}$ or, to two decimal places, as 3.14. This improves upon the value of 3 given in the Bible (I Kings 7:23), which describes a basin as 10 cubits from one brim to the other, "and a line of 30 cubits did compass it round about."

Archimedes inscribed and circumscribed regular polygons about a circle, determined their perimeters, and then used these values as upper and lower bounds to the length of the circumference. He kept doubling the number of sides until the perimeters of polygons of 96 sides were obtained. With these perimeters he established the value of π to be $3\frac{10}{71} < \pi < 3\frac{1}{7}$.

Archimedes' method, known as the *classical method* of computing π, was the one used for hundreds of years after Archimedes. This method can yield approximations for π to as many decimal places as we wish. In 1579, François Viète used polygons with 393,216 sides to find π correct to nine decimal places.

In 1610, Ludolph van Ceulen worked on π almost to the day of his death at the age of 70, and computed π to 35 decimal places by using regular polygons with 2^{62} sides. Before he died, he requested that the 35 digits of π be engraved on his tombstone. This was actually done. The tombstone seems now to be lost.

About a hundred years after Archimedes, Ptolemy computed the value of π as 3.1416 from the lengths of the chords of a circle subtended by central angles of 1°. If the length of the chord of the central angle is multiplied by 360, and the result divided by the length of the diameter of the circle, this value for π is obtained.

Algebraic Methods for Computing π

In the seventeenth century the classical method of Archimedes was abandoned and the algebraic methods of convergent infinite series, products, and continued fractions became the method for computing the value of π. For instance,

1. John Wallis (1616–1703) expressed the value of π this way:

$$\frac{\pi}{2} = \frac{2}{1} \times \frac{2}{3} \times \frac{4}{3} \times \frac{4}{5} \times \frac{6}{5} \times \frac{6}{7} \times \frac{8}{7} \times \frac{8}{9} \times \cdots.$$

2. Lord Brouncker (1620–1684) expressed the value of π this way:

$$\pi = \cfrac{4}{1 + \cfrac{1^2}{2 + \cfrac{3^2}{2 + \cfrac{5^2}{2 + 7^2 \cdots}}}}.$$

3. The simplest and most elegant expression for the value of π was given by G. W. von Leibnitz (1646–1716):

$$\frac{\pi}{4} = 1 - \frac{1}{3} + \frac{1}{5} - \frac{1}{7} + \frac{1}{9} - \frac{1}{11} + \frac{1}{13} - \frac{1}{15} + \cdots.$$

About 100 years ago, William Shanks computed π to 707 places. In 1945 his result was proven wrong starting with the 528th place!

In 1961, the value of π was calculated to 100,265 decimal places by an IBM 7090 computer. The job was completed in 8.7 hours.

In 1967, π was computed on a CDC 6600 computer to 500,000 decimal places by Jean Guilloud in Paris.

Why This Unusual Interest in π?

The preoccupation with calculating π to so many decimal places is all the more remarkable in light of at least one authoritative opinion that if π, expressed to only 10 decimal places, were used to compute the circumference of the earth, the result would be correct to inches!

There are several reasons, however, for this astonishing interest. It was suggested earlier that the appearance of π in a context different from geometry has contributed to this interest. For example, the equation

$$y = \frac{N}{\sigma\sqrt{2\pi}} e^{-x^2/2\sigma^2}$$

defines the normal-distribution curve, which is most essential in inferential statistics. It is this curve that enables us to make predictions about entire populations on the basis of small samples.

Another example of the appearance of π in a nongeometric context is in the famous Buffon Needle Problem. In 1760, G. L. Leclerc, Comte de Buffon, investigated the probability of dropping at random a uniform needle of length l onto a plane ruled with parallel lines d distance part ($l < d$) in such a way that the needle will touch one of the ruled lines (FIG. 12-21). He found this probability to be $2l/\pi d$.

If the length of the needle is half the distance between the parallel lines, that is, $l = \frac{1}{2}d$, then the probability of success becomes $1/\pi$. This provides a good laboratory-type procedure for obtaining an approximation of π.

A second reason for the exceptional interest in calculating π to a large number of decimal places is the desire to test the "regularity" or "normalcy" of π. By "normalcy" we mean the following: A real number is said to be *simply normal* if in its infinite decimal expression all digits occur with equal frequency. It is said to be *normal* if all blocks of digits of the same length occur with equal frequency.

Starting in 1949, computers were used to find out by actual count whether π is *normal* or *simply normal*. The results seem to indicate that π is *normal*.

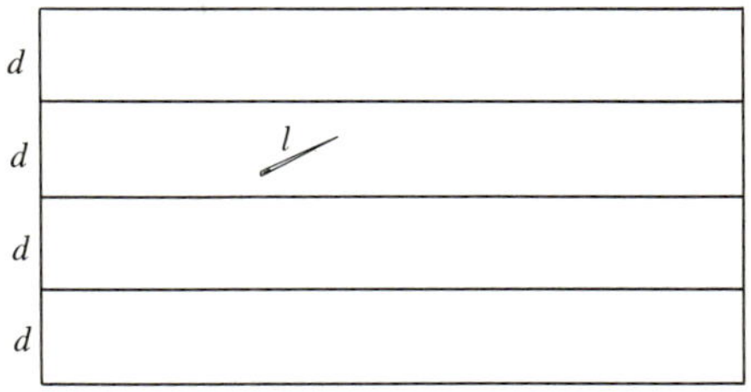

FIGURE 12-21

13 COORDINATE GEOMETRY

INTRODUCING THE CHAPTER TO STUDENTS

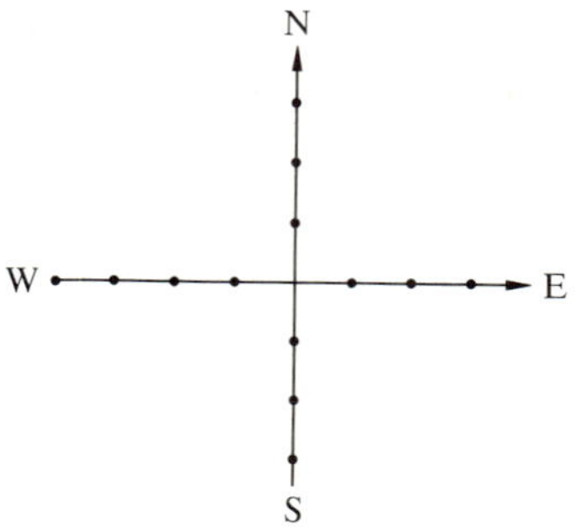

FIGURE 13-1

The basic idea of *coordinate geometry* is so simple that one wonders how such an idea can be regarded, as it is, as one of the greatest contributions of all time to mathematics.

Draw any two intersecting lines in a plane (FIG. 13-1). To keep things simple, draw them perpendicular to each other so that one line runs north–south and the other runs east–west. We shall refer to these two lines as the "axes" and call the point of intersection of the axes the "origin."

Let us further imagine that a city is laid out by this plan, with the "avenues" numbered consecutively along the north–south axis and the "streets" numbered consecutively along the east–west axis. It is obvious that on such a map we can locate instantly any location in the city. We only need to give the pair of numbers that measure its east–west and north–south distances from the axes. These two numbers are called the *coordinates* of the point with respect to the axes.

So far we have developed a convenient system for locating a point on a plane. The importance of the system lies in how it is used in mathematics.

When we combine this system with an algebraic equation it becomes possible to obtain a geometrical portrait of the equation. Conversely, we can now use algebraic methods and equations to study geometric problems. By expressing geometrical ideas in algebraic terms and algebraic ideas in geometrical terms we frequently add clarity and gain greater insight into abstract ideas. In this chapter we shall see how this is done with straight lines and circles. But the method of coordinate geometry becomes particularly valuable when we start with more complicated equations and then interpret their properties geometrically.

We can extend the coordinate system to any number of dimensions. For the plane we need *two* coordinates. For three-dimensional space we need *three* coordinates. For example, any position on the earth can be specified by its latitude, longitude, and height above sea level. For *n*-dimensional space we need *n* coordinates.

The method of coordinate geometry is so powerful that high school students today can use it to solve problems that would have baffled the greatest geometers of antiquity. The method is powerful because it permits us to bring to bear on the solution of a geometrical problem our knowledge of algebra and the properties of the real number system. We can, thereby, transform a geometrical problem into a corresponding, and often simpler, algebraic problem.

Although coordinates were used by the ancient Egyptians and Romans in surveying and by the Greeks in map making, most historians agree that the decisive contributions to the creation of coordinate geometry were made by René Descartes (1596–1650) and Pierre de Fermat (1601–1665).

According to one legend, the idea of coordinate geometry first came to Descartes as he was watching a fly crawl on the ceiling of his room. He surmised that the path of the fly might be described in terms of the distances from the two adjacent walls.

Descartes began with a curve and used coordinate geometry to find the equation that corresponds to that curve. Fermat started with an equation and then studied its corresponding curve. We shall see in this chapter that these are the two fundamental ideas of coordinate geometry.

CONTENT REVIEW

PREREQUISITES FOR LEARNING THIS CHAPTER

Knowledge and understanding of: Ruler Postulate; Cartesian product; similarity of triangles; parallelism and perpendicularity; negative reciprocal; Distance Postulate; Pythagorean Theorem; absolute value; basic algebraic skills required in the solution of first-degree and second-degree equations.

COORDINATE SYSTEMS IN A PLANE[1]

We can establish a one-to-one correspondence between the set of points of the Euclidean plane and the set of ordered pairs of real numbers in the following way:

1. We first set up a coordinate system on a line X (FIG. 13-2). We call this line the *x axis* (or axis of *abscissas*).

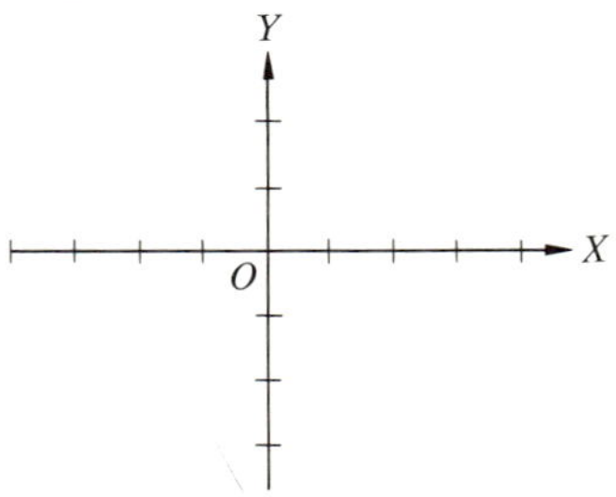

FIGURE 13-2

2. We then set up a coordinate system on a line Y perpendicular to the x axis through the point with coordinate 0. We make the zero point on Y the same as the zero point on X. We call this line the *y axis* (or axis of *ordinates*).

3. The point where the x axis intersects the y axis is called the *origin*. We call the x axis and the y axis the *coordinate axes for the plane.*

[1] By the Ruler Postulate (page 35) we established a coordinate system for a line. Under this system, every real number is associated with a unique point on a line, and every point on a line is assocated with a unique real number.

We shall now develop a coordinate system for a plane in terms of the coordinate system for a line. But in a plane, a point is associated not with a single number but with an ordered pair of numbers. More specifically, we shall associate the points of a plane with the Cartesian product $R \times R$, R being the set of real numbers.

4. We can now establish a one-to-one correspondence between the points of a plane and ordered pairs of real numbers in the following way:

   a. Let us start with any point P on the plane and draw through it lines perpendicular to the coordinate axes (FIG. 13-3).

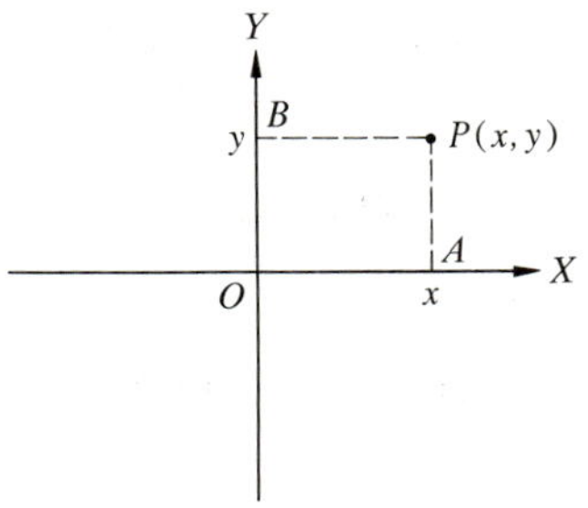

FIGURE 13-3

There is only one perpendicular from P to each of the coordinate axes. Let the foot of the perpendicular to the x axis be point A, whose coordinate on line X is x. Let the foot of the perpendicular to the y axis be point B, whose coordinate on line Y is y. The number x is called the *x coordinate* (or *abscissa*) of P, and the number y is called the *y coordinate* (or *ordinate*) of P. We indicate that P has these coordinates by writing "$P(x, y)$." Note that (x, y) is an ordered pair of real numbers whose first component is the x coordinate and whose second component is the y coordinate. Thus, with every point P we associate a unique pair of real numbers.

   b. Conversely, let us start with any ordered pair of real numbers (x, y). Let x be the coordinate of point A on line X, and let y be the coordinate of point B on line Y (FIG. 13-4). If we draw a perpendicular to the x axis

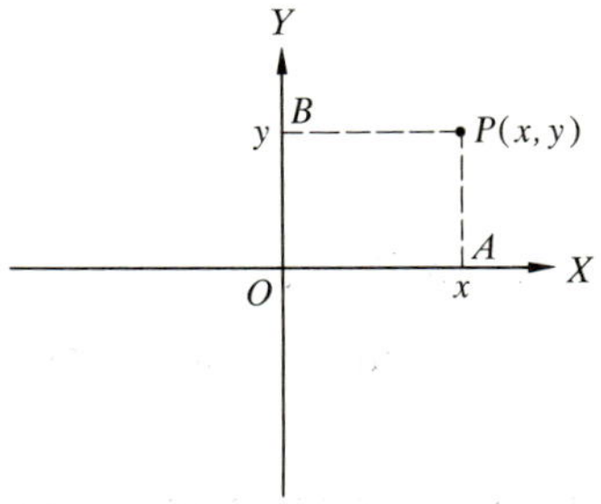

FIGURE 13-4

through A and a perpendicular to the y axis through B, the two perpendiculars will intersect in a unique point P. This point has the coordinates (x, y) and is labeled $P(x, y)$. Thus, with every ordered pair of real numbers we associate a unique point of the plane.

5. Such a one-to-one correspondence is called a *coordinate system.*

The system described above leads to the following:

Definitions

1. The x *coordinate* of a point P is the coordinate of the foot of the perpendicular from P to the x axis.
2. The y *coordinate* of P is the coordinate of the foot of the perpendicular from P to the y axis.
3. If P has an x coordinate of a and a y coordinate of b, then we write $P(a, b)$.

If P_1 is on the x axis, then P_1 has coordinates $(x, 0)$; if P_2 is on the y axis, then P_2 has coordinates $(0, y)$ (FIG. 13-5).

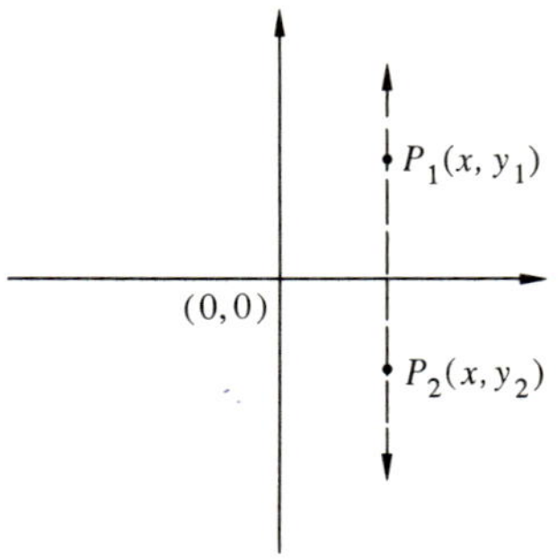

FIGURE 13-6

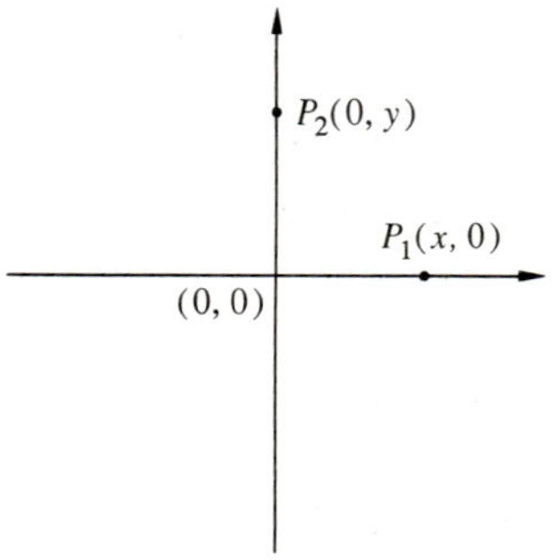

FIGURE 13-5

The origin has coordinates (0, 0) (FIG. 13-5).

Any two points with the same first coordinate are on the same line parallel to the y axis (FIG. 13-6).

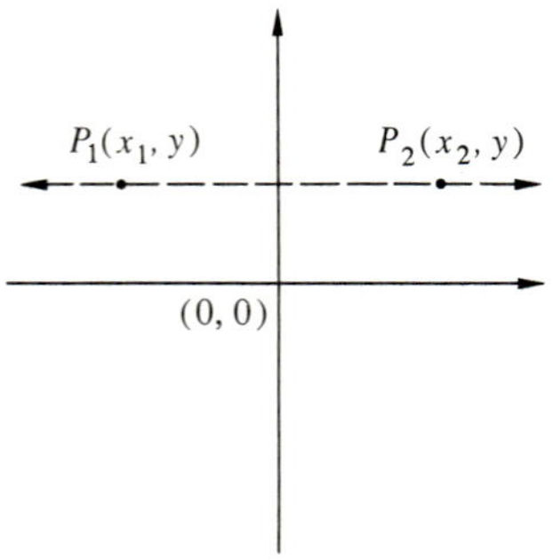

FIGURE 13-7

Any two points with the same second coordinate are on the same line parallel to the x axis (FIG. 13-7).[2]

[2] 1. In drawing the coordinate axes we usually put an arrowhead on the x axis and on the y axis to specify the positive direction on each axis. This is particularly useful when dealing with *directed segments* and *directed distances*.

2. Although we chose the coordinate axes to be perpendicular, this is not essential. For some purposes it is just as easy to use oblique axes, as in FIG. 13A. Starting from the origin O, we reach the point $P(x, y)$ by moving a distance x along the x axis, and then a distance y along a line parallel to the y axis.

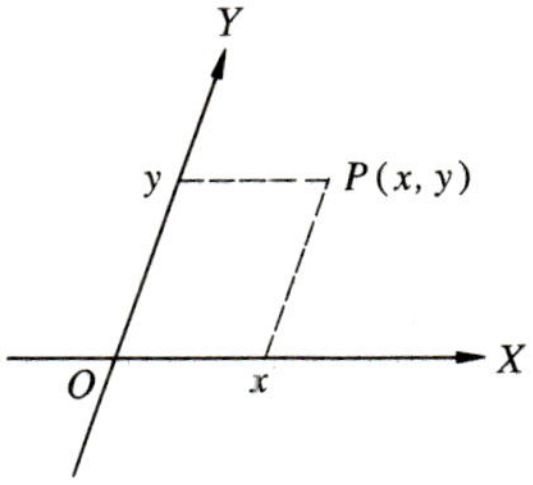

FIGURE 13A

3. The unit of length used on the y axis need not be the same as the unit of length used on the x axis if the scales measure different kinds of things. But in geometry, the scales need to be the same for both axes in order not to distort geometric figures. For instance, a circle will look like an ellipse if different scales are used.

4. For the sake of convenience, we shall refer freely to points by their coordinates and to coordinates by their corresponding points. For example, we shall speak of "the point $(1, -4)$" and "$P = (-2, 5)$."

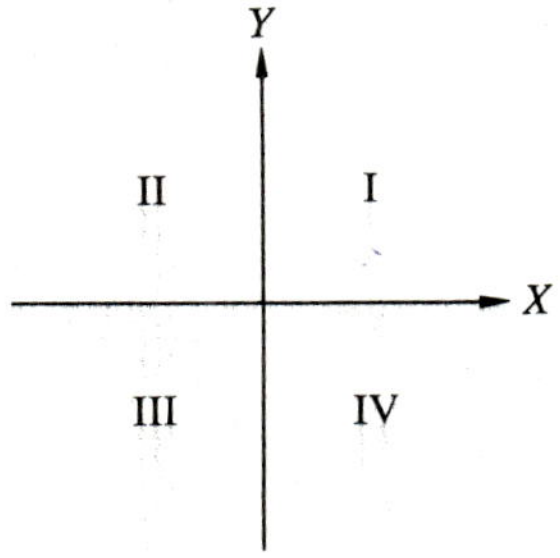

FIGURE 13-8

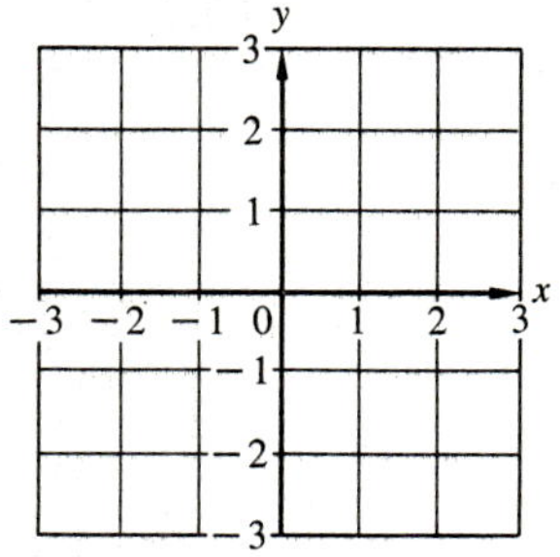

FIGURE 13-9

The coordinate axes separate the rest of the plane into four regions called *quadrants*. These quadrants are numbered I, II, III, and IV, as shown in FIG. 13-8.

To show a coordinate system on graph paper, we draw the axes and indicate the scale (FIG. 13-9).

[3] The following are immediate consequences of the definition of slope:

1. The slope of $\overline{P_1P_2}$ is the same as the slope of $\overline{P_2P_1}$, since

$$\frac{y_2 - y_1}{x_2 - x_1} = \frac{-(y_1 - y_2)}{-(x_1 - x_2)} = \frac{y_1 - y_2}{x_1 - x_2}.$$

   But the order in which y_2 and y_1 are written in the formula must be the same as that of x_2 and x_1. For instance, $m = (y_2 - y_1)/(x_2 - x_1)$ cannot be written as $m = (y_2 - y_1)/(x_1 - x_2)$.

2. The slope of a *vertical* segment is undefined since the denominator, $x_2 - x_1$, is equal to 0.

3. The slope of a *horizontal* segment is 0 since the numerator, $y_2 - y_1$, is equal to 0 (FIG. 13B).

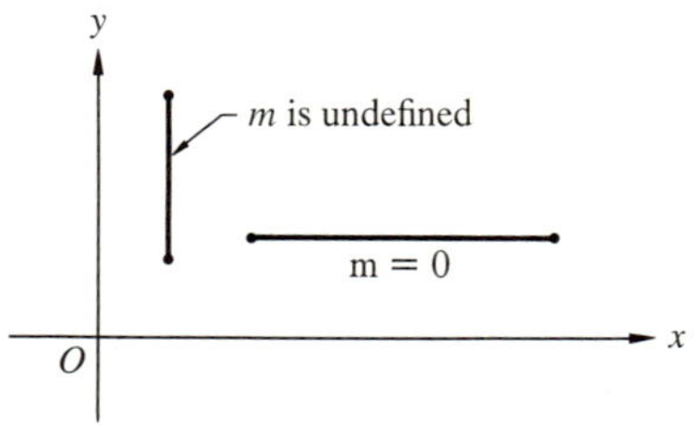

FIGURE 13B

4. If a segment rises from left to right, its slope is positive, since the numerator and denominator are both positive (FIG. 13C). If the segment drops from left to right, its slope is negative since the numerator is negative while the denominator remains positive.

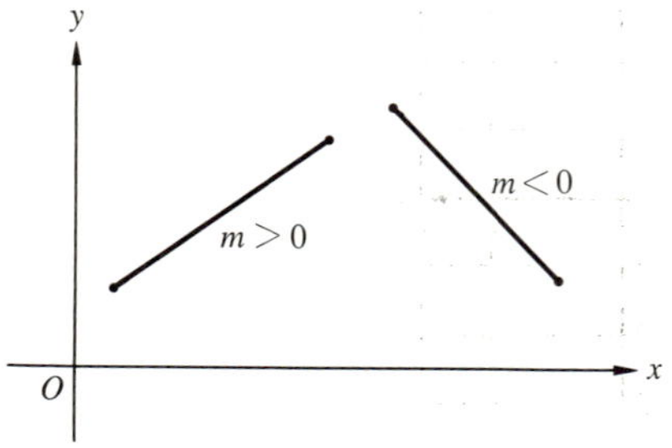

FIGURE 13C

1. We usually draw the x axis horizontally with the coordinates increasing from left to right; and the y axis vertically, with the coordinates increasing from bottom to top. The axes are labeled x and y rather than X and Y as a reminder that the coordinates are going to be denoted by these lowercase letters.

2. We label the origin "O."

3. We then mark a unit length to the right of O on the x axis, and the same unit length above O on the y axis. That is, we label the points with coordinates (1, 0) and (0, 1) in order to indicate the unit to be used.

THE SLOPE OF A LINE

Definition

If $P_1 = (x_1, y_1)$ and $P_2 = (x_2, y_2)$ and $\overline{P_1P_2}$ is nonvertical, then the *slope* of $\overline{P_1P_2}$ is the number

$$m = \frac{y_2 - y_1}{x_2 - x_1}$$

(FIG. 13-10).[3]

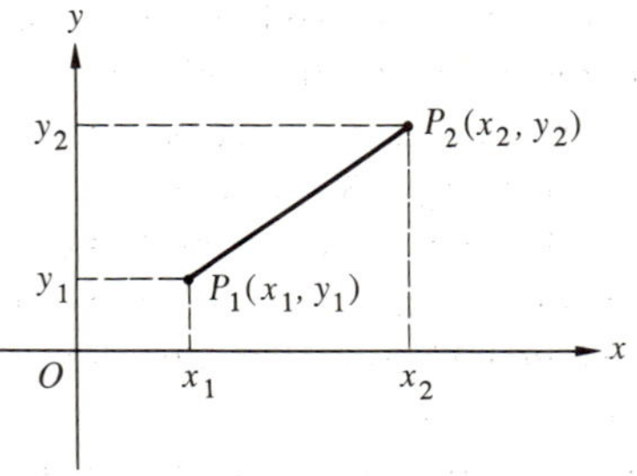

FIGURE 13-10

The following theorem enables us to define the slope of a nonvertical *line* by the slope of any of its segments.

THEOREM 13-1 *On a nonvertical line, all segments have the same slope.*

Proof

1. The theorem is obviously true for a horizontal line, since all its segments have a slope equal to 0.

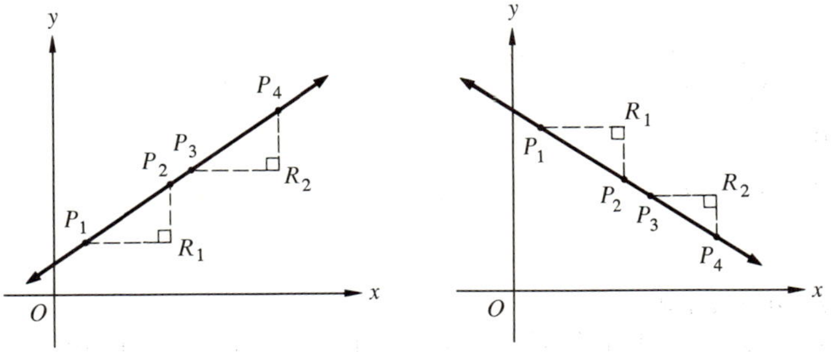

FIGURE 13-11

2. In the other two cases (FIG. 13-11), we have

$\triangle P_1R_1P_2 \sim \triangle P_3R_2P_4$ (since corresponding angles are congruent)

and

$$\frac{R_1P_2}{P_1R_1} = \frac{R_2P_4}{P_3R_2} \quad \text{or} \quad \frac{y_2 - y_1}{x_2 - x_1} = \frac{y_4 - y_3}{x_4 - x_3}.$$

Therefore, $\overline{P_1P_2}$ and $\overline{P_3P_4}$ have the same slope.

Definition
The *slope* of a line is the number that is the slope of every segment of the line.[4]

PARALLEL AND PERPENDICULAR LINES

We use the concept of slope to determine whether lines are parallel.

THEOREM 13-2 Two distinct nonvertical lines are parallel if and only if they have the same slope.

That is, (1) If $l_1 \parallel l_2$, then $m_1 = m_2$; and (2) If $m_1 = m_2$, then $l_1 \parallel l_2$.

Proof (Part 1)
1. If $l_1 \parallel l_2$ (FIG. 13-12), then $\triangle P_1R_1P_2 \sim \triangle P_3R_2P_4$, and

2. $$\frac{R_1P_2}{P_1R_1} = \frac{R_2P_4}{P_3R_2},$$

or

[4] 1. In the slope formula, the change in y, $y_2 - y_1$, may be represented by the notation Δy (read "delta y"), and the change in x may be represented by Δx (read "delta x") (FIG. 13D). Δy may then be regarded as the "rise" and Δx as the "run" from P_1 to P_2 along the line l. The slope of l would then be defined as

$$\text{slope} = \frac{\Delta y}{\Delta x} = \frac{\text{rise}}{\text{run}},$$

that is, *the rate of rise per unit of run.* If in the equation $m = \Delta y/\Delta x$ we multiply both sides by Δx, we get $\Delta y = m\,\Delta x$. This means that as a point moves along a line, the change in y is proportional to the change in x, the slope m being the proportionality factor.

2. The fact that the slope of a line is unique distinguishes lines from all other types of curves.

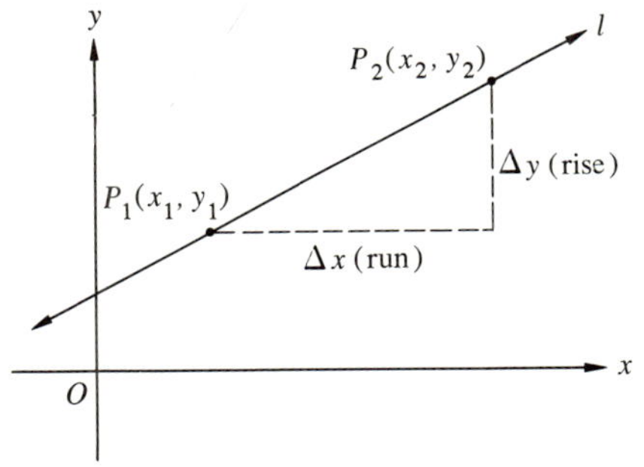

FIGURE 13D

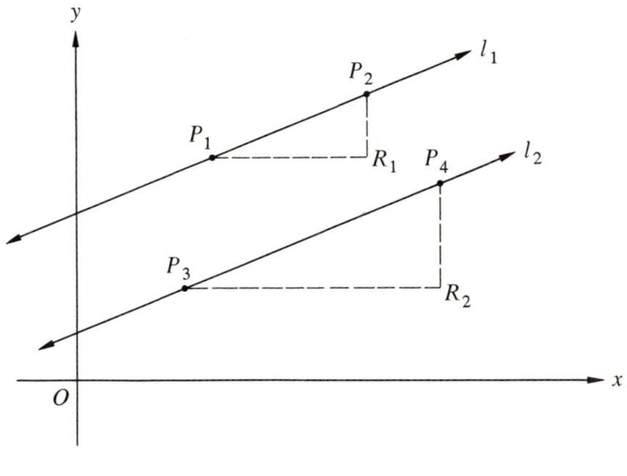

FIGURE 13-12

$$\frac{y_2 - y_1}{x_2 - x_1} = \frac{y_4 - y_3}{x_4 - x_3},$$

or

3. $m_1 = m_2$.

Proof (*Part 2*)

1. Assume that $l_1 \not\parallel l_2$, that is, that l_1 and l_2 intersect at $P(x_0, y_0)$. Choose points $P_1(x_1, y_1)$ and $P_2(x_1, y_2)$, as shown in FIG. 13-13.

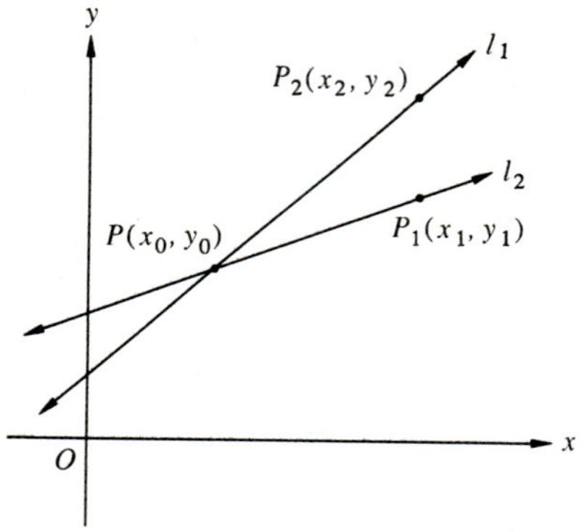

FIGURE 13-13

2. The slopes of l_1 and l_2 are

$$m_1 = \frac{y_1 - y_0}{x_1 - x_0} \quad \text{and} \quad m_2 = \frac{y_2 - y_0}{x_1 - x_0}.$$

3. Since the denominators are the same but the numerators are different, $m_1 \neq m_2$.
4. But step 3 contradicts the hypothesis.
5. Therefore, $l_1 \parallel l_2$.

We use the concept of slope to determine whether lines are perpendicular.

THEOREM 13-3 *Two nonvertical lines are perpendicular if and only if their slopes are negative reciprocals of each other.*

That is,

1. If $l_1 \perp l_2$, then $m_1 = -1/m_2$ (or $m_1 m_2 = -1$).
2. If $m_1 = -1/m_2$, then $l_1 \perp l_2$.

Proof (Part 1)

1. Let l_1 and l_2 be perpendicular and intersect at P (FIG. 13-14).

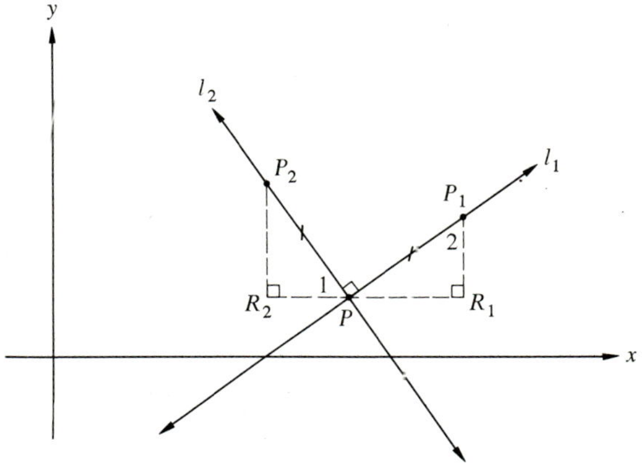

FIGURE 13-14

2. Choose P_1 above and to the right of P, and draw right $\triangle PR_1P_1$.
3. Choose P_2 above and to the left of P so that $PP_2 = PP_1$, and draw right $\triangle P_2R_2P$.
4. Since $\angle 1 \cong \angle 2$, $\triangle PR_1P_1 \cong \triangle P_2R_2P$ (by SAA), making $R_1P_1/PR_1 = R_2P/P_2R_2$.
5. The slope of l_1 is $m_1 = R_1P_1/PR_1$, and the slope of l_2 is $m_2 = -P_2R_2/R_2P$.
6. Therefore, by step 4, $m_1 = -1/m_2$.

Proof (Part 2)

1. Given that $m_1 = -1/m_2$. Draw $\triangle PR_1P_1$ as before (FIG. 13-15).
2. Choose R_2 so that $R_2P = R_1P_1$ and then draw right $\triangle P_2R_2P$ with P_2 on l_2.

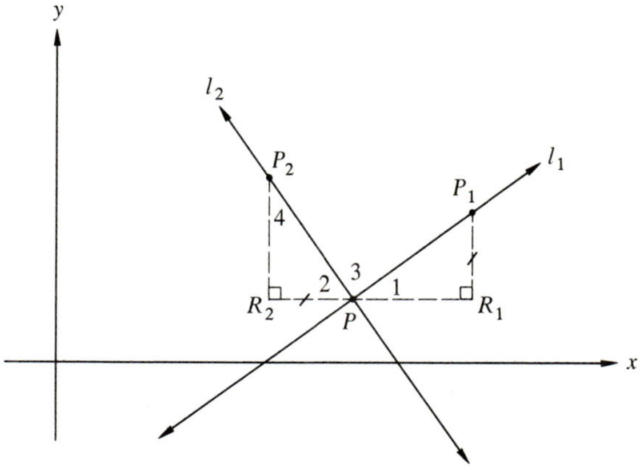

FIGURE 13-15

[5] 1. Theorems 13-2 and 13-3 deal with nonvertical lines. If l_1 is vertical, then

a. For parallelism, l_2 must also be vertical.

b. For perpendicularity, l_2 must be horizontal.

2. Note that Theorem 13-2 deals with *lines*. To show that *segments* are parallel, we need to show not only that they have the same slope but also that they are not collinear.

[6] By the Distance Postulate (page 35), the distance between any two points is determined if we know the coordinates of the points. By the Ruler Postulate we have a way to compute the distance between two points on a horizontal or a vertical line. Theorem 13-4 gives us a way to calculate the distance between two points on *any* line in terms of the coordinates of the two points.

3. Since $m_1 = -1/m_2$,

$$\frac{P_2R_2}{R_2P} = \frac{PR_1}{R_1P_1}.$$

4. Since the denominators in step 3 are equal, the numerators must also be equal; that is, $P_2R_2 = PR_1$.

5. Therefore, $\triangle PR_1P_1 \cong P_2R_2P$ (by SAS), and $\angle 1 \cong \angle 4$.

6. $\angle 1$ and $\angle 2$ are complementary since $\angle 4$ and $\angle 2$ are complementary.

7. Therefore, $\angle 3$ is a right angle and $l_1 \perp l_2$.[5]

THE DISTANCE FORMULA[6]

THEOREM 13-4 (The Distance Formula) The distance between the points (x_1, y_1) *and* (x_2, y_2) *is*

$$d = \sqrt{(x_2 - x_1)^2 + (y_2 - y_1)^2}.$$

Proof

If $\overline{P_1P_2}$ is horizontal or vertical, the formula is easily verified. If $\overline{P_1P_2}$ is neither horizontal nor vertical, then the proof is as follows:

1. Let A_1 and A_2 be the feet of the perpendiculars from $P_1(x_1, y_1)$ and $P_2(x_2, y_2)$ to the x axis, and let B_1 and B_2 be the feet of the perpendiculars from P_1 and P_2 to the y axis (FIG. 13-16). Also, let C be the point of intersection of the horizontal line through P_1 and the vertical line through P_2.

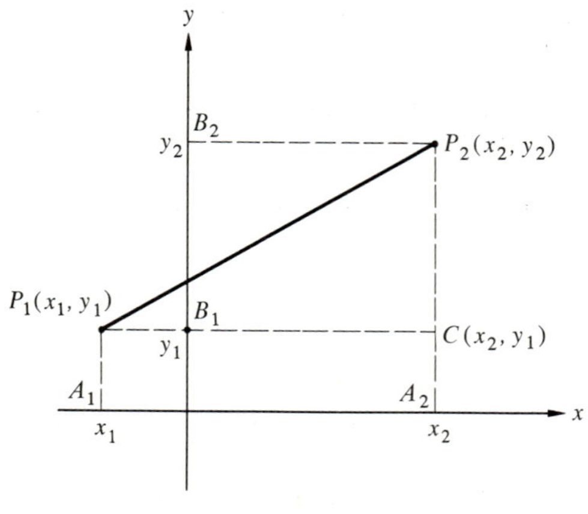

FIGURE 13-16

2. Then $(P_1P_2)^2 = (P_1C)^2 + (CP_2)^2$, by the Pythagorean Theorem.

3. Since $P_1CA_2A_1$ is a rectangle,

   $P_1C = A_1A_2 = |x_2 - x_1|$

   and

   $CP_2 = B_1B_2 = |y_2 - y_1|$, by the Ruler Postulate.

4. Therefore, $(P_1P_2)^2 = |x_2 - x_1|^2 + |y_2 - y_1|^2$.

5. But $(P_1P_2)^2 = (x_2 - x_1)^2 + (y_2 - y_1)^2$, since the square of a number is the same as the square of its absolute value.

6. Since $P_1P_2 \geq 0$, we get

   $$P_1P_2 = \sqrt{(x_2 - x_1)^2 + (y_2 - y_1)^2}.$$ [7]

[7] In using the Distance Formula it makes no difference which point is considered to be (x_1, y_1), since

$$(x_1 - x_2)^2 = (x_2 - x_1)^2$$

and

$$(y_1 - y_2)^2 = (y_2 - y_1)^2.$$

THE MIDPOINT FORMULA[8]

[8] Theorem 13-5 gives us a way to find the coordinates of the midpoint of a segment in terms of the coordinates of the endpoints of the segment. The formula is often used for proving geometric theorems, like those involving medians of a triangle, by coordinate geometry.

THEOREM 13-5 (The Midpoint Formula) *Let* $P_1 = (x_1, y_1)$ *and* $P_2 = (x_2, y_2)$. *The midpoint of* $\overline{P_1P_2}$ *is the point*

$$P = \left(\frac{x_1 + x_2}{2}, \frac{y_1 + y_2}{2}\right).$$

Proof

1. Consider a segment $\overline{P_1P_2}$ on the x axis, with the coordinates as in FIG. 13-17 and with $x_1 < x_2$.

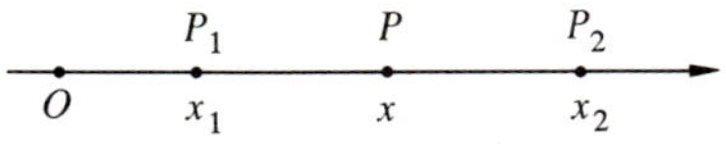

FIGURE 13-17

   a. If P is the midpoint of $\overline{P_1P_2}$, then $P_1P = PP_2$.
   b. $P_1P = |x - x_1| = x - x_1$, and $PP_2 = |x_2 - x| = x_2 - x$.
   c. From parts a and b we get

   $$x - x_1 = x_2 - x$$

   or

   $$x = \frac{x_1 + x_2}{2}.$$

d. Similarly, if $\overline{P_1P_2}$ is on the y axis, we get

$$y = \frac{y_1 + y_2}{2} \quad \text{for the midpoint.}$$

2. To get the midpoint of an oblique segment, we note (FIG. 13-18) that

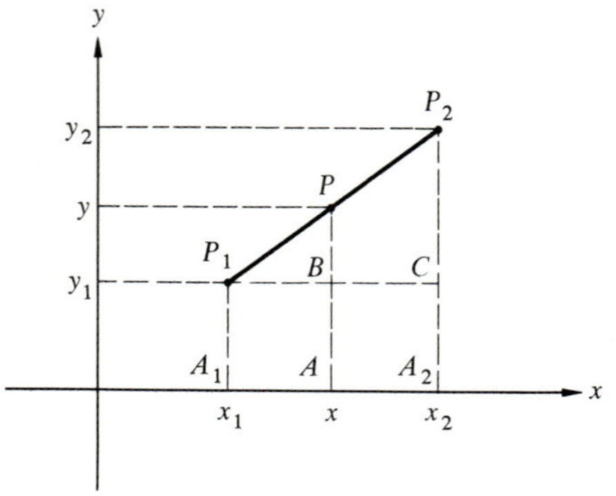

FIGURE 13-18

a. If P is the midpoint of $\overline{P_1P_2}$, then A is the midpoint of $\overline{A_1A_2}$ because $\triangle P_1PB \sim \triangle P_1P_2C$.

b. Therefore, $x = (x_1 + x_2)/2$ and, in a similar way, we get $y = (y_1 + y_2)/2$.
That is,

$$P = \left(\frac{x_1 + x_2}{2}, \frac{y_1 + y_2}{2}\right).$$ [9]

[9] 1. The Midpoint Formula also holds when $x_2 < x_1$ because if we interchange x_1 with x_2 in the derivation of the formula, we end up with the same result.

2. Note that the x and y coordinates of the midpoint are the arithmetic averages of the x and y coordinates of the endpoints.

3. We next consider a more general problem: the problem of finding the coordinates of a point that divides a segment in the ratio $r:1$, where r is any positive real number. Finding the coordinates of the midpoint is a special case of this more general problem, with $r = 1$.

THEOREM 13-6 *If P is between P_1 and P_2, and $P_1P/PP_2 = r$, where r is any positive real number, then*

$$P = \left(\frac{x_1 + rx_2}{1 + r}, \frac{y_1 + ry_2}{1 + r}\right).$$

Proof

1. Consider a segment $\overline{P_1P_2}$ on the x axis, with $x_1 < x_2$ (FIG. 13-19).

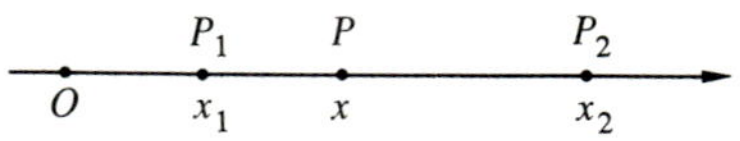

FIGURE 13-19

a. From the hypothesis,

$$\frac{P_1P}{PP_2} = r \quad \text{or} \quad P_1P = rPP_2.$$

b. $x - x_1 = r(x_2 - x)$; that is, $x + rx = x_1 + rx_2$ or

$$x = \frac{x_1 + rx_2}{1 + r}.$$

c. Similarly, if $\overline{P_1P_2}$ is on the y axis, we get

$$y = \frac{y_1 + ry_2}{1 + r} \qquad \text{for the midpoint.}$$

d. Therefore,

$$P = \left(\frac{x_1 + rx_2}{1 + r}, \frac{y_1 + ry_2}{1 + r}\right).$$

2. In the case of an oblique segment, we note that
   a. If $P_1P/PP_2 = r$, then

$$\frac{A_1A}{AA_2} = r$$

(FIG. 13-20). (Why?)

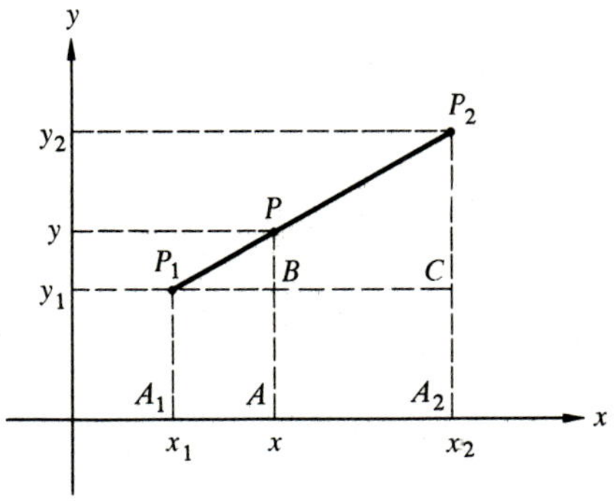

FIGURE 13-20

b. Therefore, by step 1,

$$x = \frac{x_1 + rx_2}{1 + r},$$

$$y = \frac{y_1 + ry_2}{1 + r}.$$

That is,

$$P = \left(\frac{x_1 + rx_2}{1 + r}, \frac{y_1 + ry_2}{1 + r}\right).$$

[10] 1. When we prove a theorem by using coordinate systems we call such a proof an *analytical* proof. The proofs we used in earlier chapters are called *synthetic* proofs.

2. We shall now see how we can use coordinate systems to prove geometric theorems.

ANALYTICAL PROOFS FOR GEOMETRIC THEOREMS[10]

Selecting a Coordinate System

For any given problem we must first decide upon a *suitable* coordinate system. We usually have a number of choices, but some choices make our solution easier than other choices. For example, suppose we wish to prove a theorem relating to an isosceles triangle. Three ways in which we can set up our coordinate system are as shown in FIG. 13-21. Note that in system (1) there is no special

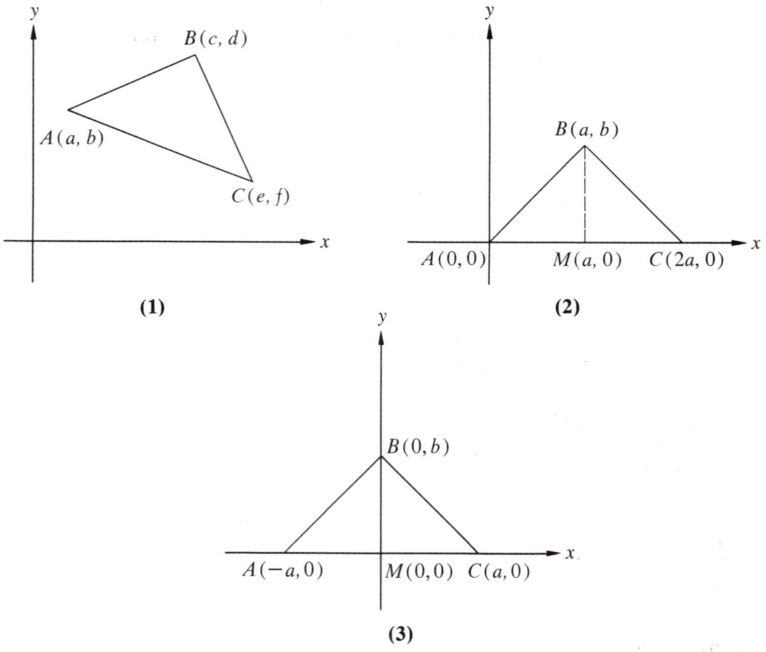

FIGURE 13-21

relationship between it and the isosceles triangle $\triangle ABC$. Systems (2) and (3), however, are so set up as to reflect the fact that the altitude $\overline{BM}$ bisects the base, making $AM = MC$. We use this knowledge to designate the coordinates of points A, B, and C. Also note that in systems (2) and (3) we use the zero coordinate as often as possible.

If we compare the Distance Formula for AB under each of the three systems, we get

(1) $AB = \sqrt{(c - a)^2 + (d - b)^2}$,

(2) $AB = \sqrt{a^2 + b^2}$,

(3) $AB = \sqrt{a^2 + b^2}$.

Obviously, systems (2) and (3) are much simpler than system (1).

A good way to place the axes when proving theorems about parallelograms is shown in FIG. 13-22. Note that after placing point

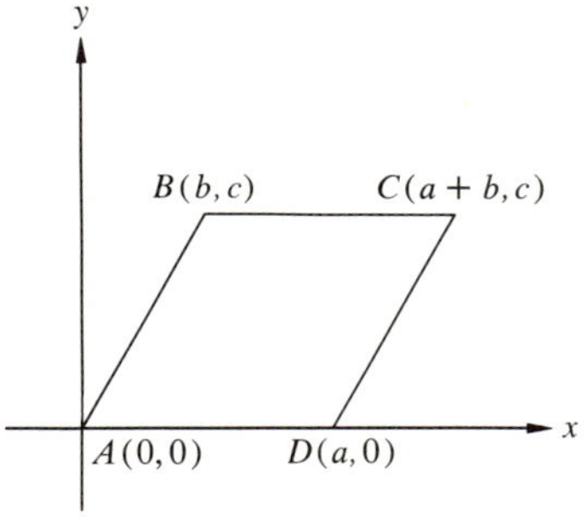

FIGURE 13-22

A at the origin and point D on the x axis, we use our knowledge about the properties of a parallelogram to compute the coordinates of point C in terms of the other coordinates.

In using our knowledge from synthetic geometry to help us set up our coordinate system, we must be careful not to assume any fact that we are required to prove. For instance, if we had to prove that the altitude to the base of an isosceles triangle bisects the base, we could not set up our coordinate system the way we did in systems (2) and (3).

Illustrative Proofs Using Coordinate Systems

THEOREM 1 *If a triangle is isosceles, then the medians to the two congruent sides are congruent* (FIG. 13-23).

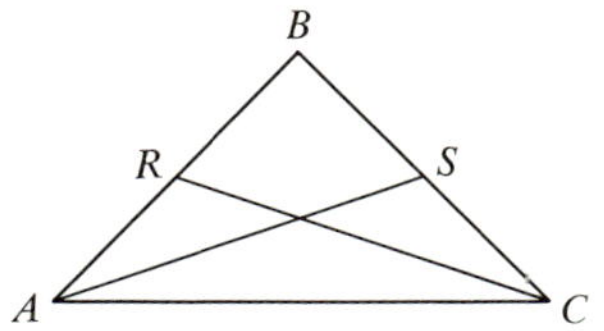

FIGURE 13-23

That is, if $\overline{AB} \cong \overline{BC}$ and R is the midpoint of $\overline{AB}$ and S is the midpoint of $\overline{BC}$, then $AS = CR$.

Analytical Proof

1. Choose the coordinate system shown in FIG. 13-24.

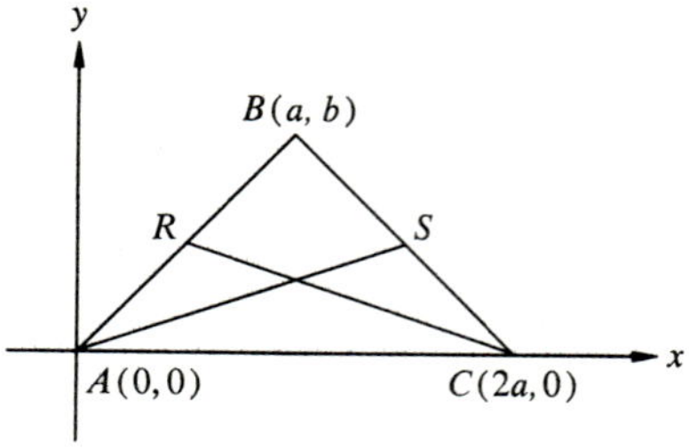

FIGURE 13-24

2. By the Midpoint Formula,

   $R = (a/2, b/2)$,

   $S = (3a/2, b/2)$.

3. By the Distance Formula,

$$AS = \sqrt{\left(\frac{3a}{2} - 0\right)^2 + \left(\frac{b}{2} - 0\right)^2} = \frac{1}{2}\sqrt{9a^2 + b^2},$$

$$CR = \sqrt{\left(2a - \frac{a}{2}\right)^2 + \left(0 - \frac{b}{2}\right)^2} = \frac{1}{2}\sqrt{9a^2 + b^2}.$$

4. Therefore, the medians AS and CR are congruent.

THEOREM 2 The diagonals of a parallelogram bisect each other (FIG. 13-25).

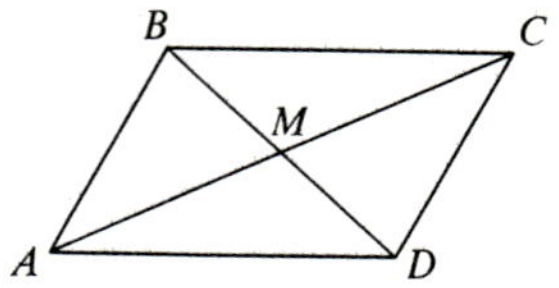

FIGURE 13-25

That is, M is the midpoint of $\overline{AC}$ and $\overline{BD}$.

Analytical Proof

1. Choose the coordinate system shown in FIG. 13-26.

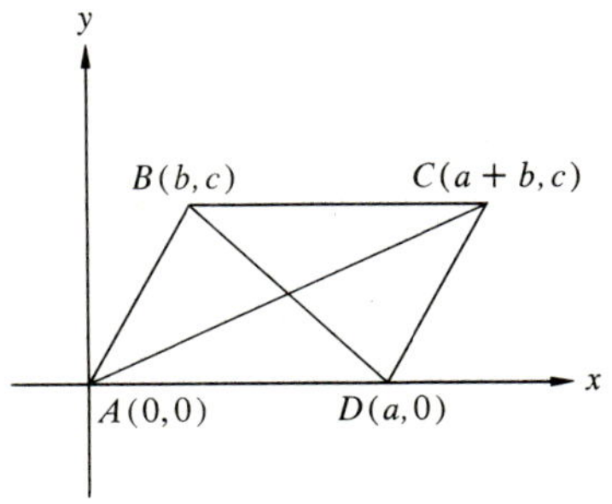

FIGURE 13-26

2. By the Midpoint Formula,

$$\text{midpoint of } \overline{AC} = \left(\frac{a+b}{2}, \frac{c}{2}\right),$$

$$\text{midpoint of } \overline{BD} = \left(\frac{a+b}{2}, \frac{c}{2}\right).$$

3. Therefore, the midpoint of $\overline{AC}$ is the same as the midpoint of $\overline{BD}$.

EQUATIONS OF A STRAIGHT LINE

Introduction

We can think of a curve such as a circle or a straight line as a set of points that satisfy certain conditions. These conditions may be stated in words, or they may be stated as equations or as inequalities. The set of all points of the plane that satisfy a given condition is called the *graph of the condition.*

For instance, the graph of the condition that the set of points in a plane be at a given distance (r) from a point (P) in the plane is a circle (FIG. 13-27).

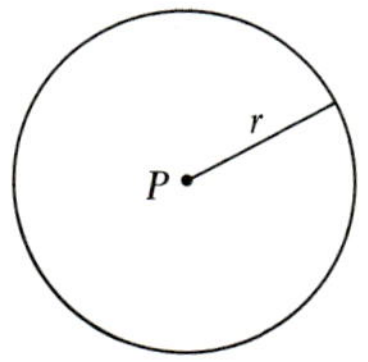

FIGURE 13-27

The graph of the condition $x = 2$ is the vertical line 2 units to the right of the y axis (FIG. 13-28).

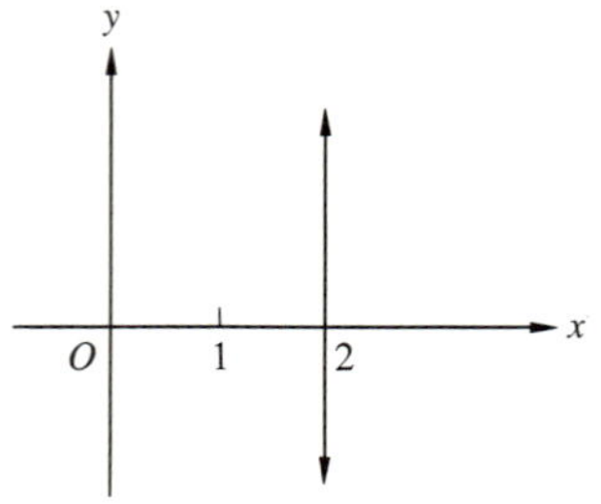

FIGURE 13-28

The graph of the condition $y > 1$ is the set of points more than 1 unit above the x axis (FIG. 13-29).

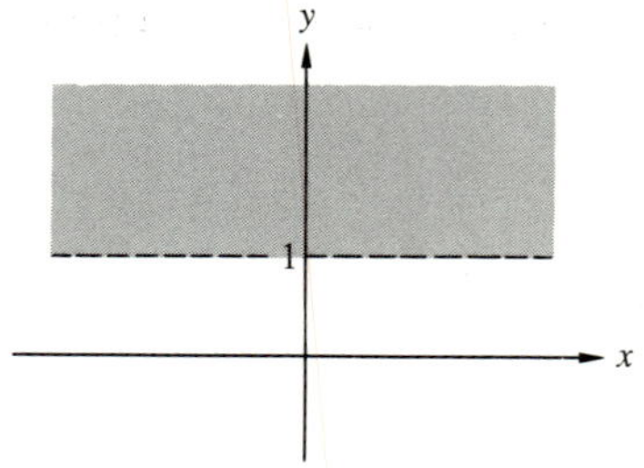

FIGURE 13-29

In the graph–equation relationship we can start with an equation and find its graph, or we can start with a graph and find its equation. A graph and equation correspond (1) if the coordinates of each point in the graph satisfy the equation, and (2) if each point whose coordinates satisfy the equation lies on the graph.

We shall start with the simplest graph—the straight line—and investigate its relationship to its corresponding equation. We shall see that the equation of an oblique straight line may be expressed in any one of several forms, each more appropriate for certain purposes.

Equations of Lines Parallel to the Axes

THEOREM 13-7 The equation of the straight line parallel to the y axis and a units from it is

$$x = a \qquad (13\text{-}1)$$

(FIG. 13-30). If a is positive, the line is to the right of the y axis; if a is negative then the line is to the left of the y axis. In either case, every point on the line will be a units from the y axis.

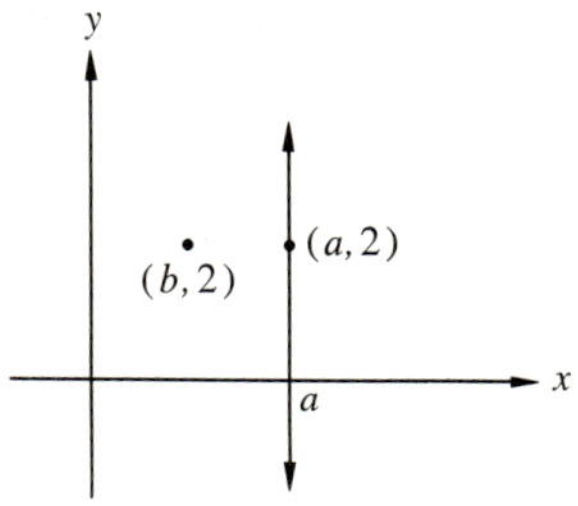

FIGURE 13-30

Equation (13-1) is satisfied by the coordinates of each point on the given line, but by no other point. For example, the point $(a, 2)$, which lies on the line, satisfies equation (13-1), since $a = a$. However, the point $(b, 2)$, which is not on the line, does not satisfy the equation since $b \neq a$.

The points satisfying the equation $x = a$ lie on the given line since the equation is satisfied by any point (a, y) where y is any real number, and every such point lies on the given line.

THEOREM 13-8 The equation of the straight line parallel to the x axis and b units from it is

$$y = b \qquad (13\text{-}2)$$

(FIG. 13-31). If b is positive, the line is above the x axis; if b is negative, then the line is below the x axis. In either case, every point on the line will be b units from the x axis.

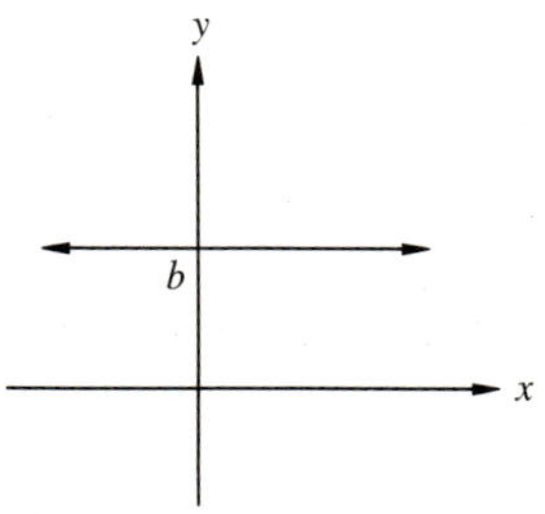

FIGURE 13-31

The proof of this theorem is similar to that of Theorem 13-7.

Equations of Oblique Lines

1. *Point-slope form of the equation*
Consider the line l, which passes through the point (x_1, y_1) and has slope m (FIG. 13-32). For every point (x, y) of l, the equation

$$\frac{y - y_1}{x - x_1} = m$$

or

$$y - y_1 = m(x - x_1)$$

is satisfied. This equation is called the *point-slope form* of the equation of the line.

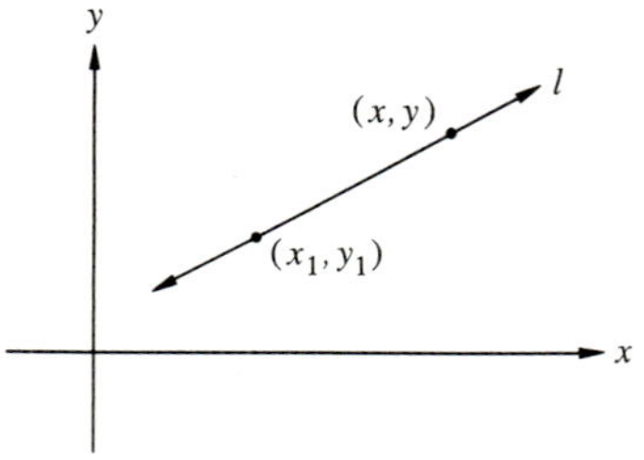

FIGURE 13-32

THEOREM 13-9 The equation of the straight line passing through the point (x_1, y_1) *with slope m is*

$$y - y_1 = m(x - x_1). \qquad (13\text{-}3)$$

To prove that equation (13-3) is the equation of line l, we prove two things:

1. Every point on l satisfies the equation.
2. Every point that satisfies the equation is on l.

We have just shown 1. To show 2 we shall select a point *not* on l and prove that it does *not* satisfy equation (13-3).

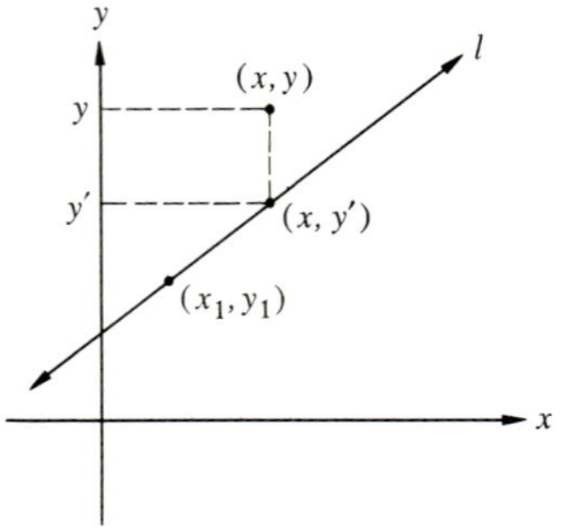

FIGURE 13-33

1. Let (x, y) be a point not on l. Then there is a point (x, y'), $y \neq y'$, which is on l (FIG. 13-33).

2. By Theorem 13-9,

$$\frac{y' - y_1}{x - x_1} = m$$

or

$$y' = y_1 + m(x - x_1).$$

3. Since $y' \neq y$, we conclude from step 2 that

$$y \neq y_1 + m(x - x_1)$$

or

$$y - y_1 \neq m(x - x_1).$$

That is, a point not on l does not satisfy equation (13-3).

2. *Slope-intercept form of the equation*
Every oblique line intersects the y axis in some point whose x coordinate is 0. The y coordinate of this point is called the *y intercept* of the line and is often denoted by b. For every point (x, y) of l (FIG. 13-34), we get

$$\frac{y - b}{x - 0} = m$$

or

$$y - b = m(x - 0)$$

or

$$y - b = mx$$

or

$$y = mx + b.$$

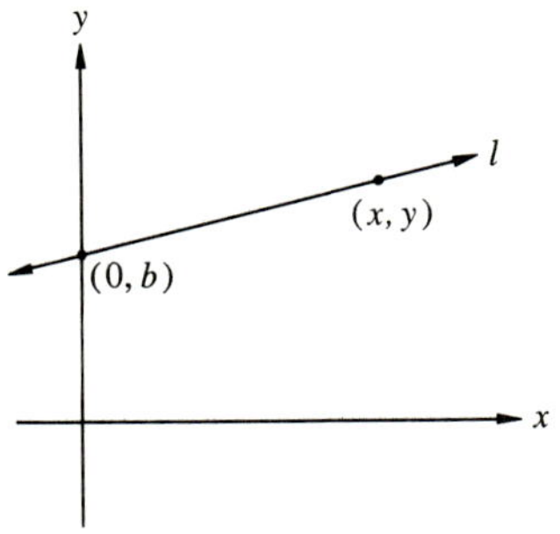

FIGURE 13-34

This equation is called the *slope-intercept form* of the equation of the line.

Definitions

1. The point where the line crosses the y axis is called the *y intercept*.
2. The point where the line crosses the x axis is called the *x intercept*.

THEOREM 13-10 *The equation of the straight line with slope m and y intercept b is*

$$y = mx + b.^{11} \tag{13-4}$$

[11] The slope-intercept form gives us a way of finding the slope of a straight line from its equation: If we solve for y, then m is the coefficient of x. For example, in the equation $2x + 3y = 5$, $y = -\frac{2}{3}x + \frac{5}{3}$. The slope of the line is, therefore, $-\frac{2}{3}$.

3. *Intercept form of the equation*
If the x and y intercepts of a line are a and b, respectively, with $a \neq 0$ and $b \neq 0$, the line passes through the points $(a, 0)$ and $(0, b)$ and has the slope $m = -b/a$ (FIG. 13-35). Using the point-slope form, the equation of the line is

$$y - 0 = -\frac{b}{a}(x - a)$$

or

$$ay = -bx + ab$$

or

$$bx + ay = ab.$$

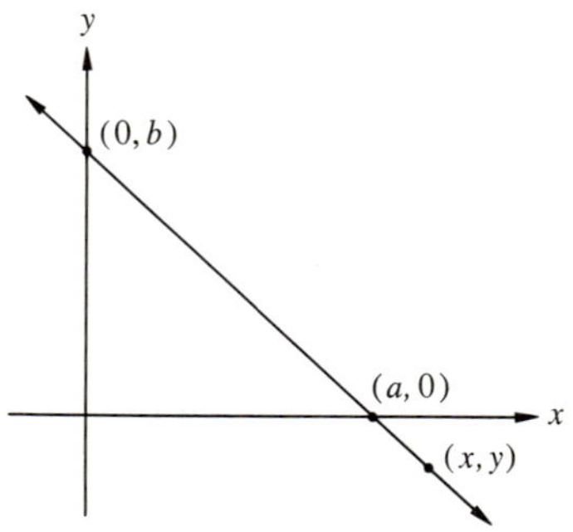

FIGURE 13-35

Dividing both members of the last equation by ab, we get

$$\frac{x}{a} + \frac{y}{b} = 1.$$

This equation is called the *intercept form* of the equation of the line.

THEOREM 13-11 *The equation of the straight line with x intercept a and y intercept b is*

$$\frac{x}{a} + \frac{y}{b} = 1. \qquad (13\text{-}5)$$

First-Degree Equations and Their Graphs

Definition

An equation of the form

$$Ax + By + C = 0 \qquad (13\text{-}6)$$

is called a *first-degree equation* in x and y if A, B, and C are real numbers and A and B are not both zero.

1. We have already seen that every straight line has a first-degree equation. Each of these equations is a special case of equation (13-6). Consider, for instance, the equations of lines parallel to the axes. The equation $x = a$ is a special case of equation (13-6) with $A = 1$, $B = 0$, and $C = -a$. In the equation $y = b$, $A = 0$, $B = 1$, and $C = -b$. In the equation for an oblique line, $y = mx + b$, we have $A = m$, $B = -1$, and $C = b$.

2. We now show that equation (13-6), with A and B not both zero, represents a *straight line.*
   a. If $B = 0$ and $A \neq 0$, then $Ax + C = 0$, and $x = -C/A$. The last equation represents a vertical line parallel to the y axis, $-C/A$ units from it.
   b. If $B \neq 0$, then

   $$y = -\frac{A}{B}x - \frac{C}{B}.$$

   This represents a straight line with slope $-A/B$ and y intercept $-C/B$.

A fundamental relationship between algebra and geometry is expressed by the following theorem.

THEOREM 13-12 *All straight lines have first-degree equations in two unknowns, and, conversely, all first-degree equations in two unknowns represent straight lines.*[12]

[12] 1. Because of the relationship established in Theorem 13-12, equations of the first degree are called *linear equations,* and the equation $Ax + By + C = 0$ is called the *general linear equation.*

2. Theorem 13-12 says, in effect, that all points whose coordinates satisfy the equation $Ax + By + C = 0$ lie on a line, and that the coordinates of any point on the line satisfy the equation.

3. Two different equations may represent the same line. For example, the two equations $y = x$ and $3y - 3x = 0$ represent the same line. We call such equations *equivalent equations.*

THE EQUATION OF A CIRCLE

Consider the circle with center at the origin and radius r (FIG. 13-36).

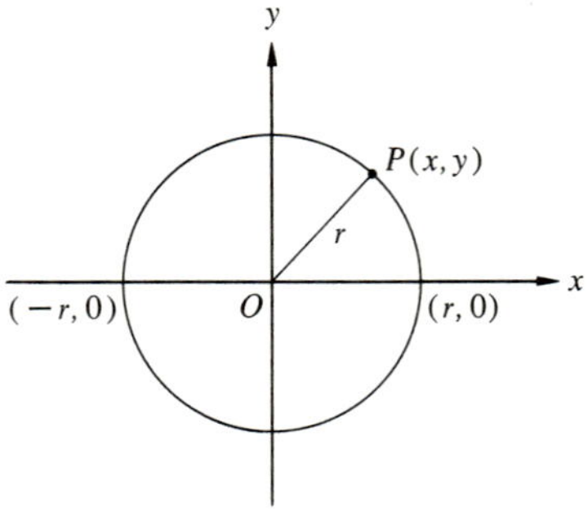

FIGURE 13-36

1. Since $OP = r$, by the Distance Formula we obtain

   $$\sqrt{(x-0)^2 + (y-0)^2} = r$$

   or

   $$x^2 + y^2 = r^2. \tag{13-7}$$

   Equation (13-7) says that if $P(x, y)$ is a point of the circle, then $x^2 + y^2 = r^2$.

2. We shall now show that if $x^2 + y^2 = r^2$, then $P(x, y)$ is a point of the circle. If $x^2 + y^2 = r^2$ and $r > 0$, then

   $$\sqrt{(x-0)^2 + (y-0)^2} = r.$$

   That is, $OP = r$, or P is a point of the circle.

THEOREM 13-13 The equation of the circle with center at the origin and radius r is

$$x^2 + y^2 = r^2. \tag{13-7}$$

We now consider a circle with center not at the origin but at (h, k) (FIG. 13-37). Then

$$(x-h)^2 + (y-k)^2 = r^2. \tag{13-8}$$

Equation (13-8) is the standard form of the equation of the circle with center (h, k) and radius r.

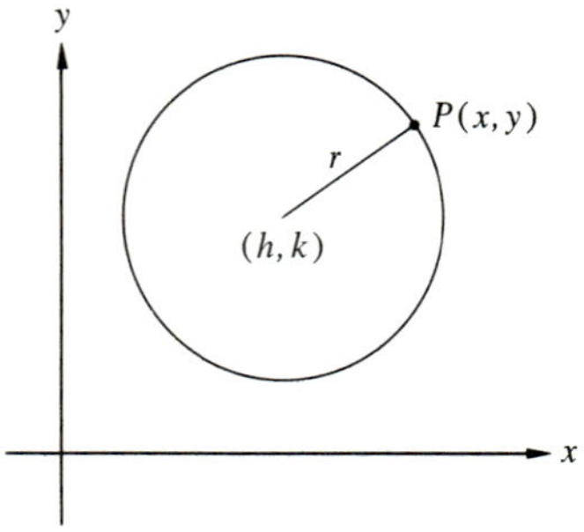

FIGURE 13-37

THEOREM 13-14 *The equation of the circle with center (h, k) and radius r is*

$$(x - h)^2 + (y - k)^2 = r^2. \qquad (13\text{-}8)$$

If we expand equation (13-8) we get

$$x^2 - 2hx + h^2 + y^2 - 2ky + k^2 = r^2$$

or

$$x^2 + y^2 + Dx + Ey + F = 0, \qquad (13\text{-}9)$$

where $D = -2h$, $E = -2k$, and $F = h^2 + k^2 - r^2$.

THEOREM 13-15 *Every circle is the graph of an equation of the form*

$$x^2 + y^2 + Dx + Ey + F = 0.^{13} \qquad (13\text{-}9)$$

[13] 1. Equation (13-7) is a special case of equation (13-8), where $(h, k) = (0, 0)$.

2. The converse of Theorem 13-15 is not true. That is, it is not true that every equation of form (13-9) has a circle as its graph. For example, the equation $x^2 + y^2 = 0$ (where D, E, and F are all zero) is satisfied only when x and y are both zero. The graph of this equation is, therefore, not a circle but a single point, (0, 0).

LOCUS

We have seen that a circle is a set of points that satisfies certain conditions:

1. That every point of the circle is a given distance from a given point in the plane.
2. That every point that satisfies this condition is on the circle.

A curve, or graph, that is associated with a given condition in this way is sometimes called a *locus* of that condition. (The plural of *locus* is *loci*.)

Definition
A set of points S is the *locus* of a condition if and only if *both* of the following statements are true:

1. Every point of S satisfies the condition.
2. Every point that satisfies the condition belongs to S.

Examples of Loci

1. The locus of points at a given distance d from a given line l is a pair of lines parallel to l, one on each side of l (FIG. 13-38).

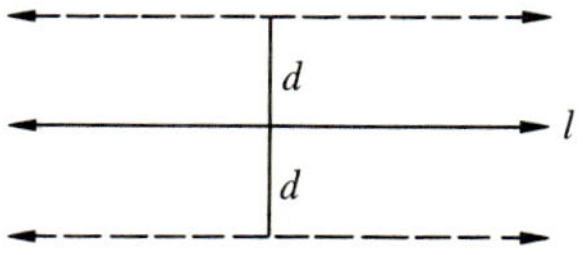

FIGURE 13-38

2. The locus of points equidistant from the sides of an angle is the set of points on the bisector of the angle (FIG. 13-39).

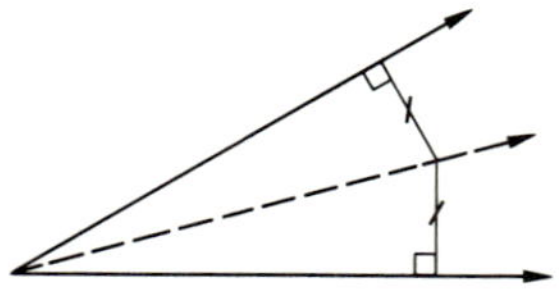

FIGURE 13-39

3. The locus of points equidistant from two given points A and B is the set of points on the perpendicular bisector of the segment $\overline{AB}$ (FIG. 13-40).

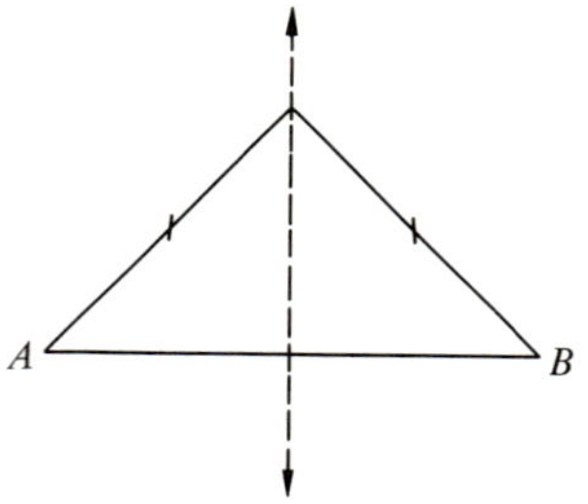

FIGURE 13-40

The last locus theorem can also be stated this way: The perpendicular bisector of a segment is the locus of all points of the plane that are equidistant from the endpoints of the segment.

TEACHING SUGGESTIONS AND ACTIVITIES

VISUAL AIDS

Visual 1

PURPOSE

String construction of an ellipse.

CONSISTS OF

1. Piece of string.
2. Two thumbtacks.
3. Stiff paper and pencil.

HOW TO USE

1. Tie the ends of the string to the thumbtacks (FIG. 13-41). Designate the length of the string left free between the thumbtacks by $2a$.
2. Press the thumbtacks into the paper at the foci F and F'.
3. Place a pencil at a point P on the paper in such a way that the string lies taut in the position F to P to F' (FIG. 13-42).
4. Point P lies on the ellipse, since $PF + PF' = 2a$.
5. If the pencil is now moved, always keeping the string taut, the ellipse will be drawn.

Note: A slight variation of this visual is to place a loop of string about the two foci and then take up the slack with a moving pencil. The pencil will trace an ellipse (FIG. 13-42).

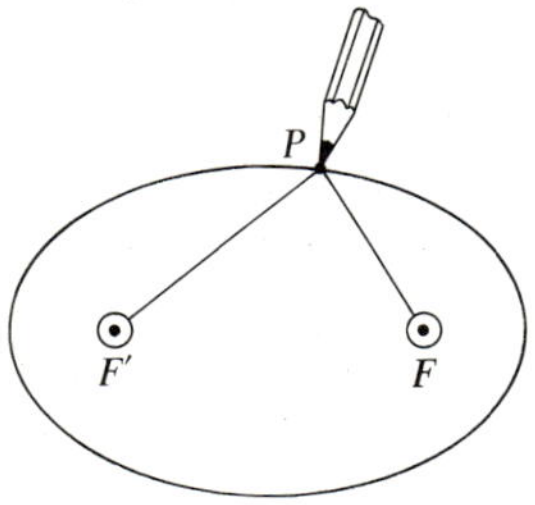

FIGURE 13-41

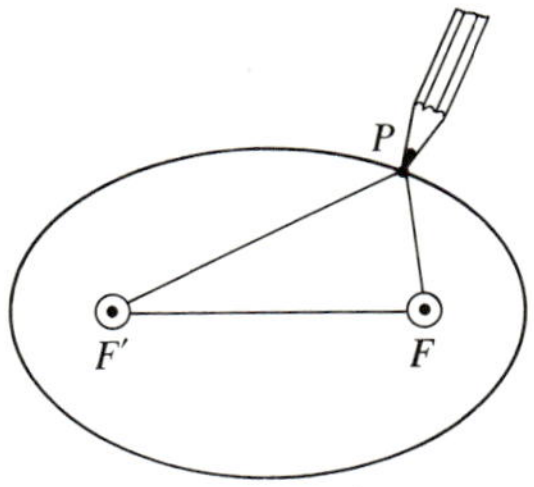

FIGURE 13-42

Visual 2

PURPOSE

Construction of a parabola by curve-stitching.[14]

CONSISTS OF

An angle drawn on a stiff piece of cardboard. The sides of the angle are labeled with equally spaced points, as shown in FIG. 13-43.

[14] Curve-stitching is an old method of constructing mathematical designs—some simple, others more intricate—by means of stitching on cards. The principle on which the method is based is to construct a curve by making an envelope of stitches which are tangents to the curve.

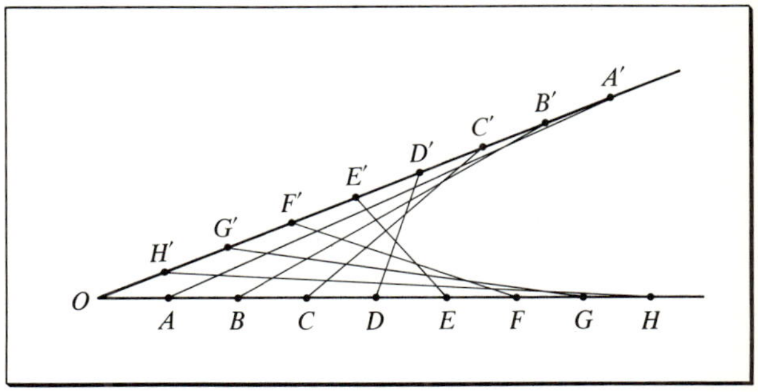

FIGURE 13-43

[15] The parabola was enveloped by taking $OA + OA' = OB + OB' = \cdots =$ a constant. If we take $(OA)(OA') = (OB)(OB') = \cdots =$ a constant, we envelop a hyperbola. We can form such an envelope by starting with a circle and locating a point O outside the circle (FIG. 13E). If secant lines are drawn from point O intersecting the circle in points A and A', B and B', etc., then $(OA)(OA') = (OB)(OB') = \cdots =$ a constant. The point O is one focus of the hyperbola, and the other focus is a corresponding point O' lying on the other side of the circle.

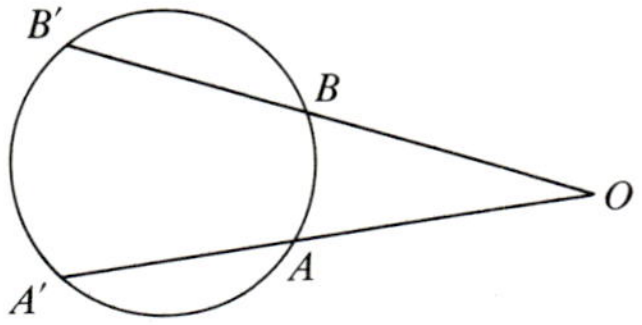

FIGURE 13E

If we choose O to be any point *inside* the circle (FIG. 13F) and draw lines through O intersecting the circle in points A and A', B and B', etc., we envelop an ellipse with one focus at O and the other focus a corresponding point O' on the other side of the center C of the circle.

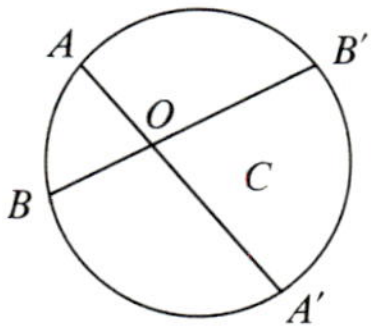

FIGURE 13F

HOW TO USE

1. The parabola is obtained by connecting with colored yarn the points in opposite directions on the sides of the angle. Stitches are made from A to A', from B' to B, from C to C', from D' to D, etc., as shown.

2. Observe the following procedures:
   a. After threading the needle, tie a double knot in the end of the yarn and then apply a drop of glue to secure the knot.
   b. Start on the "wrong" side of the card. This is the side where you always put the needle into the next hole and where you put all the knots and marks.
   c. The curve appears on the "right" side of the card.
   d. After the last stitch, again make a double knot and apply a drop of glue.

Note: A parabola can be *drawn* by joining the points on the card with line segments instead of stitches with yarn.[15]

Visual 3

Note: This visual is based on the discussion on Locating Points of an Ellipse on page 372.

PURPOSE

To provide a simple device for locating as many points of an ellipse as we wish.

CONSISTS OF

1. Three strips of any stiff material (FIG. 13-44). If the ellipse is $(x^2/a^2) + (y^2/b^2) = 1$, then strip 1 will have length a, and strip 2 and strip 3 will have lengths $a - b$.

2. The three strips are joined with eyelets to form a linkage in the following way (FIG. 13-45): At a point B, b units from one end of strip 1, attach strip 2; and at the other end of strip 1, at point A, attach strip 3.

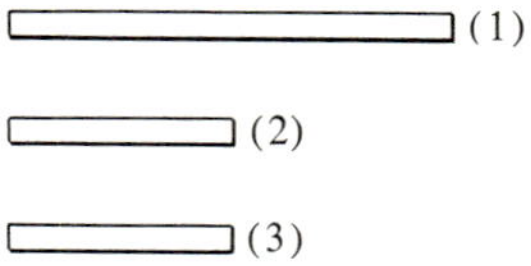

FIGURE 13-44

HOW TO USE

1. Draw coordinate axes on a sheet of paper.

2. Place point C at the origin O with strip 1 forming an angle θ with the x axis (FIG. 13-46).

3. Rotate strip 2 to a horizontal position and strip 3 to a vertical position.
4. The point P of intersection of strips 2 and 3 is a point of the ellipse $(x^2/a^2) + (y^2/b^2) = 1$.
5. By varying θ and repeating the procedure in steps 2, 3, and 4 we obtain as many points as we wish.

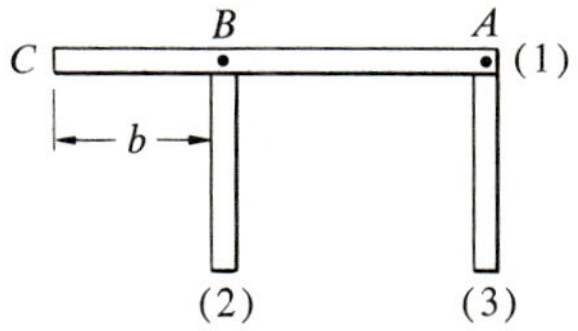

FIGURE 13-45

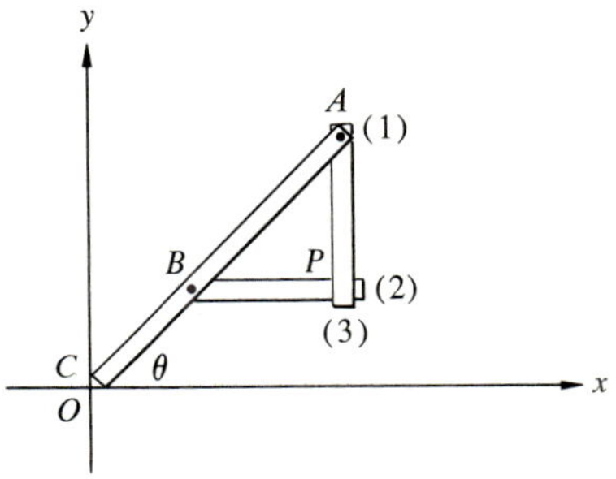

FIGURE 13-46

SLOPES OF PERPENDICULAR LINES

Because students may have difficulty accepting intuitively the relation between the slopes of perpendicular lines, that is, $m_1 = -1/m_2$, it would be well to draw several pairs of perpendicular lines, estimate the slope of each line, and then compare the slopes.

Students will quickly see that if one slope is positive, the other is negative; and if the absolute value of one slope is greater than 1, the absolute value of the other is less than 1. It will also become apparent that if one slope is zero (that is, the line is parallel to the x axis), the other slope is not defined (that is, the line is parallel to the y axis). The converse will likewise become apparent; that is, if one slope is not defined, then the other slope will be zero.

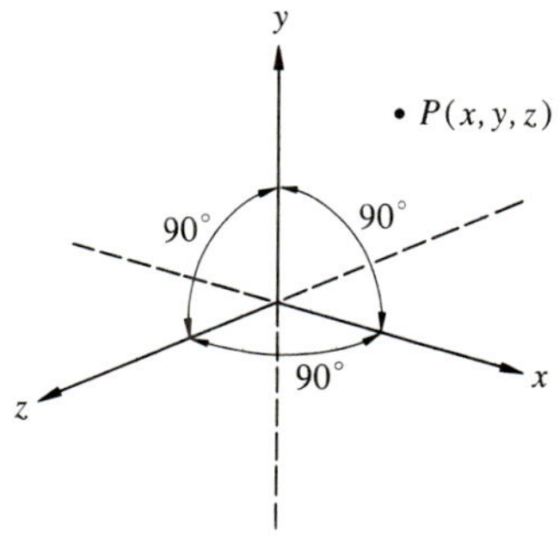

FIGURE 13-47

A THREE-DIMENSIONAL COORDINATE SYSTEM

The idea of a two-dimensional coordinate system can be extended to three dimensions. A three-dimensional coordinate system can be formed by considering three mutually perpendicular axes as shown in FIG. 13-47. The y axis represents a line perpendicular to the plane of the paper.

Taken in pairs, the three axes determine three planes called the yz plane, the xz plane, and the xy plane. These planes are called the *coordinate planes*.

A point $P(x, y, z)$ is located by its three coordinates. The first coordinate, x, represents the distance from the yz plane to P. The coordinates y and z represent the distances from the xz and xy planes to P, respectively.

Just as there is a one-to-one correspondence between the set of all points in a plane and the set of all ordered pairs of real numbers, so is there a one-to-one correspondence between the set of all points in a three-dimensional coordinate system and the set of all ordered triples of real numbers. In other words, there is one and only one point $P(a, b, c)$.

Distance

The distance between two points $P_1(x_1, y_1, z_1)$ and $P_2(x_2, y_2, z_2)$ in a three-dimensional coordinate system is

$$d = \sqrt{(x_2 - x_1)^2 + (y_2 - y_1)^2 + (z_2 - z_1)^2}.$$

The coordinates of the midpoint of the segment $\overline{P_1P_2}$ are

$$x_0 = \frac{x_1 + x_2}{2}, \qquad y_0 = \frac{y_1 + y_2}{2}, \qquad z_0 = \frac{z_1 + z_2}{2}.$$

The graph of an equation of the form

$$Ax + By + Cz + D = 0,$$

where A, B, and C are constants and not all zero, is a plane.

Can we extend the idea of a three-dimensional system to a four-dimensional or an n-dimensional system? Although we cannot visualize such a system, we find it convenient to say that an ordered set of *four real numbers* represents a point in *four dimensions*, and an ordered set of *n real numbers*, called an *n-tuple*, represents a point in *n dimensions*. So the set of numbers (1, 5, 8, 0) is thought of as a point in four-dimensional space, and the set of numbers (−1, 2, 6, −3, −2) is referred to as a point in five-dimensional space. The geometry of n dimensions, where n is greater than 3, is completely a creation of our imagination and can be portrayed by equations but not by pictures or some other physical experience.

The concept of distance is extended accordingly. The distance between two points $A(a_1, a_2, a_3, a_4, a_5)$ and $B(b_1, b_2, b_3, b_4, b_5)$ in five-dimensional space is

$$d = \sqrt{(a_1 - b_1)^2 + (a_2 - b_2)^2 + (a_3 - b_3)^2 + (a_4 - b_4)^2 + (a_5 - b_5)^2}$$

and the distance between the two points $(a_1, \ldots, a_n)$ and $(b_1, \ldots, b_n)$ in n-dimensional space is

$$d = \sqrt{(a_1 - b_1)^2 + \cdots + (a_n - b_n)^2}.$$

LOCATING POINTS OF AN ELLIPSE

The following is a simple procedure for locating as many points of an ellipse as we wish:

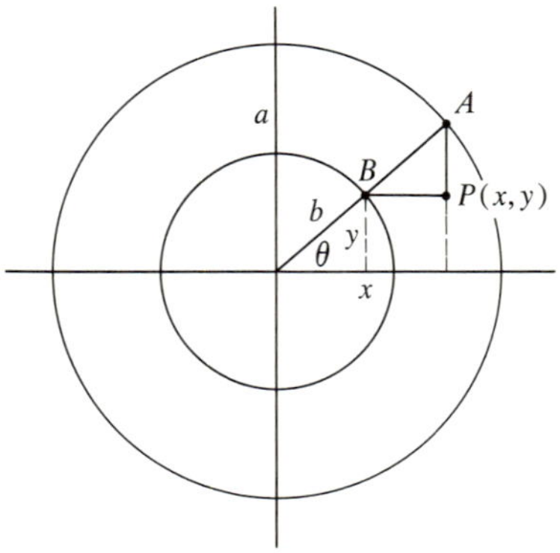

FIGURE 13-48

1. Draw concentric circles of radius a and b, $a > b$, with center at the origin (FIG. 13-48).

2. Draw a line through the origin intersecting the smaller circle at B and the larger at A.

3. Through B draw a horizontal line and through A a vertical line, the two lines intersecting at a point $P(x, y)$.

4. $P(x, y)$ is a point of the ellipse.

5. We can locate other points of the ellipse by drawing other lines through the origin intersecting both circles, and then following the procedure given in step 3.

There are two ways of showing that the locus of $P(x, y)$ is an ellipse. (See page 378.)

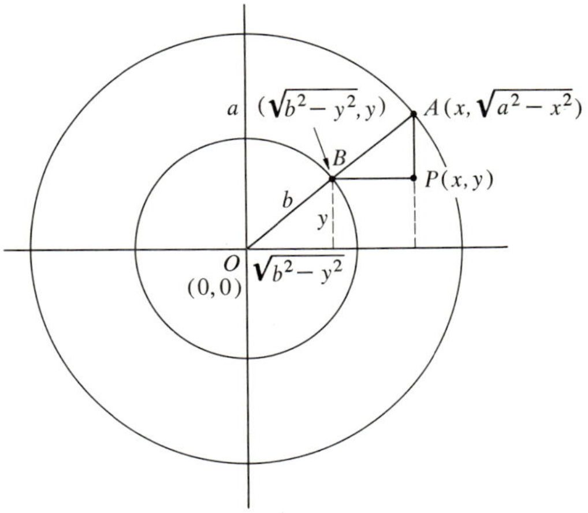

FIGURE 13-49

1. In FIG. 13-48,

$$\frac{x}{a} = \cos\theta, \qquad \frac{y}{b} = \sin\theta,$$

$$\frac{x^2}{a^2} + \frac{y^2}{b^2} = \cos^2\theta + \sin^2\theta$$
$$= 1,$$

which is the equation of an ellipse with center at the origin and axes along the coordinate axes.

2. In FIG. 13-49, the slope of $\overline{BO}$ is the same as the slope of $\overline{AB}$. Therefore,

$$\frac{y}{\sqrt{b^2 - y^2}} = \frac{\sqrt{a^2 - x^2} - y}{x - \sqrt{b^2 - y^2}}$$

or simplifying,

$$b^2x^2 + a^2y^2 = a^2b^2.$$

Dividing both sides by a^2b^2, we get

$$\frac{x^2}{a^2} + \frac{y^2}{b^2} = 1.$$

QUESTIONS

1. A straight line passes through the points (0, 0) and (2, 2). What acute angle does it make with the positive x axis?

2. A line passes through the points (2, 3) and (0, −2). Give the coordinates of any other point of the line.

3. Does the line through the points (−2, −2) and (2, 3) pass through the origin? Give a reason for your answer.

4. The vertices of a triangle are $A(-2, 0)$, $B(2, 1)$, and $C(1, 5)$.

   a. Show that $\triangle ABC$ is a right triangle, and name the vertex of the right angle.
   b. Find the perimeter of the triangle.
   c. Find the center and radius of the circumscribed circle.

5. Prove that the diagonals of a rectangle are congruent.

6. Prove that an isosceles triangle has two congruent medians.

7. In which quadrant is $P(x, y)$ located if

   a. $x < 0$ and $y > 0$. b. $x < 0$ and $y < 0$.
   c. $x > 0$ and $y > 0$. d. $x > 0$ and $y < 0$.

8. If a line, different from and not parallel to either axis, does not pass through the origin, in how many quadrants does it lie?

9. What must be the value of r if

   a. The line through $(0, -2)$ and $(3, r)$ is horizontal?
   b. The line through (r, y_1) and (x_2, y_2) is vertical?
   c. The line through $(-3, r)$ and $(2, -3)$ is horizontal?
   d. The line through $(-2, 0)$ and $(3, r)$ is to have a slope of $\frac{2}{3}$?

10. The vertices of a triangle are $A(0, 0)$, $B(6, 2)$, and $C(3, -2)$. Find

    a. The slopes of its sides.
    b. The slopes of its altitudes.
    c. The lengths of its medians.

11. Prove that the midpoint of the hypotenuse of a right triangle is equidistant from the three vertices of the triangle.

12. Prove that every point equidistant from the ends of a segment lies on the perpendicular bisector of the segment.

13. Prove that the altitudes of a triangle are concurrent.

14. Show that the points $(4, \frac{3}{2})$, $(3, 2)$, and $(-1, 4)$ are collinear.

15. Prove that if the diagonals of a parallelogram are perpendicular, then the parallelogram is a rhombus.

16. Find the center and radius of the following circles:

 a. $x^2 + y^2 - 4x + 10y - 20 = 0$.

 b. $\frac{x^2}{5} + \frac{y^2}{5} = 2$.

17. Find the equation of the circle with diameter $\overline{AB}$, where $A = (-5, 3)$ and $B = (3, 9)$.

18. Prove that the line $4x + 3y = 25$ is tangent to the circle $x^2 + y^2 = 25$. (*Hint:* Show that there is only *one* point of intersection.)

PUZZLE

How is it possible to draw an ellipse by using only three strips of wood (or any other stiff material) S_1, S_2, and S_3?

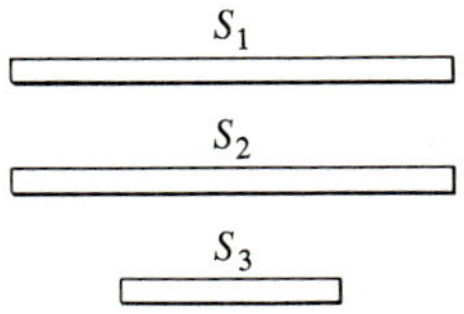

SOLUTION

Using S_1 and S_2 as coordinate axes and S_3 as a movable piece, slide S_3 in such a way that the same end of S_3 is always in contact with S_1 and the other end is always in contact with S_2.

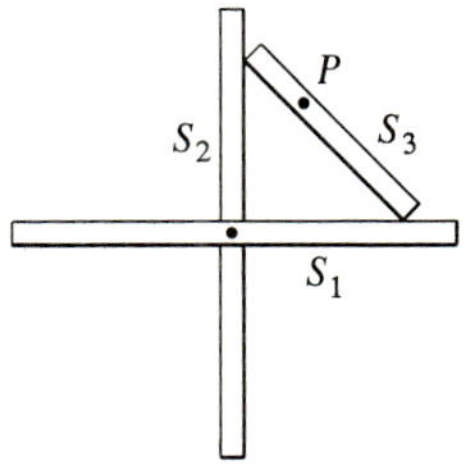

Select any point P on S_3 and insert a pencil through it. As S_3 moves through all four quadrants, the pencil will trace an ellipse.

(continued)

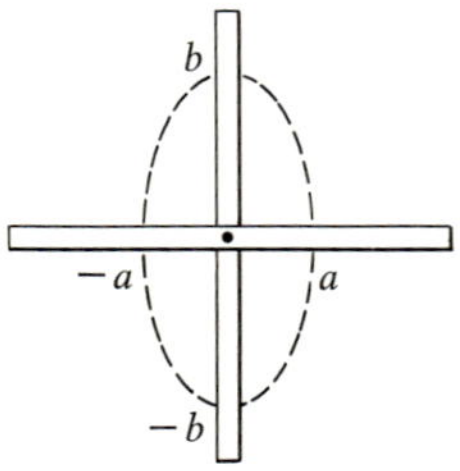

The reason that the locus of P is an ellipse is that

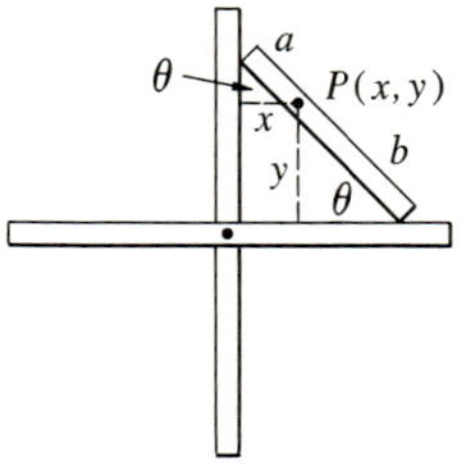

1. If P divides S_3 into lengths a and b, then

   $$\frac{x}{a} = \cos\theta \qquad \text{and} \qquad \frac{y}{b} = \sin\theta.$$

2. Therefore,

   $$\frac{x^2}{a^2} + \frac{y^2}{b^2} = 1.$$

   This is called the *trammel construction* of an ellipse.

Note:

1. The locus of P can easily be shown to be an ellipse without using trigonometric functions. (See page 373.)
2. This puzzle can be the basis of a physical device for drawing an ellipse.

ENRICHMENT

EQUATIONS OF THE CONIC SECTIONS

We have seen that the equation of a circle with center (h, k) and radius r is

$$(x - h^2) + (y - k)^2 = r^2 \tag{13-10}$$

(FIG. 13-50). We shall now derive the equations of the other conic sections.

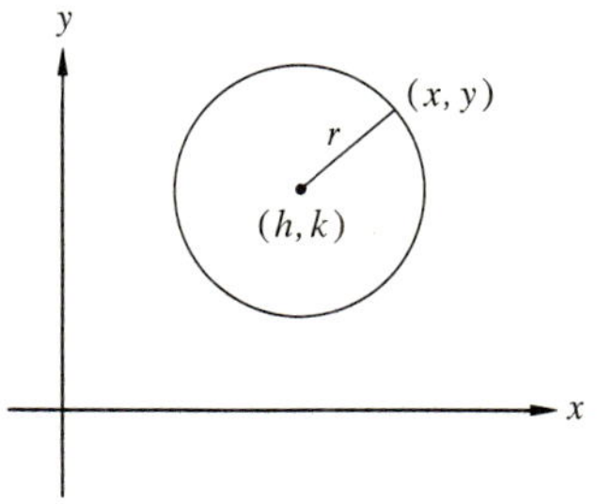

FIGURE 13-50

The Parabola

In Chapter 6 we defined the *parabola* as the set of all points equidistant from a fixed line l (called the *directrix*) and a fixed point F (called the *focus*).

Consider a parabola with focus $F(a, 0)$ and directrix $x = -a$, as shown in FIG. 13-51.

If $P(x, y)$ is any point on the parabola, then by the definition of a parabola, $FP = AP$; that is,

$$\sqrt{(x - a)^2 + y^2} = x + a$$

or

$$y^2 = 4ax. \tag{13-11}$$

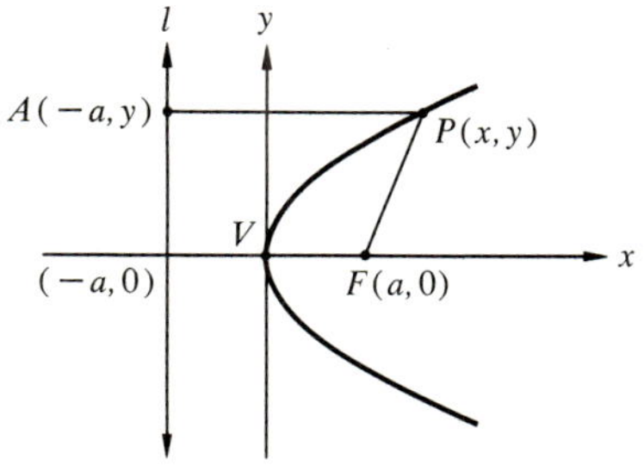

FIGURE 13-51

Conversely, it can be shown that if a point $P(x, y)$ satisfies the equation $y^2 = 4ax$, then the point is on the parabola.

The point V lies halfway between the focus and the nearest point on the directrix and is called the *vertex*. The vertex in FIG. 13-51 is the origin and, since it is equidistant from the focus and the directrix, is a point on the parabola. The line $\overleftrightarrow{VF}$ is called the *axis of symmetry*. The axis of any parabola is the line that contains the focus and is perpendicular to the directrix.

The equations of the parabolas with vertices at the origin and foci at $(-a, 0)$, $(0, a)$, and $(0, -a)$, respectively, are shown in FIG. 13-52.

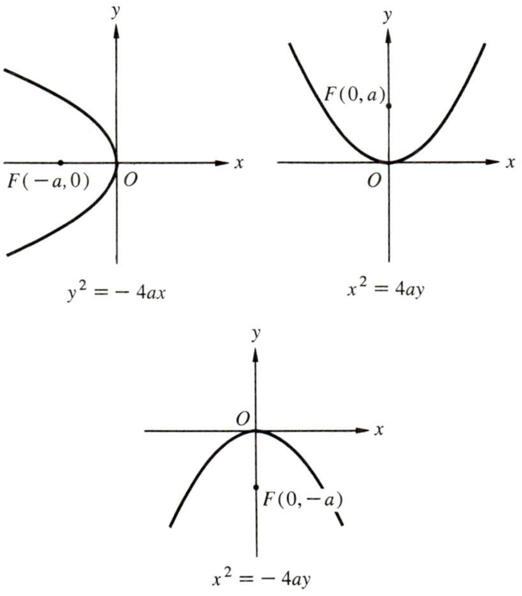

FIGURE 13-52

Not all parabolas have their vertices at the origin, nor do their axes always coincide with the coordinate axes. By the same method used to derive equation (13-11), we can show that the equation of a parabola with vertex at $V(h, k)$ and axis parallel to the x axis is

$$(y - k)^2 = 4a(x - h). \tag{13-12}$$

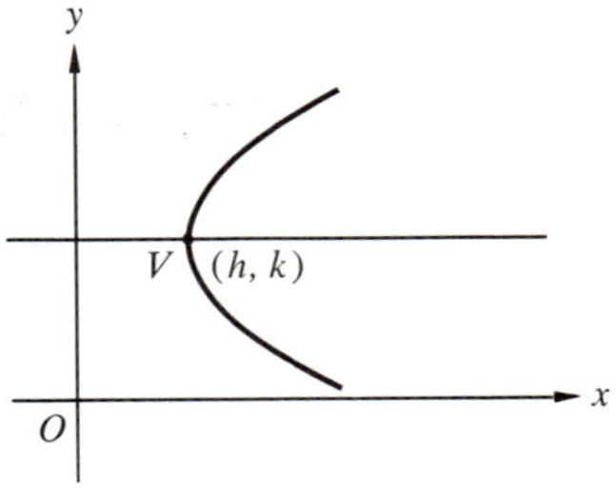

FIGURE 13-53

This parabola (FIG. 13-53) opens to the right if $a > 0$ and to the left if $a < 0$.

The equation of a parabola with vertex at $V(h, k)$ and axis parallel to the y axis is

$$(x - h)^2 = 4a(y - k). \tag{13-13}$$

This parabola (FIG. 13-54) opens upward if $a > 0$ and downward if $a < 0$.

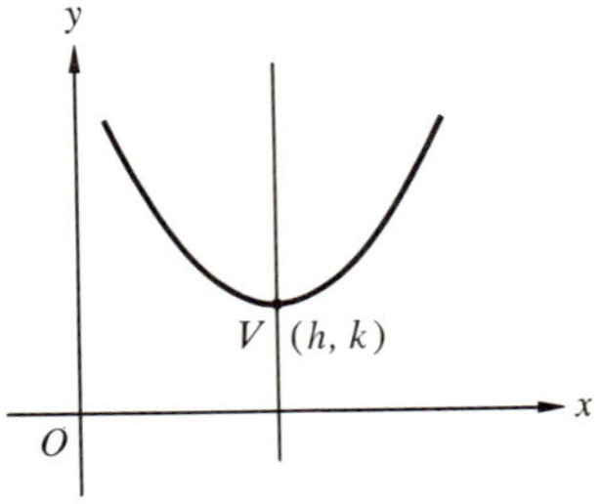

FIGURE 13-54

The Ellipse

In Chapter 6 we defined the ellipse as the set of all points in a plane such that for each point the *sum* of its distances from two fixed points F and F' in the plane is constant (FIG. 13-55). F and F' are called the *foci* of the ellipse.

Consider the ellipse whose center is at the origin and whose foci lie on the x axis and are $2c$ units apart, as shown in FIG. 13-56.

Let $P(x, y)$ be any point on the ellipse and let the sum of its distances to F and F' be the constant distance $2a$.

Since $2a > 2c$ (why?), $a > c$. Then

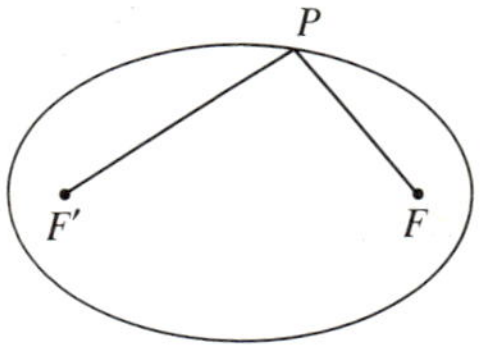

FIGURE 13-55

1. $F'P + FP = 2a$. Applying the Distance Formula, we get
2. $\sqrt{(x + c)^2 + y^2} + \sqrt{(x - c)^2 + y^2} = 2a.$
3. Transposing, squaring, and simplifying, we get

   $$cx - a^2 = -a\sqrt{(x - c)^2 + y^2}.$$
4. Squaring again and simplifying, we get

   $$(a^2 - c^2)x^2 + a^2y^2 = a^2(a^2 - c^2).$$

If we let $a^2 - c^2 = b^2$ and then divide both sides by a^2b^2, we get

$$\frac{x^2}{a^2} + \frac{y^2}{b^2} = 1, \tag{13-14}$$

as the equation of the ellipse shown in FIG. 13-56.

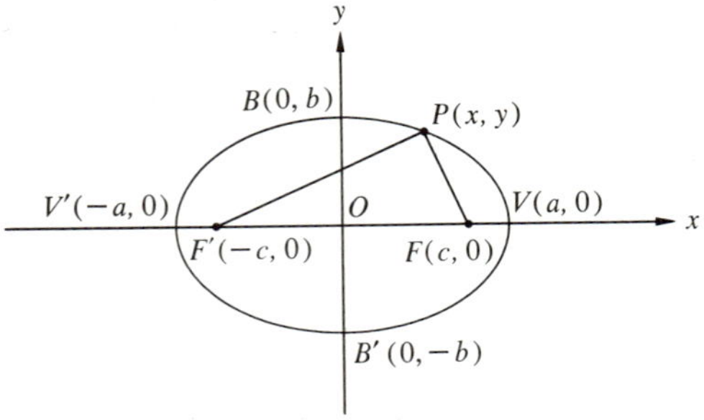

FIGURE 13-56

The curve is symmetric with respect to *both* axes. Segment $\overline{V'V}$ is called the *major axis* of the ellipse, and $\overline{B'B}$ is called the *minor axis* of the ellipse. The point of intersection of the major and minor axes is the *center* of the ellipse. V and V' are called the *vertices* and B and B' are called the *covertices* of the ellipse.

If we let $y = 0$ in equation (13-14), we find that $V'V = 2a$. If we let $x = 0$, we get $B'B = 2b$.

When $c = 0$, the foci F' and F coincide and equation (13-14) reduces to $x^2 + y^2 = a^2$, which is the equation of a circle. When $c = a$, the sum of the two distances in the definition of the ellipse is equal to the distance between the foci, and the curve becomes the straight-line segment $\overline{F'F}$.

When the coordinates of the foci are $(0, c)$ and $(0, -c)$ as in FIG. 13-57, the equation becomes

$$\frac{y^2}{a^2} + \frac{x^2}{b^2} = 1. \tag{13-15}$$

Equations (13-14) and (13-15) are the *intercept form* of the equation for an ellipse.

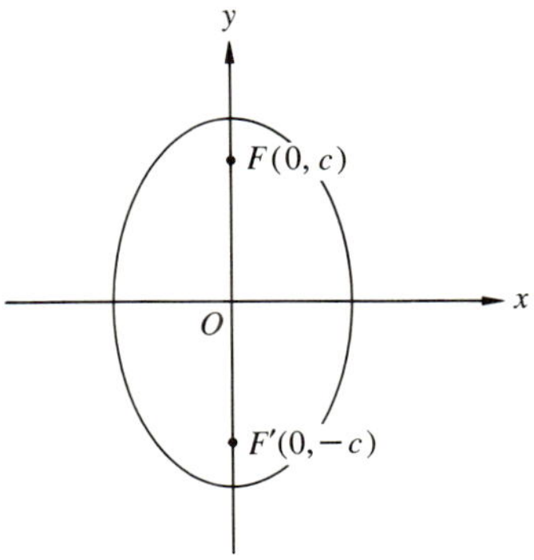

FIGURE 13-57

When the coordinates of the center are (h, k) and the major axis is parallel to the x axis, as in FIG. 13-58, we get the equation

$$\frac{(x-h)^2}{a^2} + \frac{(y-k)^2}{b^2} = 1. \tag{13-16}$$

When the coordinates of the center are (h, k) and the major axis is parallel to the y axis, as in FIG. 13-59, we get the equation

$$\frac{(y-k)^2}{a^2} + \frac{(x-h)^2}{b^2} = 1. \tag{13-17}$$

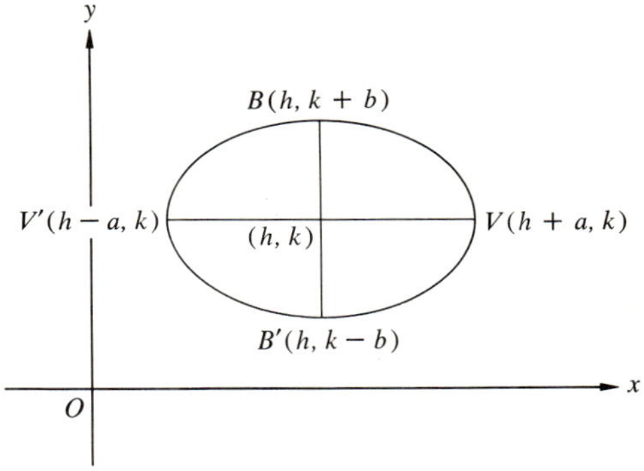

FIGURE 13-58

In both equations (13-16) and (13-17), a, b, and c are the distances from the center to V, B, and F, respectively (FIG. 13-60). The three distances are related by the equation

$$a^2 = b^2 + c^2.$$

The *eccentricity* of the ellipse,

$$e = \frac{c}{a} \qquad (e < 1),$$

helps to describe the shape of the ellipse. If two ellipses have the same eccentricity, their constants, a, b, and c, will be proportional and the ellipses will be "similar." If both ellipses are centered at the origin with major axes in the same direction, it will be possible to associate with each point P on one ellipse (FIG. 13-61) a point P' on the second ellipse such that $OP' = kOP$, k being the same constant for all pairs of points on the ellipses.

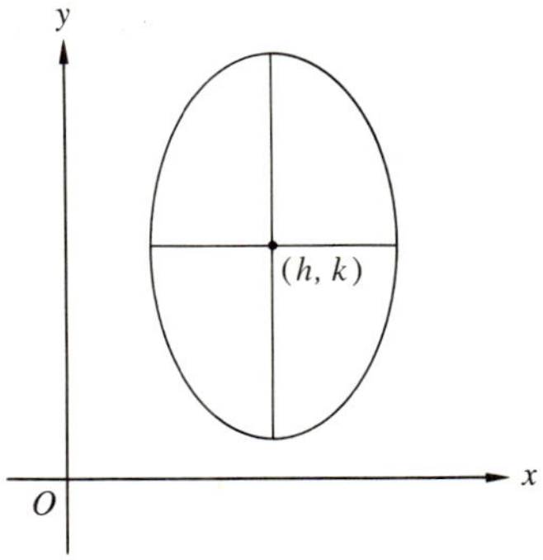

FIGURE 13-59

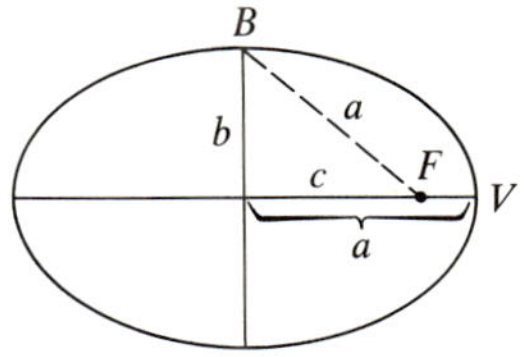

FIGURE 13-60

The Hyperbola

In Chapter 6 we defined the hyperbola as the set of all points in a plane such that for each point the *difference* of its distances from two fixed points F and F' is constant. F and F' are called the *foci* of the hyperbola.

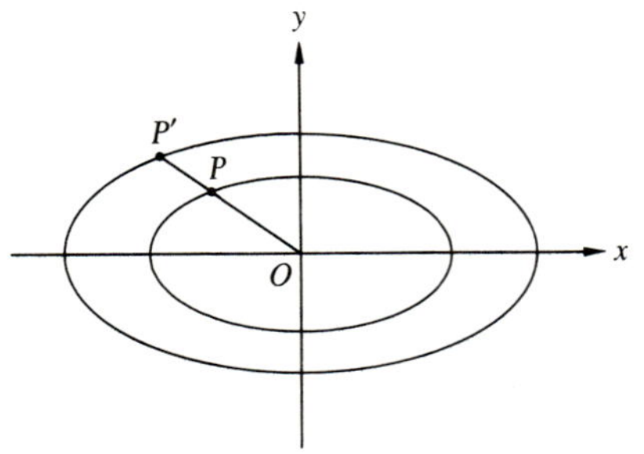

FIGURE 13-61

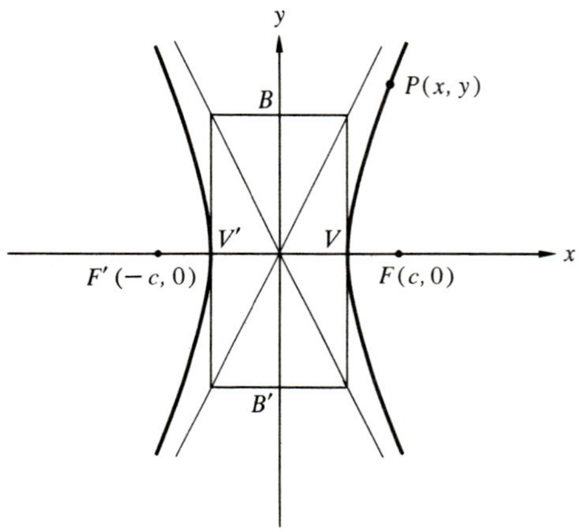

FIGURE 13-62

Choose the coordinates of the two points to be $F(c, 0)$ and $F'(-c, 0)$ as in FIG. 13-62, and let $P(x, y)$ be any point on the hyperbola and $2a$ be the constant difference of the distances from P to F and F'. Then

1. $|F'P - FP| = 2a$. Applying the Distance Formula, we get
2. $|\sqrt{(x + c)^2 + y^2} - \sqrt{(x - c)^2 + y^2}| = 2a.$
3. Squaring and simplifying step 2 leads to

   $(c^2 - a^2)x^2 - a^2y^2 = a^2(c^2 - a^2).$

If we let $c^2 - a^2 = b^2$ and then divide both sides of step 3 by a^2b^2, we get

$$\frac{x^2}{a^2} - \frac{y^2}{b^2} = 1. \tag{13-18}$$

Setting $y = 0$, we find the coordinates of the vertices to be $V(a, 0)$ and $V'(-a, 0)$. The coordinates of the covertices are $B(0, b)$ and $B'(0, -b)$. The segment $\overline{VV'}$ is called the *transverse axis* and the segment $\overline{BB'}$ the *conjugate axis.*

If the foci are on the y axis, the equation of the hyperbola becomes

$$\frac{y^2}{a^2} - \frac{x^2}{b^2} = 1. \tag{13-19}$$

Equations (13-18) and (13-19) are the *intercept form* of the equation for a hyperbola.

Note that FIG. 13-62 contains a rectangle. The lines containing the diagonals of this rectangle are

$$y = \pm\frac{b}{a}x.$$

These lines are called the *asymptotes* of the hyperbola. Roughly speaking, as we move along each line, the distance between the hyperbola and the line, measured perpendicular to the line, approaches zero. Of all the conic sections, only the hyperbola has asymptotes and only the hyperbola has two branches.

If the center of the hyperbola is at (h, k), equations (13-18) and (13-19) become

$$\frac{(x - h)^2}{a^2} - \frac{(y - k)^2}{b^2} = 1, \tag{13-20}$$

$$\frac{(y-k)^2}{a^2} - \frac{(x-h)^2}{b^2} = 1. \qquad (13\text{-}21)$$

In both equations (13-20) and (13-21), a, b, and c are the distances from center to vertex, cocenter, and focus, respectively, and the constants are related by the equation

$$c^2 = a^2 + b^2.$$

As in the case of the ellipse, the eccentricity of the hyperbola,

$$e = \frac{c}{a} \qquad (e > 1),$$

is related to the shape of the hyperbola. When e is close to 1, the hyperbola fits between two asymptotes that meet at a small angle. If e is large, the hyperbola fits between asymptotes that meet at a large angle. When $e = \sqrt{2}$, the asymptotes meet at a right angle and the hyperbola is called a *rectangular hyperbola*, or an *equilateral hyperbola.*[16]

SECOND-DEGREE CURVES

The circle, parabola, ellipse, and hyperbola are curves whose equations are all special cases of the second-degree general equation

$$Ax^2 + Bxy + Cy^2 + Dx + Ey + F = 0, \qquad (13\text{-}22)$$

where at least one of the coefficients A, B, and C is different from zero.

For example, if we start with the equation of the circle

$$(x-h)^2 + (y-k)^2 = r^2 \qquad (13\text{-}23)$$

and then expand and simplify (13-23), we get

$$x^2 - 2xh + h^2 + y^2 - 2yk + k^2 = r^2 \qquad (13\text{-}24)$$

or

$$x^2 + y^2 - 2xh - 2yk + h^2 + k^2 - r^2 = 0. \qquad (13\text{-}25)$$

Equation (13-25) is a special case of equation (13-22) with

$$A = 1,\ B = 0,\ C = 1,\ D = -2h,\ E = -2k,$$

$$F = h^2 + k^2 - r^2.$$

Similarly, the equation of parabola $(y-k)^2 = 4a(x-h)$ is a special case of equation (13-22) with $A = 0$, $B = 0$, $C = 1$, $D = -4a$, $E = -2k$, and $F = k^2 + 4ah$.

It is possible to tell directly from the equation

$$Ax^2 + Bxy + Cy^2 + Dx + Ey + F = 0 \qquad (13\text{-}22)$$

[16] It is interesting to note that in a given triangle, with base fixed in magnitude and position, and with fixed area, the locus of the points of intersection of the bisectors of the angles is an ellipse (FIG. 13G). The major axis of the ellipse is the base of the triangle. [We obtain a complete ellipse if $\triangle ABC$ (FIG. 13G) can appear on both sides of the base $\overline{AB}$.]

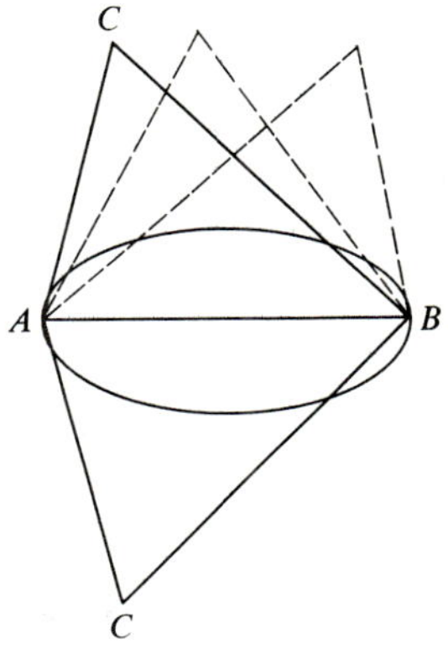

FIGURE 13G

The locus of the points of intersection of the altitudes is one branch of a hyperbola (FIG. 13H).

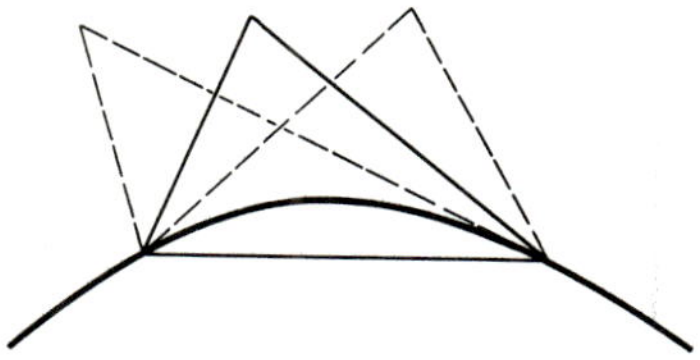

FIGURE 13H

For proofs of these theorems see A. Bakst, "Conic Sections Formed by Some Elements in a Plane Triangle," *The Mathematics Teacher*, **24** (1931), pp. 28–30.

whether it is an ellipse, parabola, or hyperbola according to whether $B^2 - 4AC$ is less than, equal to, or greater than zero. The expression $B^2 - 4AC$ is called the *discriminant* of equation (13-22).

It can be shown that

$$B^2 - 4AC < 0 \text{ for ellipses,}$$

$$B^2 - 4AC = 0 \text{ for parabolas,}$$

and

$$B^2 - 4AC > 0 \text{ for hyperbolas.}$$

It can likewise be shown that equation (13-22) always represents a circle, parabola, ellipse, or hyperbola except for four special cases where the curve degenerates into a point (for example, $x^2 + y^2 = 0$); a line (for example, $y^2 = 0$); a pair of lines, either parallel or intersecting (for example, $x^2 - y^2 = 0$); or fails to exist (for example, $x^2 + y^2 + 5 = 0$).

APPENDIX

TABLES FOR REFERENCE

- Linear Measure
- Square Measure
- Cubic Measure
- Time
- Counting
- Liquid Measure
- Dry Measure
- Avoirdupois Weight
- Practical Equivalents
- Metric Equivalents
- Squares and Square Roots
- Sines, Cosines, and Tangents

FORMULAS

COORDINATE GEOMETRY FORMULAS

REFERENCE MATERIALS

- Books
- Periodicals and Journals
- Films

DEFINITIONS

POSTULATES

THEOREMS

TABLES FOR REFERENCE

Linear Measure

12 inches (in.) = 1 foot (ft)
3 feet = 1 yard (yd)
$5\frac{1}{2}$ yards = 1 rod
$16\frac{1}{2}$ feet = 1 rod
320 rods = 1 mile
1760 yards = 1 mile
5280 feet = 1 mile
1 nautical mile = 1 knot = 1.15 miles

Square Measure

144 sq in. = 1 sq ft
9 sq ft = 1 sq yd
$30\frac{1}{4}$ sq yd = 1 sq rod
160 sq rod = 1 acre
640 acres = 1 sq mile

Cubic Measure

1728 cu in. = 1 cu ft
27 cu ft = 1 cu yd
128 cu ft = 1 cord of wood
1 cu yd = 1 load of dirt

Time

60 seconds = 1 minute
60 minutes = 1 hour
24 hours = 1 day
7 days = 1 week
52 weeks = 1 year
12 months = 1 year
366 days = 1 leap year
100 years = 1 century
$4\frac{1}{3}$ weeks = 1 month
360 days = 1 business year

Counting

12 units = 1 dozen
12 dozen = 1 gross
20 units = 1 score

Liquid Measure

2 pints = 1 quart
4 quarts = 1 gallon
$31\frac{1}{2}$ gallons = 1 barrel
1 gallon = 231 cu in.

Dry Measure

2 pints = 1 quart
8 quarts = 1 peck
4 pecks = 1 bushel
1 bushel = $1\frac{1}{4}$ cu ft

Avoirdupois Weight

16 ounces = 1 pound
100 pounds = 1 hundred weight (cwt)
20 cwt = 1 ton
2000 pounds = 1 ton

Practical Equivalents

1 bushel = $1\frac{1}{4}$ cu ft
1 gallon = 231 cu in.
1 cu ft = $7\frac{1}{2}$ gallons
1 cu ft water = $62\frac{1}{2}$ pounds
1 gallon water = $8\frac{1}{3}$ pounds

Metric Equivalents

1 centimeter = 0.39 inch
1 meter = 39.37 inches or 1.1 yards
1 kilometer = 0.62 mile
1 liter = 1.06 liquid quarts
1 gram = 0.035 ounce
1 kilogram = 2.2 pounds

TABLE OF SQUARES AND SQUARE ROOTS

No.	Square	Square Root	No.	Square	Square Root	No.	Square	Square Root
1	1	1.	**51**	2,601	7.14	**101**	10,201	10.05
2	4	1.41	**52**	2,704	7.21	**102**	10,404	10.10
3	9	1.73	**53**	2,809	7.28	**103**	10,609	10.15
4	16	2.	**54**	2,916	7.35	**104**	10,816	10.20
5	25	2.24	**55**	3,025	7.42	**105**	11,025	10.25
6	36	2.45	**56**	3,136	7.48	**106**	11,236	10.30
7	49	2.65	**57**	3,249	7.55	**107**	11,449	10.34
8	64	2.83	**58**	3,364	7.62	**108**	11,664	10.39
9	81	3.	**59**	3,481	7.68	**109**	11,881	10.44
10	100	3.16	**60**	3,600	7.75	**110**	12,100	10.49
11	121	3.32	**61**	3,721	7.81	**111**	12,321	10.54
12	144	3.46	**62**	3,844	7.87	**112**	12,544	10.58
13	169	3.61	**63**	3,969	7.94	**113**	12,769	10.63
14	196	3.74	**64**	4,096	8.	**114**	12,996	10.68
15	225	3.87	**65**	4,225	8.06	**115**	13,225	10.72
16	256	4.	**66**	4,356	8.12	**116**	13,456	10.77
17	289	4.12	**67**	4,489	8.19	**117**	13,689	10.82
18	324	4.24	**68**	4,624	8.25	**118**	13,924	10.86
19	361	4.36	**69**	4,761	8.31	**119**	14,161	10.91
20	400	4.47	**70**	4,900	8.37	**120**	14,400	10.95
21	441	4.58	**71**	5,041	8.43	**121**	14,641	11.
22	484	4.69	**72**	5,184	8.49	**122**	14,884	11.05
23	529	4.80	**73**	5,329	8.54	**123**	15,129	11.09
24	576	4.90	**74**	5,476	8.60	**124**	15,376	11.14
25	625	5.	**75**	5,625	8.66	**125**	15,625	11.18
26	676	5.10	**76**	5,776	8.72	**126**	15,876	11.22
27	729	5.20	**77**	5,929	8.77	**127**	16,129	11.27
28	784	5.29	**78**	6,084	8.83	**128**	16,384	11.31
29	841	5.39	**79**	6,241	8.89	**129**	16,641	11.36
30	900	5.48	**80**	6,400	8.94	**130**	16,900	11.40
31	961	5.57	**81**	6,561	9.	**131**	17,161	11.45
32	1,024	5.66	**82**	6,724	9.06	**132**	17,424	11.49
33	1,089	5.74	**83**	6,889	9.11	**133**	17,689	11.53
34	1,156	5.83	**84**	7,056	9.17	**134**	17,956	11.58
35	1,225	5.92	**85**	7,225	9.22	**135**	18,225	11.62
36	1,296	6.	**86**	7,396	9.27	**136**	18,496	11.66
37	1,369	6.08	**87**	7,569	9.33	**137**	18,769	11.70
38	1,444	6.16	**88**	7,744	9.38	**138**	19,044	11.75
39	1,521	6.24	**89**	7,921	9.43	**139**	19,321	11.79
40	1,600	6.32	**90**	8,100	9.49	**140**	19,600	11.83
41	1,681	6.40	**91**	8,281	9.54	**141**	19,881	11.87
42	1,764	6.48	**92**	8,464	9.59	**142**	20,164	11.92
43	1,849	6.56	**93**	8,649	9.64	**143**	20,449	11.96
44	1,936	6.63	**94**	8,836	9.70	**144**	20,736	12.
45	2,025	6.71	**95**	9,025	9.75	**145**	21,025	12.04
46	2,116	6.78	**96**	9,216	9.80	**146**	21,316	12.08
47	2,209	6.86	**97**	9,409	9.85	**147**	21,609	12.12
48	2,304	6.93	**98**	9,604	9.90	**148**	21,904	12.17
49	2,401	7.	**99**	9,801	9.95	**149**	22,201	12.21
50	2,500	7.07	**100**	10,000	10.	**150**	22,500	12.25

TABLE OF SINES, COSINES, AND TANGENTS

Angle	Sine	Cosine	Tangent	Angle	Sine	Cosine	Tangent
1°	.0175	.9998	.0175	46°	.7193	.6947	1.0355
2°	.0349	.9994	.0349	47°	.7314	.6820	1.0724
3°	.0523	.9986	.0524	48°	.7431	.6691	1.1106
4°	.0698	.9976	.0699	49°	.7547	.6561	1.1504
5°	.0872	.9962	.0875	50°	.7660	.6428	1.1918
6°	.1045	.9945	.1051	51°	.7771	.6293	1.2349
7°	.1219	.9925	.1228	52°	.7880	.6157	1.2799
8°	.1392	.9903	.1405	53°	.7986	.6018	1.3270
9°	.1564	.9877	.1584	54°	.8090	.5878	1.3764
10°	.1736	.9848	.1763	55°	.8192	.5736	1.4281
11°	.1908	.9816	.1944	56°	.8290	.5592	1.4826
12°	.2079	.9781	.2126	57°	.8387	.5446	1.5399
13°	.2250	.9744	.2309	58°	.8480	.5299	1.6003
14°	.2419	.9703	.2493	59°	.8572	.5150	1.6643
15°	.2588	.9659	.2679	60°	.8660	.5000	1.7321
16°	.2756	.9613	.2867	61°	.8746	.4848	1.8040
17°	.2924	.9563	.3057	62°	.8829	.4695	1.8807
18°	.3090	.9511	.3249	63°	.8910	.4540	1.9626
19°	.3256	.9455	.3443	64°	.8988	.4384	2.0503
20°	.3420	.9397	.3640	65°	.9063	.4226	2.1445
21°	.3584	.9336	.3839	66°	.9135	.4067	2.2460
22°	.3746	.9272	.4040	67°	.9205	.3907	2.3559
23°	.3907	.9205	.4245	68°	.9272	.3746	2.4751
24°	.4067	.9135	.4452	69°	.9336	.3584	2.6051
25°	.4226	.9063	.4663	70°	.9397	.3420	2.7475
26°	.4384	.8988	.4877	71°	.9455	.3256	2.9042
27°	.4540	.8910	.5095	72°	.9511	.3090	3.0777
28°	.4695	.8829	.5317	73°	.9563	.2924	3.2709
29°	.4848	.8746	.5543	74°	.9613	.2756	3.4874
30°	.5000	.8660	.5774	75°	.9659	.2588	3.7321
31°	.5150	.8572	.6009	76°	.9703	.2419	4.0108
32°	.5299	.8480	.6249	77°	.9744	.2250	4.3315
33°	.5446	.8387	.6494	78°	.9781	.2079	4.7046
34°	.5592	.8290	.6745	79°	.9816	.1908	5.1446
35°	.5736	.8192	.7002	80°	.9848	.1736	5.6713
36°	.5878	.8090	.7265	81°	.9877	.1564	6.3138
37°	.6018	.7986	.7536	82°	.9903	.1392	7.1154
38°	.6157	.7880	.7813	83°	.9925	.1219	8.1443
39°	.6293	.7771	.8098	84°	.9945	.1045	9.5144
40°	.6428	.7660	.8391	85°	.9962	.0872	11.4301
41°	.6561	.7547	.8693	86°	.9976	.0698	14.3007
42°	.6691	.7431	.9004	87°	.9986	.0523	19.0811
43°	.6820	.7314	.9325	88°	.9994	.0349	28.6363
44°	.6947	.7193	.9657	89°	.9998	.0175	57.2900
45°	.7071	.7071	1.0000	90°	1.0000	.0000	

FORMULAS

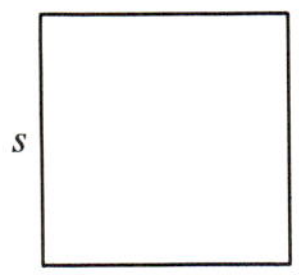

Square

$A = s^2$ (p. 295)

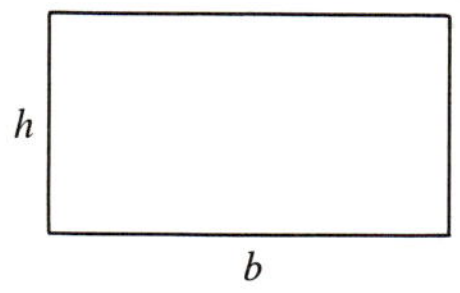

Rectangle

$A = bh$ (p. 296)

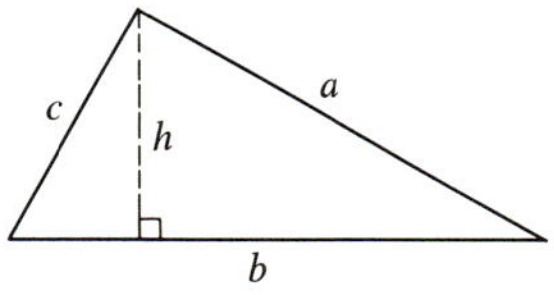

Triangle

$A = \frac{1}{2}bh$ (p. 297)

$s = \frac{1}{2}(a + b + c)$

$A = \sqrt{s(s - a)(s - b)(s - c)}$ (p. 314)

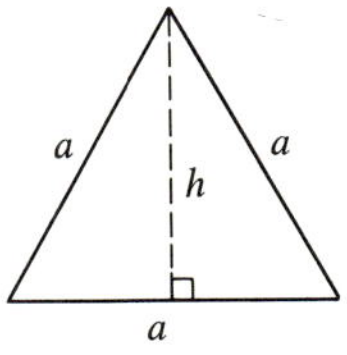

Equilateral Triangle

$$h = \frac{a}{2}\sqrt{3}$$

$$A = \frac{a^2}{4}\sqrt{3}$$

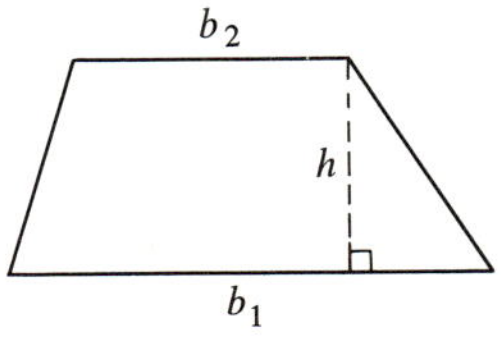

Trapezoid

$A = \frac{1}{2}h(b_1 + b_2)$ (p. 300)

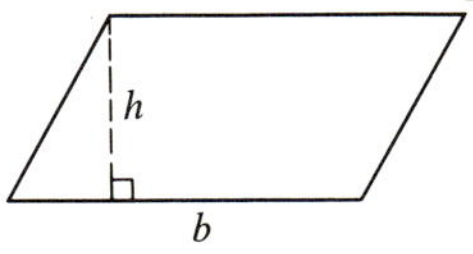

Parallelogram

$A = bh$ (p. 301)

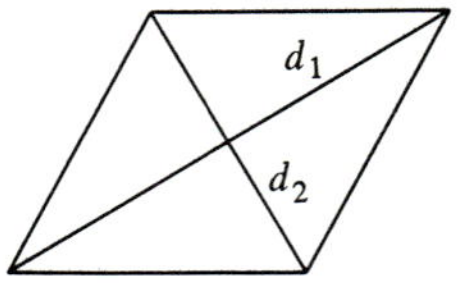

Rhombus

$A = \frac{1}{2}d_1d_2$ (p. 312 #20)

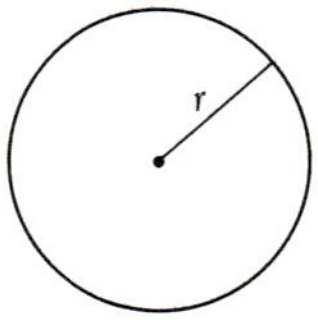

Circle

$C = 2\pi r = \pi d$ (p. 323)
$A = \pi r^2$

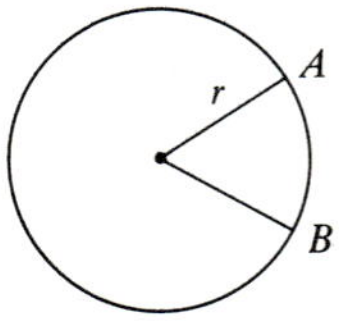

Arc

$L = \frac{q}{180} \cdot \pi r$ (p. 325)

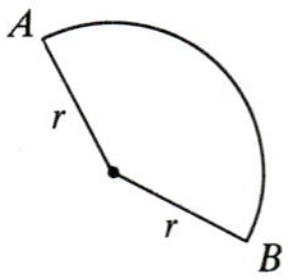

Sector

$A = \frac{q}{360} \cdot \pi r^2$ (p. 326)

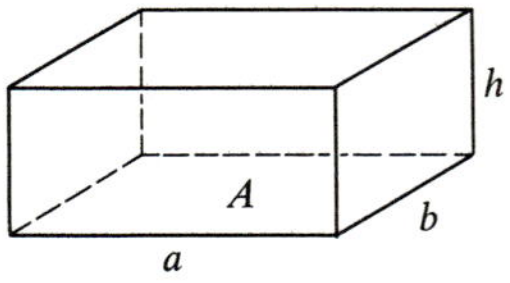

Rectangular Parallelepiped

$V = Ah = abh$ (p. 302)

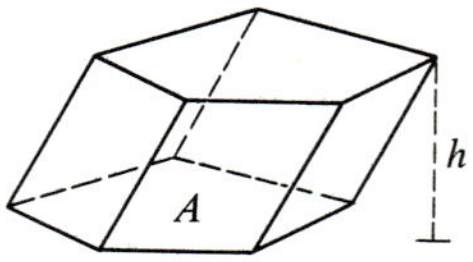

Prism

$V = Ah$ (p. 304)

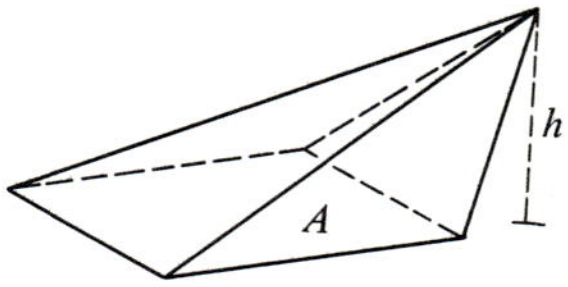

Pyramid

$V = \frac{1}{3}Ah$ (p. 305)

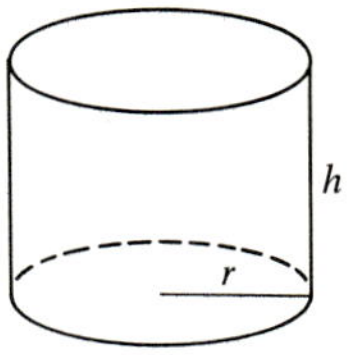

Right Circular Cylinder

$S = 2\pi rh$ (p. 326)

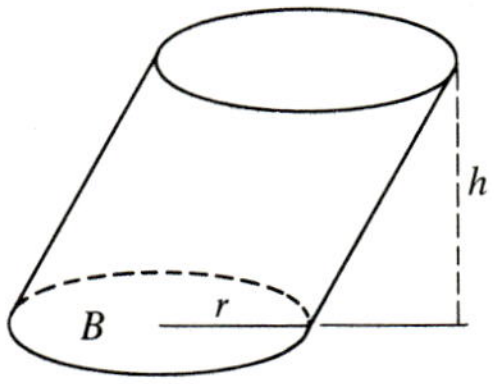

Circular Cylinder

$V = Bh = \pi r^2 h$ (p. 327)

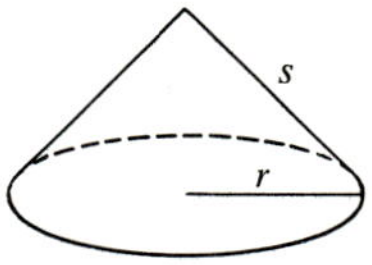

Right Circular Cone

$S = \pi rs$ (p. 327)

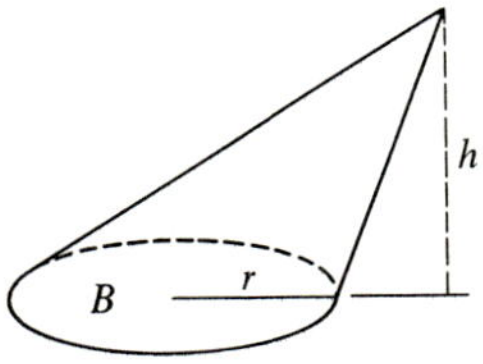

Circular Cone

$V = \frac{1}{3}Bh$

$= \frac{1}{3}\pi r^2 h$ (p. 327)

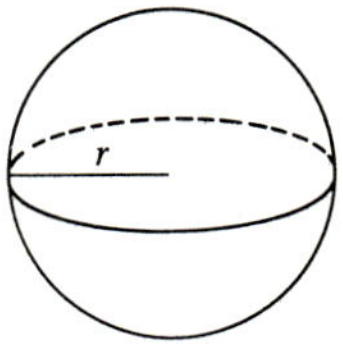

Sphere

$S = 4\pi r^2$

$V = \frac{4}{3}\pi r^3$ (p. 328)

COORDINATE GEOMETRY FORMULAS

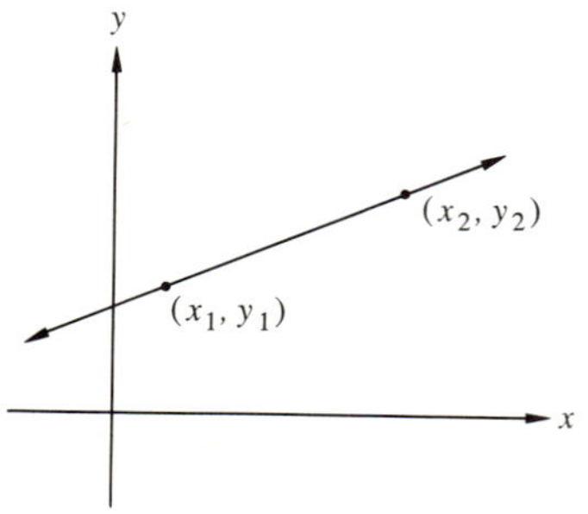

Slope of a Line

$$m = \frac{y_2 - y_1}{x_2 - x_1} \text{ (p. 348)}$$

Distance

$$d = \sqrt{(x_2 - x_1)^2 + (y_2 - y_1)^2} \text{ (p. 352)}$$

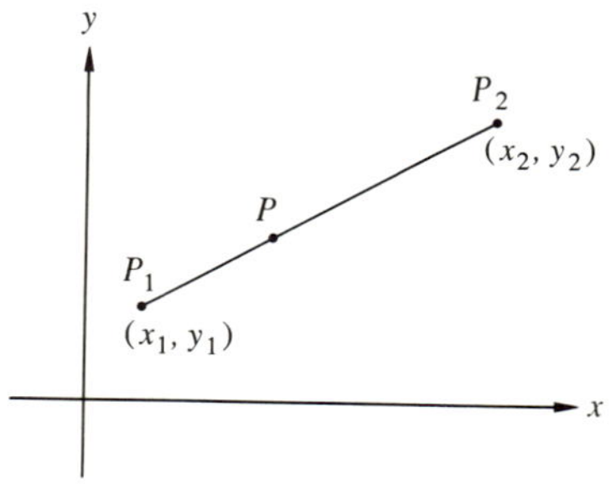

Midpoint

$$P = \left(\frac{x_1 + x_2}{2}, \frac{y_1 + y_2}{2}\right) \text{ (p. 353)}$$

If

$$\frac{P_1P}{PP_2} = r$$

where r is any positive real number, then

$$P = \left(\frac{x_1 + rx_2}{1 + r}, \frac{y_1 + ry_2}{1 + r}\right) \text{ (p. 354)}$$

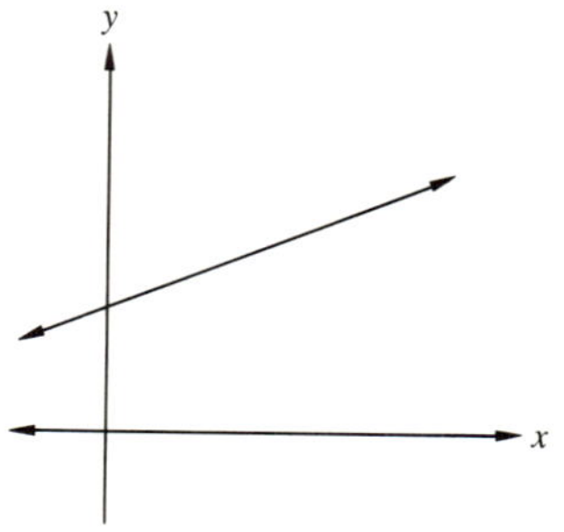

Straight Line

$x = a$ Parallel to y axis (p. 360)

$y = b$ Parallel to x axis (p. 361)

$y - y_1 = m(x - x_1)$ Point-Slope Form (p. 362)

$y = mx + b$ Slope-Intercept Form (p. 364)

$\frac{x}{a} + \frac{y}{b} = 1$ Intercept Form (p. 365)

$Ax + By + c = 0$ General Form (p. 365)

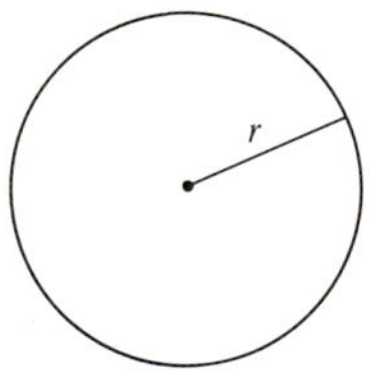

Circle

$x^2 + y^2 = r^2$ (p. 366)

$(x - h)^2 + (y - k)^2 = r^2$ (p. 367)

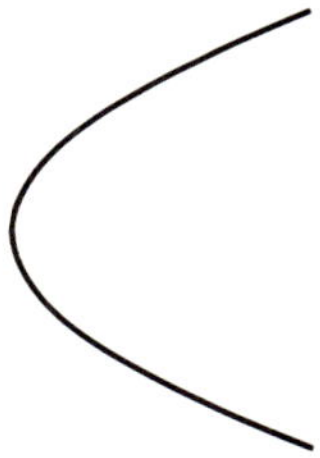

Parabola

$y^2 = 4ax$ (p. 377)

$$\left.\begin{array}{l} y^2 = -4ax \\ x^2 = 4ay \\ x^2 = -4ay \\ (y - k)^2 = 4a(x - h) \\ (x - h)^2 = 4a(y - k) \text{ (p. 378)} \end{array}\right\} \text{(p. 377)}$$

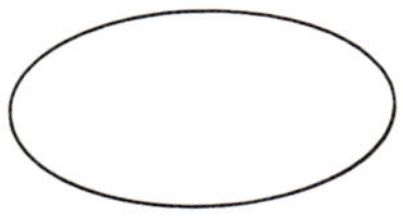

Ellipse

$\dfrac{x^2}{a^2} + \dfrac{y^2}{b^2} = 1$ (p. 378)

$$\left.\begin{array}{l} \dfrac{y^2}{a^2} + \dfrac{x^2}{b^2} = 1 \\ \dfrac{(x - h)^2}{a^2} + \dfrac{(y - k)^2}{b^2} = 1 \\ \dfrac{(y - k)^2}{a^2} + \dfrac{(x - h)^2}{b^2} = 1 \end{array}\right\} \text{(p. 379)}$$

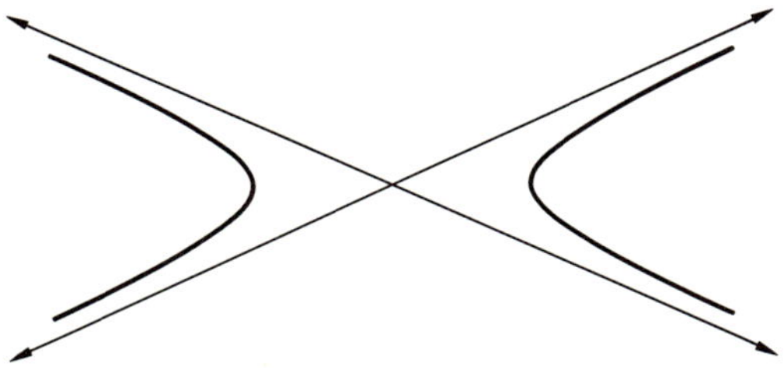

Hyperbola

$\dfrac{x^2}{a^2} - \dfrac{y^2}{b^2} = 1$ (p. 380)

$\dfrac{y^2}{a^2} - \dfrac{x^2}{b^2} = 1$ (p. 380)

$\dfrac{(x - h)^2}{a^2} - \dfrac{(y - k)^2}{b^2} = 1$ (p. 380)

$\dfrac{(y - k)^2}{a^2} - \dfrac{(x - h)^2}{b^2} = 1$ (p. 381)

Asymptotes of a Hyperbola

$y = \pm\frac{b}{a}x$ (p. 380)

Second-Degree Curves

$Ax^2 + Bxy + Cy^2 + Dx + Ey + F = 0$ General Form (p. 381)

REFERENCE MATERIALS

BOOKS

Abbott, E., *Flatland*. New York: Dover Publications, Inc., 1952.

Adler, C. F., *Modern Geometry*. New York: McGraw-Hill Book Company, Inc., 1958.

Adler, Irving, *The Adler Book of Puzzles and Riddles*. New York: The John Day Company, Inc., 1962.

Adler, Irving, *A New Look at Geometry*. New York: New American Library of World Literature, Inc., 1966.

Allendoerfer, C. B., and C. O. Oakley, *Principles of Mathematics*. New York: McGraw-Hill Book Company, Inc., 1963.

Altshiller-Court, Nathan, *College Geometry*. New York: Barnes & Noble, Inc., 1952.

Bakst, A., *Mathematical Puzzles and Pastimes*. New York: Van Nostrand Reinhold Company, 1954.

Bakst, A., *Mathematics: Its Magic and Mastery*. New York: Van Nostrand Reinhold Company, 1952.

Birkhoff, G. D., and Ralph Beatley, *Basic Geometry*. New York: Chelsea Publishing Company, 1959.

Bonola, Roberto, *Non-Euclidean Geometry*. New York: Dover Publications, Inc., 1955.

Breuer, J., *Introduction to the Theory of Sets*. Englewood Cliffs, N.J.: Prentice-Hall, Inc., 1958.

Courant, R., and H. Robbins, *What Is Mathematics*? New York: Oxford University Press, 1941.

Coxeter, H. S., *Introduction to Geometry*. New York: John Wiley & Sons, Inc., 1961.

Cundy, H. M., and A. P. Rollett, *Mathematical Models*. New York: Oxford University Press, 1961.

Eves, Howard, *An Introduction to the History of Mathematics*. New York: Holt, Rinehart and Winston, Inc., 1969.

Eves, Howard, *A Survey of Geometry*. Boston: Allyn and Bacon, Inc., 1963.

Fishback, W. T., *Projective and Euclidean Geometry*. New York: John Wiley & Sons, Inc., 1962.

Gamow, G., *One, Two, Three, Infinity*. New York: The Viking Press, Inc., 1947.

Harkin, D., *Fundamental Mathematics*. Englewood Cliffs, N.J.: Prentice-Hall, Inc., 1941.

Heath, Sir T. L., *The Thirteen Books of Euclid's Elements*. New York: Dover Publications, Inc., 1956.

Hess, A. L., *Mathematics Project Handbook*. Boston: D. C. Heath and Company, 1962.

Hilbert, David, *The Foundations of Geometry*. La Salle, Ill.: The Open Court Publishing Company, 1950.

Johnson, D. A., *Paper Folding for the Mathematics Class*. Washington, D.C.: National Council of Teachers of Mathematics, 1957.

Kasner, E., and J. Newman, *Mathematics and the Imagination.* New York: Simon and Schuster, Inc., 1940.

Kemeny, J. G., et al., *Introduction to Finite Mathematics.* Englewood Cliffs, N.J.: Prentice-Hall, Inc., 1966.

Kline, Morris, *Mathematics, A Cultural Approach.* Reading, Mass.: Addison-Wesley Publishing Company, Inc., 1962.

Kline, Morris, *Mathematics in Western Culture.* New York: Oxford University Press, 1953.

Kraitchik, M., *Mathematical Recreations.* New York: Dover Publications, Inc., 1942.

Lindgren, H., *Geometric Dissections.* New York: Van Nostrand Reinhold Company, 1964.

Moise, E., *Elementary Geometry from an Advanced Standpoint.* Reading, Mass.: Addison-Wesley Publishing Company, Inc., 1963.

National Council of Teachers of Mathematics, *Enrichment Mathematics for the Grades.* Twenty-seventh Yearbook. Washington, D.C.: National Council of Teachers of Mathematics, 1962.

National Council of Teachers of Mathematics, *Enrichment Mathematics for High School.* Twenty-eighth Yearbook. Washington, D.C.: National Council of Teachers of Mathematics, 1963.

National Council of Teachers of Mathematics, *The Growth of Mathematical Ideas.* Twenty-fourth Yearbook. Washington, D.C.: National Council of Teachers of Mathematics, 1959.

National Council of Teachers of Mathematics, *Historical Topics for the Mathematics Classroom.* Thirty-first Yearbook. Washington, D.C.: National Council of Teachers of Mathematics, 1969.

National Council of Teachers of Mathematics, *Insights into Modern Mathematics.* Twenty-third Yearbook. Washington, D.C.: National Council of Teachers of Mathematics, 1957.

National Council of Teachers of Mathematics, *Multi-Sensory Aids in the Teaching of Mathematics.* Eighteenth Yearbook. Washington, D.C.: National Council of Teachers of Mathematics, 1958.

Newman, James, *The World of Mathematics.* New York: Simon and Schuster, Inc., 1956.

Norton, M. S., *Geometric Constructions.* New York: McGraw-Hill, Inc. (Webster Division), 1963.

Polya, George, *How To Solve It.* Garden City, New York: Doubleday & Company, Inc., 1957.

Polya, George, *Mathematical Discovery.* New York: John Wiley & Sons, Inc., 1962.

Prenowitz, W., and M. Jordan, *Basic Concepts of Geometry.* Waltham, Mass.: Blaisdell Publishing Company, 1965.

Ravielli, A., *An Adventure in Geometry.* New York: The Viking Press, Inc., 1957.

Reid, Constance, *A Long Way from Euclid.* New York: Thomas Y. Crowell Company, 1963.

Row, T. S., *Geometric Exercises in Paper Folding.* New York: Dover Publications, Inc., 1966.

Sawyer, W. W., *Prelude to Mathematics.* Baltimore: Penguin Books, Inc., 1955.
School Mathematics Study Group, *Mathematics for High School, Geometry.* New Haven: Yale University Press, 1961.
Steinhaus, Hugo, *Mathematical Snapshots.* New York: Oxford University Press, 1960.
Taylor, E. H., and G. C. Bartoo, *An Introduction to College Geometry.* New York: The Macmillan Company, 1949.
Wilder, R. L., *Introduction to the Foundations of Mathematics.* New York: John Wiley & Sons, Inc., 1952.
Wolfe, H. E., *Introduction to Non-Euclidean Geometry.* New York: Holt, Rinehart and Winston, Inc., 1948.
Wylie, C. R., *Foundations of Geometry.* New York: McGraw-Hill Book Company, Inc., 1964.

PERIODICALS AND JOURNALS

American Mathematical Monthly. Mathematical Association of America, 1225 Connecticut Avenue, N.W., Washington, D.C. 20036.
The Arithmetic Teacher. Washington, D.C.: National Council of Teachers of Mathematics, 1201 Sixteenth Street, N.W., Washington, D.C. 20036.
Australian Mathematics Teacher. Teacher's College, Newton, Sidney, Australia.
The Fibonacci Quarterly. St. Mary's College, California 94575.
International Study Group for Mathematics Learning Bulletin. Z. P. Dienes, University of Sherbrooke, Sherbrooke, Quebec, Canada.
Mathematical Gazette. Mathematical Association, Gorden House, 29 Gorden Square, London, W.C.1., England.
The Mathematics Student Journal. Washington, D.C.: National Council of Teachers of Mathematics, 1201 Sixteenth Street, N.W., Washington, D.C. 20036.
The Mathematics Teacher. Washington, D.C.: National Council of Teachers of Mathematics, 1201 Sixteenth Street, N.W., Washington, D.C. 20036.
Mathematics Teaching. Association of Teachers of Mathematics, Vine Street Chambers, Nelson, Lancashire, England.
School Science and Mathematics. Menasha, Wis.: The Central Association of Science and Mathematics Teachers, Inc., P.O. Box 108, Bluffton, Ohio 45817.
Scientific American. 415 Madison Avenue, New York, New York 10017.
Scripta Mathematica. Yeshiva University, Amsterdam Avenue and 186 Street, New York, New York 10033.

FILMS

Area Under a Curve (14 minutes)
(Modern Learning Aids)
Conic Sections (11 minutes)
(Bailey-Film Associates)

Developing the General Equation of the Circle (12 minutes)
(Bailey-Film Associates)
Discovering Solids Series (5 films, Cenco Educational Films)
Solids in the World Around Us (5 minutes)
Volumes of Cubes, Prisms, Cylinders (5 minutes)
Volumes of Pyramids, Cones, and Cylinders (15 minutes)
Surface Areas of Solids, Parts I and II (15 minutes each part)
Donald in Mathmagic Land (26 minutes)
(Walt Disney Productions)
The Dynamics of a Circle (14 minutes)
(McGraw-Hill)
Geometry: Some Quadrilaterals (12 minutes)
(Bailey-Film Associates)
Graphs of Linear Equations (9 minutes)
(Bailey-Film Associates)
Intersection of Sets (15 minutes)
(McGraw-Hill)
Let Us Teach Guessing (Two 30-minute reels; a demonstration with George Polya)
(Modern Learning Aids)
Locus (13 minutes)
(McGraw-Hill)
Man Is the Measure (23 minutes)
(Ford Motor)
Mathematics of the Honeycomb (13 minutes)
(Moody Institute of Science)
Meaning of Area (12 minutes)
(McGraw-Hill)
The Meaning of Pi (12 minutes)
(Coronet Instructional Films)
Notes on a Triangle (5 minutes)
(International Film Bureau)
Possibly So, Pythagoras (14 minutes)
(International Film Bureau)
Prisms: Some Properties (12 minutes)
(Bailey-Film Associates)
Pythagorean Theorem: Proof by Area ($5\frac{1}{2}$ minutes)
(Coronet Instructional Films)
Seven Bridges of Koenigsberg (4 minutes)
(International Film Bureau)
Similar Triangles in Use (11 minutes)
(International Film Bureau)
Stretching Imagination (Discusses notions of topology; 30 minutes)
(Association Films)
Topology (30 minutes)
(Modern Learning Aids)

Volume by Shells (10 minutes)
(Modern Learning Aids)
Volume of a Solid of Revolution (9 minutes)
(Modern Learning Aids)
What Is Area? (20 minutes)
(Modern Learning Aids)
What's the Angle? (30 minutes)
(Association Films)

Association Films, Inc., 347 Madison Avenue, New York, New York 10017
Bailey-Film Associates, 11559 Santa Monica Boulevard, Los Angeles, California 90025
Cenco Educational Films, 1700 Irving Park Road, Chicago, Illinois 60613
Coronet Instructional Films, 65 East South Water Street, Chicago, Illinois 60601
Ford Motor Company, Film Library, Dearborn, Michigan 48121
International Film Bureau, 322 South Michigan Avenue, Chicago, Illinois 60604
McGraw-Hill, Inc., Text-Film Division, 330 West 42 Street, New York, New York 10036
Modern Learning Aids, 3 East 54 Street, New York, New York 10022
Moody Institute of Science, 11428 Santa Monica Boulevard, Los Angeles, California 90025
Walt Disney Productions, Educational Films Division, Burbank, California 91500

DEFINITIONS

ANGLE

An *angle* is the union of two distinct rays that have the same endpoint but do not lie on the same line. (p. 77)

The *interior of* $\angle ABC$ is the intersection of the two half-planes $\overleftrightarrow{AB}/C$ and $\overleftrightarrow{BC}/A$. (p. 78)

The *exterior of* $\angle ABC$ is the set of all points that are in the plane of the angle but neither on the angle nor in the interior of the angle. (p. 78)

A *right angle* is an angle whose measure is 90. (p. 83)

An *acute angle* is any angle whose measure is less than 90. (p. 83)

An *obtuse angle* is any angle whose measure is greater than 90 and less than 180. (p. 83)

If the sum of the measures of any two angles is 90, then the angles are *complementary*, and each angle is said to be the complement of the other. (p. 84)

If $\overrightarrow{AB}$ and $\overrightarrow{AC}$ are opposite rays and $\overrightarrow{AD}$ is any third ray, then $\angle DAB$ and $\angle DAC$ form a *linear pair*. (p. 84)

If the sum of the measures of two angles is 180, then the angles are called *supplementary*, and each angle is said to be the supplement of the other. (p. 85)

Two angles in a plane are said to be *adjacent* if they have a side in common and their interiors do not intersect. (p. 85)

Two angles are *vertical angles* if their union is the union of two intersecting lines. (p. 85)

ARCS AND INSCRIBED ANGLES

A *central angle* of a given circle is an angle whose vertex is the center of the circle. (p. 168)

In the central angle $\angle APB$, the set of points consisting of A and B together with all points of the circle that are in the interior of $\angle APB$ is called a *minor arc* of the circle. The set consisting of A, B, together with all points of the circle that are in the exterior of $\angle APB$, is called a *major arc* of the circle. (p. 168)

If A and B are the endpoints of a diameter, then the set of points consisting of A and B

together with all points of the circle that lie on a given side of the line is called a *semicircle.* (p. 169)

An *arc* is any minor arc, major arc, or semicircle. (p. 169)

An angle *intercepts* an arc if (a) the endpoints of the arc lie on different rays of the angle, and (b) except for its endpoints, the arc lies in the interior of the angle. (p. 170)

An angle is *inscribed* in an arc if the intersection of the angle and the arc consists of three points: the vertex of the angle and the two endpoints of the arc. (p. 171)

BETWEEN

B is *between* A and C if (1) A, B, and C are distinct and collinear, and (2) $AB + BC = AC$. (p. 35)

CIRCLE

A *circle* is the set of all points in a plane at a given distance from a given point in the plane. (p. 164)

A *chord* is a segment having both endpoints on a circle. (p. 165)

If a chord contains the center of a circle, it is called a *diameter* of the circle. (p. 165)

A *tangent* to a circle is a line in the plane of the circle which intersects the circle in only one point. If the line intersects the circle at more than one point, it is called a *secant* line. (p. 165)

Circles that lie in the same plane and have the same center are called *concentric circles.* (p. 165)

The *interior of a circle* is the set of all points of the plane whose distance from the center is less than the radius. (p. 166)

The *exterior of a circle* is the set of all points of the plane whose distance from the center is greater than the radius. (p. 166)

The union of a circle and its interior is a *closed circular region.* (p. 166)

Two circles are *congruent* if and only if they have congruent radii. (p. 166)

COMPLEMENT

If $A \subseteq U$, then the elements of U that are *not* in A form another subset of U called the *complement of A.* (p. 6)

CONE

If C is a curve lying in a plane P and V is a point not in P, then C and V determine a *conical surface* that is the union of all lines each of which passes through V and a point of C. If C is a closed curve, the surface is called a *closed conical surface*. Any of the lines through the vertex and contained in the surface is called an *element* of the surface. (p. 178)

The vertex separates the conical surface into two parts called the *nappes* of the conical surface. (p. 178)

A *cone* is a solid bounded by a closed conical surface and a plane that intersects all the elements of the surface. (p. 178) If the curve C in the plane is a circle, the cone is called a *circular cone*. (p. 179)

The surface of a cone consists of two parts: (a) The *lateral surface*, which consists of those points of the cone which lie in the elements of the conical surface, and (b) the base. (p. 179)

The perpendicular segment from the vertex to the plane of the base is called the *altitude* of the cone, and its length is called the *height* of the cone. (p. 179)

If the segment from the vertex to the center of the circle C is perpendicular to the plane of C, we call the cone a *right circular cone*. (p. 179)

The segment that joins the vertex to any point on the base circle of a right circular cone is called the *slant height* of the cone. (p. 180)

If we cut a circular cone by a plane parallel to the base, we get a circle. If we cut a circular cone by other planes, we get curves known as *conic sections*. (p. 180)

CONGRUENCE

Two *segments are congruent* if they have the same length. (p. 38)

Two *angles are congruent* if they have the same measure. (p. 85)

Given $\triangle ABC$, $\triangle DEF$, and a one-to-one correspondence $ABC \leftrightarrow DEF$ between their vertices, if every pair of corresponding sides is congruent and every pair of corresponding angles is congruent, then the correspondence is a congruence and we say that *$\triangle ABC$ and $\triangle DEF$ are congruent*. (p. 108)

Two *sets of points A and B are congruent* if and only if there exists a one-to-one correspondence between their points such that the segment determined by any two points of one set is congruent to the segment determined by the corresponding points of the other set. (p. 115)

CONVEX SET

A set of points S is *convex* if for every two distinct points A and B of S, the entire segment $\overline{AB}$ is in S. (p. 39)

COORDINATE

The x *coordinate* of a point P is the coordinate of the foot of the perpendicular from P to the x axis. (p. 346)

The y *coordinate* of a point P is the coordinate of the foot of the perpendicular from P to the y axis. (p. 346)

CYLINDER

If C is a curve lying in a plane P and if l is a line not in P but intersecting P in a point of C, then the union of l and all lines parallel to l passing through C is a *cylindrical surface*. (p. 176)

If the curve C is a simple closed curve such as a circle or a rectangle, the cylindrical surface is called a *closed cylindrical surface*. Each of the lines parallel to l and passing through a point in C is called an *element* of the cylindrical surface. (p. 177)

A *cylinder* is a solid bounded by a closed cylindrical surface and two parallel planes which intersect all the elements of the cylindrical surface. That portion of the surface which lies in one of the parallel planes is called a *base* of the cylinder. That portion of the surface which lies in the closed cylindrical surface is called the *lateral surface* of the cylinder. (p. 177)

An *altitude* of a cylinder is any segment that has endpoints in the planes of the bases and is perpendicular to them. (p. 177)

A *circular cylinder* is one whose bases are circles. (p. 177)

If the line segment joining the centers of the bases of a circular cylinder is perpendicular to the bases, then the cylinder is a *right circular cylinder*. (pp. 177–178)

DISTANCE

The *distance between a line and a point not on the line* is the length of the perpendicular segment from the point to the line. The *distance between a line and a point on the line* is defined to be zero. (p. 91)

ELLIPSE

An *ellipse* is the set of all points in a plane such that for each point the *sum* of its distances from two fixed points F and F' in the plane is constant. F and F' are called the *foci* of the ellipse. (p. 189)

FIRST-DEGREE EQUATION

An equation of the form $Ax + By + C = 0$ is called a *first-degree equation* in x and y if A, B, and C are real numbers and A and B are not both zero. (p. 365)

HYPERBOLA

A *hyperbola* is the set of all points in a plane such that for each point the *difference* of its distances from two fixed points F and F' is constant. F and F' are called the *foci* of the hyperbola. (p. 189)

INTERCEPT

The point where a line crosses the y axis is called the *y intercept*. (p. 264)

The point where a line crosses the x axis is called the *x intercept*. (p. 264)

LENGTH

The *length* of a segment is the distance between its endpoints. (p. 38)

LOCUS

A set of points S is the *locus* of a condition if and only if both of the following statements are true: (1) Every point of S satisfies the condition, and (2) every point that satisfies the condition belongs to S. (p. 367)

MIDPOINT

Point B is called the *midpoint* of $\overline{AC}$ if (a) A-B-C, and (b) $AB = BC$. (p. 37)

PARABOLA

A *parabola* is the set of points in a plane such that every point is equally distant from a fixed point and a fixed line. The fixed point is called the *focus*, and the fixed line is called the *directrix*. (p. 189)

PARALLEL

Two *lines are parallel* if they are contained in the same plane and do not intersect. (p. 34)

Two *planes are parallel* if they do not intersect. (p. 34)

PERPENDICULAR

Two lines are *perpendicular* to each other if their intersection forms a right angle. (p. 83)

POLYGON

Let $A_1, A_2, A_3, \ldots, A_n$ be a sequence of n distinct coplanar points, with $n \geq 3$. Let the n segments $\overline{A_1A_2}, \overline{A_2A_3}, \ldots, \overline{A_nA_1}$ have the following properties: (1) No two segments intersect, except for common endpoints, and (2) no two segments with a common endpoint are collinear. Then the union of the n segments is a *polygon*. (p. 158)

A *diagonal* of a polygon is a segment joining two vertices that are not endpoints of the same side of the polygon. (p. 159)

A polygon is called a *regular polygon* if all its sides are congruent and all its angles are congruent. (p. 159)

A polygon is a *convex polygon* if and only if each side lies on a line determining a half-plane that contains the remainder of the polygon. (p. 159)

The intersection of the interiors of all the angles of a convex polygon is the *interior of the polygon*. (p. 159)

The union of a polygon and its interior is called a *closed polygonal region*. (p. 159)

POLYHEDRON

A *polyhedron* is a solid whose surface consists entirely of polygonal regions. (p. 172)

PRISM

A *prism* is a polyhedron of which two faces, called *bases*, are congruent polygons in parallel planes, and the other faces, called *lateral faces*, are parallelograms. Each edge determined by the intersection of two lateral faces is called a *lateral edge*. (p. 173)

An *altitude* of the prism is any segment that has endpoints in the two bases and is perpendicular to them. We call the length of any altitude the *height* of the prism. (p. 173)

If the lateral edges of a prism are perpendicular to its bases, the prism is called a *right prism*. (p. 173)

A *parallelepiped* is a prism whose faces are parallelograms. If each of the faces is a rectangle, then the prism is a *rectangular parallelepiped* or a *rectangular solid* or box. (p. 174)

PROPORTIONAL

Given two sequences of numbers $a, b, c, \ldots, a', b', c', \ldots$, if $a/a' = b/b' = c/c' = \cdots$, then we say that the two sequences are *proportional*. When a proportionality involves only four numbers, such as $a/b = c/d$, we refer to such a proportionality as a *proportion*. (p. 219)

If a, b, and c are positive numbers and $a/b = b/c$, then b is called the *geometric mean* between a and c. Another name for geometric mean is *mean proportional*. (p. 219)

PYRAMID

A *pyramid* is a polyhedron of which one face is called the *base*, and the other faces, called *lateral faces*, are triangles having a common vertex called the *apex*. (p. 175)

If the base of a pyramid is a regular polygon, the line through the vertex of the pyramid and the center of the base is the *axis* of the pyramid. If the base of a pyramid is a regular polygon and if its axis is perpendicular to its base, then the pyramid is called a *regular pyramid*. (p. 175)

If a plane parallel to the base of a pyramid intersects all its lateral edges, then the polyhedron formed between the parallel plane and the base is called a *frustum*. (p. 175)

QUADRILATERAL

A polygon with four sides is called a *quadrilateral*. (p. 158)

A quadrilateral that has at least one pair of parallel sides is called a *trapezoid*. (p. 163)

A quadrilateral that has two pairs of parallel sides is called a *parallelogram.* (p. 163)

A *rhombus* is a parallelogram all of whose sides are congruent. (p. 164)

A parallelogram containing at least one right angle is called a *rectangle.* (p. 164)

A rhombus with at least one right angle is called a *square.* (p. 164)

RAY

Ray AB, denoted by $\overrightarrow{AB}$, where A and B are distinct points, is the union of two sets of points: (a) the line segment $\overline{AB}$, and (2) all the points C such that point B is between points A and C. (p. 37)

SEGMENT

If A and B are two distinct points, then the set consisting of A, B, and all points between A and B is called *segment AB* and is denoted by $\overline{AB}$. (p. 37)

SEPARATION

Point P *separates* line l into the nonempty sets L_1 and L_2 if the following conditions are satisfied: (1) $l = L_1 \cup L_2 \cup P$, where L_1, L_2, and P are disjoint. (2) P is between each point of L_1 and each point of L_2. (3) P is not between two points of L_1 or of L_2. (p. 41)

SETS

If every element of set A is also an element of set B, then A is a *subset* of B. (p. 5)

If A is a subset of B but B is not a subset of A, then we say that A is a *proper subset* of B. (p. 5)

If $A \subseteq U$, then the elements of U that are not in A form another subset of U called the *complement of A*. (p. 6)

Two sets are *equal* if each is a subset of the other. (p. 7)

Two sets are *equivalent* if there exists a one-to-one correspondence between their elements. (p. 7)

The *union* of two sets is the set of all elements belonging to either of the two sets or to both of them. (p. 8)

The *intersection* of two sets is the set of elements belonging to both sets. (p. 10)

If A and B are sets, then the set $A \times B$, called the *Cartesian product of A and B*, is the set of all possible ordered pairs such that the first component of an ordered pair is an element of A and the second component is an element of B. (p. 12)

SIMILARITY

Given a correspondence between two triangles, if corresponding angles are congruent and corresponding sides are proportional, then the correspondence is called a *similarity*, and the triangles are said to be similar. (p. 219)

Two *figures F and F′ are similar* if a one-to-one correspondence exists between the points of F and those of F' such that for any pair of points A, B of F and the corresponding pair of points A', B' of F', we have $AB/A'B'$ = a constant k. (p. 224)

SKEW LINES

Two lines are said to be *skew lines* if and only if they are not in the same plane. (p. 34)

SLOPE

If $P_1 = (x_1, y_1)$ and $P_2 = (x_2, y_2)$ and $\overline{P_1P_2}$ is nonvertical, then the *slope* of $\overline{P_1P_2}$ is the number

$$m = \frac{y_2 - y_1}{x_2 - x_1}.$$

(p. 348)

The *slope of a line* is the number that is the slope of every segment of the line. (p. 349)

SPHERE

A *sphere* is the set of all points at a given distance from a given point in space. The given distance is called the *radius* of the sphere; the given point is called the *center* of the sphere. If A is any point of the sphere and P is the center of the sphere, then the segment $\overline{PA}$ also is called a *radius* of the sphere. (p. 180)

Any segment through the center and having endpoints in the sphere is a *diameter* of the sphere. The two endpoints of the diameter are called *antipodal points* of the sphere. (p. 180)

A *great circle* on a sphere is any intersection of the sphere with a plane through the center of the sphere. All circles on a sphere which are not great circles are called *small circles*. (p. 180)

A plane is *tangent* to a sphere if it intersects the sphere in exactly one point. (p. 181)

A point is in the *interior of a sphere* if its distance from the center of the sphere is less than the radius of the sphere. (p. 181)

A point is in the *exterior of a sphere* if its distance from the center of the sphere is greater than the radius of the sphere. (p. 181)

TRANSVERSAL

A line that intersects two other lines in exactly two points is called a *transversal* of the other two lines. (p. 132)

TRIANGLE

If A, B, and C are three noncollinear points, then the set $\overline{AB} \cup \overline{BC} \cup \overline{AC}$ is called a *triangle*. (p. 86)

The *interior of a triangle ABC* is the intersection of the half-planes $\overleftrightarrow{AB}/C$, $\overleftrightarrow{BC}/A$, and $\overleftrightarrow{AC}/B$. (p. 88)

The *exterior of a triangle ABC* is the set of all points in the plane of the triangle that are not on the triangle and not in the interior of the triangle. (p. 88)

An *exterior angle of a triangle* is an angle that forms a linear pair with one of the angles of the triangle. (p. 89)

The two angles of a triangle not adjacent to an exterior angle are called the *remote interior angles*. (p. 89)

A *right triangle* is a triangle that has one right angle. The side opposite the right angle is called the *hypotenuse*. The sides that include the right angle are called the *legs*. (p. 199)

POSTULATES

Postulate 2-1	All lines and planes are sets of points. (p. 30)
Postulate 2-2	Given any two different points, there is exactly one line containing them. (p. 30)
Postulate 2-3	Three different noncollinear points are contained in exactly one plane. (p. 31)
Postulate 2-4	If a plane contains two distinct points, then the line containing them lies in the plane. (p. 31)
Postulate 2-5	If two distinct planes intersect, then their intersection is a line. (p. 31)
Postulate 2-6	Every line contains at least two distinct points. Every plane contains at least three noncollinear points. Space contains at least four noncoplanar points. (p. 31)
Postulate 2-7	(*Distance Postulate*) To each pair of distinct points there corresponds a unique positive number called the *distance* between the two points. (p. 35)
Postulate 2-8	(*Ruler Postulate*) The points of a line can be placed in correspondence with the real numbers in such a way that 1. To every point on the line there corresponds exactly one real number. 2. To every real number there corresponds exactly one point of the line. 3. The distance between two points equals the absolute value of the difference between the corresponding numbers. (p. 35)
Postulate 2-9	(*Plane-Separation Postulate*) Given a line l and a plane X containing it, the set of all points of the plane that do not lie on l is the union of two disjoint sets. (p. 41)
Postulate 2-10	(*Space-Separation Postulate*) Given a plane in space, the set of all points that do not lie in the plane is the union of two disjoint sets S_1 and S_2. (p. 43)
Postulate 3-1	(*Angle Measurement Postulate*) For every angle there is one and only one real number n, called its *measure*. When the angle is measured in degrees, n is between 0 and 180; that is, $0 < n < 180$. (p. 83)
Postulate 3-2	(*Angle Construction Postulate*) Let $\overrightarrow{AB}$ be a ray on the edge of the half-plane H. For every number n such that $0 < n < 180$, there is one and only one ray $\overrightarrow{AP}$ with P in H such that $m\angle PAB = n$. (p. 84)

Postulate 3-3 (*Angle Addition Postulate*) If D is in the interior of $\angle ABC$, then $m\angle CBA = m\angle CBD + m\angle DBA$. (p. 84)

Postulate 3-4 If two angles form a linear pair, then they are supplementary. (p. 85)

Postulate 4-1 (*Side Angle Side Postulate*) Given a correspondence between two triangles (or between a triangle and itself), if two sides and the included angle of the first triangle are congruent to the corresponding parts of the second triangle, then the correspondence is a congruence. (p. 111)

Postulate 5-1 (*Parallel Postulate*) Through any point not on a given line, there is no more than one line parallel to the given line. (p. 132)

Postulate 9-1 Two nonparallel lines lying in a plane intersect in a point. (p. 238)

Postulate 9-2 A line and a circle intersect in two points if the line contains a point inside the circle. (p. 238)

Postulate 9-3 Let C and C' be circles of radius a and b, and let c be the distance between their centers. If each of the numbers a, b, and c is less than the sum of the other two, then C and C' intersect in two places. (p. 239)

Postulate 10-1 (*Comparison Postulate*) We assume that two continuous quantities of the same kind may be *compared* to determine whether they have the same size, or which one is the larger. (p. 275)

Postulate 10-2 (*Motion Postulate*) We assume that an object may be *moved* from one position to another without changing its size or shape. (p. 275)

Postulate 10-3 (*Matching Postulate*) We assume that two continuous quantities have the same size if every part of one can be *matched* to a part of the second in such a way that no part of either is left out. (p. 275)

Postulate 10-4 (*Subdivision Postulate*) We assume that a quantity may be *subdivided* without producing a change in the size of the whole. (p. 275)

Postulate 10-5 (*Archimedean Postulate*) If $\overline{AB}$ and $\overline{XY}$ are any segments, and on $\overline{AB}$ we choose points $P_1, P_2, P_3, P_4, \ldots$ such that $\overline{AP_1}, \overline{P_1P_2}, \overline{P_2P_3}, \overline{P_3P_4}, \ldots$ are all congruent to $\overline{XY}$, then eventually we shall find points P_n and P_{n+1} for which $B = P_n$ or B is between P_n and P_{n+1}. (p. 289)

Postulate 11-1 (*Area Postulate*) To each polygonal region there corresponds a unique positive number which is called the *area* of the region. (p. 295)

Postulate 11-2 (*Congruence Postulate for Area*) If two triangles are congruent, then their triangular regions have the same area. (p. 295)

Postulate 11-3 (*Area Addition Postulate*) If a polygonal region R is the union of two non-overlapping polygonal regions R_1 and R_2, then the area of $R =$ the area of $R_1 +$ the area of R_2. (p. 295)

Postulate 11-4 (*Unit Area Postulate*) The area of a square region is the square of the length of its side. (p. 295)

Postulate 11-5 (*Unit Volume Postulate*) The volume of a rectangular parallelepiped is the product of the altitude and the area of its base. (p. 302)

Postulate 11-6 (*Cavalieri's Principle*) Given two solids and a plane, suppose that every plane parallel to the given plane, intersecting one of the two solids, also intersects the other in such a way that the two cross sections have equal areas. Then the two solids have the same volume. (p. 303)

THEOREMS

Theorem 2-1	If two distinct lines intersect, then their intersection contains exactly one point. (p. 32)
Theorem 2-2	If a line intersects a plane not containing it, then the intersection is a single point. (p. 32)
Theorem 2-3	If *l* is a line and *P* is a point not on *l*, then there is one and only one plane containing *l* and *P*. (p. 32)
Theorem 2-4	If two distinct lines intersect, then there is exactly one plane containing them. (p. 33)
Theorem 2-5	If a plane intersects two parallel planes, then it intersects them in two parallel lines. (p. 34)
Theorem 2-6	If *A*-*B*-*C*, then *C*-*B*-*A*. (That is, if *B* is between *A* and *C*, then it is also between *C* and *A*.) (p. 36)
Theorem 2-7	Of any three distinct points on a line, exactly one is between the other two. (That is, if *A*-*B*-*C*, then neither *B*-*A*-*C* nor *A*-*C*-*B* is true.) (p. 36)
Theorem 2-8	Given four distinct collinear points, it is possible to name them *A*, *B*, *C*, and *D* in such a way that *A*-*B*-*C*, *A*-*B*-*D*, *A*-*C*-*D*, and *B*-*C*-*D*. (p. 36)
Theorem 2-9	If *A* and *B* are any two distinct points, then there is a point *C* such that *A*-*B*-*C*, and there is a point *D* such that *A*-*D*-*B*. (p. 37)
Theorem 2-10	If *A*-*B*-*C*, then *A*, *B*, and *C* are three different points of the same line. (p. 37)
Theorem 2-11	Congruence of segments is an equivalence relation. (p. 39)
Theorem 2-12	The intersection of any two convex sets is also a convex set. (p. 40)
Theorem 3-1	Every angle is contained in a unique plane. (p. 79)
Theorem 3-2	For any two points of an angle not on the same ray of the angle, every point between these two points is in the interior of the angle. (p. 79)
Theorem 3-3	Two distinct intersecting lines form exactly four angles. (p. 80)
Theorem 3-4	The interior of an angle is a convex set. (p. 80)
Theorem 3-5	Congruence of angles is an equivalence relation. (p. 86)

Theorem 3-6	Any two right angles are congruent. (p. 86)
Theorem 3-7	If two angles are both congruent and supplementary, then each of them is a right angle. (p. 86)
Theorem 3-8	Supplements of congruent angles are congruent. (p. 86)
Theorem 3-9	Complements of congruent angles are congruent. (p. 86)
Theorem 3-10	Vertical angles are congruent. (p. 86)
Theorem 3-11	The interior of a triangle is always a convex set. (p. 89)
Theorem 3-12	If a line in the plane of a triangle intersects one side of the triangle, then the line intersects at least one other side of the triangle. (p. 89)
Theorem 3-13	The measure of an exterior angle of a triangle is greater than the measure of either remote interior angle. (p. 90)
Theorem 3-14	If the lengths of two sides of a triangle are unequal, the measures of the angles opposite these sides are unequal, and the angle with the larger measure is opposite the longer side. (p. 90)
Theorem 3-15	If the measures of two angles of a triangle are unequal, the lengths of the sides opposite these angles are unequal, and the longer side is opposite the angle with the greater measure. (p. 90)
Theorem 3-16	The shortest segment joining a point to a line is the perpendicular segment. (p. 91)
Theorem 3-17	The sum of the lengths of two sides of a triangle is greater than the length of the third side. (p. 91)
Theorem 4-1	Congruence between triangles is an equivalence relation. (p. 110)
Theorem 4-2	(*Isosceles Triangle Theorem*) If two sides of a triangle are congruent, then the angles opposite them are congruent. (p. 111)
Theorem 4-3	(*Angle Side Angle Theorem*) Given a correspondence between two triangles (or between a triangle and itself), if two angles and the included side of the first triangle are congruent to the corresponding parts of the second, then the correspondence is a congruence. (p. 112)
Theorem 4-4	(*Side Side Side Theorem*) Given a correspondence between two triangles (or between a triangle and itself), if all three pairs of corresponding sides are congruent, then the correspondence is a congruence. (p. 112)
Theorem 4-5	Every segment has one and only one midpoint. (p. 113)
Theorem 4-6	(*Angle Bisector Theorem*) Every angle has exactly one bisector. (p. 113)

Theorem 5-1	If l is any line and P any point not on l, then there exists at least one line that contains P and is parallel to l. (p. 131)
Theorem 5-2	If two parallel lines are cut by a transversal, then any pair of alternate interior angles is congruent. (p. 133)
Theorem 5-3	(*Converse of Theorem* 5-2) If two lines are cut by a transversal and if a pair of alternate interior angles is congruent, then the lines are parallel. (p. 133)
Theorem 5-4	If two parallel lines are cut by a transversal, then any pair of corresponding angles is congruent. (p. 133)
Theorem 5-5	(*Converse of Theorem* 5-4) If two lines are cut by a transversal and if a pair of corresponding angles is congruent, then the lines are parallel. (p. 134)
Theorem 5-6	If two parallel lines are cut by a transversal, interior angles on the same side of the transversal are supplementary. (p. 134)
Theorem 5-7	The sum of the measures of the angles of any triangle is 180. (p. 134)
Theorem 5-8	If three parallel lines intercept congruent segments on one transversal, then they intercept congruent segments on any other transversal. (p. 135)
Theorem 5-9	Two parallel lines lie in one and only one plane. (p. 136)
Theorem 5-10	If a line is perpendicular to one of two parallel planes, then it is perpendicular to the other. (p. 136)
Theorem 5-11	Two lines perpendicular to the same plane are parallel. (p. 136)
Theorem 5-12	Planes parallel to the same plane are parallel to each other. (p. 137)
Theorem 5-13	Planes perpendicular to the same line are parallel. (p. 137)
Theorem 6-1	The interior of every convex polygon is a convex set. (p. 159)
Theorem 6-2	The segment joining the midpoints of two sides of a triangle is parallel to the third side, and the length of the segment is half the length of the third side. (p. 161)
Theorem 6-3	If a line is parallel to one side of a triangle and contains a midpoint of another side, then the line contains the midpoint of the third side. (p. 161)
Theorem 6-4	The bisectors of the angles of a triangle intersect in one point which is equidistant from the three sides. (p. 161)
Theorem 6-5	The perpendicular bisectors of the sides of a triangle intersect in one point which is equidistant from the three vertices. (p. 161)

Theorem 6-6	The medians of a triangle intersect in one point which is two thirds of the distance from each vertex to the midpoint of the opposite side. (p. 161)
Theorem 6-7	The altitudes of a triangle intersect in one point. (p. 162)
Theorem 6-8	If two internal angle bisectors are congruent, the triangle is isosceles. (p. 162)
Theorem 6-9	The sum of the distances from any point within an equilateral triangle to its sides is a constant equal to the length of the altitude. (p. 162)
Theorem 6-10	In a parallelogram, any two opposite sides are congruent. (p. 163)
Theorem 6-11	Any two opposite angles of a parallelogram are congruent. (p. 163)
Theorem 6-12	Any two consecutive angles of a parallelogram are supplementary. (p. 163)
Theorem 6-13	The diagonals of a parallelogram bisect each other. (p. 163)
Theorem 6-14	If in a given quadrilateral, both pairs of opposite sides are congruent, then the quadrilateral is a parallelogram. (p. 163)
Theorem 6-15	If two sides of a quadrilateral are both parallel and congruent, then the quadrilateral is a parallelogram. (p. 163)
Theorem 6-16	If the diagonals of a quadrilateral bisect each other, then the quadrilateral is a parallelogram. (p. 163)
Theorem 6-17	The diagonals of a rhombus are perpendicular to each other and bisect each other. (p. 164)
Theorem 6-18	(*Converse of Theorem* 6-17) If the diagonals of a quadrilateral bisect each other and are perpendicular to each other, then the quadrilateral is a rhombus. (p. 164)
Theorem 6-19	If a quadrilateral has four right angles, it is a rectangle. (p. 164)
Theorem 6-20	If a line is perpendicular to a radius of a circle at its outer end, then the line is a tangent. (p. 166)
Theorem 6-21	(*Converse of Theorem* 6-20) Every tangent to a circle is perpendicular to the radius drawn to the point of contact. (p. 167)
Theorem 6-22	Any perpendicular from the center of a circle to a chord bisects the chord. (p. 167)
Theorem 6-23	The perpendicular bisector of a chord passes through the center. (p. 167)
Theorem 6-24	In the same or congruent circles, chords equidistant from the center are congruent. (p. 167)

Theorem 6-25	In the same circle or in congruent circles, any two congruent chords are equidistant from the center. (p. 168)
Theorem 6-26	If $\widehat{AB}$ and $\widehat{BC}$ are arcs of the same circle, having only the point B in common, then $m\widehat{AB} + m\widehat{BC} = m\widehat{AC}$. (p. 170)
Theorem 6-27	The measure of an inscribed angle is half the measure of its intercepted arc. (p. 171)
Theorem 6-28	(*Two-Circle Theorem*) Let C and C' be two coplanar circles with radii a and b, and let d be the distance between their centers. 1. C and C' intersect in exactly two points if and only if each of the numbers a, b, and d is less than the sum of the other two. 2. If C and C' intersect in exactly two points, then these points lie on opposite sides of the line of centers. (p. 171)
Theorem 6-29	The lateral edges of a prism are parallel and congruent. (p. 174)
Theorem 6-30	The lateral faces of a right prism are rectangles. (p. 174)
Theorem 6-31	An altitude of a right prism is congruent to each of its lateral edges. (p. 174)
Theorem 6-32	The lateral edges of a regular pyramid are congruent. (p. 176)
Theorem 6-33	The lateral faces of a regular pyramid are congruent isosceles triangles. (p. 176)
Theorem 6-34	The altitudes of the lateral faces of a regular pyramid are congruent. (p. 176)
Theorem 6-35	The diameter of a sphere is twice its radius. (p. 181)
Theorem 6-36	All radii of the same sphere are congruent. (p. 181)
Theorem 6-37	All diameters of the same sphere are congruent. (p. 181)
Theorem 6-38	Every section of a sphere made by a plane that intersects the sphere but is not tangent to it is a circle. (p. 181)
Theorem 6-39	Circles of a sphere made by planes at equal distances from the center of the sphere have equal radii. (p. 181)
Theorem 6-40	The planes of circles with equal radii on a sphere are equally distant from the center. (p. 181)
Theorem 6-41	A plane that is perpendicular to a radius of a sphere at its endpoint on the sphere is tangent to the sphere. (p. 181)
Theorem 6-42	If a plane intersects a sphere in more than one point, then the intersection of the plane and the sphere is a circle. (p. 181)
Theorem 6-43	Through any two points on a sphere that are not antipodal points, there is one and only one great circle on the sphere. (p. 181)

Theorem 6-44	Through any two antipodal points there are infinitely many great circles on the sphere. (p. 181)
Theorem 7-1	(*Pythagorean Theorem*) In any right triangle, the square of the length of the hypotenuse is equal to the sum of the squares of the lengths of the legs. (p. 199)
Theorem 7-2	(*Converse of the Pythagorean Theorem*) If the square of the length of one side of a triangle is equal to the sum of the squares of the lengths of the other two sides, then the triangle is a right triangle whose right angle is opposite the longest side. (p. 201)
Theorem 7-3	(30-60 *Triangle Theorem*) In a right triangle whose acute angles have measures of 30 and 60, the hypotenuse is twice as long as the shorter leg, and the longer leg is $\sqrt{3}$ times the shorter leg. (p. 201)
Theorem 7-4	(*Isosceles Right Triangle Theorem*) In a right triangle whose acute angles have measures of 45 and 45, the hypotenuse is $\sqrt{2}$ times the length of a leg. (p. 202)
Theorem 7-5	(*Hypotenuse-Leg Theorem*) If the hypotenuse and a leg of one right triangle are congruent to the hypotenuse and a leg of another, then the triangles are congruent. (p. 202)
Theorem 8-1	(*Basic Proportionality Theorem*) If a line parallel to one side of a triangle intersects the other two sides in distinct points, then it cuts off segments that are proportional to these sides. (p. 220)
Theorem 8-2	(*Converse of Theorem* 8-1) If a line intersects two sides of a triangle, and cuts off segments proportional to these sides, then it is parallel to the third side. (p. 221)
Theorem 8-3	(*AAA Similarity Theorem*) Given a correspondence between two triangles, if corresponding angles are congruent, then the correspondence is a similarity. (p. 221)
Theorem 8-4	(*SAS Similarity Theorem*) Given a correspondence between two triangles, if two pairs of corresponding sides are proportional, and the included angles are congruent, then the correspondence is a similarity. (p. 222)
Theorem 8-5	(*SSS Similarity Theorem*) Given a correspondence between two triangles, if corresponding sides are proportional, then the correspondence is a similarity. (p. 222)
Theorem 8-6	In any right triangle, the altitude to the hypotenuse separates the triangle into two triangles which are similar to each other and to the original triangle. (p. 223)

Theorem 8-7 Given a right triangle and the altitude to the hypotenuse,

1. The altitude is the geometric mean of the segments into which it separates the hypotenuse.
2. Each leg is the geometric mean of the hypotenuse and the segment of the hypotenuse adjacent to the leg. (p. 223)

Theorem 8-8 Similarity between triangles is an equivalence relation. (p. 223)

Theorem 8-9 If two triangles are similar, then the altitudes drawn to corresponding sides are proportional. (p. 224)

Theorem 8-10 If three or more parallel lines are cut by two transversals, then the lengths of the segments intercepted on one transversal are proportional to the lengths of the corresponding segments intercepted on the other transversal. (p. 224)

Theorem 8-11 The perimeters of two similar polygons have the same ratio as any two corresponding sides. (p. 224)

Theorem 11-1 The area of a rectangle is the product of its base and its altitude. (p. 296)

Theorem 11-2 The area of a right triangle is half the product of its legs. (p. 297)

Theorem 11-3 The area of a triangle is half the product of any base and the altitude to that base. (p. 297)

Theorem 11-4 If two triangles have the same base and the same altitude, then they have the same area. (p. 298)

Theorem 11-5 If two triangles have the same altitude, then the ratio of their areas is equal to the ratio of their bases. (p. 299)

Theorem 11-6 If two triangles have the same base, then the ratio of their areas is equal to the ratio of their corresponding altitudes. (p. 299)

Theorem 11-7 If two triangles are similar, then the ratio of their areas is the square of the ratio of any two corresponding sides. (p. 300)

Theorem 11-8 The area of a trapezoid is half the product of its altitude and the sum of its bases. (p. 300)

Theorem 11-9 The area of a parallelogram is the product of any base and the corresponding altitude. (p. 301)

Theorem 11-10 (*Prism Cross-Section Theorem*) All cross sections of a prism have the same area. (p. 303)

Theorem 11-11 (*Pyramid Cross-Section Theorem*) If two pyramids have the same base area and the same altitude, then cross sections equidistant from the vertices have the same area. (p. 304)

Theorem 11-12 The volume of any prism is the product of the altitude and the area of the base. (p. 304)

Theorem 11-13 If two pyramids have the same base area and the same height, then they have the same volume. (p. 305)

Theorem 11-14 The volume of a pyramid is one third the product of its altitude and its base area. (p. 305)

(*Basic Theorem on Limits*) If two variables are always equal and each approaches a limit, then these limits are equal. (p. 320)

Theorem 12-1 The ratio $C/2r$ of the circumference to the diameter is the same for all circles. (p. 321)

Theorem 12-2 The area of a circle of radius r is πr^2. (p. 323)

Theorem 12-3 If two arcs have equal radii, then their lengths are proportional to their measures. (p. 324)

Theorem 12-4 If an arc has measure q and radius r, then its length is

$$L = \frac{q}{180}\pi r.$$

(p. 325)

Theorem 12-5 The area of a sector is half the product of its radius and the length of its arc. (p. 326)

Theorem 12-6 The area of a sector of radius r and arc measure q is

$$A = \frac{q}{360}\pi r^2.$$

(p. 326)

Theorem 12-7 The lateral surface area of a right circular cylinder is $S = 2\pi rh$, where r is the radius of the base and h is the height of the cylinder. (p. 326)

Theorem 12-8 The volume of a circular cylinder is the product of its altitude and the area of its base. (p. 327)

Theorem 12-9 The lateral area of a right circular cone is equal to half the product of the circumference of the base and the slant height. (p. 327)

Theorem 12-10 The volume of a circular cone is one third the product of its altitude and the area of its base. (p. 327)

Theorem 12-11 The volume of a sphere of radius r is $V = \frac{4}{3}\pi r^3$. (p. 328)

Theorem 12-12 The surface area of a sphere of radius r is $S = 4\pi r^2$. (p. 328)

Theorem 13-1 On a nonvertical line, all segments have the same slope. (p. 348)

Theorem 13-2 Two distinct nonvertical lines are parallel if and only if they have the same slope. (p. 349)

Theorem 13-3 Two nonvertical lines are perpendicular if and only if their slopes are negative reciprocals of each other. (p. 350)

Theorem 13-4 (*Distance Formula*) The distance between the point (x_1, y_1) and (x_2, y_2) is

$$d = \sqrt{(x_2 - x_1)^2 + (y_2 - y_1)^2}$$

(p. 352)

Theorem 13-5 (*Midpoint Formula*) Let $P_1 = (x_1, y_1)$ and $P_2 = (x_2, y_2)$. The midpoint of $\overline{P_1P_2}$ is the point

$$P = \left(\frac{x_1 + x_2}{2}, \frac{y_1 + y_2}{2}\right)$$

(p. 353)

Theorem 13-6 If P is between P_1 and P_2, and $P_1P/PP_2 = r$, where r is any positive real number, then

$$P = \left(\frac{x_1 + rx_2}{1 + r}, \frac{y_1 + ry_2}{1 + r}\right)$$

(p. 354)

Theorem 13-7 The equation of the straight line parallel to the y axis and a units from it is $x = a$. (p. 360)

Theorem 13-8 The equation of the straight line parallel to the x axis and b units from it is $y = b$. (p. 361)

Theorem 13-9 The equation of the straight line passing through the point (x_1, y_1) with slope m is $y - y_1 = m(x - x_1)$. (p. 362)

Theorem 13-10 The equation of the straight line with slope m and y intercept b is $y = mx + b$. (p. 364)

Theorem 13-11 The equation of the straight line with x intercept a and y intercept b is

$$\frac{x}{a} + \frac{y}{b} = 1.$$

(p. 365)

Theorem 13-12 All straight lines have first-degree equations in two unknowns, and, conversely, all first-degree equations in two unknowns represent straight lines. (p. 365)

Theorem 13-13	The equation of the circle with center at the origin and radius r is $x^2 + y^2 = r^2$. (p. 366)
Theorem 13-14	The equation of the circle with center (h, k) and radius r is $(x - h)^2 + (y - k)^2 = r^2$. (p. 367)
Theorem 13-15	Every circle is the graph of an equation of the form $x^2 + y^2 + Dx + Ey + F = 0$. (p. 367)

INDEX